ADVANCES IN
LIPOBIOLOGY

Volume 2 • 1997

ADVANCES IN LIPOBIOLOGY

Editor: RICHARD W. GROSS
Department of Bioorganic Chemistry
and Molecular Pharmacology
Washington University School of Medicine
St. Louis, Missouri

VOLUME 2 • 1997

JAI PRESS INC.

Greenwich, Connecticut *London, England*

55 Old Post Road No. 2
Greenwich, Connecticut 06836

JAI PRESS LTD.
38 Tavistock Street
Covent Garden
London WC2E 7PB
England

ISBN: 0-7623-0205-4

Transferred to digital printing 2006

CONTENTS

LIST OF CONTRIBUTORS

Darrell D. Belke	Cardiovascular Disease Research Group Faculty of Medicine The University of Alberta Edmonton, Alberta, Canada
Merle L. Blank	Oak Ridge Associated Universities Medical Sciences Division Oak Ridge, Tennessee
L. Maximilian Buja	Department of Pathology and Laboratory Medicine The University of Texas Medical School Houston, Texas
Patrick C. Choy	Department of Biochemistry and Molecular Biology Faculty of Medicine The University of Mannitoba Winnipeg, Mannitoba, Canada
David L. DeWitt	Department of Biochemistry Michigan State University East Lansing, Michigan
William Dowhan	Department of Biochemistry and Molecular Biology The University of Texas Medical School Houston, Texas
Math J.H. Geelen	Department of Veterinary Basic Sciences Division of Biochemistry Utrecht University Utrecht, The Netherlands

Howard Goldfine Department of Microbiology
University of Pennsylvania School of Medicine
Philadelphia, Pennsylvania

Yusuf A. Hannun Department of Medicine
Duke University Medical Center
Durham, North Carolina

D. Corinne Harrison-Shostak Department of Cell Biology and Anatomy
University of North Carolina, Chapel Hill
Chapel Hill, North Carolina

Grant M. Hatch Department of Biochemistry and
Molecular Biology
Faculty of Medicine
The University of Mannitoba
Winnipeg, Mannitoba, Canada

Brian Herman Department of Cell Biology and Anatomy
University of North Carolina, Chapel Hill
Chapel Hill, North Carolina

Supriya Jayadev Department of Medicine
Duke University Medical Center
Durham, North Carolina

Ten-ching Lee Oak Ridge Associated Universities
Medical Sciences Division
Oak Ridge, Tennessee

John J. Lemasters Department of Cell Biology and Anatomy
University of North Carolina, Chapel Hill
Chapel Hill, North Carolina

Gary D. Lopaschuk Cardiovascular Disease Research Group
Faculty of Medicine
The University of Alberta
Edmonton, Alberta, Canada

Ricky Y.K. Man Department of Biochemistry and
Molecular Biology
Faculty of Medicine
The University of Mannitoba
Winnipeg, Mannitoba, Canada

Janet H. Mar
Department of Medicine
The University of Texas Health Science Center
Houston, Texas

Jeanie B. McMillin
Department of Pathology
and Laboratory Medicine
The University of Texas
Health Science Center
Houston, Texas

Joseph C. Miller
Department of Pathology
and Laboratory Medicine
The University of Texas Medical School
Houston, Texas

Anna Liisa Nieminen
Department of Cell Biology and Anatomy
University of North Carolina, Chapel Hill
Chapel Hill, North Carolina

William L. Smith
Department of Biochemistry
Michigan State University
East Lansing, Michigan

Lilian B.M. Tijburg
Unilever Research Laboratory
Vlaardingen
Vlaardingen, The Netherlands

Fred Snyder
Oak Ridge Associated Universities
Medical Sciences Division
Oak Ridge, Tennessee

Lambert M.G. van Golde
Department of Veterinary Basic Sciences
Division of Biochemistry
Utrecht University
Utrecht, The Netherlands

P. Sebastiaan Vermeulen
Department of Veterinary Basic Sciences
Division of Biochemistry
Utrecht University
Utrecht, The Netherlands

Moseley Waite | Department of Biochemistry
Wake Forest University Medical Center
Winston-Salem, North Carolina

Haichao Wang | Department of Cell Biology and Anatomy
University of North Carolina, Chapel Hill
Chapel Hill, North Carolina

Xue Feng Wang | Department of Cell Biology and Anatomy
University of North Carolina, Chapel Hill
Chapel Hill, North Carolina

PREFACE

During the last several years substantial growth in the research of lipid metabolism and lipid second messengers has occurred. In large part, these advances have resulted from the application of a wide variety of molecular biologic techniques to the examination of the kinetics of lipid metabolic enzymes and the study of structure-activity relationships on cellular physiologic function. Our knowledge of the importance of covalent modification of lipid synthetic and metabolizing enzymes in intact cells has expotentially expanded and studies on the role of specific regulatory aspects of lipid metabolism can now be routinely performed in molecular detail. Through the cloning, engineering, and insightful execution of critical experiments, the field has witnessed the refinement of old concepts and the emergence of new ideas and new paradigms. As always, it is through the careful testing of these new hypotheses and the reexamination of prior dogma that the next decade will witness an ever greater penetrance of the depth of our scientific understanding of lipid metabolism, membrane structure and function, and the mechanisms of generation of lipid second messengers in intact cells after cellular stimulation. While a change in the political, economic, and social climate has indeed refocused and restructured many practical aspects of this process of investigation, the emergence of new technologies has facilitated our dealing with these pressures to result in a collective enterprise whose progress in many respects surpasses that manifest in years past. The power of the scientific approach is ever increasingly fueled by the force of novel technologies and the innovation of human spirit. The field of membrane biochemistry and lipid metabolism continues to expand in new directions thrust upon us by a new generation of students and investigators and it is this infusion of new thoughts and new ideas that will assure the continued growth of our field in the years to come.

Richard W. Gross
Editor

RELATIONSHIP OF LIPID ALTERATIONS AND IMPAIRED CALCIUM HOMEOSTASIS DURING MYOCARDIAL ISCHEMIA

L. Maximilian Buja and Joseph C. Miller

Advances in Lipobiology
Volume 2, pages 1-28.

ISBN 0-7623-0205-4

I. INTRODUCTION

Coronary heart disease can lead to a major reduction in myocardial blood flow which induces severe myocardial ischemia due to oxygen and substrate deprivation (Reimer and Jennings, 1986). These profound metabolic alterations lead to damage of the cardiac myocyte (Figure 1). Progressive alterations in myocardial electrolytes, including deranged calcium homeostasis, are characteristic of the pathogenesis of ischemic injury in the heart (Buja et al., 1988). There is also considerable evidence that progressive derangements in lipid metabolism are involved in the development of membrane damage and altered membrane function (Buja, 1991). In addition, changes in the cytoskeleton lead to destabilization of the membranes making them susceptible to rupture from cell swelling or other mechanisms (Buja et al., 1993). The damage may be produced either by abnormal actions of hydrolytic enzymes, such as phospholipases and/or proteases, by chemical or physical disruption, or by a combination of these factors. From the cumulative body of clinical and experimental observations, one must conclude that damage to cell membranes plays a key role in the transition from reversible to irreversible myocardial injury (Reimer and Jennings, 1986). Unlike apoptosis, the critical determinant of necrotic cell death produced by ischemia involves irreversible damage to the integrity and function of cellular membranes (Buja et al., 1993). In contrast, apoptotic cell death involves primary intracellular events, including activation of proteases and nuclear damage due to activation of an endonuclease and subsequent double stranded DNA cleavage (Buja et al., 1993; Patel et al., 1996).

Thus, a general relationship is apparent between altered intracellular calcium homeostasis and altered lipid metabolism in the cells of ischemic tissue. The purpose of this presentation is to review evidence for the nature of the alterations in calcium and lipid metabolism in ischemia and related forms of injury, and discuss the possible interrelationships between these two phenomena.

A. Myocardial Calcium Regulation

Intracellular calcium ion concentration is a key regulator of myocardial function. Electrical stimulation leads to calcium influx across the sarcolemma, calcium-induced calcium release from the sarcoplasmic reticulum, and an increase in calcium concentration at the myofibrils which triggers myocyte contraction during systole;

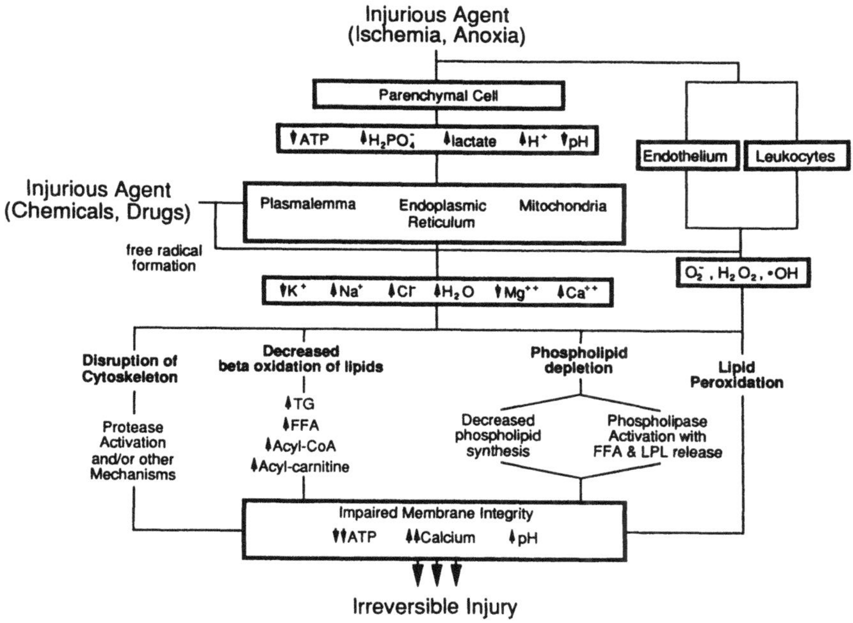

Figure 1. Scheme of the pathogenesis of irreversible cell injury. ATP, adenosine triphosphate; CoA, coenzyme A; FFA, free fatty acids; LPL, lysophospholipids; TG, triglycerides; H_2PO_4, phosphoric acid; H_2O_2, hydrogen peroxide; OH•, hydroxyl radical; H_20, water; Ca^{++}, calcium ion; Mg^{++}, magnesium ion; Cl^-, chlorine ion; Na^+, sodium ion; and K^+, potassium ion. From Buja et al. (1993).

reuptake of calcium ions (Ca^{2+}) into the sarcoplasmic reticulum and relaxation occurs during diastole (Braunwald, 1982; Barry and Bridge, 1993). Intracellular, free calcium ion concentrations ($[Ca^{2+}]_i$) normally range from approximately 50-100 nM during diastole to peaks of 500-1000 nM during systole with each contraction cycle (Figure 2). Precise documentation of $[Ca^{2+}]_i$ in normal tissue and cells, in myocardial ischemia, and related experimental models has come from technical advances in methods for measurement of calcium (Barry et al., 1987; Morris et al., 1989; Lee et al., 1988; Steenbergen et al., 1990; Koretsume and Marban, 1990; Doeller and Wittenberg, 1990; Thandroyen et al., 1992). These approaches include electron probe x-ray microanalysis to determine total cellular calcium levels and microspectrofluoroscopy using fluorescent probes, such as fura-2 and indo-1, for determination of intracellular, free calcium ion concentration, $[Ca^{2+}]_i$. These technical advances have allowed correlation of intracellular $[Ca^{2+}]_i$ levels with the severity of cell injury as discussed in the following section.

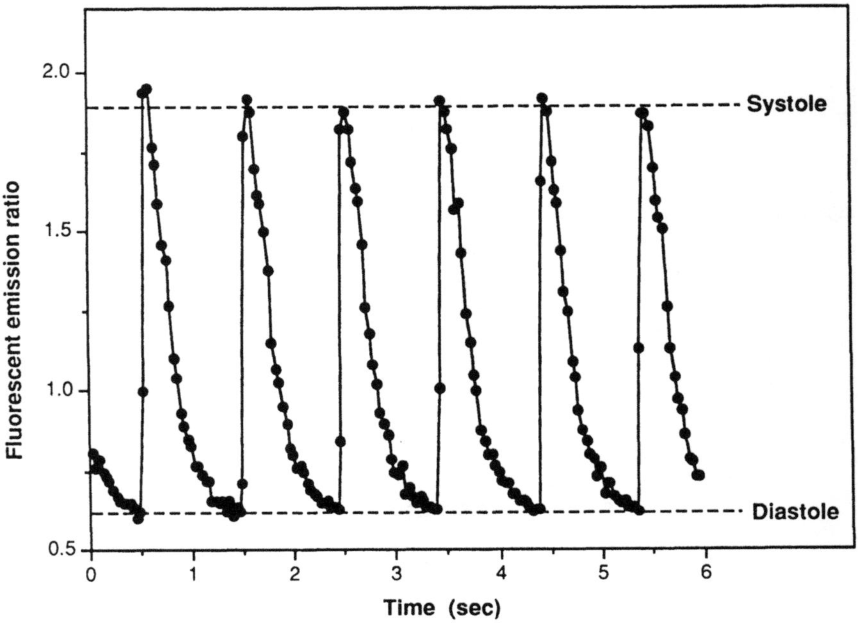

Figure 2. Normal intracellular, free calcium transients measured with fura-2 microspectrofluorometry. The fluorescence emission exceeding 470 nm produced from an alternating of 340 and 380 nm excitation was measured in cultured neonatal myocytes loaded with fura-2/AM. The 340/380 emission ratio increases proportionally with $[Ca^{2+}]_i$ associated with each contraction (systole) in the mycoytes then decreases during relaxation (diastole).

$[Ca^{2+}]_i$ represents only a very small fraction of total cellular calcium in the normal cell (Buja et al., 1988). Electron probe measurements determine that total cellular calcium concentrations range from approximately 5-10 mmol/kg tissue dry weight (Jones et al., 1989). Assuming the dry/wet tissue weight ratio of approximately 20% and the specific density of 1 g/ml, we can estimate that total calcium concentrations of 1-2 ug/ml, or 1-2 mM in normal myocytes. Therefore, the cytoplasmic, free calcium pool represents only 0.05-0.10% of the total calcium. This 20-fold range represents the minimum to maximum $[Ca^{2+}]_i$ in myocytes during the contraction cycle, with regulation of the contractile cycle being the primary physiological function of shifts in $[Ca^{2+}]_i$. Accordingly, the myocyte has tremendous capacity to quickly move calcium ions between the various intracellular and extracellular compartments.

The very rapid calcium fluxes in contracting myocytes are mediated by ion pumps and channels located in the cellular and organellar membranes (Braunwald, 1982; Barry and Bridge, 1993). Loss of calcium homeostasis can produce myocardial dysfunction. Function of these membrane proteins can be influenced by the properties of the membrane phospholipids which encompass them (Spector and Yorek, 1985; Philipson and Ward, 1985; Katz and Messineo, 1981; Corr et al., 1984).

B. Myocardial Phospholipid Metabolism

Cellular membranes are fluid mosaic structures composed primarily of phospholipids arranged into a molecular bilayer containing proteins which mediate membrane functions (Singer and Nicolson, 1972). Maintenance of membrane integrity is absolutely critical for cell viability; loss of either structural or functional integrity leads to irreversible cell injury (Katz and Messineo, 1981; Corr et al., 1984). Membrane homeostasis represents a balance between phospholipid anabolism and catabolism (Figure 3). Synthesis of phospholipids can occur by *de novo* synthesis or by reacylation pathways where the acyl moieties may be drawn from extracellular or intracellular sources (Katz and Messineo, 1981; Corr et al., 1984; Buja, 1991). Phospholipid catabolism in myocardium is mediated by phospholipases, lysophospholipases, and lipases yielding a diverse group of products (Buja, 1991; Gross, 1992; Wolf and Gross, 1992; Hazen and Gross, 1992a and 1992b).

The myocardium contains phospholipases in different subcellular fractions which can be distinguished functionally by differences in their substrate specificities, pH optima, and dependence on calcium for activation (Gross, 1992; Hazen and Gross, 1992b). Phospholipases of the A_1 and A_2 classes hydrolyze the fatty acyl moieties from the sn-1 and sn-2 positions of the phospholipid molecule, respectively, producing a free fatty acid and a lysophospholipid. Cells normally maintain very low levels of the monoacyl lipid species, presumably to minimize their potent detergent effects. These lipophilic products are susceptible to further degradation or may be used for resynthesis of phospholipids. Specifically, the lysophospholipids can be further hydrolyzed by lysophospholipases, and the resultant fatty acids can undergo beta-oxidation in the mitochondria to meet energy needs of the myocytes. Alternatively, regeneration of the intact phospholipid molecule can occur by the enzymatic reacylation of a high-energy fatty acyl-CoA to the lysophospholipid by a lysophospholipid acyl transferase. A deacylation-reacylation cycle is formed by the combined action of phospholipase A and lysophospholipid acyl transferase. In addition to diacyl phospholipids, myocardial membranes contain large amounts of plasmalogens (Gross, 1992). These lipids have an ether-linked fatty acid in the sn-1 position. Choline and ethanolamine plasmalogens with an arachidonyl moiety in the sn-2 position are particularly abundant species. The myocardium contains phospholipases with a selective affinity for plasmalogens (Wolf and Gross, 1985; Hazen and Gross, 1992b).

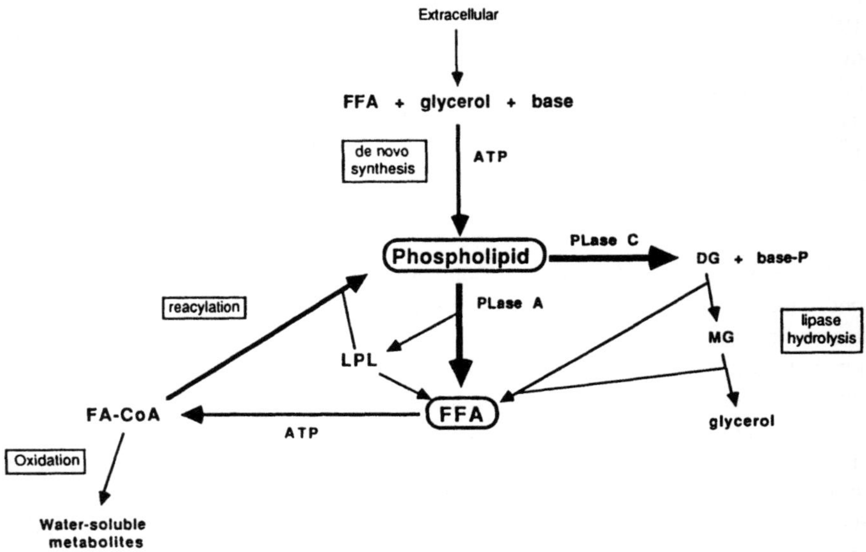

Figure 3. Major pathways of phospholipid metabolism. Phospholipid catabolism may be mediated by phospholipase A or C. Anabolic pathways involve reacylation and *de novo* synthesis. *DG*, diglyceride; *FFA*, free fatty acid; and *MG*, monoglyceride. From Buja (1991).

Myocardium also contains cytosolic and membrane-associated phospholipase C activities (Buja, 1991). These enzymes cleave the phosphate-base head group from the diacylglycerol backbone of phospholipids resulting in the formation of a diacylglycerol and a water-soluble phosphate-base molecule. In the case of phosphatidylinositol-specific phospholipase C, both the inositol-polyphosphate and diacylglycerol products can be potent second messengers. The diacylglycerol molecule can be degraded further by diacylglycerol and monoacylglycerol lipases, thereby releasing free fatty acids. Activities of these enzymes are influenced by their associated receptors, respective receptor agonists or antagonists, and G-proteins. Changes in signal transduction complexes have been implicated in a variety of models of injury, such as myocardial ischemic damage discussed below.

II. ALTERATIONS IN MYOCARDIAL ISCHEMIA AND MYOCARDIAL CELL INJURY

Essentially, ischemic injury represents the detrimental downstream effects of abnormal cellular metabolism due to oxygen depletion. If adequate oxygen levels could be maintained in myocardial tissue, then initiation of the pathological biochemical sequelae would be effectively prevented. The circular nature of this truism focuses on the critical point of research on ischemic injury. Since we know the most basic factor initiating the pathogenesis of ischemic injury, what happens subsequent to oxygen depletion? At what points may we develop therapeutics to prevent or delay progression of the pathological cascade? The answers to these questions have led to a diverse body of research seeking to delineate the biochemical etiology at the cellular level. It seems implicit that oxygen depletion leads to alterations in the metabolic status of the myocyte which then produces tertiary effects on a wide variety of cellular processes. It has been proposed that myocyte viability is linked to maintenance of glycolysis and that inhibition of glycolysis may lead to loss of cellular calcium homeostasis (Opie, 1993). Alterations in myocardial calcium homeostasis and membrane damage represent two of the most consistent aspects of many studies seeking to identify the critical, downstream effects within the metabolically compromised myocyte.

A. Increased Phospholipid Degradation

There is abundant evidence that severe derangements in lipid metabolism occur early in myocardial ischemic injury and contribute to the progression of injury (Buja, 1991; Gross, 1992; Buja et al., 1993). These alterations are characterized by phospholipid degradation and accumulation of free fatty acids, lysophospholipids, acyl carnitines, and other lipid species, including triacylglycerols. In a canine model of regional ischemia produced by coronary occlusion, we observed time-dependent reductions of 10-20% in total phospholipids and individual phospholipid species after 3 hours (Chien et al., 1981). In addition, there were 2- to 4-fold increases in saturated and unsaturated free fatty acid levels during the first hour of ischemia (Figure 4) (Chien et al., 1984). The fatty acid accumulation included a prominent increase in free arachidonic acid, a fatty acid normally present exclusively in membrane phospholipids (Figure 5); the accumulation of free arachidonic acid has proven to be a sensitive indicator of phospholipid degradation (Hseuh et al., 1979; Chien et al., 1981; Chien et al., 1984). Similar findings have been demonstrated in other models (Shaikh et al., 1981; van der Vusse et al., 1987; Burton et al., 1986; van Bilsen et al., 1989).

In addition to accumulation of free fatty acids, ischemic myocardium may exhibit a transient increase in lysophospholipids (Corr et al., 1984). Elevated lysophospholipid levels have been associated with significant electrophysiological alterations which can lead to arrhythmias. Various studies have shown a 50-100% increase in lysophospholipids after 10 minutes of coronary occlusion whereas no increases were detected after 30 minutes of coronary occlusion (Corr et al., 1984; Chien et al., 1981; Chien et al., 1984; Shaikh et al., 1981). These results suggest that the increase in lysophospholipids is a transient phenomenon which may be explained by the sequential activities of phospholipases and lysophospholipases.

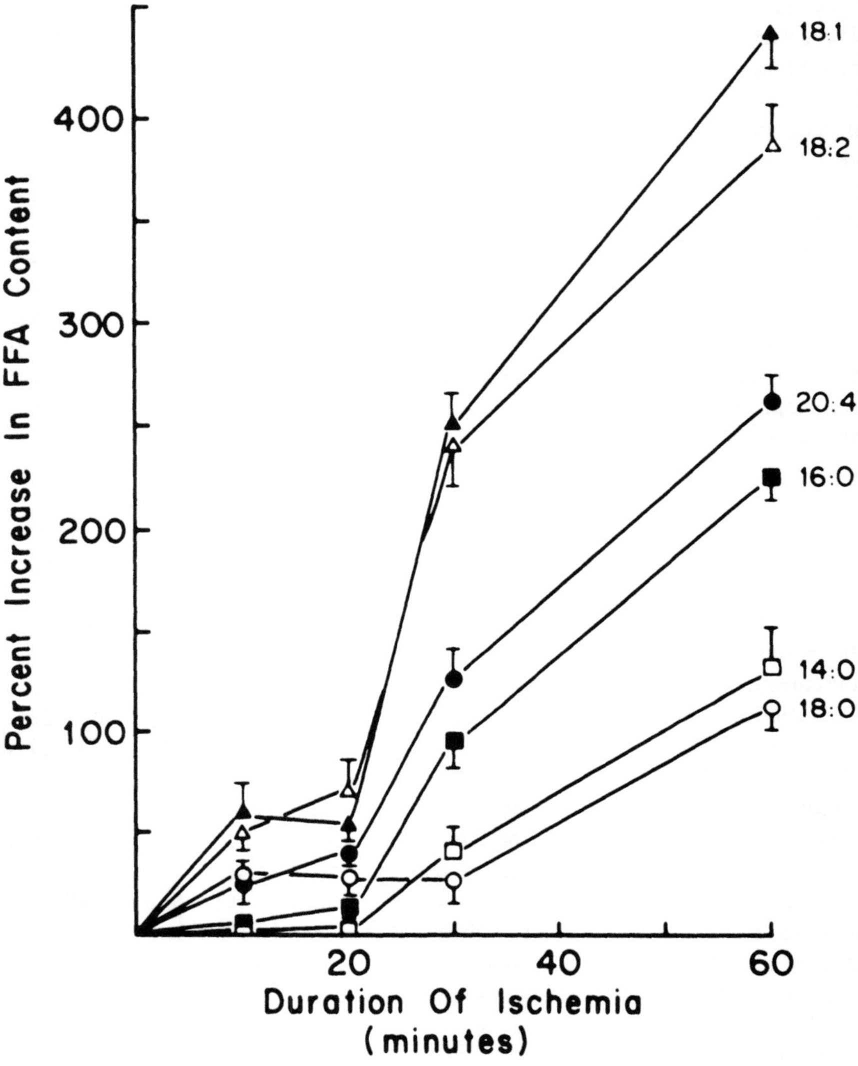

Figure 4. Time course of accumulation of unesterified fatty acids in ischemic canine myocardium produced by ligation of the left anterior descending coronary artery. The unesterified fatty acids were isolated by thin-layer chromatography and were quantified by high-performance liquid chromatography. The data are expressed as the percent increase in the ischemic tissues versus sham control tissues. From Chien et al. (1984).

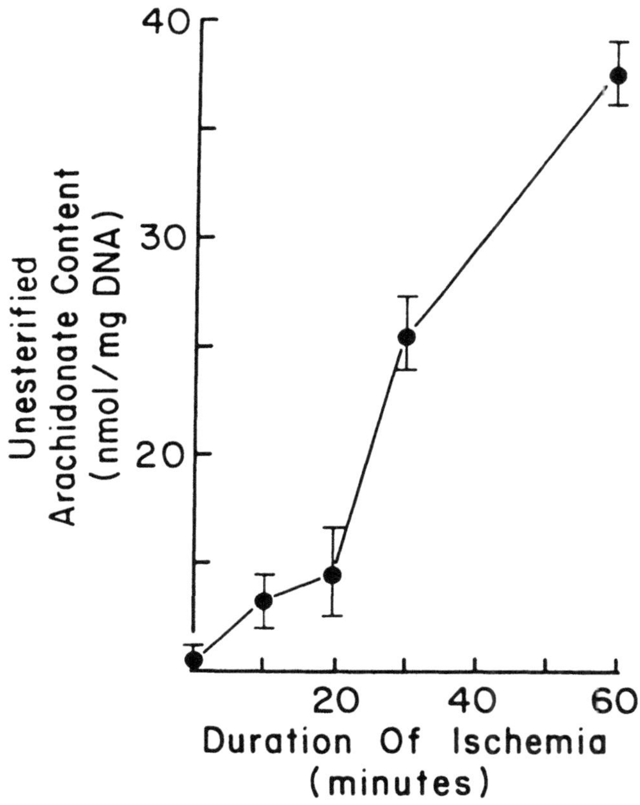

Figure 5. Time course of accumulation of unesterified arachidonatic acid in ischemic canine myocardium. Unesterified arachidonic acid concentrations were measured in lipid extracts of canine heart tissue after various periods of ischemia after separation by thin-layer chromatography and quantified by reverse phase HPLC. From Chien et al. (1984).

Ischemia also introduces alterations in various signal transduction pathways. A significant body of evidence demonstrates changes in adrenergic receptor-mediated processes in ischemic myocardium and cellular models (Brachmann and Schömig, 1989; Muntz et al., 1994). Briefly, systemic and myocardial catecholamine levels increase and adenylate cyclase activity decreases as ischemia progresses. Comparable changes may be occurring in alpha-adrenergic receptor-stimulated phospholipase C-mediated phosphatidylinositol turnover. In myocytes prelabelled with ^{3}H-inositol, phenylephrine-stimulated phosphatidylinositol hydrolysis was reduced under hypoxic conditions during a 60 minute incubation as compared to a normoxic atmosphere (Muntz et al., 1993). Phenylephrine, an alpha$_1$-adrenergic receptor agonist, produces a biphasic response in neonatal rat cardiac myocytes stimulated from 0 to 60 minute (Figure 6). Under hypoxic conditions, phosphatidylinositol turnover was decreased and exhibited a monophasic pattern. Although we can not identify the specific mechanism producing this pattern, it seems that at least two processes are occurring in normal cells and that one becomes inactive during hypoxia. Two phospholipase C enzymes or two separate substrate pools are logical possibilities. We must consider that catecholamines released during myocardial ischemia can bind to myocytes and activate various cellular processes leading to altered phospholipase C activity that varies with the duration of ischemia. Furthermore, the inositol-phosphates and diacylglycerols produced regulate a diverse range of cellular processes, such as calcium regulation. These data suggest that phospholipases may contribute to the injury in multiple ways, including aspects of intracellular regulation as well as structural defects.

B. Changes In Calcium and Other Electrolytes

Early electrolyte changes in myocardial cell injury are characterized by increases in inorganic phosphate and hydrogen ions, loss of K^+, and decreased Mg^{2+} following an early transient increase (Thandroyen et al., 1992). These electrolyte changes are accompanied by lactate accumulation (Braunwald, 1982). Progressive intracellular acidosis occurs with lowering of the intracellular pH from 7.2 to 6.5 or below (Braunwald, 1982; Buja et al., 1993). These changes in electrolytes and lactate are related to progressive derangements in high energy phosphate metabolism and decreasing ATP. An early K^+ loss occurs without change in Na^+ and appears to involve activation of ATP-sensitive K+ channels secondary to a decreasing energy charge (ratio of ATP to ADP plus AMP) or some other factor. The early electrolyte changes also are accompanied by an increase in cytosolic, free calcium ion concentration, $[Ca^{2+}]_i$ (Barry et al., 1987).

A subsequent phase involves an increase in Na^+ accompanied by a further decrease in K^+ presumably due to inhibition of the Na^+, K^+-ATPase (Thandroyen et al., 1992). The late phase of electrolyte alterations is characterized by progressive derangement in calcium homeostasis leading to an increase in total cellular calcium levels as well as $[Ca^{2+}]_i$ (Buja et al., 1993; Buja et al., 1988; Thandroyen et al.,

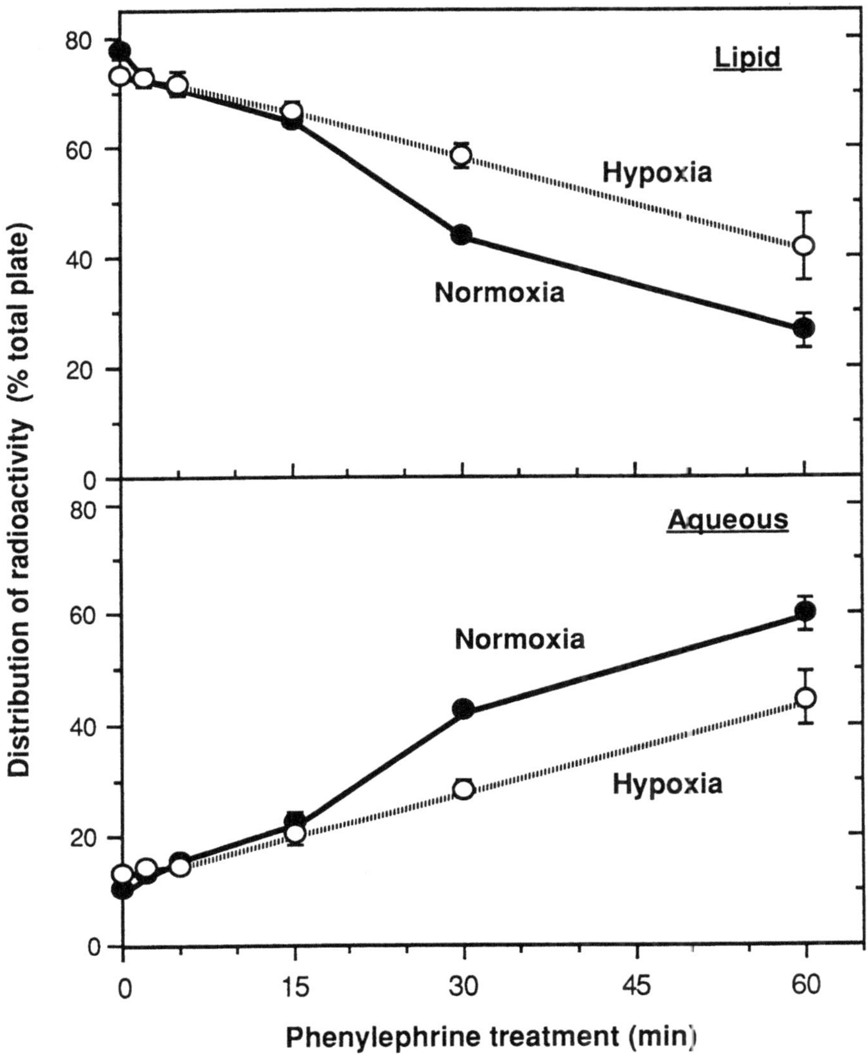

Figure 6. Phosphatidylinositol turnover in normoxic and hypoxic myocytes. Neonatal rat heart myocytes isolated were prelabelled overnight with ^{3}H-inositol then incubated for 1 h under either normoxic or hypoxic conditions; phenylephrine (100 uM final) was added from 0-60 minutes before terminating the 1 h incubation period. Radioactivity was determined in the lipid, water-soluble, and protein fractions. Adapted from Muntz et al. (1993).

1992). The increase in total calcium is accompanied by severe derangements in the other electrolytes. These changes reflect a non-specific increase in membrane permeability which cannot be reversed by calcium channel blocking agents. When severe, the calcium loading results in formation of calcium phosphate deposits in the mitochondria (Buja et al., 1988). The magnitude of calcium accumulation is influenced by the availability of extracellular calcium and is particularly prominent in the setting of myocardial reperfusion following ischemia. These observations are consistent with a biphasic pattern of cell injury, an initial increase in $[Ca^{2+}]_i$ from the extracellular milleu due to a metabolic imbalance, followed by an increase in bound calcium compartments via the free cytosolic pool leading to intracellular precipitation of calcium salts due to loss of membrane integrity.

Differences between these two phases has been demonstrated in cultured cardiac myocyte models (Figures 7 and 8). A reversible increase in $[Ca^{2+}]_i$ has been demonstrated in response to metabolic inhibition (Morris et al., 1989). More prolonged hypoxia or metabolic inhibition then leads to a phase of irreversible calcium loading. An early reversible increase in $[Ca^{2+}]_i$ also has been demonstrated with true ischemia in isolated heart models (Lee et al., 1988; Steenbergen et al., 1990; Koretsune et al., 1990). Total cytosolic calcium measured by electron probe microanalysis increases to approximately 53 mmol/kg tissue dry weight and mitochondrial levels reach nearly 800 mmol/kg (Jones et al., 1989). These concentrations are approximately 10- to 1000-fold higher than those in control cells.

C. Relationship Between Phospholipid and Calcium Alterations

We have utilized cardiac myocytes isolated from neonatal rat hearts subjected to selective components of ischemia, namely metabolic inhibition and hypoxia, to study the progression of injury (Chien et al., 1985; Gunn et al., 1985; Buja et al., 1990). This model has permitted study of interrelationships among changes in membrane phospholipids, calcium, and ATP levels in a controlled system which allows more precise measurement and correlation of these parameters than in intact heart preparations. In these experiments, myocytes are prelabelled with tritiated arachidonic acid for 18-24 hours and then subjected to metabolic inhibition with a variety of agents, including deoxyglucose and cyanide, deoxyglucose and oligomycin, or iodoacetate; hypoxia also is utilized (Chien et al., 1985; Gunn et al., 1985; Buja et al., 1990; Thandroyen et al., 1992). An important relationship was established between accelerated arachidonic acid release and the degree of ATP depletion. A threshold of 50% ATP reduction is required before arachidonate release occurs and, thereafter, an exponential amount of arachidonate release occurs as ATP concentration is reduced from 25% to negligible levels (Figure 9). Other studies showed that the increases in radioactivity in the fatty acid fraction (^{3}H-FFA) were reflective of increases in absolute levels of arachidonate (Revtyak et al., 1990). In the same model an early increase in free cytosolic calcium was demonstrated using the fura-2 microspectrofluoroscopy technique (Morris et al., 1989). Progressive

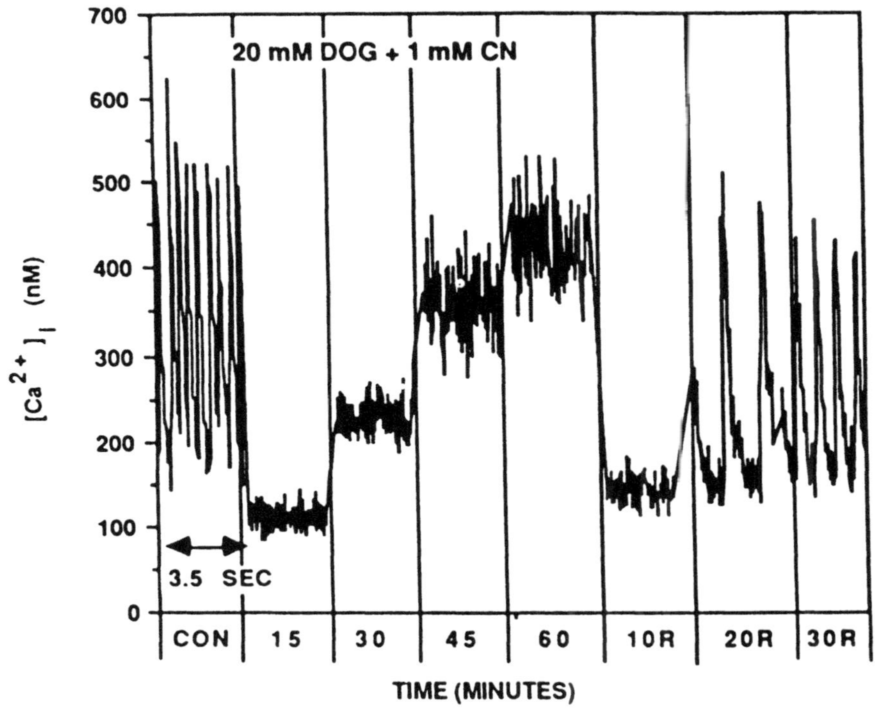

Figure 7. Reversible effects of metabolic inhibition on intracellular, free calcium transients in neonatal rat heart myocytes. $[Ca^{2+}]_i$ was measured using microspectrofluorometry in ventricular myocytes loaded with fura-2 then incubated from 0-60 minutes with 20 mM 2-deoxy-D-glucose (DOG) plus 1 mM sodium cyanide (CN); after incubation with DOG/CN 60 minutes, the cells were returned to control medium for 0-30 minutes. From Morris et al. (1989).

accumulation of arachidonic acid was associated with the development of severe cell injury characterized by hypercontraction, blebbing, and marked electrolyte derangements, including calcium loading (Chien et al., 1985; Gunn et al., 1985; Buja et al., 1990). These alterations in lipid metabolism have been confirmed in other myocyte models of cell injury (Hagve et al., 1990; Miyazaki et al., 1990). The myocyte model also has provided evidence that the progression of membrane and cellular injury also can be influenced by cytokines, inflammatory mediators and stress-induced proteins (Wang et al., 1996).

D. Effects of Phospholipase Inhibition

It is important to recognize that the severity and nature of the insult influences the magnitude of lipid alterations. Further insights into ischemic pathogenesis have

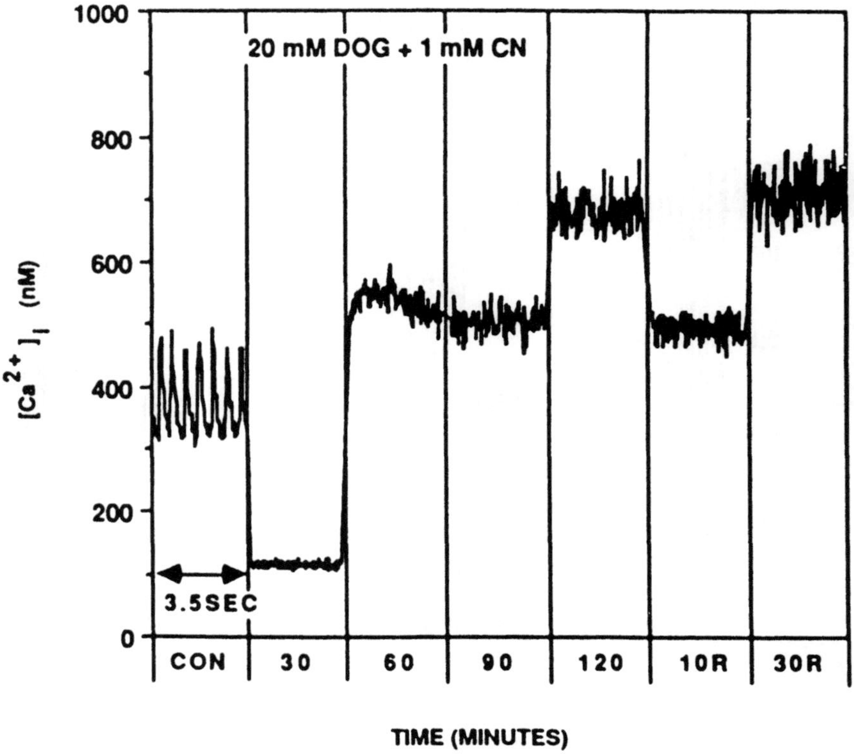

Figure 8. Irreversible effects of metabolic inhibition on intracellular, free calcium transients in neonatal rat heart myocytes. $[Ca^{2+}]_i$ was measured using microspectrofluorometry in ventricular myocytes loaded with fura-2 then incubated from 0-120 minutes with 20 mM 2-deoxy-D-glucose (DOG) plus 1 mM sodium cyanide (CN); after incubation with DOG/CN 60 min, the cells were returned to control medium for 0-30 min. From Morris et al. (1989)

been obtained using phospholipase inhibitors in attempts to block the progression of injury. In isolated heart models and intact dog models, agents which inhibit phospholipase A, including chlorpromazine and mepacrine, have been shown to protect against membrane damage and myocardial ischemic injury (Das et al., 1986; Otani et al., 1989; Hostetler et al., 1989). In addition, the anti-arrhythmic agent, amiodarone, has been shown to inhibit phospholipid degradation (Shaikh and Downar, 1987; Shaikh, 1992).

Phospholipase inhibitory agents reduced radioactive fatty acid release from isolated cardiac myocytes exposed to iodoacetic acid (IAA) to produce ATP-depletion (Table 1). Treatment of myocytes with mepacrine partially prevented io-

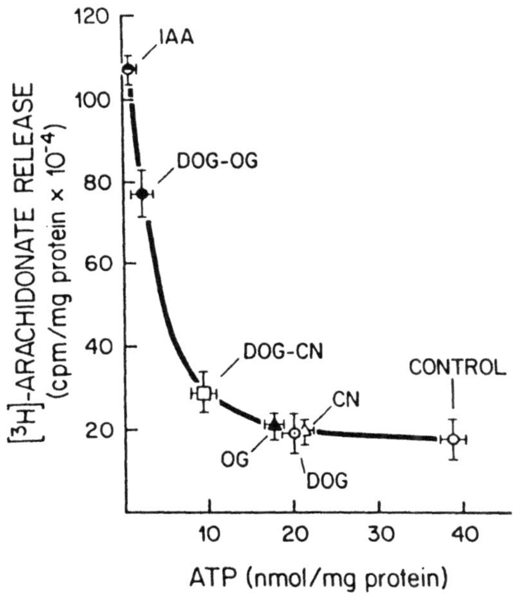

Figure 9. Inverse relationship between arachidonic acid release and ATP levels in metabolically inhibited myocytes. ATP concentrations and ^{3}H-arachidonic acid release were measured in neonatal rat heart myocytes incubated with various inhibitors. ATp was determined in perchloric acid extracts and fatty acid radioactivity measured in chloroform/methanol extracts separated by thin-layer chromatography. From Gunn et al. (1985)

doacetate induced cell damage (Jones et al., 1989). But this alkyl acridine was only partially effective as a protective agent in this cellular model, being approximately 25% effective when presented as a pretreatment and up to 80% effective as a continuous treatment. Prevention of morphological damage and calcium loading was similarly limited.

We have demonstrated that another phospholipase inhibitor (U26,384, N-[3-(dimethylamino) prophl]-3-methoxy-N-methyl estra 2,5-(10)-dien-17β amine developed by the Upjohn Company) much more effectively prevented increased release of unesterifed fatty acids in metabolically-inhibited cardiac myocytes (Table 1). In addition, this steroidal diamine also blocked calcium overload as measured by electron probe x-ray microanalysis and gross morphological damage without diminishing the degree of ATP depletion (Table 2) (Sen et al., 1988; Jones et al., 1989). *In vitro* studies have documented that this steroidal diamine effectively inhibits pancreatic phospholipase A_2 and platelet aggregation.

Additional studies are being undertaken to determine the effects of U26,384 on intracellular, free calcium concentration ($[Ca^{2+}]_i$) in this model under identical

Table 1. Effect of Phospholipase Inhibition on Arachidonic Acid Distribution in ATP-Depleted Neonatal Rat Cardiac Myocytes[a]

	Lipid radioactivity		
	PL	FFA	
Treatment	*(% total)*	*(% total)*	*(cpm/mg protein)*
Medium only[b]			
Control (n = 14)	81.2 ± 2.2	5.6 ± 1.1	(418 ± 55)
Iodoacetate (n-14)	71.2 ± 1.8	22.0 ± 1.8	(2119 ± 433)
IAA plus 5 uM U26,384 (n-11)	6.1 ± 1.6	5.2 ± 0.9	(229 ± 23)
Control (n = 15)	79.1 ± 2.1	6.9 ± 1.2	(368 ± 56)
Iodoacetate (n = 15)	70.8 ± 1.8	21.5 ± 2.2	(1692 ± 269)
IAA plus 50 uM mepacrine (n = 6)	86.8 ± 0.7	7.1 ± 1.0	(926 ± 275)
Medium plus cells[c]			
Control (n = 8)	93.2 ± 0.6	0.7 ± 0.1	
Iodoacetate (n = 8)	83.2 ± 0.7	2.4 ± 0.1	
IAA plus 5 uM U26,384 (n = 10)	92.1 ± 0.3	1.0 ± 0.1	
IAA plus 10 uM bromenol lactone (n = 4)	83.2 ± 0.4	1.5 ± 0.2	

Notes: [a] Values (mean ± SEM) represent the percentage of total radioactivity or as cpm per mg protein isolated in the free fatty acid (FFA) and phospholipid (PL) fractions from either medium only[b] or medium plus cells[c]. Data are given for five difference treatment groups: control; treatment with 30 μM iodoacetate for 3 h; pretreatment with 5 μM U26,384 for 60 or 90 min followed by treatment with 30 μM iodoacetate; pretreatment with 50 μM mepacrine for 90 min followed by treatment with 30 μM iodoacetate; pretreatment with 10 μM bromenol lactone for 60 min followed by treatment with 30 μM iodoacetate;

[b,c] Adapted from Jones et al., (1989)[b] and (Miller et al. 1993)[c].

Table 2. Effect of the Phospholipase Inhibitor, U26,384, on Electrolyte Concentrations in ATP-Depleted, Neonatal Rat, Cardiac Myocytes[a]

	Concentration (mmol/kg tissue, dry weight)				
Treatment	*Na*	*Mg*	*Cl*	*K*	*Ca*
Control (n = 42)	109 ± 14	46 ± 3	243 ± 11	613 ± 18	10 ± 3
Iodoacetate (n = 56)	498 ± 30[b]	14 ± 4[b]	433 ± 32[b]	78 ± 15[b]	242 ± 41[b]
Iodoacetate plus U26,384 (n = 49)	155 ± 17[c]	28 ± 3	255 ± 15	529 ± 29[c]	12 ± 2

Notes: [a] Values are given as mean ± SEM of elemental concentrations (mmol/kg dry weight) measured in individual cells (n) from 2-4 separate experiments. The electrolyte concentrations were measured by electron probe microanalysis. Data are given for three groups: control; treatment with 30 μM iodoacetate for 3 h; pretreatment with 5 μM U26,384 for 90 min followed by treatment, with 30 μM iodoacetate and an additional dose of 5 μM U26,384 for 3 h. From Jones et al. (1989).

[b] Significantly different from control.

[c] Significantly different from control (analysis of variance and multiple range test).

conditions (Katayama et al., 1992). Using the fluorescent $[Ca^{2+}]_i$ indicator, fura-2, measurements were performed in cultured, neonatal rat, cardiac myocytes following treatment with 40 uM iodoacetate (IAA) and 5 uM U26,384. Spontaneous cellular contractions decreased progressively then ceased following the introduction of IAA, whereas cells on the control plates contracted with normal $[Ca^{2+}]_i$ transients throughout the 180 minutes experimental period. The $[Ca^{2+}]_i$ transients, calculated from the fluorescence emission ratio of fura-2 (340/380 nm excitation), initially ranged from approximately 100 to 600 nM during each contraction. $[Ca^{2+}]_i$ increased rapidly 15-90 minutes after introducing IAA into the culture medium and remained elevated thereafter exceeding 1000 nM, coincident with the cessation of cellular contractions. Treatment with U26,384 did not alter the increase in $[Ca^{2+}]_i$. Therefore, $[Ca^{2+}]_i$ still increased under conditions where phospholipid degradation, unesterified fatty acid accumulation, morphological damage, and gross calcium overload were prevented. These data suggest that U26,384 prevents cell damage and gross calcium loading in cardiac myocytes by limiting hydrolysis of membrane phospholipids in spite of an early increase in intracellular free calcium levels. Whether or not this hydrolytic activity is induced by increased $[Ca^{2+}]_i$ remains unclear; all the data would be consistent with increased activity of either a calcium-dependent or independent phospholipase A_2.

Using the same cultured, neonatal rat myocyte model with metabolic inhibition produced by iodoacetic acid (IAA), we also have determined the effects of a bromoenol lactone (BrEL, from Monsanto). This compound was designed to specifically inhibit the calcium-independent, plasmalogen-selective phospholipase A_2 isolated from heart (Hazen et al., 1991b; Miller et al., 1993). Pretreatment with 0.5 to 10 uM BrEL before incubation with IAA progressively but only partially reduced increased ^{3}H-FFA levels from 2.1 ± 0.2 to 1.5 ± 0.2% (controls, 0.7 ± 0.1%). However, myocytes exhibited severe morphological damage with IAA exposure, irrespective of BrEL treatment. Comparatively, pretreatment with 5 uM steroidal diamine U26-384 prevented the morphological damage and reduced ^{3}H-FFA release to 1.0 ± 0.1%. Incubation of myocytes with 1 uM 3H-BrEL shows that approximately 8-10% becomes associated with the cell pellet within 15 min. These data suggest that BrEL may limit PLA_2-mediated phospholipid hydrolysis and the accumulation of unesterified fatty acids but not the morphological damage in myocytes, in contrast to the steroidal diamine. The highly reactive nature of BrEL may limit its cellular permeability thereby reducing the effective intracellular concentration of the agent. Still the partial effectiveness of this compound further implicates the involvement of a phospholipase A_2 in the early events of myocardial ischemia, consistent with our studies using U26, 384 and mepacrine.

III. MECHANISMS OF MYOCARDIAL CELL INJURY

Potential mechanisms for the early increase in $[Ca^{2+}]_i$ include a sustained and unregulated influx of calcium ions following the primary event in an ischemic episode, (i.e., oxygen depletion due to interrupted blood flow). Subsequent metabolic changes most likely increase calcium movement across the sarcolemma due to altered function of the slow calcium channel and other calcium transport systems plus the release of calcium from the sarcoplasmic reticulum and mitochondria (Braunwald et al., 1982; Barry et al., 1993; Buja et al., 1993). Interestingly, arachidonic acid can stimulate a transient increase in $[Ca^{2+}]_i$ by modulating sarcolemmal ion channels and/or calcium release from the sarcoplasmic reticulum (Damron and Bond, 1993). Whether a similar phenomenon is operative in ischemic myocardium is unknown. Note that extracellular calcium concentrations greatly exceed cytoplasmic levels in normal cells, millimolar versus submicromolar. Also, calcium *per se* does not injure myocytes, but extended elevation in $[Ca^{2+}]_i$ should be considered as a triggering mechanism for other cellular processes. The progressive increase in $[Ca^{2+}]_i$ may have a number of deleterious effects, including activation of ATPases which accelerate ATP depletion, mitochondrial calcium accumulation which inhibits ATP production, and activation of various phospholipases and proteases. The increase in $[Ca^{2+}]_i$ also can lead to the induction of ventricular arrhythymias (Thandroyen et al., 1991) The critical transition to irreversible injury appears to be loss of the ability to maintain compartmentalization, especially across the plasma membrane. Protease activation is likely important in cytoskeletal damage which destabalizes the support of the plasma membrane and contributes to membrane dysfunction (Van Winkle et al., 1995). Therefore, one must consider changes in membrane integrity as the critical aspect of any mechanism leading to cell death.

A. Phospholipase Activation

Phospholipase-mediated phospholipid degradation is considered to play an important role in the lipid alterations in ischemic myocardium (Hseuh et al., 1979; Chien et al., 1981; Chien et al., 1984; Shaikh et al., 1981; van der Vusse et al., 1987; Burton et al., 1986; Ford and Gross, 1989). In fact, it is difficult to conceive a pathogenic scheme not involving phospholipases, particularly a phospholipase A. The progressive accumulation of free fatty acids coupled with the transient increase in lysophospholipids is consistent with activation of a phospholipase A. However, contributions from other pathways to the production of free fatty acids must be considered. The sequential activation of a phospholipase C and diacylglycerol and monoacylglycerol lipases could lead to the release of arachidonic acid and other fatty acids. Phospholipase C production of diacylglycerols and inositol-phosphates, as well as phospholipase A_2 production of arachidonic acid leading to its potent biologically active metabolites, may have significant effects leading to cellular dysfunction and subsequent loss of structural integrity. The protective effects of various phospholipase inhibitors reinforce the involvement of a phospholipase. Thus, all evidence to date is consistent with impairment of phospholipid metabolism during ischemia.

A key issue is the nature of the early events initiating phospholipid alterations. Activation of a calcium-dependent phospholipase A secondary to an increase in cytosolic calcium has long been proposed as an initiating mechanism (Buja, 1991). The temporal changes in calcium metabolism and fatty acid accumulation form the basis of this hypothesis. The apparent dependency of cell damage on the availability of calcium also contributes to the potential of this interaction.

However, recent work by Gross and colleagues has implicated a calcium-independent, plasmalogen-selective phospholipase A_2 in the lipid alterations associated with myocardial ischemia (Hazen and Gross, 1991; Hazen, Ford, and Gross, 1991; Hazen et al., 1991). The relatively high concentration of plasmalogens in heart seems well-matched for the substrate selectivity of this enzyme. In Langendorff-perfused hearts, its activity in the microsomal fraction increased 10-fold after 15 minutes of ischemia without additional translation or transcription. Interestingly, ATP activates this enzyme and lowers its rate of thermal denaturation and chemical modification of an essential thiol residue (Hazen and Gross, 1991). The ATP regulation stems from the formation of a complex with phosphofructokinase. Clearly, ischemic myocardium would be particularly vulnerable to increased activity of this phospholipase due to its plasmalogen-selectivity.

Activation of phospholipases by other metabolic alterations can not overlooked. The potential problems stemming from receptor-mediated responses is very unclear at this time. The effects of altered phospholiase C production of diacylglycerol and inositol-phosphates can not be fully interpreted without additional information in a variety of models. Effects of changes in membrane lipid composition on other signal transduction systems, such as adenylate cyclase, also would reflect abnormal lipid metabolism, albeit indirectly. The potential impact of platelet activating factor released from other cells present in the ischemic myocardium, including platelets and endothelial cells, could also contribute to myocyte injury directly or via secondary processes, such as neutrophil infiltration (Lefer 1989).

B. Imbalance Between Phospholipid Synthesis and Degradation

The free fatty acids which accumulate in ischemic myocardium are derived in part from exogenous uptake and internal sources, including degradation of membrane phospholipids (Buja, 1991). Another mechanism leading to phospholipid degradation may involve impairment of the reacylation and *de novo* synthesis pathways (Hseuh et al., 1979; Chien et al., 1981; Chien et al., 1984; Shaikh et al., 1981; van der Vusse et al., 1987; Burton et al., 1986). Both of these pathways are driven by ATP-dependent reactions. Therefore, fatty acid utilization by these pathways may become limited as ATP is depleted during states of limited oxygen availability, such as ischemia and hypoxia. Otani et al. have presented evidence that both enhanced degradation and impaired reacylation and *de novo* synthesis are operative in myocardial ischemia (Otani et al., 1989). Additionally, decreased mitochondrial beta oxidation stemming from oxygen depletion would accentuate the rate and magnitude of free fatty accumulation. The 85% decrease in the glutathione redox potential observed in ischemic myocardium may contribute to alterations in fatty acid oxidation (Pauly et al., 1987). The activity and regulation of carnitine-palmitoyl transferase reflects changes in protein thiol oxidation (see chapter by Mar and McMillin).

These alterations lead to accumulation of free fatty acids and their derivatives, acyl-CoA and acylcarnitine. Briefly, oxygen depletion leads to decreased fatty acid oxidation via beta-oxidation in the mitochondria. Subsequently, increases in the metabolites can occur at each preceding step, including the initial pool of free fatty acids. Continued uptake of extracellular fatty acids and normal rates of phospholipid hydrolysis contribute to the accumulation of unesterified fatty acids within the cell. Increased rates of phospholipid hydrolysis would accelerate the rate of accumulation of the primary fatty acid pool and of the secondary intermediate pools.

Neutral lipids in the form of triglycerides also accumulate in ischemic myocardium (Buja, 1991). Lipid accumulation is particularly prominent in states of low-flow ischemia, hypoxia, and ischemia followed by reperfusion. The accumulation of large numbers of triglyceride-rich lipid droplets in the ischemic myocardium may represent an adaptive response by the cells to decrease the excess free fatty acids by esterifying them with glycerol. This relationship between altered phospholipid metabolism and triacylglycerol accumulation has been studied in an isolated heart model of global ischemia (Burton et al., 1986). In early myocardial ischemia, arachidonic acid and other fatty acids released from membrane phospholipids were incorporated into triacylglycerols. Later in the course of ischemia, free arachidonic acid and other free fatty acids accumulated. These findings suggest that esterification mechanisms become impaired as ischemia progresses, thereby contributing to the sequential increases in free fatty acids. Thus, the progressive accumulation of monoacyl lipids during ischemia most likely represents the combined effects of increased fatty acid production by phospholipases, decreased utilization via reacylation and *de novo* synthesis, and decreased oxidation. Reoxygenation can remedy the latter two pathways, but it should have no direct impact on phospholipase activity.

C. Free Radical-mediated Injury

Superoxide anions (O_2^-), hydrogen peroxide (H_2O_2), and the highly reactive hydroxyl radicals (HO•) are produced normally in aerobic cells via enzymatic and non-enzymatic reactions (Farber 1990; Janssen 1993). These chemically reactive oxygen species may contribute to membrane phospholipid damage during myocardial ischemia. The increase in oxidized glutathione levels during ischemia may enhance this process either by increasing hydroxyl radical formation or by altering enzyme activities due to protein thiol oxidation (Ziegler, 1985). Free radicals can interact with the polyunsaturated fatty acids of phospholipids with resultant generation of lipid peroxides and subsequent degradation of the fatty acids (Buja, 1991). Free radicals also can induce increases in $[Ca^{2+}]_i$ in myocytes (Burton et al., 1990). In fact, measurement of lipid peroxidation products is used to characterize oxidative stress. Peroxidized fatty acids tend to be more susceptible to phospholipase cleavage (Massey and Burton, 1989). Thus, the combination of increased lipid peroxidation and phospholipase activation synergistically may potentiate phospholipid degradation thereby expediting membrane damage. Scavengers of free radicals and inhibitors of lipid peroxidation, such as alpha tocopherol, can retard phospholipid degradation and membrane damage in ischemic myocardium (Massey and Burton, 1989).

IV. CONCLUSIONS AND PERSPECTIVES

Myocardial necrosis following a transient ischemic episode represents the most common cause of disability and death in our society today. The progression of localized injury initiated by impaired blood flow exhibits characteristic biochemical changes. Increased phospholipid degradation with unesterified fatty acid accumulation and increased cellular calcium concentrations represent three such alterations, in addition to ATP depletion and acidosis. Cytoskeletal damage also occurs. The ultimate endpoint of this process is complete loss of cellular membrane integrity, both functional and structural, in the irreversibly injured cell. Any intervention which can prevent the onset of this succession, or delay its rate of progression, would lead to a significant improvement of the patient's prognosis. Although this presentation has focused on myocardial ischemia, it must be noted that similar observations and conclusions can be considered in other tissues, including cerebral and renal ischemia (Siesjö and Katsura, 1992; Weinberg 1991). Therefore, it is critical that we clearly determine the specific biochemical aberrations which occur in ischemic tissue, only then can we develop therapeutic interventions to reduce the debilitating and lasting effects of ischemia.

A. Consequences of Altered Lipid Metabolism

Phospholipid degradation contributes to alterations in membrane integrity and function in several ways. First, phospholipid hydrolysis directly alters the composition of cell membranes. With continued degradation, the basic phospholipid bilayer structure will cease to exist. Implicitly, release of fatty acids from cellular phospholipids requires the action of a phopholipase. The accumulating monoacyl products, including free fatty acids, lysophospholipids, acyl-carnitines, and acyl-CoAs, are amphipathic lipids which can incorporate into cell membranes (Katz et al., 1981; Corr et al., 1984). Increased concentrations of these lipids leads to changes in membrane lipid composition and fluidity. These related processes lead to loss of the structural and functional integrity of the cellular membranes. Subsequently, compartmentalization of the cell itself, and of intracellular processes, can not be maintained. This defines the point when the cell becomes irreversibly damaged and non-viable.

Many studies have documented that the lipid composition of the membrane microenvironment containing membrane-bound enzymes and other proteins can have profound effects on their function (Spector et al., 1985; Philipson et al., 1985; Katz et al., 1981; Katz et al., 1981; Corr et al., 1984; Miller and Weinhold, 1981). Amphipathatic lipids can alter the function of key transport systems, including the Na^+, K^+-ATPase and calcium transporters. In addition, elevation of lysophospholipid levels can induce significant electrophysiological alterations which can lead to arrhythmias (Corr et al., 1984). Treatment of ischemic hearts with carnitine palmitoyl-CoA transferase inhibitors can provide reduction in ischemic injury (Buja, 1991). This appears to be mediated both by a reduction in long-chain acylcarnitines and a shift to increased glucose utilization (Lopaschuk et al., 1993).

The elegant work of Gross and colleagues suggests involvement of the calcium-independent, plasmalogen-selective phospholipase A_2 in ischemic injury (Wolf and Gross, 1985; Hazen, Ford, and Gross, 1991; Hazen et al., 1991; Hazen and Gross, 1992a, 1992b; Gross, 1992). Its substrate specificity alone makes it a likely candidate in the plasmalogen-rich heart tissue. Additionaly, its activation by ATP via the phosphofructokinase complex may provide the triggering mechanism early in ischemia without an increase in $[Ca^{2+}]_i$. Hydrolysis of membrane phospholipids may lead to progressive loss of membrane function with regard to calcium homeostasis and finally calcium overload with cell death. However, some inconsistencies must be considered regarding this hypothesis. This is an ATP-activatible enzyme, yet ATP levels are decreasing during ischemia. One might consider different ATP pools, however phosphofructokinase is a cytosolic enzyme, presumably regulated by the cytoplasmic ATP pool which would be decreasing during ischemia, although an early adaptive response may produce a transient increase in ATP levels. The implications of its regulation via the complex with phosphofructokinase also remain unclear. Perhaps many other degradative enzymes form regulatory complexes with critical glycolytic enzymes thereby providing a overall level of regulation related to cellular energy charge. Furthermore, the accumulation of fatty acids occurs progressively during ischemia yet with decreasing ATP levels this enzyme should become progressively less active. Therefore, another phospholipase would be necessary for the continued hydrolytic release of fatty acids from membrane phospholipids. The data currently available, including the effects of phospholipase inhibitors, are consistent with either a calcium-dependent or-independent phospholipase.

B. Consequences of Increased Intracellular Calcium Levels

Clearly, calcium levels increase within the ischemic myocardium and myocytes used in cellular models used to study this pathogenic phenomenon. The early increase in $[Ca^{2+}]_i$ most likely represents a subtle and reversible impairment of the complex system of ion channels and pumps which maintain intracellular calcium homeostasis. The myocytes can easily recover from short periods of this nature.

However, it must be noted that elevated calcium concentrations *per se* are not detrimental to myocytes. These cells are uniquely capable of clearing large amounts on calcium within very brief periods of time. Therefore, any consideration of calcium as a trigger precipitating injury must be related to the intracellular dose of free calcium ions. Increased $[Ca^{2+}]_i$ could increase the duration of increased activity for calcium-dependent enzymes indirectly leading to a calcium-dependent injury.

The calcium overload that appears with longer periods of ischemia represents the transition into a state of irreversible injury. The transformation most likely stems from the loss of membrane integrity, both structural and functional. Once the cell is unable to maintain the calcium concentration gradient across its plasma membrane, the cell can not remain viable. Therefore, the appearance of calcium phosphate precipitates characterizes the terminal stage of pathogenesis in myocardial ischemia.

C. Potential Therapeutic Approaches to Minimize Ischemia-induced Damage

Clearly, inhibition of phospholipase-mediated degradation of membrane phospholipids during an ischemic episode would seem to be the most likely approach to preventing, or at least delaying, loss of membrane integrity. Any time gained before the ischemic tissue makes the transition from reversible to irreversible injury represents an opportunity for recovery. Regaining full function in the ischemic zone is the primary goal, second only to prevention of the ischemia in the first place. However, to develop an efficacious therapeutic agent requires that we fully understand the biochemical etiology of this problem. The current body of data strongly implicates the role of some phospholipase; however, we currently are unable to definitively identify a specific enzyme upon which to focus our efforts. Based on the evidence currently available, we must assume that future studies will more clearly identify the specific enzyme involved and lead to a method of abating its detrimental effects during a limited period of ischemia.

The increased rate of release of fatty acids from phospholipids implicitly demonstrates the catalytic action mediated by some phospholipase. The protective effects observed in cellular models of injury using the steroidal diamine and mepacrine further support this point. The calcium-independent, plasmalogen-selective phospholipase A_2 elegantly characterized by Gross and colleagues seems to an excellent candidate (Hazen et al., 1991a). But, as discussed earlier, there remain some significant points of uncertainty regarding the role of this enzyme when ATP levels are decreasing. Most data suggest that membrane damage is inversely related to intracellular ATP concentration yet the activity and half-life of this hydrolase is dependent on ATP. This contradiction appears to be intractable and most likely represents a gap in our understanding of all the aspects of its regulation. Our preliminary work demonstrating the relative ineffectiveness of the bromoenol lactone in one cellular model of injury can be rationalized most easily by the highly reactive nature of this compound. It may not be able to inactivate the calcium-independent phospholipase A_2 because it reacts with something else before gaining access to the enzyme within the cell. Perhaps some future derivative will be more membrane permeable or less non-specifically reactive and subsequently will be more efficacious. The ability of the steroidal diamine to prevent phospholipid degradation, calcium overload, and morphological damage in metabolically inhibited myocytes strongly reinforces the potential of a phospholipase inhibitor. Whether this compound mediates its effects via a calcium-dependent or -independent phospholipase does not alter the primary conclusion; these hydrolytic enzymes appear to play a critical role in the development of irreversible injury and cell death.

It may be possible to mute the detrimental effects of increased unesterified fatty acid concentrations which occur during ischemia without preventing increased phospholipid degradation. Evidence suggests that a relative imbalance between glycolysis and glucose oxidation during ischemia may potentiate cell damage (Opie, 1993; Lopaschuk, 1993). Agents which intervene at metabolic regulatory points, not directly related to preventing phospholipid degradation, may provide a means to limit the degree to which these metabolic pathways become disproportionate during ischemia. Activation of pyruvate dehydrogenase complex by dichloroacetate can enhance acetyl-CoA production and lead to decreased carnitine-palmitoyl transferase I activity by increased malonyl-CoA production via acetyl-CoA carboxylase (Saddik, 1993). This leads to a relative increase in glycolysis despite increased phospholipid degradation and myocardial protection. Although this could afford some protection metabolically, phospholipid degradation would continue ultimately leading to loss of membrane integrity. A multiple inhibitor therapy may be able to delay cell injury effectively by preserving both metabolic balance and membrane phospholipids.

The role of calcium in the initiation of injury will not be fully resolved until we know the specific biochemical processes producing the cellular damage. The possibility remains that increased calcium concentrations may increase the hydrolytic activity of at least one phospholipase. If this proves to be the case, then one might consider approaches to block the activation site, a competive inhibitor. Our studies suggest that the accumulation of fatty acids is dependent on calcium. However, reducing $[Ca^{2+}]_i$ in the presence of membrane dysfunction does not seem feasible since intracellular concentrations are submicromolar and extracellular levels are millimolar. Bringing extracellular calcium concentrations to effective levels would impede normal myocardial function even if possible. Therefore, it seems likely that any attempt to moderate irreversible injury by removing calcium *per se* will be untenable. This leaves open the possibility of developing inhibitors of the activation site which would be operative in spite of the early increase in $[Ca^{2+}]_i$.

The temporal relationship of calcium and phospholipase in ischemic injury seems reflective of the "chicken versus the egg" metaphor, however the solution to preventing myocardial cell damage always leads back to the phospholipase. This point should remain clear to those seeking to moderate the detrimental effects of ischemia on cellular membrane structure and function.

ACKNOWLEDGMENTS

We wish to thank Kailas Patel for technical assistance, Donna Buja for preparing the cultured myocytes, Drs. John E. Bleasdale and Roy A. Johnson of The Upjohn Company for graciously donating the U26,384 for us to undertake these studies, and Dr. Randy Weiss of the Monsanto Company for providing the bromoenolactone. Our research was supported by NIH grants HL-47462 and HL-17669, a grant from the American Heart Association, Texas Affiliate (88R074), and a Materials Testing Agreement from the Monsanto Company.

REFERENCES

Barry, W. H., & Bridge, J.H.B. (1993). Intracellular calcium homeostasis in cardiac myocytes. Circ. Res. 87, 1806-1815.

Barry, W.H., Peeters, G.A., Rasmussen, C.A.F., & Cunningham, M.J. (1987). Role of changes in $[Ca^{2+}]_i$ in energy deprivation contracture. Circ. Res. 61, 726-734.

Brachmann, J., & Schömig, A. (eds). (1989). Adrenergic System and Ventricular Arrhythmias in Myocardial Infarction Springer-Verlag.

Braunwald, E. (1982). Mechanism of action of calcium channel blocking agents. N. Eng. J. Med. 307, 1618-1627.

Buja, L.M., Fattor, R.A., Miller, J.C., Chien, K.R., & Willerson, J.T. (1990). Effects of calcium loading and impaired energy production on metabolic and ultrastructural features of cell injury in cultured neonatal rat cardiac myocytes. Lab. Invest. 63, 320-331.

Buja, L.M., Eigenbrodt, M., & Eigenbrodt, E. (1993). Apotosis and necrosis: Basic types and mechanisms of cell death. Arch. Pathol. Lab. Med. 117, 1208-1214.

Buja, L.M., Hagler, H.K., & Willerson, J.T. (1988). Altered calcium homeostasis in the pathogenesis of myocardial ischemic and hypoxic injury. Cell Calcium 9, 205-217.

Buja, L.M. (1991). Lipid abnormalities in myocardial cell injury. Trends Cardiovascular Med. 1, 40-45.

Burton, K.P., Buja, L.M., Sen, A., Willerson, J.T., & Chien, K.R. (1986). Accumulation of arachidonate in triacylglycerols and unesterified fatty acids during ischemia and reflow in the isolated rat heart. Am. J. Pathol. 124, 238-245.

Burton, K.P., Morris, A.C., Massey, K.D., Buja, L.M., & Hagler, H.K. (1990). Free radicals alter ionic calcium levels and membrane phospholipids in cultured rat ventricular myocytes. J. Molec. Cell Cardiol. 22, 1035-1047.

Chien, K.R., Sen, A., Reynolds, R., Chang, A., Kim, Y., Gunn, M.D., Buja, L.M., Willerson, J.T. (1985). Release of arachidonate from membrane phospholipids in cultured myocardial cells during ATP depletion: Correlation with progression of cell injury. J. Clin. Invest. 75,1770-1780.

Chien, K.R., Reeves, J.P., Buja, L.M., Bonte, F., Parkey, R.W., & Willerson, J.T. (1981). Phospholipid alterations in canine ischemic myocardium. Circ. Res. 48, 711-719.

Chien, K.R., Han, A., Sen, A., Buja, L.M., & Willerson, J.T. (1984). Accumulation of unesterified arachidonic acid in ischemic canine myocardium. Circ. Res. 54, 313-322 .

Corr, P.B., Gross, R.W., & Sobel, B.E. (1984). Amphipathic metabolites and membrane dysfunction in ischemic myocardium. Circ. Res. 55, 135-154.

Damron, D.S., & Bond, M. (1993). Modulation of Ca^{2+} cycling in cardiac myocytes by arachidonic acid. Circ. Res. 72, 376-386.

Das, D.K., Engelman, R.M., Rousou, J.A., Breyer, R.H., Otani, H., & Lemeshow, S. (1986). Role of membrane phospholipids in myocardial injury induced by ischemia and reperfusion. Am. J. Physiol. 251, H71-H79.

Doeller, J.E., & Wittenberg, B.A. (1990). Intracellular calcium and high-energy phosphates in isolated cardiac myocytes. Am. J. Physiol. 259, Hl85l-Hl859.

Farber, J.L., Kyle, M.E., & Coleman, J.B. (1990). Biology of disease: Mechanisms of cell injury by activated oxygen species. Lab. Invest. 62, 670-679.

Ford, D.A., & Gross, R.W. (1989). Differential accumulation of diacyl and plasmalogenic diglycerides during myocardial ischemia. Circ. Res. 64, 173-177.

Gross, R.W. (1992). Myocardial phospholipases A_2 and their membrane substrates. Trends Cardiovasc. Med. 2, 115-121.

Gunn, M.D., Sen, A., Chang, A., Willerson, J.T., Buja, L.M., & Chien, K.R. (1985). Mechanisms of accumulation of arachidonic acid in cultured myocardial cells during ATP depletion. Am. J. Physiol. 249, H1188-H1194.

Hagve, T.A., Sprecher, H., & Hohl, C.M. (1990). The effect of anoxia on lipid metabolism in isolated rat cardiac myocytes. J. Mol. Cardiol. 22, 1467-1475.

Hazen, S.L., & Gross, R.W. (1992a). Identification and characterization of human myocardial phospholipase A2 from transplant recipients suffering from end-stage ischemic heart disease. Circ. Res. 70, 486-495.

Hazen, S.L., & Gross, R.W. (1992b). Principles of membrane biochemistry and their applications to the pathophysiology of cardiovascular disease. In The Heart and Cardiovascular System, 2 ed., (Fozzard, H.A. et al., Eds.), pp. 839-860. Raven Press, New York.

Hazen, S.L., Ford, D.A., & Gross, R.W. (1991). Activation of a membrane-associated phospholipase A2 during myocardial ischemia which is highly selective for plasmalogen substrate. J. Biol. Chem. 266, 5629-5633.

Hazen, S.L., & Gross, R.W. (1991). ATP-dependent regulation of rabbit myocardial cytolsolic calcium-independent phospholipase A_2. J. Biol. Chem. 266, 14526-14534.

Hazen, S.L., Zupan, L.A., Weiss, R.H., Getman, D.P., & Gross, R.W. (1991). Suicide inhibition of canine myocardial cytosolic calcium-independent phospholipase A_2. J. Biol. Chem. 266, 7227-7232.

Hostetler, K.Y., & Jellison, E.J. (1989). Role of phospholipases in myocardial ischemia: Effect of cardioprotective agents on the phospholipases A of heart cytosol and sarcoplasmic reticulum *in vitro*. Mol. Cell Biochem. 88, 77-82.

Hseuh, W., & Needleman, P. (1979). Cardiac and renal lipases and prostaglandin biosynthesis. Lipids 14, 236-240.

Janssen, Y.M., van Houten, B., Borm, P.J.A., & Mossman, B.T. (1993). Biology of disease: Cell and tissue responses to oxidative damage. Lab. Invest. 69, 261-274.

Jones, R.L., Miller, J.C., Hagler, H.K., Chien, K.R., Willerson, J.T., & Buja, L.M. (1989). Association between inhibition of arachidonic acid release and prevention of calcium loading during ATP depletion in cultured rat cardiac myocytes. Am. J. Pathol. 135, 541.

Katayama, A., Buja, L.M., & Miller, J.C. (1992). Phospholipase inhibition prevents cell damage in metabolically-inhibited cardiac myocytes despite increased intracellular calcium. FASEB J. 6, A163.

Katz, A.M., & Messineo, F.C. (1981). Lipid-membrane interactions and the pathogenesis of ischemic damage in the myocardium. Circ. Res. 48, 1-16.

Koretsune, Y., & Marban, E. (1990). Mechanism of ischemic contracture in ferret hearts: relative roles of $[Ca^{2+}]_i$, elevation and ATP depletion. Am. J. Physiol. 258, H9-Hl6.

Lee, H.C., Monhabir, R., Smith, N., Franz, M.R., & Clusin, W.T. (1988). Effect of ischemia on calcium-dependent fluorescence transients in rabbit hearts containing indo-1. Circulation 78, 1047-1059.

Lefer, A.M. (1989). Platelet activating factor (PAF) and its role in cardiac injury. In: Prostaglandins in Clinical Research: Cardiovascular System, (Schrör, K., & Sinzinger, H., Eds.) pp. 53-60. Alan R. Liss, New York.

Lopaschuk, G.D., Wambolt, R.B., & Barr, R.L. (1993). An imbalance between glycolysis and glucose oxidation is a possible explanation for the detrimental effects of high levels of fatty acids during aerobic reperfusion of ischemic hearts, J. Pharm. Exp. Therap. 264, 135-144.

Massey, K.D., & Burton, K.P. (1989). Alpha-tocopherol attenuates myocardial membrane-related alterations resulting from ischemia and reperfusion. Am. J. Physiol. 256, H1192-H1199.

Miller, J.C., Buja, L.M., Patel, K.D., & Weiss, R.H. (1993). Effects of phospholipase inhibition with bromoenol lactone in metabolically-inhibited, neonatal rat, cardiac myocytes. Circulation 88(Suppl. I), I-489.

Miller, J.C., & Weinhold, P.A. (1981). Cholinephosphotransferase in rat lung: The *in vitro* synthesis of dipalmitoylphosphatidylcholine from dipalmitoylglycerol, J. Biol. Chem. 259, 12662-12665.

Miyazaki, Y., Gross, R.W., Sobel, B.E., & Saffitz, J.E. (1990). Selective turnover of sarcolemmal phospholipids with lethal cardiac myocyte injury. Am. J. Physiol. 259, C325-C331.

Morris, A.C., Hagler, H.K., Willerson, J.T., & Buja, L.M. (1989). Relationship between calcium loading and impaired energy metabolism during Na^+, K^+ pump inhibition in cultured neonatal rat cardiac myocytes. J. Clin. Invest. 83, 1876-1877.

Muntz, K.H., Zhao, M., & Miller, J.C. (1994). Downregulation of myocardial β-adrenegic receptors: Receptor subtype selectivity. Circ. Res. 74, 369-375.

Muntz, K.H., Neyman, S.L., & Miller, J.C. (1993). Alterations in $alpha_1$-adrenergic receptor-mediated phosphatidylinositol turnover in hypoxic myocytes, J. Mol. Cell Biol. 25, 1187-1202.

Opie, L.H. (1993). The mechanism of myocyte death in ischemia, Eur. Heart J. 14G, 31-33.

Otani, H., Renuka, M., Jones, R.M., & Das, D.K. (1989). Mechanism of membrane phospholipid degradation in ischemic-reperfused rat hearts. Am. J. Physiol. 257, H252-H258.

Patel, T., Gores, G.J., & Kaufmann, S.H. (1996). The role of proteases during apoptosis. FASEB J. 10, 587-597.

Pauly, D.F., Yoon, S.B., & McMillin, J.B. (1987). Carnitine-acylcarnitine translocase in ischemia: Evidence for sulfhydryl modification. Am. J. Physiol. 253, H1152-H1565.

Philipson, K.D., & Ward, R. (1985). Effects of fatty acids on Na^+-$Ca2^+$ exchange and Ca^{2+} permeability of cardiac sarcolemmal vesicles. J. Biol. Chem. 260, 9666-9671.

Reimer, K.A., & Jennings, R.B. (1986). Myocardial ischemia, hypoxia, and infarction. In: The Heart and Cardiovascular System (Fozzard, H.A., Haber, E., Jennings, R.B., Katz, A.M., & Morgan, H.E., Eds.), pp. 1133-1201.

Revtyak, G.E., Buja, L.M., Chien, K.R., & Campbell, W.B. (1990). Reduced arachidonate metabolism in ATP-depleted myocardial cell occurs early in cell injury. Am. J. Physiol. 259, H582-H591.

Saddik, M., Gamble, J., Witters, L.A., & Lopaschuk, G.D. (1993). Acetyl-CoA carboxylase regulation of fatty acid oxidation in the heart, J. Biol. Chem. 268, 25836-25845.

Sen, A., Miller, J.C., Reynolds, R., Willerson, J.T., Buja, L.M., & Chien, K.R. (1988). Inhibition of the release of arachidonic acid prevents the development of sarcolemmal membrane defects in cultured rat myocardial cells during adenosine triphosphate depletion. J. Clin. Invest. 82, 1333.

Shaikh, N.A., & Downar, E. (1981). Time course of changes in porcine myocardial phospholipid levels during ischemia. Circ. Res. 49, 316-325.

Shaikh, N.A., & Downar, E. (1987). Effects of chronic amiodarone treatment on cat myocardial phospholipid content and on *in vitro* phospholipid catabolism, Mol. Cell Biochem. 78, 17-25.

Shaikh, N.A. (1992). Effect of amiodarone therapy on the time course of myocardial phospholipid hydrolysis during *in vitro* total ischaemia in cat hearts, J. Molec. Cell. Cardiol. 24, 507-521.

Siesjö, B.K., & Katsura, K. (1992). Ischemic brain damage: Focus on lipids and lipid messengers. In: Neurobiology of Essential Fatty Acids, (Bazan, N.G., Muphy, M.G., & Toffano, G., Eds.), pp. 41-56. Plenum Press, New York.

Singer, S.S., & Nicolson, G.L. (1972). The fluid mosaic model of the structure of cell membranes. Science 175, 720-731.

Spector, A.A., & Yorek, M.A. (1985). Membrane lipid composition and cellular function. J. Lipid Res. 26, 1015-1035.

Steenbergen, C., Murphy, E., Watts, J.A., & London, R.E. (1990). Correlation between cytosolic free calcium, contracture, ATP, and irreversible ischemic injury in perfused rat heart. Circ. Res. 66, 135-146.

Thandroyen, F.T., Bellotto, D., Katayama, A., Hagler, H.K., Willerson, J.T., & Buja, L.M. (1992). Subcellular electrolyte alterations during progressive hypoxia and following reoxygenation in isolated neonatal rat ventricular myocytes. Circ. Res. 71, 106-119.

Thandroyen, F.T., Morris, A.C., Hagler, H.K., Ziman, B., Pai, L., Willerson, J.T., & Buja, L.M. (1991). Intracellular calcium transients and arrhythmias in isolated cells. Circ. Res. 69, 810-819.

van Bilsen, M., van der Vusse, G.J., Willemsen, P.H.M., Coumans, W.A., Roemen, T.H.M., & Reneman, R.S. (1989). Lipid alterations in isolated, working rat hearts during ischemia and reperfusion: Its relation to myocardial ischemia. Circ. Res. 64, 304-314.

van der Vusse, G.J., Prinzen, F.W., van Bilsen, M., Engels, W., & Reneman, R.S. (1987). Accumulation of lipids and lipid-intermediates in the heart during ischemia. Basic Res. Cardiol. 82, 157-167.

Van Winkle, W.B., Snuggs, M., & Buja, L.M. (1995). Hypoxia-induced alterations in cytoskeleton coincide with collagenase expression in cultured neonatal rat cardiomyoeytes. J. Molec. Cell. Cardiol. 27, 2532-2542.

Wang, D., McMillin, J.B., Bick, R., & Buja, L.M. (1996). Response of the neonatal rat cardiomyocyte in culture to energy depletion: Effects of cytokines, nitric oxide, and heat shock proteins. Lab. Invest. 75, 809-818.

Weinberg, J.M. (1991). The cell biology of ischemic renal injury. Kidney Int. 39, 476-500.

Wolf, R.A., & Gross, R.W. (1985). Identification of neutral active phospholipase C which hydrolyzes choline glycerolphospholipids and plasmalogen selective phospholipase A_2 in canine myocardium. J. Biol. Chem. 260, 7295-7303.

Ziegler, D.M. (1985). Role of reversible oxidation-reduction of enzyme thiols-disulfides in metabolic regulation. Ann. Rev. Biochem. 54, 305-329.

FATTY ACID METABOLISM IN THE REPERFUSED ISCHEMIC HEART

Darrell D. Belke and Gary D. Lopaschuk

ABSTRACT

Elevated serum levels of fatty acids can be seen in most clinically relevant situations of myocardial ischemia. During an episode of mild to moderate ischemia these high levels of fatty acid have been shown to contribute to contractile dysfunction. Recent evidence suggests that high levels of fatty acids also depress the recovery of contractile

Advances in Lipobiology
Volume 2, pages 29-46.

ISBN 0-7623-0205-4

function following a transient episode of severe ischemia. The mechanisms by which fatty acid are detrimental during reperfusion of severely ischemic hearts appears to differ from the mechanism by which fatty acids enhance injury during ischemia itself. Evidence from our laboratory suggests that a marked depression of glucose oxidation during reperfusion may be primarily responsible for the detrimental effects of high levels of fatty acids. This is supported by the observation that interventions which overcome fatty acid inhibition of glucose oxidation will significantly improve post-ischemic functional recovery. The exact reason(s) why stimulating glucose oxidation during reperfusion is beneficial has yet to be determined. As will be discussed in this paper, we speculate that stimulation of glucose oxidation improves the coupling between glycolysis and glucose oxidation in the immediate reperfusion period. By doing so, a reduction in H^+ production from glucose metabolism will occur during the actual reperfusion period. This will result in a lessening of the myocardial H^+ load during the critical period in which the heart is recovering from the acidosis that has occurred during the actual ischemic episode. Inhibition of glucose oxidation by fatty acids may not contribute to injury following a period of mild to moderate ischemia, since the decrease in myocardial pH during ischemia is less dramatic. As a result, the heart is not overwhelmed with a high intracellular H^+ load during reperfusion.

I. INTRODUCTION

The role of fatty acid metabolism in mediating the damage induced by ischemia-reperfusion has been the subject of great conjecture since early studies showed a positive correlation between plasma fatty acid levels and the degree of myocardial injury following ischemia (Oliver et al., 1968; Rutenberg et al., 1969; Severeid et al., 1969; Kjekshus and Mjos, 1972; Russo and Margolis, 1972, Mueller and Ayres, 1978). The clinical relevance of this phenomenon can be observed as the dramatic rise in plasma fatty acid concentrations following acute myocardial infarction or cardiac surgery (Oliver et al., 1968; Allison et al., 1969; Opie, 1975; Svensson et al., 1990; Lopaschuk et al., 1994), likely a result of increased plasma catecholeamine concentrations and the use of anticoagulants such as heparin. Although clinically relevant to man, detrimental effects of high circulating fatty acid concentrations following ischemia have also been shown to occur in a number of animal models, including dog, pig, rabbit and rat hearts (Mjos, 1978; Liedtke et al., 1978; Lopaschuk et al., 1990; Johnson and Lewandowski, 1991).

Several different experimental models of ischemia have been used to examine the relationship between energy substrate metabolism and mechanical function during ischemia and during reperfusion following ischemia. The degree of ischemia used in the different studies varies, such as a 50 to 60% reduction in coronary flow used in both the in situ pig heart (Liedtke et al., 1978; Liedtke et al., 1988) and isolated working rat heart (Lopaschuk and Spafford, 1990), or the severe no flow of very low flow ischemia used in isolated Langendorff (Görge et al., 1991; Benzi and Lerch,1992) or working rat hearts (Lopaschuk et al, 1988; Lopaschuk et al, 1990). As may be expected, the differing models lead to different results as to the nature of the detrimental effects of fatty acids during ischemia and during reperfusion. Under low flow conditions in pig hearts, Liedtke and colleagues (1978) found that high levels of fatty acids depressed contractile function during ischemia itself, an observation we also observed in isolated rat hearts subjected to a 50% reduction in coronary flow (Lopaschuk and Spafford, 1989). Of interest, is the observation that high levels of fatty acids do not appear to have a detrimental effect during reperfusion following ischemia of moderate severity (Liedtke et al., 1988). In contrast, high concentrations of fatty acids are detrimental to the recovery of mechanical function during reperfusion following a severe episode of no flow ischemia (Ichihara and Neely, 1985; Lopaschuk et al., 1988; Johnson and Lewandowski, 1991). In fact, in the same study, Ichihara and Neely (1985) using working rat hearts demonstrated a detrimental effects of 1.2 mM palmitate on functional recovery following global no-flow ischemia, an effect that was not seen during reperfusion following low flow ischemia. As a result, the effects of fatty acids on reperfusion recovery of heart function is clearly dependent on the severity of the preceding ischemia.

Several theories have emerged as to the mechanism by which increased extracellular and intracellular fatty acids exert their detrimental effects during ischemia. The detrimental effects of fatty acid metabolism have previously been linked to: (1) perturbations in myocyte electrophysiology leading to the development of severe arrhythmia's during reperfusion (Bricknell and Opie, 1978), (2) the disruption of sarcolemmal and sarcoplasmic reticulum membrane integrity and transmembrane protein function secondary to the production of amphipathic metabolites such as acyl-CoA and acylcarnitine (see Katz and Messineo, 1981; Corr et al., 1984 and Van der Vusse et al., 1992 for reviews), or (3) the development of the ATP wasting futile cycle of triacylglycerol resynthesis and breakdown during ischemia and early reperfusion due to increased fatty acid accumulation (Van Bilsen et al., 1989). However, considerable debate continues as to whether the detrimental effects of high fatty acid concentrations are related primarily to the physical characteristics of the fatty acids and their metabolites (e.g., membrane perturbations), (4) or whether they occur secondary to an overeliance of the heart on fatty acids as a source of ATP production. The effects of ischemia-reperfusion and increased fatty acid concentrations on heart function and energy substrate metabolism are shown in Table 1. Although the detrimental effects of amphipathic fatty acids have been described in a number of studies (see Corr et al., 1984), others have dissociated the correlation between myocardial levels of these amphipathic compounds and the loss of myocardial function (Ichihara and Neely, 1985; Lopaschuk and Spafford, 1989: Lopaschuk et al., 1988). This paper will not debate these issues, but rather concentrate on the contribution of high levels of fatty acids to myocardial injury during reperfusion following a transient period of severe ischemia.

Table 1. Energy Substrate Metabolism Following Ischemia-Reperfusion

Study (Species)	*Substrate(s) Studied*	*Perfusion Conditions*	*Function Change*	*Results*
Schweiger et al. (1985a) (Dog)	$[^{11}C]$Palmitate	-*In situ* hearts subjected to 20 min of LAD occlusion. -Isotope injected after LAD occlusion.	-Depressed contractile recovery in affected area.	-Decreased fatty acid metabolism.
Myears et al. (1988) (Dog)	$[^{3}H]$Glucose, $[^{14}C]$Palmitate	-*In situ* hearts subjected to 1 hr of LAD occlusion. -Isotope injected after LAD occlusion.	-Coronary flow depressed.	-Glucose use increased. -Palmitate oxidation decreased.
Liedtke et al. (1988) (Pig)	$[^{14}C]$Palmitate	-*In situ* hearts subjected to 45 min of reduced LAD blood flow. -Isotope infused prior to and following ischemia. -Blood glucose > 90 mg/dl. -Low fat group - 0.5 mM -High fat group -1.4 mM.	-Depressed contractile recovery in affected area. -High fat worse than low fat group.	-Increased palmitate oxidation during reperfusion.
Lopaschuk et al. (1990) (Rat)	$[^{14}C]$Glucose $[^{14}C]$Palmitate	-Isolated working hearts. -25 min global no-flow ischemia. -11 mM glucose, 1.2 mM palmitate.	-Decreased mechanical functionduring reperfusion.	-Decreased Glucose oxidation. -Palmitate oxidation unchanged.
Johnston & Lewandowski (1991) (Rabbit)	$[^{13}C]$Pyruvate	-Langendorff perfused hearts -10 or 20 min global ischemia. -5 mM glucose, 2 mM palmitate.	-Fatty acid depresses contractility during reperfusion.	-Palmitate decreases pyruvate oxidation during reperfusion.

Benzi and Lerch (1992) (Rat)	[^{14}C]Glucose [^{14}C]Palmitate	-Langendorff perfused hearts -60 min of no-flow ischemia. -11 mM glucose, 0.4 mM palmitate	-Increased diastolicpressure. -Decreased LV pressure development.	-Increased glucose oxidation. -Decreased Palmitate oxidation.
Saddik & Lopaschuk (1992) (Rat)	[^{3}H]Palmitate [^{14}C]Palmitate	-Isolated working hearts. -30 min no-flow ischemia. -11 mM glucose, 1.2 mM palmitate	-Depressed functional recovery with fat. -Little depression in functional recovery without fat.	-Exogenous fatty acids preferred substrate in reperfusion. -Initial triacylglycerol synthesis increased.
Liedtke et al. (1993) (Pig)	[^{14}C]Palmitate [5-^{3}H]Glucose	-*In situ* hearts subjected to1 hr reduced LAD blood flow. -Plasma glucose 7-8 mM, plasma fatty acid 0.4 mM. -Function studies 4 days post surgery.	-Initial decrease in contractility. -Contractile recovery at 4 days.	-Decreased fatty acid oxidation. -Increased glucose usage.

II. FATTY ACID METABOLISM IN THE REPERFUSED ISCHEMIC HEART

The contribution of fatty acid oxidation as a source of ATP production during reperfusion following ischemia has recently been a source of debate. Early studies established that fatty acid oxidation is the primary substrate for energy metabolism in the aerobic heart, which supplies over 90% of total ATP production when circulating fatty acid levels are high (Bing, 1965; Neely and Morgan, 1974). The reliance of the heart on fatty acids as a source of energy is illustrated by the observation that even if hearts are perfused in the absence of any exogenous fatty acid supply, they will continue to oxidize endogenous triacylglycerol stores to supply upwards of 40% of overall ATP production (Saddik and Lopaschuk, 1991).

During ischemia fatty acid oxidation decreases secondary to a decrease in O_2 supply. During reperfusion fatty acid oxidation rapidly recovers and remains the primary substrate for energy metabolism (Myears et al., 1987; Liedtke et al., 1988, Lopaschuk et al., 1990; Saddik and Lopaschuk, 1992). In most of these studies, fatty acid oxidation was measured directly using ^{14}C labeled palmitate. An example of this is the study of Liedtke et al (1988) which demonstrated an increase in fatty acid oxidation during reperfusion of pig hearts following ischemia. These results, showing increased fatty acid oxidation in parallel with a decrease in mechanical function, differ from results obtained from Positron Emission Tomography (PET). Using ^{11}C-palmitate, these studies suggest that a delayed clearance of the positron following ischemia is consistent with an impairment of fatty acid β-oxidation during reperfusion (Schwaiger et al., 1985a,b, Bergmann et al., 1983). However, in these studies the decreased clearance of ^{11}C-palmitate from the myocardium may be due to differences in the backdiffusion of ^{11}C-palmitate (Fox et al., 1985; Nellis et al., 1991), or the result of label trapped within the triacylglycerol pool, as triacylglycerol synthesis is increased in rat hearts immediately following ischemia-reperfusion (Saddik and Lopaschuk, 1992).

The increase in fatty acid oxidation following ischemia-reperfusion is accompanied by a concomitant decrease in glucose oxidation. In isolated working rat hearts subjected to global no-flow ischemia, the addition of 1.2 mM palmitate to the perfusion buffer leads to a decrease in glucose oxidation during reperfusion, and a reduction in mechanical function (Lopaschuk et al., 1990). Stimulation of glucose oxidation during reperfusion with the carnitine palmitoyltransferase I inhibitor, etomoxir, or the pyruvate dehydrogenase complex (PDC) activator, dichloroacetate, will improve recovery of mechanical function in isolated working rat hearts following ischemia (Lopaschuk et al., 1988; McVeigh and Lopaschuk, 1990). The reason why fatty acid inhibition of glucose oxidation is deleterious to the recovery of mechanical function following a period of ischemia-reperfusion has yet to be determined.

III. POTENTIAL MECHANISMS BY WHICH FATTY ACIDS DEPRESS THE RECOVERY OF MECHANICAL FUNCTION FOLLOWING ISCHEMIA

As mentioned, a number of different mechanisms have been proposed to explain the detrimental effects of fatty acids during and following ischemia. One potential mechanism involves the disruption of energy metabolism by interfering with ATP transport due to the accumulation of the fatty acid intermediate, long chain acyl CoA. Direct effects of long chain acyl-CoAs on the inhibition of mitochondrial ADP/ATP translocase have been proposed to interfere with cellular energy production by limiting ADP access to the mitochondrial matrix (Chua and Shrago, 1977; Shrago, 1978; Shug et al., 1978). Acyl-CoA has been proposed to act as a competitive inhibitor with adenine nucleotides for binding to the translocase on the cytoplasmic side of the inner mitochondrial membrane. However, it is uncertain whether acyl-CoA exerts a similar negative effect on adenine nucleotide translocase from the matrix side of the inner mitochondrial membrane (Lanoue et al., 1981), where the majority of acyl-CoA is thought to be localized (Idell-Wenger et al., 1978). Similarly, the protective effect of the CPT I inhibitor, etomoxir, is not closely correlated with alterations in myocardial acyl CoA levels (Lopaschuk et al., 1988, Lopaschuk et al, 1990). These results suggest that the decrease in functional recovery following ischemia in the presence of high concentrations of fatty acids may not be due to insufficient mitochondrial adenine nucleotide transport. Similarly, this theory can not explain why a specific increase in glucose oxidation would be beneficial to functional recovery even when associated with an increase in myocardial long chain acyl-CoA content (Lopaschuk et al., 1988).

Another possible mechanism which may explain the deleterious effects of high levels of fatty acids is related to studies suggesting that a decreased pyruvate flux through PDC decreases the responsiveness of mitochondrial energy production to cellular energy requirements. High levels of fatty acids markedly reduce glucose oxidation during reperfusion, due to a decrease in PDC activity (Lopaschuk et al., 1989; Lopaschuk et al., 1990, Renstrom et al., 1989). Recent evidence suggests that PDC activity is required to re-establish cytosolic phosphorylation potentials, allowing ADP to drive mitochondrial respiration at the critical moment of reperfusion following hypoxia or ischemia (Zimmer et al., 1989; Mallet et al., 1990). If ischemic guinea pig hearts are reperfused with perfusion medium containing both glucose and pyruvate, a rapid restoration of cytosolic phosphorylation potential, oxygen consumption and ventricular work occurs, as well as a rapid decrease intracellular phosphate concentration (Bunger et al., 1989; Mallet et al., 1990). Omission of glucose, or the substitution of pyruvate by lactate or acetate during reperfusion fails to restore phosphorylation potential upon reperfusion and leads to a depression in oxygen consumption and contractile force. The effect of pyruvate on protecting pyruvate flux through PDC may relate to the ability of pyruvate to act as an inhibitor of the pyruvate dehydrogenase E_1 kinase, preventing the phosphorylation and inactivation of PDC (Patel and Roche, 1990). In contrast, high levels of fatty acids will activate this kinase, secondary to an increase in intramitochondrial acetyl CoA/CoA. Theoretically, this could slow the recovery of oxidative metabolism during reperfusion.

Studies on rat hearts using ^{31}P-NMR have shown that stimulating PDC activity couples energy demand with energy use. In Langendorf perfused rat hearts, stimulation of mitochondrial dehydrogenases following ischemia increases mitochondrial NADH production, allowing intramitochondrial NADH production to proceed at a high rate so that ATP production is limited only by ADP availability (Zimmer et al., 1989). This should theoretically make mitochondrial energy production more responsive to cellular energy demand. Under these conditions, an increased flux through the PDC during reperfusion following ischemia should facilitate the recovery of oxidative metabolism and mechanical function, and improve the coupling between energy production and energy demand. The converse of this hypothesis is that inhibition of PDC activity, as observed in the presence of high concentrations of fatty acids, may be deleterious to the recovery of mechanical function by suppressing oxidative metabolism and uncoupling the requirement of ATP synthesis from ADP concentrations. However, although glucose oxidation is depressed in ischemic hearts reperfused in the presence of high levels of fatty acids (1.2 mM palmitate) overall ATP production does not appear to be limited (Lopaschuk et al., 1988; Lopaschuk and Saddik, 1992), and may be slightly enhanced following reperfusion (150 mmol ATP produced/g dry/min vs. 129 mmol/g dry/min prior to ischemia). This is because fatty acid oxidation rates are greater than those observed prior to ischemia (Lopaschuk and Saddik, 1992). Although oxidative metabolism could not be accurately measured during the initial moment of reperfusion, these results suggest that it is unlikely that overall oxidative metabolism is depressed during reperfusion. This is supported by the studies of Lerch's group (Görge et al., 1991, Benzi and Lerch, 1992), who demonstrate that in the presence of fatty acids an uncoupling between contactile function and both O_2 consumption and mitochondrial oxidation occurs during reperfusion. This suggests that recovery of mitochondrial metabolism precedes the recovery of mechanical function.

While the beneficial effect of dichloroacetate on mechanical function following ischemia-reperfusion suggest that flux through PDC is important during reperfusion (McVeigh and Lopaschuk, 1990; Lopaschuk et al., 1993), it is unlikely to occur secondary to an increase in the phosphorylation potential. This is because we have recently demonstrated that the increase in glucose oxidation following DCA administration is accompanied by a parallel decrease in fatty acid oxidation (Saddik et al, 1993). As a result, is unlikely that DCA increases overall ATP production during reperfusion secondary to an increase in intramitochondrial NADH concentrations. Recent experiments by Bunger and Mallet (1993) in isolated working guinea pig hearts show that dichloroacetate is able to stimulate pyruvate oxidation without effecting myocardial O_2 consumption if hearts are perfused in the presence of octanoic acid. Under these conditions, it appears likely that oxidative metabolism is not suppressed following reperfusion of ischemic myocardium, and that the beneficial effects of increasing glucose oxidation is likely due to factors other than increases in intramitochondrial NADH concentrations and oxidative phosphorylation.

IV. POTENTIAL EFFECTS OF FATTY ACIDS ON INTRACELLULAR ION HOMEOSTASIS

Another mechanism explaining how fatty acid oxidation and its accompanying decrease in glucose oxidation may lead to an increase in the susceptibility of myocardium to ischemia-reperfusion is by altering the relationship between glucose oxidation and glycolysis. During severe no-flow ischemia, anaerobic glycolysis results in the development of intracellular acidosis, through the production of H^+ from the hydrolysis of ATP derived from glycolysis uncoupled from glucose oxidation (Gevers, 1977; Hochachka and Mommsen, 1983; Opie, 1990). We hypothesize that during reperfusion, glycolysis uncoupled from glucose oxidation continues to be a major source of intracellular H^+ production during the actual reperfusion period (see Figure 1).

The intracellular acidosis accompanying global no-flow ischemia plays a role in the etiology of intracellular Ca^{2+} overload which develops upon reperfusion. Experiments by Neely's group (Neely and Grotyohann, 1984; Tani and Neely, 1989) suggest that H^+ ions produced through anaerobic glycolysis lead to the development of an intracellular Ca^{2+} overload upon reperfusion by exchanging for Na^+ through the Na^+-H^+ exchanger and subsequently exchanging the Na+ for Ca^{2+} through the Na^+-Ca^{2+} exchanger. These studies tested the hypothesis of Lazdunski et al. (1985) that intracellular regulation of H^+, Na^+ and Ca^{2-} ions are coupled through the actions of ion exchangers. Tani and Neely (1989) demonstrated a slow rise in intracellular Na^+ during the ischemic period, with little or no change in intracellular Ca^{2+}. During reperfusion, a sharp rise and then fall in the level of intracellular Na+ coincided with a large increase in Ca^{2+} uptake by the rat heart. This was strongly correlated with decreased ventricular function. Furthermore, a decrease in H^+ ion production by decreasing glycolytic flux during ischemia, or by inhibiting Na^+-H^+ ion exchange with exchange inhibitors prevents the intracellular Ca^{2+} overload leading to cell death (Tani and Neely, 1989; du Toit and Opie, 1993; Meng et al., 1993).

While much attention has focussed on alterations in the severity of acidosis during ischemia, relatively little attention has focussed on H+ production during the actual reperfusion period. According to the concepts outline above, any increase in H^+ ion production during reperfusion could be detrimental to functional recovery, as it would contribute to the increase in intracellular Ca^{2+} overload during reperfusion. While the anaerobic metabolism of glucose produces 2 ATP and 2 lactate, the hydrolysis of this ATP results in the production of 2 H^+ (Hochachka et al., 1983, Opie, 1990). However, the complete oxidation of glucose through to CO_2 and H_2O is H^+ neutral (equations 1 and 2).

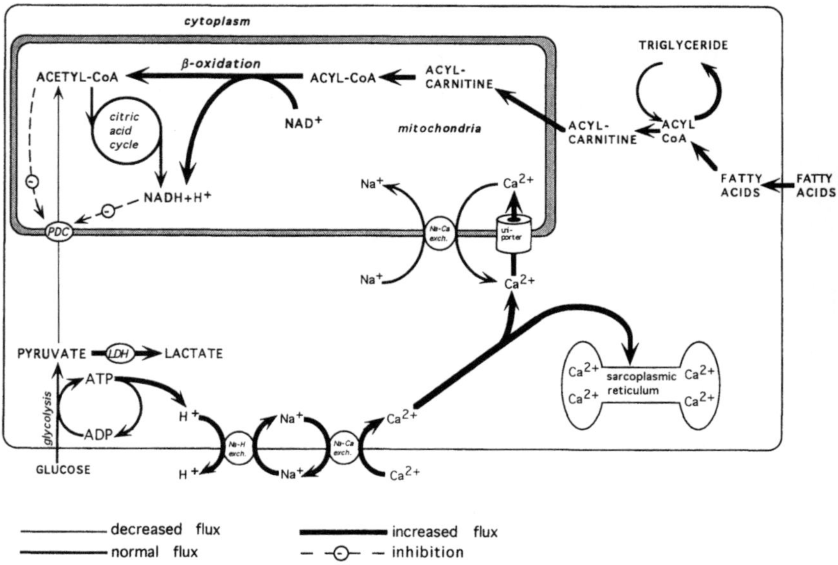

Figure 1. Hypothetical model of how high levels of fatty acids contribute to ischemic injury during reperfusion of hearts following a severe episode of ischemia. This figure depicts the potential consequences of alterations in energy substrate utilization during reperfusion of previously ischemic hearts. Thick lines represent increased flux through the indicated pathways compared to aerobic conditions, while thin lines represent decreased flux through these pathways. High levels of fatty acids seen during and following ischemia lead to accelerated rates of fatty acid oxidation during reperfusion. High rates of fatty acid oxidation, secondary to high circulating levels of fatty acids, markedly decrease pyruvate dehydrogenase complex (PDC) activity, resulting in a decrease in glucose oxidation. Normal glycolytic rates result in an uncoupling between glycolysis and glucose oxidation, resulting in a greater amount of glycolytically derived pyruvate being converted to lactate. The uncoupling of glycolysis from glucose oxidation also increases the production of H^+ derived from glycolytic ATP hydrolysis. As a result, a significant amount of H^+ is produced from glucose metabolism during this critical period of reperfusion, at a time when the myocardium in attempting to clear the H^+ that accumulated during ischemia. This extra H^+ produced may exchange with Na^+ to increase intracellular Na^+, which can then exchange with Ca^{2+}. The increase in cytosolic Ca^{2+} is partly sequestered by the sarcoplasmic reticulum and the mitochondria.

Glycolysis uncoupled from glucose oxidation (1)

$$\text{glucose} + 2\ \text{Pi} + 2\ \text{ADP} \longrightarrow 2\ \text{lactate} + 2\ \text{ATP} + 2\ H_2O$$

$$2\ \text{ATP} \longrightarrow 2\ \text{ADP} + 2\ \text{Pi} + 2\ H^+$$

Net H^+ production per glucose = 2

Glycolysis coupled to glucose oxidation (2)

$$\text{glucose} + 38\ \text{ADP} + 38\ \text{Pi} + 38\ H^+ + 6\ O_2 \longrightarrow 6\ CO_2 + 38\ \text{ATP} + 42\ H_2O$$

$$38\ \text{ATP} \longrightarrow 38\ \text{ADP} + 38\ \text{Pi} + 38\ H^+$$

Net H^+ production per glucose = 0

As a result, the dissociation of glycolysis from glucose oxidation in the presence of high concentrations of fatty acids increases the H^+ burden the myocyte must cope with in order to maintain intracellular pH and ion homeostasis. In the presence of high levels of fatty acids, fatty acid oxidation recovers quickly following ischemia-reperfusion (Liedtke et al., 1988; Lopaschuk et al., 1990; Gorge et al., 1991; Benzi and Lerch, 1992; Saddik and Lopaschuk, 1992). This leads to inhibition of glucose oxidation, resulting in a dramatic uncoupling of glycolysis from glucose oxidation (Lopaschuk et al., 1993). If fatty acids are excluded from the perfusion buffer the ratio of glycolysis to glucose oxidation in isolated working rat hearts is approximately 2:1 (Kobayashi and Neely, 1979; Saddik and Lopaschuk, 1991), while in hearts perfused in the presence of 1.2 mM palmitate the ratio of glycolysis to glucose oxidation increases to approximately 13:1 (Saddik and Lopaschuk, 1991; Lopaschuk et al., 1993). The reason for this differential effect of fatty acid oxidation on glycolysis and glucose oxidation appears to be related to the sensitivity of the key regulatory enzymes phosphfructokinase and PDC to inhibition by the products of oxidative metabolism. Fatty acid oxidation inhibits PDC through an increase in mitochondrial acetyl-CoA/CoA and $NADH_2$/NAD ratios (Kerby et al., 1976; Dennis et al., 1979; Weiss et al., 1989, Patel and Roche, 1990), while PFK is inhibited to a lesser extent by an increase in citrate levels (Newsholme et al., 1962; Neely and Morgan, 1974). Thus, the suppression of glucose oxidation during reperfusion has the potential to contribute to intracellular Ca^{2+} overload. As a result, improving glucose oxidation should be beneficial by decreasing H^+ production from glycolysis uncoupled from glucose oxidation. An overall scheme relating the deleterious effects of changes in energy substrate metabolism to Ca^{2+} overload is illustrated in Figure 1.

This hypothesis is supported by observations that pharmacological interventions which increase glucose oxidation relative to glycolysis will improve the functional recovery in isolated working hearts perfused with high concentrations of fatty acids (see Table 2). Etomoxir, a carnitine palmitoyltransferase I inhibitor, protects mechanical function in reperfused rat hearts at either 10^{-8} or 10^{-5} M, concentrations which increase glucose oxidation during reperfusion (Lopaschuk et al., 1989,

Table 2. Pharmacological Alteration of Glucose and Fatty Acid Oxidation During Reperfusion Following Ischemia

Study	*Drug (Class)*	*Perfusion Conditions*	*Results*
Lopaschuk et al. (1990)	Etomoxir (CPT I inhibitor) -(10^{-6} M) added during reperfusion.	-Isolated working rat heart -11 mM glucose, 1.2 mM palmitate. -25 min of no-flow ischemia.	-Increased functional recovery. -Increased glucose oxidation.
McVeigh et al., (1990)	Dichloroacetate (PDC activator) -(1 mM) added during reperfusion.	-Isolated working rat heart -11 mM glucose, 1.2 mM palmitate -30 min of no-flow ischemia.	-Increased functional recovery. -Increased glucose oxidation.
Benzi and Lerch, (1992)	Ruthenium red (mitochondrial Ca^{2+} uniporter inhibitor, and SR Ca^{2+} release inhibitor) -(6 μM) added during initial aerobic perfusion.	-Langendorff perfused rat heart. -11 mM glucose, 0.4 mM palmitate. -60 min of no-flow ischemia.	-Increased functional recovery. -Decreased creatine kinase release. -Improved efficiency of O_2 consumption and energy substrate metabolism.
Broderick et al., (1993)	Carnitine -(10 mM) added during initial aerobic perfusion.	-Isolated working rat heart. -11 mM glucose, 1.2 mM palmitate. -35 min of no-flow ischemia.	-Increased functional recovery. -Increased glucose oxidation.

Lopaschuk et al., 1990). Recently, we have shown that treating isolated working rat hearts with L-carnitine will also increase glucose oxidation (Broderick et al., 1992, Broderick et al., 1993). The increase in glucose oxidation probably occurs secondary to free carnitine reacting with intramitochondrial acetyl-CoA to form acetylcarnitine via the action of carnitine acetyltransferase (Lysiak et al., 1988). The acetylcarnitine is subsequently transported out of the mitochondria lowering the acetyl-CoA/CoA ratio and stimulating PDC activity (Uziel et al., 1988). The increase in glucose oxidation stimulated by L-carnitine treatment is beneficial in preserving mechanical function in hearts undergoing ischemia-reperfusion (Broderick et al., 1993). As previously mentioned, dichloroacetate, a compound which stimulates glucose oxidation directly at the level of the PDC by inhibiting the action of PDH kinase (Stacpoole, 1989), also improves mechanical function following ischemia (McVeigh and Lopaschuk, 1990; Lopaschuk et al., 1993). Dichloroacetate decreases the ratio of glycolysis to glucose oxidation from 13.6 to 3.7 in isolated working rat hearts reperfused following ischemia with 1.2 mM palmitate. This improves the coupling of glycolysis to glucose oxidation and results in a marked reduction in H^+ production from glucose metabolism during the actual reperfusion period (Lopaschuk et al, 1993).

Increased myocardial intracellular Ca^{2+} can have a number of adverse effects on the heart, which range from a decrease in compliance within the ventricular walls, to the activation of phospholipases and membrane degradation (see Van der Vusse et al., 1990). Alterations in ion homeostasis, such as Ca^{2+} accumulation, increases the requirement of ATP for non-contractile purpose (i.e. such as re-establishing sarcolemmal ion gradients). It has already been shown that during reperfusion following ischemia a dissociation of contractile function from O_2 consumption and oxidative metabolism occurs (Benzi and Lerch, 1992). Although myocardial oxygen consumption recovers rapidly following ischemia-reperfusion, this occurs well before the recovery of left ventricular pressure development (Gorge et al., 1991). The decrease in the efficiency of energy substrate metabolism and oxygen consumption may be related to an increase in energy utilization by processes not directly involved in contractile function, or it may represent an uncoupling of oxidative phosphorylation. Ruthenium red is able to improve the coupling between mechanical function and oxidative metabolism (Benzi and Lerch, 1992). This suggests that alterations in intracellular Ca^{2+} transport may be involved in the uncoupling of oxidative metabolism and mechanical function, as ruthenium red inhibits Ca^{2+} release from sarcoplasmic reticulum and blocks mitochondrial Ca^{2+} uptake via the Ca^{2+} uniporter (Henry et al., 1977; Chamberlain et al., 1984). The marked increase in diastolic tension noted in hearts during and following ischemia is generally associated with a pronounced cytosolic and mitochondrial Ca^{2+} overload (Steenbergen et al., 1990; Henry et al., 1977). A number of studies have shown that ruthenium red is able to decrease intramitochondrial Ca^{2+} accumulation during reperfusion (Peng et al., 1980; Ferrari et al., 1982). Although these previous studies were primarily concerned with examining total mitochondrial Ca^{2+} accumulation, recent studies using hypoxic-reoxygenated rat hearts have shown that intramitochondrial free Ca^{2+} increases upon reoxygenation of the heart (Allen et al., 1993). As intramitochondrial free calcium increases, a greater proportion of H^+ ions transported out of the mitochondria by the cytochromes must be diverted from ATP production to lowering intramitochondrial calcium concentration through the actions of the mitochondrial Na^+-Ca^{2+} exchanger and Na^+-H^+ exchanger (McCormack et al., 1992). The extent to which increased concentrations of fatty acids in the perfusate contribute to the uncoupling of oxidative phosphorylation in the mitochondria of myocardial tissue following ischemia reperfusion remains to be explored.

V. CONCLUSIONS

Despite observations 25 years ago demonstrating that the severity of myocardial damage during ischemia-reperfusion is linked to high plasma fatty acids levels, the nature of how fatty acids exert their deleterious effect remains uncertain. Recent studies have focused on the role fatty acids play in balancing energy substrate metabolism pathways within the heart. Fatty acids inhibit glucose oxidation to a much greater extent then glycolysis, which can result in an increased H^+ load on the myocyte. This potentially contributes to an increased intracellular Ca^{2+} overload during reperfusion. High levels of fatty acids are present in the blood in most clinically relevant situations of reperfusion following myocardial ischemia. As a result, the omission of perfusate fatty acids in studies examining ischemia-reperfusion in the isolated perfused heart may be eliminating a critical determinant of injury during and following ischemia.

ACKNOWLEDGMENT

This work was supported by a grant from the Medical Research Council of Canada. DDB is a trainee of the Heart and Stroke Foundation of Canada. GDL is a Alberta Heritage Foundation for Medical Research Scholar and a Medical Research Council of Canada Scientist.

REFERENCES

Allison, S. P., Chamberlain, M. J., & Hinton, P. (1969). Intravenous glucose tolerance, insulin, glucose and free fatty acid levels after myocardial infarction. Br. Med. J. 4, 776-778.

Allen, S. P., Darley-Usmar, V. M., McCormack, J. G., & Stone, D. (1993). Changes in mitochondria matrix free calcium in perfused rat hearts subjected to hypoxia-reoxygenation. J. Mol. Cell. Cardiol. 25, 949-958.

Benzi, R. H., & Lerch, R. (1992). Dissociation between contractile function and oxidative metabolism in postischemic myocardium. Circ. Res. 71, 567-576.

Bing, R. J. (1965) Cardiac metabolism. Physiol. Rev. 45, 171-213.

Bricknell, O. L., & Opie, L. H., (1978). Effects of substrates on tissue metabolic changes in the isolated rat heart during underperfusion and on release of lactate dehydrogenase and arrhythmias during reperfusion. Circ. Res. 43, 102-115.

Broderick, T. L., Quinney, H. A., & Lopaschuk, G. D. (1992). Carnitine stimulation of glucose oxidation in the fatty acid-perfused isolate working rat heart. J. Biol. Chem. 267, 3758-3763.

Broderick, T. L., Quinney, H. A., Barker, C. C. & Lopaschuk, G. D. (1993). Beneficial effect of carnitine on mechanical recovery of rat hearts reperfused after a transient period of global ischemia is accompanied by a stimulation of glucose oxidation. Circ. Res. 87, 972-981.

Bunger, R., Mallet, R. T., & Hartman, D. A. (1989). Pyruvate-enhanced phosphorylation potential and inotropism in normoxic and post ischemic isolated working heart. Eur. J. Biochem. 180, 221-233.

Bunger, R., & Mallet, R. T. (1993). Mitochondrial pyruvate transport in working guinea-pig heart. Work-related vs. carrier-mediated control of pyruvate oxidation. Biochim. Biophys. Acta 1151, 223-236.

Chamberlain, B. K., Volpe, P., & Fleischer, S. (1984). Inhibition of calcium induced calcium release from purified cardiac sarcoplasmic reticulum vesicles. J. Biol. Chem. 259, 7547-7553.

Chua, B. H., & Shrago, E. (1977). Reversible inhibition of adenine nucleotide translocation by long chain acyl CoA esters in bovine heart mitochondria and inverted submitochondrial particles. J. Biol. Chem. 252, 6711-6714.

Corr, P. B., Gross, R. W., & Sobel, B. E. (1984). Amphipathic metabolites and membrane dysfunction in ischemic myocardium. Circ. Res. 55, 135-154.

Dennis, S., Debuysere, M. S., & Olson, M. S. (1979). Studies on the regulation of pyruvate dehydrogenase in the isolated perfused rat heart. J. Biol. Chem. 254, 1252-1258.

du Toit, E. F., & Opie, L. H. (1993). Role for Na^+/H^+ exchanger in reperfusion stunning in isolated perfused rat heart. J. Cardiovasc. Pharmacol. 22, 877-883.

Ferrari, R., Di Lisa, F., Raddino, R., & Visioli, O. (1982). The effects of ruthenium red on mitochondrial function during postischemic reperfusion. J. Mol. Cell. Cardiol. 14, 737-740.

Fox, K.A.A., Abendschein, D.R., Ambos, H.D., Sobel, B.E. (1985) Efflux of metabolized and non-metabolized fatty acid from canine myocardium: Implications for quantifying myocardial metabolism tomographically. Circ. Res. 57, 232-243.

Gevers, W. (1977). Generation of protons by metabolic processes in heart cells. J. Mol. Cell. Cardiol. 9, 867-874.

Görge, G., Chatelain, P., Schaper, J., & Lerch, R. (1991). Effect of increasing dergrees of ischemic injury on myocardial oxidative metabolism early after reperfusion in isolated rat hearts. Circ. Res. 68, 1681-1692.

Henry, P. D., Shuchleib, R., Davis, J., Weiss, E. S., & Sobel, B. E. (1977). Myocardial contracture and accumulation of mitochondrial calcium in ischemic rabbit heart. Am. J. Physiol. 233, H677-H684.

Hochachka, P. W., & Mommsen, T. P. (1983). Protons and anerbiosis. Science 219, 1491-1397.

Ichihara, K., & Neely, J. R. (1985). Recovery of ventricular function in reperfused ischemic rat hearts exposed to fatty acids. Am. J. Physiol. 249, H492-H497.

Idell-Wenger, J. A., Grotyohann, L. W., & Neely, J. R. (1978). Coenzyme A and Carnitine distribution in normal and ischemic hearts. J. Biol. Chem. 253, 4310-4318.

Johnson, D. L., & Lewandowski, E. D. (1991). Fatty acid metabolism and contactile function in the reperfused myocardium: Multinuclear NMR studies of isolated rabbit hearts Circ. Res. 68:714-725.

Katz, K. M., & Messineo, F. C. (1981). Lipid-membrane interactions and the pathogenesis of ischemic damage in the myocardium. Circ. Res. 48, 1-16.

Kerbey, A. L., Randle, P. J., Cooper, R. H., Whitehouse, S., Pask, H. T., & Denton, R. M. (1976). Regulation of pyruvate dehydrogenase in rat heart. Biochem. J. 154, 327-348.

Khandoudi, N., Bernard, M., Cossone, P., & Feuvray, D. (1990). Intracellular pH and role of Na^+/H^+ exchange during ischemia and reperfusion of normal and diabetic rat hearts. Cardiovasc. Res. 11, 873-878.

Kjekshus, J. K., & Mjos, O. D. (1972). Effect of free fatty acids on myocardial function and metabolism in the ischemic dog heart. J. Clin. Invest. 51, 1767-1776.

Kobiashi, K., & Neely, J. R. (1979). Control of maximum rates of glycolysis in rat cardiac muscle. Circ. Res. 44, 166-175.

Lazdunski, M., Frelin, C., & Vigne, P. (1985). The sodium/hydrogen exchange system in cardiac cells: Its biochemical and pharmacological properties and its role in regulating internal concentrations of sodium and internal pH. J. Mol. Cell. Cardiol. 17, 1029-1042.

Lanoue, K. F., Watts, J. A., & Koch, C. D. (1981). Adenine nucleotide transport during cardiac ischemia. Am. J. Physiol. 241, H663-H671.

Liedtke, A.J., Nellis, S.H., & Neely, J.R. (1978) Effects of excess free fatty acids on mechanical and metabolic function in normal and ischemic myocardium of swine. Circ. Res. 343, 652-661.

Liedtke, A. J., Demaison, L., Eggleston, A. M., Cohen, L. M., & Nellis, S. H. (1988). Changes in substrate metabolism and effects of excess fatty acids in reperfused myocardium. Circ. Res. 62, 535-542.

Liedtke, A. J., Renstrom, B., Nellis, S. H., Subramanian, R., & Woldegiorgis, G. (1993). Myocardial metabolism in chrinic reperfusion after nontransmural infarction in pig hearts. Am. J. Physiol. 265, H1614-H1622.

Lopaschuk, G. D., Wall, S. R., Olley, P.M., & Davies, N. J. (1988). Etomoxir, a carnitine palmitoyltransferase I inhibitor, protects hearts from fatty acid-induced ischemic injury independent of changes in long chain acylcarnitine. Circ. Res. 63, 1036-1043.

Lopaschuk, G. D., & Spafford, M. (1989). Response of isolated working hearts to fatty acids and carnitine palmitoyltransferase 1 inhibition during reduction of coronary flow in acutely and chronically diabetic rats. Circ. Res. 65, 378-387.

Lopaschuk, G. D., McNeil, J., & McVeigh, J. (1989). Glucose oxidation is stimulated in repefused hearts with the carnitine palmitoyltransferase I inhibitor, Etomoxir. Mol. Cell. Biochem. 88, 175-179.

Lopaschuk, G. D., Spafford, M. A., Davies, N. J., & Wall, S. R. (1990). Glucose and palmitate oxidation in isolated working rat hearts reperfused after a period of transient global ischemia. Circ. Res. 66, 546-553.

Lopaschuk, G. D., & Saddik, M. (1992). The relative contribution of glucose and fatty acids to ATP production in hearts reperfused following ischemia. Mol. Cell. Biochem. 116, 111-116.

Lopaschuk, G. D., Wambolt, R. B., & Barr, R. L. (1993). An imbalance between glycolysis and glucose is a possible explanation for the detrimental effects of high levels of fatty acids during aerobic referfusion of ischemic hearts. J. Pharm. Exp. Therap. 264, 135-144.

Lopaschuk, G.D., Collins-Nakai, R., Olley, P.M., Montague, T.J., McNeil, G., Gayle, M., Penkoske, P., & Finegan, B.A., (1994). Plasma fatty acid levels in infants and adults following myocardial ischemia. Am. Heart J. 128, 61-67.

Lysiak , W., Lilly, K., DiLisa, F., Toth, P.P., & Bieber, L.L. (1988). Quatntification of the effect of L-carnitine on the levels of acid-soluble short-chain acyl CoA and CoASH in rat heart and liver mitochondria. J. Biol. Chem. 263, 1511-1516.

Mallet, R. T., Hartman, D. A., & Bunger, R. (1990). Glucose requirement for postischemic recovery of perfused working heart. Eur. J. Biochem. 188, 481-493.

McCormack, J. G., Daniel, R. L., Osbaldeston, N. J., Rutter, G. A., & Denton, R. M. (1992). Mitochondrial Ca^{2+} transporter and the role of matrix Ca^{2+} in mammalian tissues. Biochem. Soc. Trans. 20, 153-159.

McVeigh, J. J., & Lopaschuk, G. D. (1990). Dichloroacetate stimulation of glucose oxidation improves recovery of ischemic rat hearts. Am. J. Physiol. 259, H1070-H1085.

Meng, H., Maddaford, T. G., & Pierce, G. N. (1993). Effect of amiloride and selected analogs on postischemic recovery of cardiac contractile function. Am. J. Physiol. 264, H1831-H1835.

Mjos, O. D. (1978). Effect of free fatty acids on myocardial function and oxygen consumption in intact dogs. J. Clin. Invest. 50, 1386-1389.

Mueller, H.S., Ayres, S.T. (1978) Metabolic responses of the heart in acute myocardial infarction in man. Am J. Cardiol. 42, 363-371.

Myears, D. W., Sobel, B. E., & Bergmann, S. R. (1987). Substrate use in ischemic and reperfused canine myocardium: Quantitative considerations. Am. J. Physiol. 253, H107-H114.

Neely, J. R., & GrotyoHann, L. W. (1984). Role of glycolytic products in damage to ischemic myocardium. Circ. Res. 55, 816-824.

Neely, J. R., & Morgan, H. E. (1974). Relationship between carbohydrate and lipid metabolism and the energy balance of heart muscle. Ann. Rev. Physiol. 36, 413-459.

Newsholme, E. A., Randle, P. J., & Manchester, K. L. (1962). Inhibition of the phosphofructokinase reaction in perfused rat heart by respiration of ketone bodies, fatty acids and pyruvate. Nature 193, 270-271.

Oliver, M. F., Kurien, V. A., & Greenwood, T. W. (1968). Relation between serum free-fatty acids and arrhythmia and death after myocardial infarction. Lancet 1, 710-715.

Opie, L. H. (1975). Metabolism of free fatty acids, glucose, and catecholamines in acute myocardial infarction. Am. J. Cardiol. 36, 938-953.

Opie, L. H. (1990) Myocardial ischemia–Metabolic pathways and implication of increased glycolysis. Cardiovasc. Drugs Therapy 4, 777-790.

Opie, L. H. (1988). Hypothesis: Glycolytic rates control cell viability in ischemia. J. Appl. Cardiol. 3, 407-414.

Patel, M. S., & Roche, T. E. (1990). Molecular biology of pyruvate dehydrogenase complexes. FASEB J. 4, 3224-3233.

Peng, C. F., Kane, J. J., Straub, K. D., & Murphy, M. L. (1980). Improvement of mitochondrial energy production in ischemic myocardium by *in vivo* infusion of ruthenium red. J. Cardiovasc. Pharmacol. 2, 45-54.

Renstrom, B., Nellis, S.H., Liedtke, A.J. (1989) Metabolic oxidation of glucose during early myocardila reperfusion. Circ. Res. 65, 1094-1101.

Russo, J. V., & Margolis, S. (1972). Hemodynamics effects of free fatty acid augmentation following myocardial infarction. Circ. (Suppl. II) 215, 46-46.

Rutenbeng, H. L., Pamintuan, J. C., & Soloff, L. A. (1969). Serum-free-fatty-acids and their relation to complications after acute myocardial infarction. Lancet 2, 559-564.

Saddik, M., & Lopaschuk, G. D. (1991). Myocardial triglyceride turnover and contribution to energy substrate utilization in isolated working rat hearts. J. Biol. Chem. 266, 8162-8170.

Saddik, M., & Lopaschuk, G. D. (1992). Myocardial triglyceride turnover during reperfusion of isolated rat hearts subjected to a transient period of global ischemia J. Biol. Chem. 267, 3825-3831.

Saddik, M., Gamble, J., Witters, L. A., & Lopaschuk, G. D. (1993). Acetyl-CoA carboxylase regulation of fatty acid oxidationin the heart. J. Biol. Chem. 268, 25836-25845.

Schwaiger, M., Schelbert, H. R., Keen, R., Vinten-Johansen, J., Hansen, H., Selin, C., Barrio, J., Huang, S. C., & Phelps, M. E. (1985a). Retention and clearance of C-11 palmitic acid in ischemic and reperfused canine myocardium. J. Am. Coll. Cardiol. 6, 311-320.

Schwaiger, M., Schelbert, H. R., Keen, R., Ellison, D., Hansen, H., Yeatmen, L., Vinten-johansen, J., Selin, C., Barrio, J., & Phelps, M. E. (1985b). Sustained regional abnormalities in cardiac metabolism after transient ischemia in the chronic dog model. J. Am. Coll. Cardiol. 6, 336-347.

Severeid, L., Connor, W. E., & Long, J. P. (1969). The depressant effect of fatty acids on the isolated rabbit heart. Proc. Soc. Exp. Biol. Med. 131, 1239-1242.

Shug, A.L., Thomsen, J.H., Folts, J.D., Bittar, N., Klein, M.I., Koke, J.R., Huth, P.J., (1978). Changes in tissue levels of carnitine and other metabolites during myocardial ischemia and anoxia. Biochim. Biophys. Acta 187:25-33.

Shrago, E. (1978). The effect of long chain fatty acyl CoA esters on the adenine nucleotide translocase in myocardial metabolism. Life Sci. 22, 1-6.

Stacpoole, P. W. (1989). The pharmacology of dichloroacetate. Metabolism 38, 1124-1144.

Steenbergen, C., Murphy, E., Watts, J. A., & London, R. E. (1990). Correlation between cytosolic free calcium, contracture, ATP, and irreversible ischemic injury in perfused rat heart. Circ. Res. 166, 135-146.

Svensson, S., Svedjeholm, R., Ekroth, R., Milocco, I., Nilsson, F., Sabel, K. G., & William-Olsson, G. (1990). Trauma metabolism and the heart. Uptake of substrates and effects of insulin early after cardiac operations. J. Thorac. Cardiovasc. Surg. 99, 1063-1073.

Tani, M., & Neely, J. R. (1989). Role of intracellular Na^+ and Ca^{2+} overload and depressed recovery of ventricular function of reperfused ischemic rat hearts: Possible involvement of H^+-Na^+ and Na^+-Ca^{2+} exchange. Circ. res. 65, 1045-1056.

Uziel, G., Gravaglia, B., & di Donato, S. (1988). Carnitine stimulation of pyruvate dehydrogenase complex (PDHC) in isolated human skeletal muscle mitochondria. Muscle & Nerve 11, 720-724.

Van Bilsen, M., Van der Vusse, G. J., Willemsen, P. H. M., Coumans, W.A., Roemen, T. H. M., & Reneman, R. S. (1989). Lipid alterations in isolated, working rat hearts during ischemia and reperfusion: its relation to myocardial damage. Circ. Res. 64, 304-314.

Van der Vusse, G. J., Glatz, J. F. C., Stam, H. C. G., & Reneman, R. S. (1992). Fatty acid homeostasis in the normoxic and ischemic heart. Physiol. Rev. 72, 881-940.

Van der Vusse, G. J., van Bilsen, M., Sonderkamp, T., & Reneman, R. S. (1990). Hydrolysis of phospholipids and cellular integrity. In: Pathophysiology of severe ischemic myocardial injury (Piper, H. M. ed.), pp. 167-193. Kluwer, London.

Weiss, R. G., Chacko, V. P., & Gerstenblith, G. (1989). Fatty acid regulation of glucose metabolism in the intact beating rat heart assessed by Carbon-13 NMR spectroscopy: The critical role of pyruvate dehydrogenase. J. Mol. Cell. Cardiol. 21, 469-478.

Zimmer, S. D., Ugurbil, K., Michurski, S. P., Mohanakrishnan, P., Ulstad, V. K., & Folker, J. E. (1989). From AHL: Alterations in oxidative function and respiratory regulation in the post-ischemic myocardium. J. Biol. Chem. 264, 12402-12411.

PHOSPHOLIPID BIOSYNTHESIS IN HEALTH AND DISEASE

Patrick C. Choy, Grant M. Hatch, and Ricky Y.K. Man

Advances in Lipobiology
Volume 2, pages 47-78.

ISBN 0-7623-0205-4

I. INTRODUCTION

A. Nomencleature

Phosphoglycerides are the most abundant class of lipids in all mammalian membranes (White, 1973). The phosphoglycerides can be divided into three distinct subclasses according to the type of bond at the *sn*-1 position. The 1,2-diacylphosphoglyceride is by far the most abundant type whereas 1-alkyl-2-acyl-phosphoglyceride is present only in small quantity (1-2%). The 1-alkenyl-2-acyl-phosphoglyceride is commonly known as plasmalogen and its distribution in mammalian tissues ranges from a low of 2% to a high of 40% (Arthur et al., 1985). In this chapter, our attention will be focussed on the biosynthesis of diacylphosphoglycerides in mammalian tissues in health and disease. The metabolism of 1-alkyl-2-acyl-phosphoglycerides and 1-alkenyl-2-acyl-phosphoglycerides will be discussed in other chapters.

B. Function

As the major component of the external sarcolemma, the diacylphospholipids provide the cell with a protective barrier that has selective permeability for cellular metabolism. Within the cell, these phospholipids participate in the membrane network that delineates individual organelles which constitute uniquely ordered system for intracellular metabolic reactions. In addition to their role in membrane formation, diacylphospholipids are involved in the modulation of a number of membrane bound enzymes (Coleman, 1973). In the last two decades, the participation of several phospholipids in the transmission of biological signals across the membrane has been well documented (Berridge and Irvine, 1989; Exton, 1990).

The direct involvement of phosphatidylcholine and phosphatidylethanolamine in lung surfactant, lipoprotein secretion and arachidonate production for prostaglandin biosynthesis have been clearly demonstrated (Cullis and Hope, 1991). Recently, the phosphatidic acid that accumulates in platelet-derived growth factor stimulated 3T3 cells has been identified as a potential mitogenic signal (Fukami and Takenawa, 1992). It is clear from these studies that diacylphospholipids have physiological roles beyonds their primary function as building blocks of the biological membrane.

II. BIOSYNTHESES OF PHOSPHOLIPIDS

A. The Biosynthesis of Phosphatidylcholine

Phosphatidylcholine and phosphatidylethanolamine are the major diacylphosphoglycerides in mammalian tissues. Phosphatidylcholine was first identified in 1847 and the pathways for its biosynthesis were elucidated in the early 1950s (Figure 1). The CDP-choline pathway is quantitatively the major pathway for phosphatidylcholine biosynthesis whereas the progressive methylation of phosphatidylethanolamine and the base-exchange reaction may contribute to its biosynthesis in selected organs (Hatch et al., 1989; Vance, 1991). In the CDP-choline pathway, choline is phosphorylated to phosphocholine which is converted to CDP-choline for the condensation with diacylglycerol to form phosphatidylcholine. The phosphorylation of choline is catalyzed by choline kinase which exists in multiple molecular forms in the cytosol. The purified enzyme from the liver cytosol displays a molecular mass of 160 kDa by gel filtration determination and has a subunit molecular mass of 47 kDa by SDS-PAGE analysis (Ishidate and Nakazawa, 1992; Porter and Kent, 1992). The second step of the pathway is the formation of CDP-choline which is the rate-limiting step catalyzed by CTP:phosphocholine cytidylyltransferase. The enzyme is found in both the microsomal and the cytosolic fractions, and the translocation of the enzyme from one compartment to another is an important mechanism for the regulation of the enzyme activity (Vance, 1989). The cytidylyltransferase from the rat liver cytosol has a molecular mass of 97 ± 10 kDa which consists of two identical subunits of 45 kDa (Weinhold et al., 1986; Feldman and Weinhold, 1987; Weinhold and Feldman, 1992). The purified enzyme requires lipids for maximum activity but enzyme activity is inhibited by sphingolipids (Choy and Vance, 1978; Sohal and Cornell, 1990). The final step of the CDP-choline pathway is catalyzed by CDP-choline: diacylglycerol cholinephosphotransferase. This enzyme is located on the cytoplasmic side of the endoplasmic reticulum (Cornell, 1992). Recently, the solubilization of the functional enzyme in hamster liver and heart by Triton QS-15 has been reported (O et al., 1989). The enzyme has been partially purified by DEAE-Sepharose column chromatography (O and Choy, 1990).

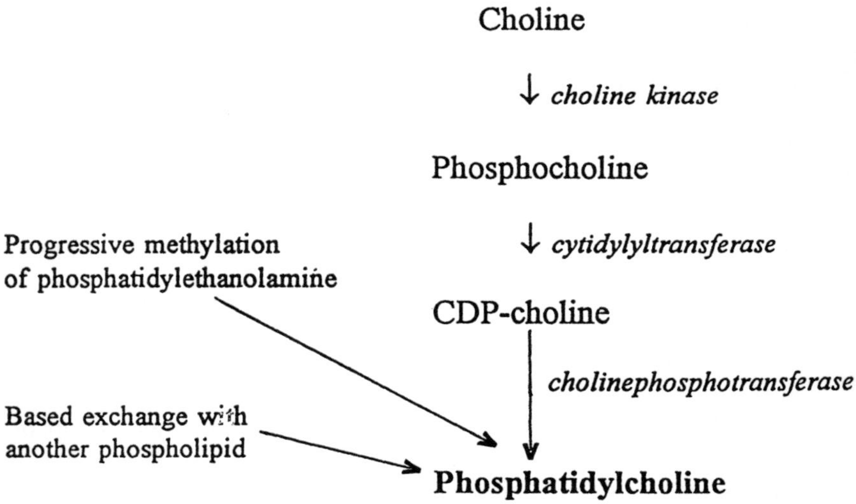

Figure 1. Pathways for the biosynthesis of phosphatidylcholine.

B. The Biosynthesis of Phosphatidylethanolamine

The CDP-ethanolamine pathway, which is corollary to the CDP-choline pathway, is the major route for phosphatidylethanolamine biosynthesis in most mammalian tissues (Vance, 1991). However, the decarboxylation of phosphatidylserine has been recognized as an important pathway for phosphatidylethanolamine biosynthesis in prokaryotes and some eukaryotic cells (Voelker, 1984). The base-exchange pathway appears to be a minor route for the quantitative biosynthesis of phosphatidylethanolamine. In the CDP-ethanolamine pathway, ethanolamine is phosphorylated to phosphoethanolamine which is converted to CDP-ethanolamine and finally to phosphatidylethanolamine. Although the rate-limiting step of this pathway appears to be the conversion of phosphoethanolamine to CDP-ethanolamine (Vance, 1991), the phosphorylation of ethanolamine can also become rate-limiting under certain conditions (Sundler, 1973). Interestingly, the identity of ethanolamine kinase is still under debate. In some studies, purified choline kinase preparations have full ethanolamine kinase activities but it is not clear if the phosphorylation of choline and ethanolamine are catalyzed by the same enzyme in all tissues (Ishidate and Nakazawa, 1992; Porter and Kent, 1992). The conversion of phosphoethanolamine for the formation of CDP-ethanolamine is catalyzed by CTP:phosphoethanolamine cytidylyltransferase which is found exclusively in the cytosol. Although the enzyme was purified by Sundler (1975) almost two decades ago, only limited information is available on its regulation (Tijburg et al., 1992).

The conversion of CDP-ethanolamine to phosphatidylethanolamine is catalyzed by the CDP-ethanolamine: diacylglycerol ethanolaminephosphotransferase. Similar to the cholinephosphotransferase, this enzyme is also located in the endoplasmic reticulum (Kanoh and Ohno, 1976; Cornell, 1992). We have shown in a recent study that the enzyme can be solubilized by Triton QS-15, and the cholinephosphotransferase activity can be partially separated from the ethanolaminephosphotransferase activity by ion exchange chromatography. The two enzymes also display some difference in their characteristics and kinetic properties (O and Choy, 1990).

C. Biosyntheses of Phosphatidylinositol and Phosphatidylserine

These phosphoglycerides are present in significant quantities in mammalian tisssues. They carry a net negative charge and are called acidic phospholipids. The pathways for the biosynthesis of phosphatidylinositol is depicted in Figure 2. The key step for the synthesis of this phospholipid is the conversion of phosphatidic acid to CDP-diacylglycerol by the action of CTP:phosphatidic acid cytidylyltransferase (E.C. 2.7.7.41). The enzyme is located in both the microsomal and the mitochondrial fractions. The microsomal form of the enzyme is activated by GTP

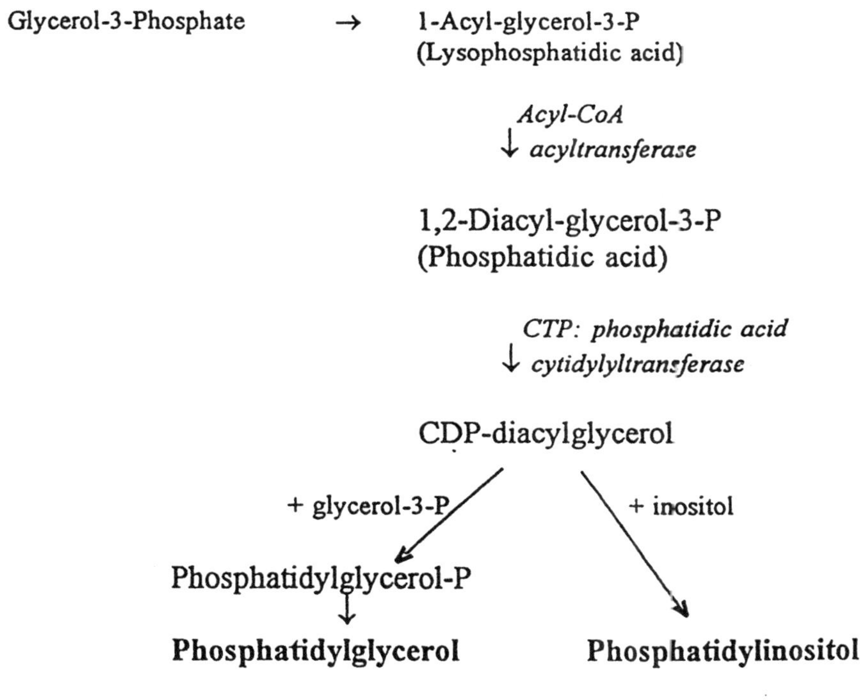

Figure 2. Pathways for the biosynthesis of phosphatidylglycerol and phosphatidylinositol.

whereas the mitochondrial form is not, suggesting that these two forms of the enzyme may be different (Mok et al., 1992). The mitochondrial enzyme has been purified from yeast (Carman and Kelley, 1992). It has a native mass of 114 kD and a subunit mass of 56 kD, and $MgCl_2$ is required for full enzyme activity. Subsequently, the CDP-diacylglycerol formed reacts with inositol for the formation of the phospholipid. The reaction is catalyzed by phosphatidylinositol synthase (EC 2.7.8.11) which has been purified to near homogeneity by Takenawa and Egawa (1977). The involvement of phosphatidylinositol in the anchoring of membrane proteins has been well-documented (Low, 1989).

In prokaryotes, phosphatidylserine is formed via the CDP-diacylglycerol pathway similar to the formation of phosphatidylinositol. In mammalian tissues, phosphatidylserine is synthesized from a base-exchange reaction in which the base group of a preexisting phospholipid is exchanged for serine. The base-exchange enzyme is located in the microsomal fraction and requires calcium for activity. However, the calcium requirement can be circumvented by the presence of ATP. The purified enzyme from brain microsomes has both ethanolamine and serine base exchange activity (Kanfer, 1992).

D. Biosyntheses of Polyglycerophospholipids

Cardiolipin is a major phospholipid in mitochondria (Hostetler et al., 1982). In the rat heart, cardiolipin comprises approximately 15% of the entire phospholipid mass (Hatch, 1994). Cardiolipin is characteristically associated with the inner mitochondrial membrane and the heart may be the only organ with significant amounts of extramitochondrial cardiolipin (Hostetler, 1982). However, this is still a point of contention since it is difficult to isolate cardiac non-mitochondrial subcellular fractions (i.e., sarcolemma and sarcoplasmic reticulum) which are completely devoid of mitochondrial enzyme activity. Cardiolipin is formed from the condensation of phosphatidylglycerol and CDP-diacylglycerol catalyzed by cardiolipin synthase in eukaryotic cells (Hostetler et al., 1971; Schlame et al., 1993). The cardiolipin synthase has recently been purified to homogeneity (Schlame and Hostetler, 1991). The enzyme appears as a 50 kD band in SDS-PAGE gel electrophoresis and requires cobalt for optimum activity.

Phosphatidylglycerol comprises 1-1.5% of the total phospholipid mass in mammalian tissues (Hostetler, 1982). In contrast to cardiolipin, phosphatidylglycerol is synthesized not only in the mitochondria but also in non-mitochondrial subcellular membranes. As depicted in Figure 2, phosphatidylglycerol is formed by the condensation of CDP-diacylglycerol and glycero-3-phosphate to form phosphatidylglycerol phosphate, catalyzed by phosphatidylglycerol phosphate synthase (E.C. 2.7.8.5). The product was dephosphorylated by phosphatidylglycerol phosphate phosphatase (E.C. 3.1.3.27) to produce phosphatidylglycerol (Kiyasu et al., 1968).

Recently, the pathways for the biosynthesis of cardiac phosphatidylglycerol and cardiolipin have been identified in isolated perfused rat hearts (Hatch, 1994). Using labeled glycerol, phosphatidylglycerol and cardiolipin were shown to be actively synthesized from the newly formed phosphatidic acid (Figure 3). The rate-limiting step of phosphatidylglycerol and cardiolipin biosynthesis in the rat heart was the conversion of phosphatidic acid to CDP-diacylglycerol.

Other polyglycerophospholipids found in mammalian tissues include bis(monoacylglycero)phosphate and acylphosphatidylglycerol which normally comprise less than 1% of the entire phospholipid mass in cells (Hostetler, 1982). An exception to this is the alveolar macrophage in which bis(monoacylglycero)phosphate may comprise 14-18% of the entire phospholipid mass. The biosynthesis of bis(monoacylglycero)-phosphate and acylphosphatidylglycerol occurs in the lysosome (Poorthius and Hostetler, 1976). Synthesis of these polyglycerophospholipids may occur via an acyl transfer from phosphatidylcholine to lysophosphatidylglycerol with formation of bis(monoacylglycero)phosphate (Huterer and Wherrett, 1989), and to bis(monoacylglycero)phosphate with formation of acyl-phosphatidylglycerol (Huterer and Wherrett, 1990).

E. The Remodeling of Diacylphosphoglycerides

Structural studies reveal that acyl groups are distributed in an asymmetrical manner in most of the diacylphosphoglycerides. Saturated fatty acids are usually esterified at the *sn*-1 position whereas unsaturated acyl groups are located at the *sn*-2 position. The distribution of molecular species are different among animal species and tissues as well as at cellular and subcellular levels (Choy and Arthur, 1989). It is clear that mechanisms exist in cells to bring about the observed distinctive and non-random distribution of acyl groups in the membrane phospholipids. The remodeling of phosphatidylcholine has been well-studied in the heart (Arthur and Choy, 1984). Since limited selectivity for the molecular species of diacylglycerol has been displayed by cholinephosphotransferase, the newly formed phospholipid must undergo extensive remodeling of the acyl group. At present, two mechanisms for the remodelling of the acyl groups in phosphatidylcholine have been identified. Lands (1960) first introduced the deacylation-reacylation of phospholipids which was based on the presence of both phospholipase A_2 and acyltransferase activities in mammalian tissues (Figure 4). Subsequently, the formation of phosphatidylcholine from lysophosphatidylcholine by an energy-independent transfer of an acyl group from a phospholipid (or lysophospholipid) was demonstrated.

The deacylation of phosphatidylcholine by the action of phospholipase A_1 or A_2 has been studied in detail (Choy and Arthur, 1989). In general, the reaction is catalyzed by the Ca^{++}-dependent enzyme which is normally associated with the cellular membrane (Hatch et al, 1989). The lysophosphatidylcholine formed can be further deacylated by lysophospholipase or reacylated to the parent phospholipid. The acylation of lysophosphatidylcholine to phosphatidylcholine by an acyl-CoA dependent process has been demonstrated in a large number of tissues.

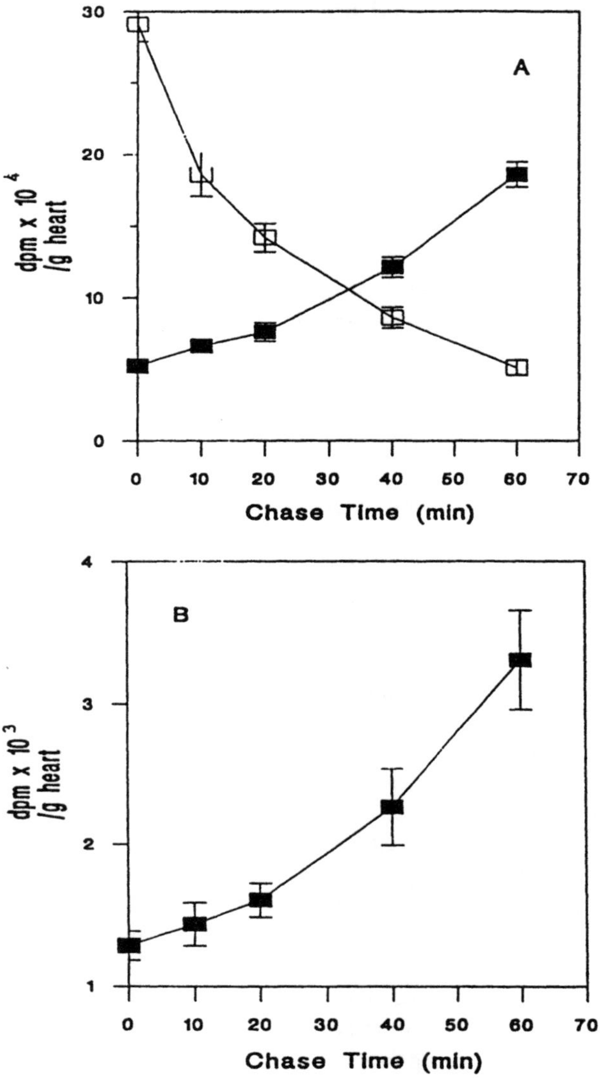

Figure 3. The rate limiting step of phosphatidylglycerol and cardiolipin biosynthesis in rat hearts. Isolated hearts were perfused for 5 min with 0.1 uM [1,3-^{3}H]glycerol, followed by perfusion for 60 min with 0.1 uM unlabeled glycerol. The amounts of labeling in cardiolipin, phosphatidylglycerol and phosphatidic acid were determined. (A) Phosphatidic acid (open symbols); phosphatidylglycerol (closed symbols). (B) Cardiolipin.

Phosphatidylcholine
(1,2-Diacylglycerophosphocholine)

Phospholipase A_2 ↓ ↑ *Acyltransferase*

Lysophosphatidylcholine
(1-Acylglycerophosphocholine)

↓ *Lysophospholipase*

Glycerophosphocholine

Figure 4. The deacylation-reacylation pathway for phosphatidylcholine.

The transfer of acyl groups to 1-acyl-glycerophosphocholine is catalyzed by acyl-CoA: 1-acyl-glycerophosphocholine acyltransferase whereas the transfer of an acyl group to 2-acyl-glycerophosphocholine is catalyzed by acyl-CoA: 2-acyl-glycerophosphocholine acyltransferase (Choy and Arthur, 1989). Indirect evidence shows that these two activities are acyl specific and catalyzed by separate enzymes (Arthur, 1989). In the last two decades, acyl-CoA: 1-acyl-glycerophosphocholine acyltransferase has been studied extensively in the liver, lung and brain and the subject has been reviewed (Choy et al., 1992). Although the purification of the enzymes from the brain (Deka et al, 1986) and bovine heart (Sanjanwala et al., 1988) have been reported, there is no further study on the purified enzyme by these investigators. At present, only limited information is available on the acyl specificity of the enzyme and its mechanism of regulation is entirely unknown.

Liver mitochondrial cardiolipin is distinguished from other phospholipids by the presence of linoleoyl in almost all molecular species. Direct evidence has been obtained for the presence of a deacylation-reacylation cycle by which a labeled linoleoyl group is incorporated into exogenous cardiolipin added to mitochondria (Schlame and Rustow, 1990). It is concluded that a cycle, comprising cardiolipin deacylation to monolysocardiolipin and its subsequent reacylation with linoleoyl-CoA, would provide a potential mechanism for the remodeling of molecular species of newly formed cardiolipin.

III. CHANGES IN PHOSPHOLIPID CONTENT IN DISEASES

A. Diseases

Under normal circumstances, deficiency or accumulation of a single phospholipid in mammalian tissues is detrimental to the proper function of that tissue and severe changes in most cases are lethal. Thus, severe alterations in the phospholipid content in mammalian tissues are rare. In this section, we have provided a synopsis on phospholipid alteration in a congenital disease as well as an acquired disease.

Respiratory distress syndrome is a condition affecting premature neonates caused by surfactant deficiency. The lung surfactant contains a composite of phospholipids, neutral lipids and proteins which coats the inner surface of the lungs with the purpose of reducing alveolar surface tension, thus allowing for the low compliance characteristics of the lung (Leff and Schumacker, 1993). Phosphatidylcholine (dipalmitoylphosphatidylcholine and oleoyl-palmitoyl-phosphatidylcholine) is the principle phospholipid component of the lung surfactant. A low level of lung surfactant is usually associated with premature birth. In premature fetal rats (delivered 1 and 2 days prematurely), the rates of incorporation of radioactive choline and phosphorus into phosphatidylcholine were low immediately after birth but were elevated 60% 3 hours postpartum (Weinhold et al., 1980). The increased incorporation of radioactivity into phosphatidylcholine was regarded as a compensatory mechanism to boost up the phosphatidylcholine content and was caused by an increase in CTP:phosphocholine cytidylyltransferase activity (Weinhold et al., 1981). Phosphatidylglycerol is another important phospholipid in pulmonary surfactant comprising approximately 5% of the total phospholipid mass. Interestingly, glucocorticosteroids and estrogens which stimulate cytidylyltransferase activity also accelerate the synthesis of phosphatidyglycerol in the lung (Post et al., 1980).

The chronic consumption of excessive alcohol may lead to a variety of symptoms. The accumulation of fat in the liver is the most common disturbance of lipid metabolism produced by alcohol. Since steatosis constitutes the most striking histological feature of the early stages of alcoholic liver injury, this initial phase is generally defined as alcoholic fatty liver (Baraona and Lieber, 1979). The increased NADH/NAD^+ ratio generated by the oxidation of ethanol via alcohol dehydrogenase also favors the production of glycero-3-phosphate from dihydroxyacetone phosphate which facilitates the formation of diacylglycerol, triacylglycerol and phospholipids. It has been shown that ethanol-fed rats incorporate significantly more labeled glycerol into hepatic phosphatidylcholine and phosphatidylethanolamine, despite ethanol-induced inhibition of glycerol uptake by splanchnic organs and the possible dilution of the hepatic glycero-3-phosphate (Mendenhall et al, 1969). In a separate study, phospholipid contents in liver microsomes and mitochondria were determined in ethanol-fed rats. The phospholipid distribution within organelles were not changed, except for a significant increase in the phosphatidylinositol content of microsomes of the ethanol-fed animals. The fatty acid composition of both microsomal and mitochondrial phospholipids were significantly altered by ethanol feeding. It was concluded that the alterations in the fatty acyl groups are suggestive of ethanol-induced changes in fatty acid desaturation activities.

In addition to altered phospholipid content and composition in various tissues, chronic ethanol feeding results in the formation of an abnormal acid phospholipid in the kidney and the brain of rats. This phospholipid was subsequently identified as phosphatidylethanol by GC-MS and other determinations (Alling et al, 1984). It is generally accepted that ethanol exerts its pharmacological effects on the lipids of the cell membrane and causes an increased fluidity and alteration in the function of integral membrane proteins (Goldstein, 1986). The most fluid membranes, including those that are low in cholesterol, are the most easily disordered by ethanol. When animals are treated chronically with ethanol, their membranes become less fluid. This can be regarded as a response that can be regarded as adaptive. Ethanol may favor the uptake of cholesterol or saturated fatty acid into membranes, thus reducing its own effect.

B. Diet and Drug Treatment

The fatty acid composition of specific phospholipids are different between tissues of various animal species. For example, phosphatidylcholine from bovine heart exhibits a high percentage of 18:2 molecular species whereas phosphatidylcholine of rat brain exhibits a high percentage of 16:0 and is devoid of 18:2 (Wheeldon et al., 1965; Skrbic and Cumings, 1970). It is well established that dietary fat may alter physiological processes in many organs including the ATPase activities in the mitochondria (Innes and Clandinin, 1981) but the mechanism is not entirely clear. The supplemenation or deficiency in dietary fatty acids may change the content of phospholipid in cellular membranes and the composition of the fatty acyl moiety within a phospholipid group and consequently, alters the physical property of the membrane. Weanling rats fed a 10% sunflower oil diet for eight weeks exhibited a reduction in fat cell plasma membrane phosphatidylethanolamine, phosphatidylcholine and sphingomyelin contents when compared to animals fed a control diet containing 1.5% sunflower oil (Khuu Thi Dinh et al., 1990). Mice fed a fat free diet supplemented with 1% safflower and 1% linolenate or 1% safflower and 1% eicosatrienoate for two weeks exhibited 36-50% reductions in 20:4 (n- 6) in liver phosphatidylcholine and liver and spleen phosphatidylethanolamine compared with mice fed 2% safflower (Berger and German, 1990). Adult rats fed with α-linolenic acid deficent diet exhibited progressive reductions in docosahexanoic acid in liver, heart and testes membranes reaching a maximum within three weeks (Bourre et al., 1992). A 50% reduction in 18:2 (n-6) molecular species in heart mitochondrial cardiolipin was observed in rats fed a hydrogenated corn oil diet for 10 days compared with animals fed a plain corn oil diet (Yamaoka et al., 1990). Interestingly, a decrease in the 18:2 molecular species of cardiolipin appears to produce a reduction in the oxygen consumption by the cardiac mitochondria. These and many other studies exemplify the fact that dietary fat has a significant role in the determination of the acyl contents of phospholipids.

Deficiency of essential vitamins or minerals may lead to an alteration of phospholipid content and composition. For example, rats fed a choline-methionine deficient diet for four days exhibited a significant decrease in liver phosphatidylcholine content (Yao et al., 1990). In this study, direct evidence was obtained to demonstrate the translocation of CTP:phosphocholine cytidylyltransferase for regulation of phosphatidylcholine biosynthesis. The translocation of the enzyme is regarded as a compensatory mechanism to maintain phosphatidylcholine biosynthesis under the dietary treatment. In another study, rats supplemented with the essential mineral selinium for three days exhibited increased cardiac phosphatidic acid and phosphatidylcholine biosynthesis (Liu et al., 1993). The mechanism for the enhanced biosynthesis of these phospholipids was caused by an increase in cardiac acyl-CoA levels and a two-fold enhancement of CDP-choline:diacylglycerol cholinephosphotransferase activity.

A number of reagents and drugs have been shown to alter the biosynthesis and content of phospholipids in tissues. Perfusion of isolated hamster hearts with the experimental anti-arrhythmic drug methyl lidocaine produced an accelerated biosynthesis of phosphatidylserine and phosphatidylinositol (Tardi et al., 1992). The mechanism for the increase in phosphatidylinositol biosynthesis was proposed to be an increase in the intracellular pool size of CDP-diacylglycerol mediated by a methyl lidocaine induced increase in CTP:phosphatidic acid cytidylyltransferase activity. Another anesthetic, diethylether, produced a 29% decrease in the biosynthesis of phosphatidylcholine in the kidney (O et al., 1988). The mechanism for this decrease in phosphatidylcholine biosynthesis was due to a reduction in microsomal membrane CTP:phosphocholine cytidylyltransferase activity. Thus, the choice of anesthetic must be carefully made when studying phospholipid biosynthesis in animal models. 3-Deazaadenosine, a structural analogue of adenosine, has been shown to dramatically reduce the biosynthesis of phosphatidylcholine via the methylation pathway (Pritchard et al, 1982). The mechanism for the reduction may be due to the competitive inhibitory activity of 3-deaza s-adenosyl homocyteine on the phosphatidylethanolamine methyltransferase. In order to compensate for the reduction in the progressive methylation of phosphatidylethanolamine, phosphatidylcholine biosynthesis via the CDP-choline pathway was stimulated (Chiang et al., 1980; Prictchard et al., 1982).

Treatment of rats with the antimalarial drug chloroquine results in a drug induced hepatic phospholipidosis (Matsuzawa and Hostetler, 1981). In particular, accumulation of the polyglycerophospholipid bis(monoacylglycero)phosphate occurs. Two mechanisms for the observed effect have been proposed. The accumulation of phospholipid may be caused by an increase in the delivery of phospholipid to lysosomes and/or an inhibition of lysosomal phospholipid catabolism.

IV. PHOSPHATIDYLCHOLINE BIOSYNTHESIS IN CELLULAR INJURY AND DISEASES

A. Introduction

The majority of phosphatidylcholine has been shown to be synthesized via the CDP-choline pathway in mammalian tissues including the heart (Zelinski et al., 1980). Alterations in its rate of biosynthesis have been detected in disease models and other pathological conditions. The relationship between phosphatidylcholine biosynthesis during pathological insult is demonstrated in the following models.

B. Cardiomyopathy

Through autosomal resessive inheritance, cardiomyopathy develops spontaneously in a strain of inbred Syrian hamsters (BIO 14.6 strain). Animals of both sexes exhibit degenerative lesions in the myocardium with a 100% incidence. Morphological and histological changes in the hearts of these animals were observed at different developmental stages (Bajusz, 1969). Biochemical changes including decreases in the rate of fatty acid oxidation, high energy phosphate levels, phospholipid content and enzyme activities were reported, suggesting that phospholipid biosyntheses in the heart of the myopathic animal are perturbed. Hence, phosphatidylcholine biosynthesis in the myopathic hamster hearts were compared with date-matched controls (Choy, 1982). Upon perfusion with labeled choline, no significant change was detected in the heart of 90-120-day-old animals, but a 22% increase in the labeling of phosphatidylcholine was observed in the hearts of 150-200-day-old myopathic hamsters (Figure 5). However, the total cardiac phosphatidylcholine remained unchanged. The cause for the increase in labeling of phosphatidylcholine during cardiomyopathy was investigated by analyzing the labeling of the intermediates in the CDP-choline pathway. The respective values of labeling and pool size of of CDP-choline was found to be 72% and 60% of those obtained from the control. The unproportional reduction in the labeling of CDP-choline caused a 20% increase in the specific radioactivity of CDP-choline and resulted in the apparent increase in the labeling of phosphatidylcholine. Based on the specific radioactivity of CDP-choline, the net amount of phosphatidylcholine synthesized was estimated to be similar between the normal and myopathic hearts. The reduction in the net CDP-choline formation was probably caused by an observed decrease in cardiac CTP concentration (34%) during the development of cardiomyopathy. Analysis of the enzyme activities in the CDP-choline pathway revealed that the phosphocholine cytidylyltransferase activity was elevated. The enhanced enzyme activity was regarded as one of the compensatory mechanisms for the myopathic heart to maintain a minimum CDP-choline level in order to prevent reduction of net phosphatidylcholine biosynthesis during the development of the disease.

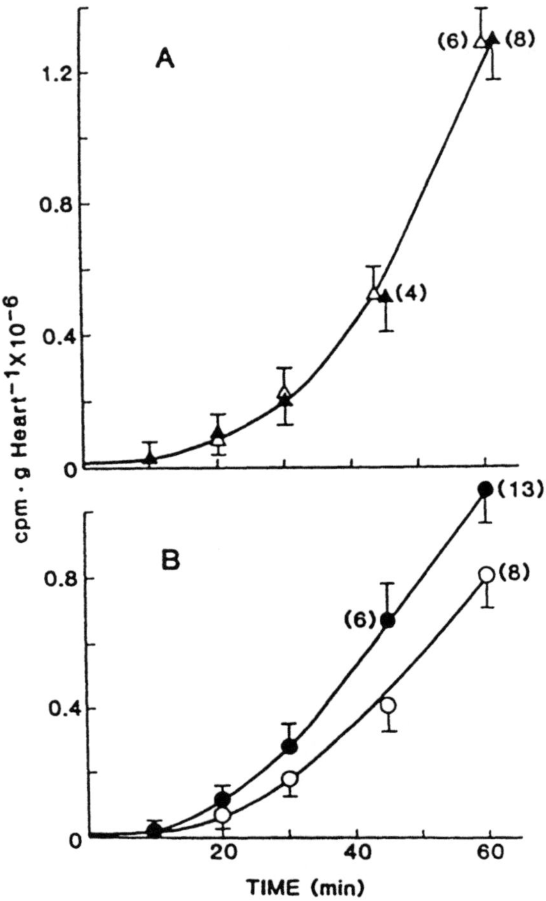

***Figure** 5*. Time course for the incorporation of labeled choline into phosphatidylcholine in normal and myopathic hamster hearts. Hearts from 90-120 day-old (A) and 150-200-day-old (B) hamsters were perfused with labeled choline for 10-60 min, and the amounts of label incorporated into phosphatidylcholine in each heart were determined. The open symbol denotes the date-matched control animals and closed symbol denotes cardiomyopathic animals. The vertical bar denotes the standard deviation for each set of values.

C. Hypoxia

The requirement for ATP and CTP in the CDP-choline pathway suggests that the intracellular levels of these compounds may affect the rate of phosphatidyl-

choline biosynthesis. In the myopathic hamster, the levels of both ATP and CTP are proportionally reduced. The reduction in CTP has been shown to cause a decrease in the formation of CDP-choline but the reduction in ATP in the myopathic heart does not affect the phosphorylation of choline.

The perfusion of the isolated heart under hypoxic conditions is an established model for decreasing the overall oxygen delivery to below the critical level required to support the metabolic ATP demands of the tissue. The ability of the heart to synthesize phosphatidylcholine under hypoxic conditions was examined (Hatch and Choy, 1990). The hamster heart was pulse-labeled with labeled choline and then chased with non-labeled choline under hypoxia for various periods. Histological examination of the ventricular tissue by electron microscopy revealed some degree of mitochondrial swelling within 60 minutes of hypoxic perfusion. The mitochondrial swelling was shown to be reversible upon reoxygenation. At 60 minutes of hypoxic perfusion, severe decreases in both ATP and CTP levels were observed with a corresponding fall in the rate of phosphatidylcholine biosynthesis (Table 1). Analysis of the choline-containing metabolites revealed that the lowered ATP level did not affect the phosphorylation of choline to phosphocholine, but the lower CTP level resulted in the decreased conversion of phosphocholine to CDP-choline which might be the cause for the observed reduction in phosphatidylcholine biosynthesis. The enzyme activities in the CDP-choline pathway were also determined. An increase in the CTP:phosphocholine cytidylyltransferase activities was detected in the hypoxic heart which might arise from the translocation of the enzyme from the cytosolic to the microsomal form. Further studies of the hamster heart under hypoxic conditions showed that the translocation of the cytidylyltransferase was caused by an increase in fatty acid content. However, careful analysis of the phosphatidylcholine content in the hypoxic heart revealed that there was no significant change between the control and the hypoxic hearts. One explanation to this apparent discrepancy was that the decrease in phosphatidylcholine biosynthesis indeed caused a reduction in the total phosphatidylcholine pool, but the change within a relatively short period (60 minutes) of hypoxic perfusion was too small to be detected.

D. Fasting

Changes in phospholipid content in the liver during fasting has been well-documented. The synthesis of phosphatidylcholine has been found to be reduced in the

Table 1. ATP and CTP Concentrations in Hamster Hearts Perfused Under Hypoxic Conditions for 60 min.

	ATP	*CTP*
	(μmol/g of heart)	(nmol/ g of heart)
Control hearts	2.54 ± 0.16 (3)	11.49 ± 2.24 (3)
Hypoxic hearts	0.99 ± 0.09 (3)	3.25 ± 1.12 (3)

Note: The values depicted are mean ± standard deviation (number of experiments).

liver of fasting animals. However, the coordination of phosphatidylcholine biosynthesis via the CDP-choline pathway with other major metabolic pathways in the liver is still undefined. Since the liver undergoes rapid adaptive changes in order to maintain the level of plasma glucose during fasting, changes in the activities of the glycolysis and gluconeogenesis pathways with alterations in the pools of major metabolites may affect the rate of phosphatidylcholine biosynthesis. In a recent study, the control of phosphatidylcholine biosynthesis in livers of hamsters fasted for 24 h and 48 h was examined (O and Choy, 1993). Under both fasting conditions, the incorporation of labeled choline into phosphatidylcholine was reduced. After 48 hours of fasting, the 52% reduction in phosphatidylcholine biosynthesis was caused by changes in several factors including a diminishing rate of choline uptake and severe reductions in the pool sizes of ATP and CTP (Table 2). The reduction in the levels of these high energy nucleotides resulted in a decrease in the pools of choline-containing metabolites. The activation of cytidylyltransferase after 48 h of fasting might be regarded as a compensatory mechanism for the maintenance of phosphatidylcholine biosynthesis. After 24 hours of fasting, a 25% reduction in phosphatidylcholine biosynthesis was observed. The ATP and CTP levels were decreased but the reduction was not severe enough to affect the choline uptake or the labeling of the phosphocholine fraction. The activities of the cytidylyltransferase remained unchanged but an accumulation of labeled CDP-choline was detected. Although cholinephosphotransferase activity was not changed in the microsomal fraction, the enzyme activity was attenuated in the postmitochondrial fraction. Further analysis revealed that cholinephosphotransferase in the liver was inhibited by an endogenous inhibitor in the cytosol which was later identified as argininosuccinate. The level of argininosuccinate was elevated during fasting and the change quantitatively accounted for the attenuation of cholinephosphotransferase activity. In addition, the level of diacylglycerol was substantially reduced in the liver during fasting.

Table 2. Effect of 48 h Fasting on the Pool Sizes of metabolites for the Biosynthesis of Phosphatidylcholine in Hamster Liver

	Control	*Fasted Animal*
	(μmol/ g liver wet weight)	
ATP	0.86 ± 0.10	0.32 ± 0.14
CTP	0.09 ± 0.01	0.03 ± 0.02
Diacylglycerol (total)	2.10 ± 0.38	0.57 ± 0.22
Diacylglycerol (microsomal) (nmol/ mg protein)	10.5 ± 1.25	6.1 ± 1.86
Argininosuccinate	0.50 ± 0.13	1.12 ± 0.32

Note: The values depicted are mean ± standard deviations of at least three experiments.

In the liver of the fasting animal, the enhancement of amino acid metabolism for gluconeogenesis causes activation of enzymes of the urea cycle (Snodgrass, 1981). The increase in argininosuccinate level after 24 h of fasting was found to produce a 10% inhibition of enzyme activity when compared with the control. Fasting for longer periods did not continue to produce substantial increases in arininiosuccinate levels. Hence, a maximum of 12-15% inhibition of cholinephosphotransferase can be produced by changes in endogenous argininosuccinate levels during fasting. Under normal conditions, the 12-15% inhibition of enzyme activity may not play a significant role in the regulation of phosphatidylcholine biosynthesis, but the importance of such an inhibition may become more prominent when diacylglycerol is in short supply. The modulation of cholinephosphotransferase activity by a key metabolite of the urea cycle serves as an attractive model to demonstrate the occurence of coordination between phospholipid and protein metabolism during gluconeogenesis. It appears that the inhibition of cholinephosphotransferase by argininosuccinate, coupled with a substantial decrease in the diacylglycerol level, would provide the hamster liver with an immediate mechanism for the transient modulation of phosphatidylcholine biosynthesis during short-term fasting.

E. Viral Infection

The rate of incorporation of labeled choline into phosphatidylcholine was found to be altered in HeLa cells infected by polio virus (Choy et al, 1980). Analysis of the choline containing metabolites in the CDP-choline pathway demonstrated that the conversion of phosphocholine to CDP-choline was elevated. However, the determination of the CTP:phosphocholine cytidylyltransferase activity under optimal assay conditions revealed that it was lower than that obtained from the mock-infected cells. In order to resolve this apparent ambiguity, a procedure was designed to assay for the enzyme activity in the postmitochondrial fraction under physiological conditions in the presence of labeled phosphocholine. When the enzyme activity was determined in this manner, an increase in enzyme activity was detected in the polio-infected cells. The increase of enzyme activity could be reproduced in the mock-infected cells by the addition of CTP. Subsequently, the CTP pool in the polio-infected cells was found to be elevated. A temporal relationship between the elevated CTP pool in the cytosol and the increase in phosphatidylcholine biosynthesis was established after polio infection. The study showed that the level of CTP in the cytosol is a key factor for the regulation of phosphatidylcholine biosynthesis in HeLa cells.

F. Hormones

The effects of hormones on phosphatidylcholine biosynthesis have been studied in several organs. The synthetic estrogen, diethylstilbestrol, has been shown to stimulate phosphatidylcholine biosynthesis in roosters (Vigo and Vance, 1981). The hormone was shown to cause a 3-fold increase in the activity of choline kinase which resulted in a higher rate of phosphorylation of choline for the formation of phosphocholine. In a subsequent study, the increase in choline kinase activity was found to be caused by an elevated amount of titratable enzyme (Vigo et al., 1981). However, the activities of other enzymes in the CDP-choline pathway were not altered during the treatment. It was concluded that the biosynthesis of choline kinase was induced by the action of diethylstilbestrol.

Glucocorticoids stimulate the synthesis of phosphatidylcholine in the lung cells. The postulated mechanism is that the hormone causes an increase in total cytidylyltransferase activities. It appears that glucocorticoids do not directly stimulate the type II alveolar cells, but rather on fetal lung fibroblasts that are stimulated to produce fibroblast-pneumonocyte factor which acts on the alveolar type II cells (Post et al., 1986). Further support to the indirect effect of the hormone was obtained by immunological studies. Antibody to this factor was found to block its effect on type II cells and caused a delay in lung maturation in intact animals (Post et al., 1984). There is ample evidence to show that the fibroblast-pneumonocyte factor causes the stimulation of the cytidylyltransferase activity in the type II cells. However, the exact mechanism for the stimulation of the enzyme activity has not been elucidated.

V. BIOSYNTHESES OF OTHER PHOSPHOLIPIDS IN DISEASES

A. Phosphatidylethanolamine

In mammalian tissues, a significant portion of phosphatidylethanolamine is synthesized via the CDP-ethanolamine pathway. The biosynthesis of this phospholipid and its relationship to cellular injury will be discussed in another chapter of this volume and our discussion will be confined to its relationship with phosphatidylcholine and phosphatidylserine. Since the CDP-ethanolamine pathway for phosphatidylethanolamine biosynthesis is similar to the CDP-choline pathway, changes in intracellular ATP and CTP levels may also affect the biosynthesis of phosphatidylethanolamine in a similar manner as in phosphatidylcholine.

Due to the structural similarity between choline and ethanolamine, the possibility of exogenous choline to regulate phosphatidylethanolamine biosynthesis was examined (Zelinski and Choy, 1982). Hamster hearts were perfused with labeled ethanolamine in the presence of up to 0.5 mM choline. The incorporation of label into phosphatidylethanolamine was decreased 28% at 0.2-0.5 mM choline. Analysis of the ethanolamine-containing metabolites of the CDP-ethanolamine pathway revealed that choline in the perfusate had no effect on the labeling of ethanolamine. However, a 28-30% reduction in the labeling of phosphoethanolamine and CDP-ethanolamine was observed. Concurrently, there was a 2-fold increase in the intracellular choline level. Although the formation of CDP-ethanolamine is regarded as the rate-limiting step of phosphatidylethanolamine biosynthesis, choline was found to inhibit ethanolamine kinase, but did not have any effect on phosphoethanolamine cytidylyltransferase. Hence, choline may provide an additional mechanism for the regulation of phosphatidylethanolamine biosynthesis in the heart.

Ethanolaminephosphotransferase catalyzes the last step of the CDP-ethanolamine pathway. Similar to the cholinephosphotransferase in the CDP-choline pathway, this enzyme is also bound to the membrane. When agents which perturb the structure and function of cytoskeletal elements were tested for effects on phospholipid metabolism in glioma cells, the biosynthesis of phosphatidylcholine and phosphatidylethanolamine were inhibited by cytochalasin B (George et al., 1991). However, this agent had no effect on the biosynthesis of other phospholipids. The effect of cytochalasin B on phospholipid biosynthesis were not due to inhibition of glucose uptake but an alteration of intracellular calcium.

Since phosphatidylethanolamine may also be produced by the decarboxylation of phosphatidylserine, the role of circulating serine on phosphatidylethanolamine biosynthesis in the hamster heart was examined (McMaster and Choy, 1992). When the hamster heart was perfused in the presence of serine, labeled ethanolamine uptake was attenuated by 0.05-10 mM serine in a non-competitive manner, and the incorporation of labeled ethanolamine into phosphatidylethanolamine was also inhibited by serine. Analysis of the ethanolamine-containing metabolites in the CDP-ethanolamine pathway revealed that the conversion of ethanolamine to phosphoethanolamine was reduced. The reduction was a result of an inhibition of ethanolamine kinase activity by an elevated pool of intracellular serine. Interestingly, serine did not cause any enhancement of phosphatidylethanolamine hydrolysis. In addition, the base-exchange reaction for phosphatidylserine formation or the decarboxylation of phosphatidylserine was not affected by exogenous serine in the perfusate. It is clear that the concentration of circulating serine has an important role in the alteration of phosphatidylethanolamine biosynthesis by regulating ethanolamine uptake. However, serine does not appear to contribute to phosphatidylethanolamine formation via the decarboxylation of phosphatidylserine pathway. In another study, some serine analogs such as isoserine and serinol cause changes in the biosynthesis of phosphatidylserine, and to a lesser extent, the synthesis of phosphatidylcholine and phosphatidylethanolamine in T cells (Pelassy et al., 1991). In serinol treated cells, the production of diacylglycerol was impaired while calcium ion mobilization remained unaffected. Serinol appeared to be a potential immunoregulatory molecule active at the level of protein kinase C regulation either through its interaction with phosphatidylserine or through diacylglycerol production.

B. Phosphatidylserine

In mammalian cells, phosphatidylserine is formed via base-exchange with another phospholipid, preferably phosphatidylcholine. The phosphatidylserine is then transported to the mitochondria where it is decarboxylated to phosphatidylethanolamine and phosphatidylserine is then regenerated possibly on the endoplasmic reticulum with the release of ethanolamine. It is clear that phosphatidylserine is generated at the expense of phosphatidylcholine and the reactions produce a molecule each of choline and ethanolamine which can be recycled for the biosynthesis of the respective phospholipids. As a consequence of these reactions, phosphatidylserine and phosphatidylethanolamine are increased in the cell. Presumably, the choline can be resynthesized into phosphatidylcholine and there should be no net decrease in phosphatidylcholine level. There is very little information on the regulation of phosphatidylserine biosynthesis in mammalian tissues, and its role in cellular injury is undefined. Recently, phosphatidylserine has been shown to suppress antigen-specific IgM production in mice (Carr et al., 1992). These animals were administered orally with phosphatidylserine and subsequently intubated intragastrically with sheep red blood cells. A significant decrease in antigen-specific IgM production by splenic lymphocytes was observed when compared with the appropriate controls. Since phosphatidylserine is released upon injury and destruction of eukaryotic cells, it was concluded that the immunosuppressive qualities of phosphatidylserine might account for the presence of an endogenous anti-inflammatory factor observed during cellular injury.

C. Polyglycerophospholipids

The effect of hypoxia on phosphatidylglycerol and cardiolipin biosynthesis has been recently examined in isolated perfused rat hearts. Hearts were perfused with labeled glycerol and subsequently chased under normoxic or hypoxic conditions. The labeled material incorporated into phosphatidylglycerol and cardiolipin was dramatically reduced in the hypoxic hearts but the activity of the phosphatidic acid: CTP cytidylyltranferase was unaltered (Cheng and Hatch, 1994). In addition, a significant reduction in the content of cardiac CTP was observed. Since CDP-diacylglycerol is required for the biosynthesis of phosphatidylglycerol and cardiolipin, a decrease in the formation of labeled CDP-diacylglycerol may be one of the factors for the reduction in phosphatidylglycerol and cardiolipin biosynthesis in hypoxic hearts.

Diabetes causes an alteration in myocardial metabolism from a carbohydrate based to a fat based metabolism (Lopaschuck, 1989). An alteration in the phospholipid composition of cardiac membranes occurs and changes in the cholesterol to phospholipid ratio following streptozotocin (STZ)-induced diabetes in rats might cause alterations in cardiac membrane enzyme function. Cardiolipin content was shown to be reduced in the sarcolemmal membranes from ventricular tissue of STZ-treated rats (Pierce et al., 1983). It has been hypothesized that such a reduction in the sarcolemmal cardiolipin may be one of the reasons for decreased calcium binding to SL membranes which causes an alteration in the contractile properties of the heart. In contrast, sarcoplasmic reticulum cardiolipin content was increased in these animals (Pierce and Dhalla, 1983). This prompted us to examine if myocardial cardiolipin biosynthesis was altered in ventricular tissue of the diabetic rat heart. The overall content of rat ventricular cardiolipin, 10.4 ± 0.5 μmoles lipid phosphorus/g freeze-dried ventricles, was unaltered in both acute (4 day) and chronic (28 day) STZ-treated diabetic rats compared with controls (Hatch et al., 1994). In addition, when hearts from these animals were perfused for short periods of time with buffer containing [^{32}P] phosphorus, the radioactivity incorporated into cardiolipin was unaltered. Thus, the observed differences in cardiolipin composition in diabetic sarcolemma and sarcoplasmic reticulum membranes were not due to an alteration in the biosynthesis of cardiolipin but likely an alteration in its membrane distribution.

Mitochondrial levels of cardiolipin appear to be regulated at least in part by thyroid hormone. A 36% decrease in the myocardial content of cardiolipin was observed in hypothyroid rats (Paradies and Ruggiero, 1990). In addition, the level of cardiolipin was found to be increased in the mitochondrial fraction of hearts prepared from hyperthyroid rats (Paradies and Ruggiero, 1988). Treatment of rats with thyroxine for 5 days days resulted in a 52% increase in liver cardiolipin synthase activity (Hostetler, 1991). We invesitaged if the increase in cardiac cardiolipin mass in hyperthyroid rats was due to increased synthesis or decreased degradation of cardiolipin. Cardiolipin biosynthesis is stimulated in the ventricles of rats treated with exogenous (250 μg/Kg) thyroxine for 5 consecutive days (Hatch, G.M., unpublished results). The mass of ventricular mitochondrial cardiolipin was increased 25% compared with controls. When hearts from hyperthyroid animals were perfused with labeled glycerol, a significant increase in the labeling of cardiolipin was observed when compared with controls, indicating an increase in cardiolipin biosynthesis.

Hypophysectomy is known to alter body growth. Hypophysectomy has been demonstrated to decrease rat liver cardiolipin levels by 25% (Clejan and Maddaiah, 1986). The normal concentration of cardiolipin was restored by treatment of these animals with growth hormone. The mechanism for the reduction in cardiolipin in hypophysectomy was not determined. Significant alterations in the mass of other phospholipids was not observed.

D. Phosphatidylinositol

Phosphatidylinositol is formed by the condensation of CDP-diacylglycerol and *myo*-inositol. There are three sources for the cellular *myo*-inositol: diet, synthesis from glucose and recycled from the phospholipid. In the proximal tubules of the kidney, the level of intracellular *myo*-inositol was found to be considerably higher than the Km for phosphatidylinositol synthase, indicating that the level of intracellular *myo*-inositol might not play a significant role in the regulation of phosphatidylinositol synthesis (Galvao and Shayman, 1990). The acyl content and composition in phosphatidylinositol are affected by diet, hormones, and certain diseases (Hrelia et al, 1992). Experiments using labeled glycerol as a precursor indicate that newly synthesized phosphatidylinositol contains mainly monoenoic and dienoic species of fatty acids which are subsequently converted to polyenoic forms through a deacylation-reacylation process (Esko and Raetz, 1983).

Phosphatidylinositol has been shown to be intimately associated with signal transduction. In this process, the phospholipid becomes phosphorylated on the inositol moiety by inositol kinase (Berridge and Irvine, 1989). The polyphosphorylated inositol in the phospholipid is subjected to phospholipase C cleavage, resulting in the production of 1,2-diacylglycerol and 1,4,5-inositol triphosphate (Rhee et al., 1989). The subject of signal transduction via the hydrolysis of phosphatidylinositol phosphate is beyond the scope of this chapter.

IV. ALTERATIONS IN REMODELING OF PHOSPHOLIPIDS

A. Reacylation of Lysophospholipids

The lysophospholipid formed from the action of phospholipase A can be acylated back to the parent phospholipid by the action of acyltransferases. The lysophosphatidylcholine: acyl-CoA acyltransferase activity is the most prominent type of acyltransferase in mammalian tissues. The enzyme is located in both the microsomal and mitochondrial fractions. The microsomal enzyme has a higher specificity towards unsaturated species of acyl-CoA than saturated species, whereas the mitochondrial enzyme is specific towards linoleoyl-CoA (Arthur et al., 1987). The acyltransferase activity in the rat lung was inhibited by high concentrations of lysophosphatidylcholine, but the inhibition by high concentrations of acyl-CoAs was not prominent (Hasegawa-Sasaki and Ohno, 1975). In an earlier study, enzyme activities in the rat liver and heart were inhibited by detergents and inhibitors of cyclic nucleotide phosphodiesterase (Shier, 1977). However, the activity of the liver microsomal enzyme was relatively unaffected by sulfhydryl-binding reagents such as iodoacetate, N-ethylmaleimide, and p-chloromercuriphenylsulfonic acid, but the activity of the partially purified enzyme was inhibited by these reagents (Hasegawa-Sasaki and Ohno, 1980). Acyltransferase activity *in vivo* was affected by long-term administration of clofibric acid and chronic administration of isoproterenol (Yashiro et al., 1989). At present, the exact mechanism for the regulation of the enzyme activity *in vivo* is unknown.

The role of acyltransferase for the regulation of lysophospholipid levels in the tissue has not been defined. In order to elicit the physiological role of the acyltransferase, hamster hearts were perfused with methyl lidocaine (Tardi et al., 1990). Methyl lidocaine is an analog of lidocaine and has been employed as an experimental drug for the treatment of cardiac arrhythmias. The drug was found to inhibit both lysophosphatidylcholine: acyl-CoA and lysophosphatidylethanolamine: acyl-CoA acyltransferase activities in the hamster. However, it has no effect on the other lysophospholipid metabolic enzymes. When the heart was perfused with 0.5 mg methyl lidocaine/ml, acyltransferase activities were attenuated but there was no change in the activities of phospholipase A or lysophospholipase. The levels of the major lysophospholipids in the heart were not altered by methyl lidocaine perfusion. When the heart was perfused with labeled lysophospholipid in the presence of methyl lidocaine, there was a reduction in the formation of the phospholipid and an increase in the release of the free fatty acid. Surprisingly, the labeling of lysophospholipid in the heart was not altered by methyl lidocaine. It is quite likely that the acylation reaction has no direct contribution to the maintenance of the lysophospholipid levels in the heart.

B. The Role of Acyl-CoA

Long chain acyl-CoA esters are intermediates involved in the acylation of lipids and proteins. The intracellular acyl-CoA content and composition have been shown to be important factors in determining acyl groups in phospholipids produced by the acylation process (Lands and Hart, 1965; Yashiro et al., 1989; Choy and Arthur, 1989). The ability to utilize acyl-CoA *in vivo* appears to be dependent on its availability, which in turn is regulated by its rate of synthesis and degradation. Since ATP is required for its synthesis, it can be predicted that the level of acyl-CoA is in part dependent on the intracellular ATP pool. In the last decade, several enzymes have been found to have the ability to hydrolyze long chain acyl-CoA. However, none of these enzymes are specific for the hydrolysis of the thioester bond, and it is not clear if any of these hydrolytic enzymes may play a role in the regulation of long chain acyl-CoA levels *in vivo*.

Recently, we have purified a rat liver carboxylesterase which has the ability to hydrolyze long chain acyl-CoA esters (Mukherjee et al., 1993). Despite the non-specific nature of most carboxylesterases, our enzyme preparation has a remarkable specificity towards the long chain acyl-CoAs. The enzyme activity is inhibited by lysophosphatidic acid but activated by lysophosphatidylcholine. Lysophosphatidic acid is an intermediate for the formation of phosphatidic acid which is a precursor of all glycerolipids. Since the long chain acyl-CoA is required for the acylation reaction, the inhibition of acyl-CoA hydrolase by lysophosphatidic acid may be an important positive control for the synthesis of glycerolipids. Alternatively, long chain acyl-CoA is also required for the acylation of other lysophospholipids including lysophosphatidylcholine. The enhancement of carboxylesterase activity by lysophosphatidylcholine may reduce the availability of long chain acyl-CoA. Hence, the enhancement of enzyme activity by lysophosphatidylcholine can be regarded as a negative control for the remodeling process and facilitates the further deacylation of the lysophospholipid during its accumulation.

VII. PHOSPHOLIPID METABOLISM IN MODIFIED LOW DENSITY LIPOPROTEIN

A. Phospholipid Content in Low Density Lipoprotein (LDL)

Plasma lipoproteins are macromolecules that are major carriers of water insoluble lipids in the blood. The outer surface of each lipoprotein consists of amphipathic phospholipids, free cholesterol and protein. The inner hydrophobic core is made up of cholesterol esters and triglycerides. In the studies of lipoprotein structure, assembly and their role in the transport of lipids, emphasis have been placed on the protein, cholesterol and triglyceride components. The phospholipid component is considered to be mainly structural and has received little attention. However, recent studies indicated that the phospholipid content in LDL can be altered and has physiological consequences.

LDL (density = 1.019-1.063) is the major carrier of cholesterol in the body. The hydrophilic outer surface of LDL is composed of a single protein (apolipoprotein B-100) together with a monolayer of phospholipids and cholesterol. The hydrophobic central core consists of mainly cholesterol esters and a small amount of triglycerides. Phosphatidylcholine and sphingomyelin make up over 95% of the total phospholipids in LDL and constitute about 20-25% of the total mass of LDL. The protein component is about 25% of the total mass whereas the free cholesterol, cholesterol esters and triglycerides make up the remaining 50-55% of the LDL particle.

B. Changes in Phospholipid Profile during LDL Modification

It is now recognized that modified form of LDL but not native (or normal) LDL may play an important role in the formation of atherosclerotic plaques (Steinberg et al., 1989). The uptake of modified LDL by macrophages is much faster than the uptake of native LDL. Foam cells are formed when macrophages are loaded with cholesterol from the modified LDL, the key step in the formation of atherosclerotic lesions. Modification of LDL can be observed when normal LDL is incubated with cultured endothelial cells, smooth muscle cells and macrophages (Henriksen et al., 1981; Morel et al., 1984; Parathasarathy et al., 1986). Transitional metals such as Cu^{++} and Fe^{++} are required (Heinecke et al., 1984). The initial step of modification appears to be the peroxidation of phospholipids in LDL (Steinbrecher et al., 1984). Subsequent changes include the production of lysophosphatidylcholine with a concomitant decrease in phosphatidylcholine (Steinbrecher et al., 1984). Of interest is that among the two major phospholipids in LDL, only phosphatidylcholine is converted to lysophosphatidylcholine and sphingomyelin remains essentially unchanged.

C. Physiological Consequences

Interest in the phospholipid component in LDL, and particularly the lysophosphatidylcholine content in modified LDL, has been greatly increased by several key observations. In the last few years, endothelium-dependent cardiovascular effects of lysophosphatidylcholine were reported. Lysophosphatidylcholine causes a slow but progressing dose-dependent relaxation in precontracted rabbit aortic strips (Saito et al., 1988). The relaxation requires intact endothelium and may involve the activation of guanylate cyclase, and it can be blocked by hemoglobin and methylene blue. These results suggest the involvement of nitric oxide in lysophosphatidylcholine-induced relaxation. More recently, the phospholipid fraction of modified LDL has been found to impair endothelium-dependent relaxation. The main component has been identified as lysophosphatidylcholine which is generated in modified LDL (Kugiyama et al., 1990; Yokoyama et al., 1990). The significance of these observations being that atherosclerotic vessels are known to show impaired endothelium-dependent relaxation. The presence of modified LDL with the associated altered phospholipids such as lysophosphatidylcholine may therefore account for the reported impairment of relaxation in atherosclerotic vessels. However, the mechanism for the impairment of endothelium-dependent relaxation by lysophosphatidylcholine in modified LDL has not been fully elucidated. Recently, the hydrolysis of phosphoinositide and the subsequent protein kinase C activation have been suggested as plausible causes (Inoue et al., 1992; Kugiyama et al., 1992). However, the differential vascular effects of lysophosphatidylcholine, such as lysophosphatidylcholine associated with modified LDL impairs relaxation while lysophosphatidylcholine itself possesses an endothelium-dependent relaxant property remain to be explained. Nevertheless, it can be envisioned that in addition to the atherogenic effect of modified LDL, the lysophosphatidylcholine associated with modified LDL may contribute to the pathogenesis of ischemic heart disease by impairing endothelium-dependent relaxation in coronary blood vessels. In spite of the important role of lysophosphatidylcholine in affecting the vascular responses of blood vessels, the mode of lysophosphatidylcholine production and its subsequent metabolism in LDL has not been clearly identified. An essential role of phospholipase A_2 activity in endothelial cells in the modification of LDL has been suggested (Parathasarathy et al., 1985). In a follow up study, the intrinsic activity of phospholipase A_2 in LDL has been demonstrated (Parathasarathy and Barnett, 1990). The possible role for lipoxygenase for the breakdown of phosphatidylcholine in LDL has also been explored (Parathasarathy et al., 1989). Alternatively, platelet-activating factor acetylhydrolase also appears to be involved in the conversion of oxidized phosphatidylcholine into lysophosphatidylcholine (Steinbrecher and Pritchard, 1989; Stremler et al., 1989).

VIII. SUMMARY AND CONCLUSION

Phospholipids are the essential components of all biological membranes which provide the cell with a protective barrier that has selective permeability for cellular metabolism. Membrane proteins and enzymes that provide essential functions for the cells such as metabolite transport and generation of second messengers are contained within the phospholipid bilayer. Hence, the survival of cells depends on the integrity and proper function of the phospholipid bilayer which is underscored by the general lack of inherited disorders in the metabolism of phospholipids.

Changes in phospholipid content and acyl composition have been detected in cellular injury and under a variety of disease states. These changes may arise from (1) alteration in the intake of precursors, (2) alteration in cellular metabolism which is associated with the biosynthesis and catabolism of the phospholipids, and (3) alteration in the energy status of the cell. From the studies described in this chapter, some changes in the content and acyl composition of the membrane phospholipids may be tolerated without noticable effects on the overall function of the cell.

It is generally accepted that changes in the acyl composition of the phospholipids may affect the fluidity of the membrane. Studies in model membranes reveal that membrane fluidity is in part dictated by chain length and the degree of saturation of the acyl group. It should be noted that the acyl composition of membrane phospholipids is not identical from one animal to another. Indeed, significant changes in the acyl composition of different phospholipid groups have been observed during cellular development, aging or dietary intake. Many of these observable changes do not appear to affect the well-being of the organism. It can be concluded that some limited changes in the acyl composition of the phospholipid groups can be tolerated as long as these changes do not significantly affect the fluidity of the membrane.

The content of phospholipid groups within a membrane system appears to be more restrictive since gross changes in the percentage of each phospholipid in the membrane is rarely detected. In order to preserve the appropriate content of each phospholipid group, the biosynthesis of individual phospholipid is under many tiers of control. The failure of one mechanism is usually compensated by another mechanism. Although changes in the rate of biosynthesis may be detected under a pathological condition, the cellular pool of the phospholipid usually remains constant. The multi-tier mechanism instilled for the maintenance of a constant pool of the phospholipid is a clear reflection on the importance to preserve the proper phospholipid content in the cell. One reason for the upkeep of the correct phospholipid content in the membrane is that phospholipids are arranged asymmetrically within the bilayer structure. Any change in their content may alter the asymmetrical distribution and disrupt the proper function of the membrane.

It is clear that the synthesis of phospholipids must be synchronized in order to achieve the proper content within a membrane system. It is obvious that the pool of each phospholipid is maintained by its rate of synthesis and degradation. An examination of the biosynthetic pathways reveals that CTP is the common precursor for the biosynthesis of all diacylphosphoglycerides. Indeed, the intracellular pool of CTP has been shown to be a key factor for the regulation of phospholipid biosynthesis under several pathological conditions. Interestingly, CTP does not participate extensively in other reactions. Hence, it is plausible that CTP is employed as a universal signal for the synchronization of phospholipid biosynthesis. We postulate that a proper intracellular CTP level must be maintained to ensure the coordinated synthesis of all phospholipids in the eukaryotic cell.

REFERENCES

Alling, C., Gustavsson, L., Mansson, J-E., Benthin, G., & Anggard, E. (1984). Phosphatidylethanol formation in rat organs after ethanol treatment. Biochim. Biophys. Acta 793, 119-122.

Arthur, G. (1989). Acylation of 2-acyl-glycerophosphocholine in guinea-pig heart microsomal fractions. Biochem. J. 261, 575-580.

Arthur, G., Page, L.L., Zaborniak, C.L., & Choy, P.C. (1987). The acylation of lysophosphoradylglycerocholines in guinea-pig heart mitochondria. Biochem. J. 242, 171-175.

Arthur, G., Mock, T., Zaborniak, C., & Choy, P.C. (1985). The distribution and fatty acid composition of plasmalogens in guinea pig heart microsomes and mitochondria. Lipids 20, 693-698.

Arthur, G., & Choy, P.C. (1984). Acyl specificity of hamster heart CDP-choline 1,2-diacylglycerol phosphocholine transferase in phosphatidylcholine biosynthesis. Biochim. Biophys. Acta 795, 221-229

Bajusz, E. (1969). Hereditary cardiomyopathy: A new disease model. Am. Heart J. 77, 686-696

Baraona, E., & Lieber, C.S. (1979). Effects of ethanol on lipid metabolism. J. Lipid Res. 20, 289-315.

Berger, A., & German, J.B. (1990). Phospholipid fatty acid composition of various mouse tissues after feeding alpha-linolenate (18:3n-3) or eicosatrienoate (20:3n-3). Lipids 25, 473-480.

Berridge, M.J., & Irvine, R.F. (1989). Inositol phosphates and cell signalling. Nature 341, 197-205.

Bourre, J.-M.E., Dumont, O.S., Piciotti, M.J., Pascal, G.A., & Durand, G.A. (1992). Dietary alpha-linolenic acid deficiency in adult rats for 7 months does not alter brain docosahexanoic acid content, in contrast to liver, heart and testes. Biochim. Biophys Acta 1124, 119-122.

Carman, G.M., & Kelley, M.J. (1992). CDPdiacylglycerol synthase from yeast. Methods in Enzymol. 209, 242-247.

Carr, D.J., Guarcello, V., & Blalock, J.E. (1992). Phosphatidylserine supresses antigen-specific IgM production by mice orally administered sheep red blood cells. Proc. Soc. Exp. Biol. Med. 200, 548-554.

Cheng, P., & Hatch, G.M. (1994). Effect of hypoxia on cardiolipin biosynthesis in the heart. Can. Fed. Biol. Soc. Abstract

Chiang, P.K., Im, Y.S., & Cantoni, G.L. (1980). Phospholipid biosynthesis by methylations and choline incorporation: effect of 3-deazaadenosine. Biochem. Biophys. Res. Comm. 94, 174-181.

Choy, P.C., Tardi, P.G., & Mukherjee, J.J. (1992). 1-acyl-glycerophosphocholine acyltransferase. Methods in Enzymol. 209, 80-86.

Choy, P.C., & Arthur, G. (1989). Phosphatidylcholine biosynthesis from lysophosphatidylcholine. In: Phosphatidylcholine Metabolism (Vance, D.E., ed.), pp. 87-101. CRC Press, Boca Raton, FL.

Choy, P.C. (1982). Control of phosphatidylcholine biosynthesis in myopathic hamster heart. J. Biol. Chem. 257, 10928-10933.

Choy, P.C., Paddon, H.P., & Vance, D.E. (1980). An increase in cytoplasmic CTP accelerates the reaction catalyzed by CTP:phosphocholine cytidylyltransferase in polio-infected HeLa cells. J. Biol. Chem. 255, 1070-1073.

Choy, P.C., & Vance, D.E. (1978). Lipid requirements for activation of CTP:phosphocholine cytidylyltransferase from rat liver. J. Biol. Chem. 253, 5163-5167.

Clejan, S., & Maddaiah, V.T. (1986). Growth hormone and liver mitochondria: Effects on phospholipid composition and fatty acyl distribution. Lipids 21, 677-683.

Coleman, R. (1973). Membrane-bound enzymes and membrane ultrastructure. Biochim. Biophys. Acta 300, 1-30.

Cornell, R.M. (1992). Cholinephosphotransferase from mammalian sources. Methods in Enzymol. 209, 267-272.

Cullis, P.R., & Hope, M.J. (1991). Physical properties and functional roles of lipids in membranes. in Biochemistry of lipids, lipoproteins and membranes (Vance, D.E., & Vance, J.E., eds.), pp. 1-41. Elsevier Science Publishers, Amsterdam.

Deka, N., Sun, G.Y., & MacQuarrie, R. (1986). Purification and properties of acyl-CoA:1-acyl-*sn*-glycero-3-phosphocholine-O-acyltransferase from bovine brain microsomes. Arch. Biochem. Biophys. 246, 554-563.

Esko, J.D., & Raetz, C.R. (1983). Synthesis of phospholipids in animal cells. In: The Enzymes (Boyer, P.D., ed.), pp. 207-253. Academic Press Inc., NY.

Exton, J.E. (1990). Signalling through phosphatidylcholine breakdown. J. Biol. Chem. 265, 1-4.

Feldman, D.A., & Weinhold, P.A. (1987). CTP:phosphorylcholine cytidylyltransferase from rat liver: isolation and characterization of the catalytic subunit. J. Biol. Chem. 262, 9075-9081.

Fukami, K., & Takenawa, T. (1992). Phosphatidic acid that accumulates in platelet-derived growth factor stimulated balb/c 3T3 cells is a potential mitogenic signal. J. Biol. Chem. 267, 10988-10993.

Gavino, V.C., & Shayman, J.A. (1990). The phosphatidylinositol synthase of proximal tubule cells. Biochim. Biophys. Acta 1044, 34-42.

George, T.P., Cook, H.W., Byers, D.M., Palmer, F.B., & Spence, M.W. (1991). Inhibition of phosphatidylcholine and phosphatidylethanolamine biosynthesis by cytochalasin B in culture glioma cells: potential regulation of biosynthesis by Ca(2+).-dependent mechanism. Biochim. Biophys Acta 1084, 185-193.

Goldstein, D.B. (1986). Effect of alcohol on cellular membranes. Annals of Emergency Medicine 15, 1013-1018.

Hasegawa-Sasaki, H., & Ohno, K. (1975). Acyltransferase activities in rat lung microsomes. Biochim. Biophys. Acta 380, 486-495.

Hasegawa-Sasaki, H., & Ohno, K. (1980). Extraction and partial purification of acyl-CoA:1-acyl-*sn*-glycero-3-phosphocholine acyltransferase from rat liver microsomes. Biochim. Biophys. Acta 617, 205-217.

Hatch, G.M., & Choy, P.C. (1990). Effect of hypoxia on phosphatidylcholine biosynthesis in the isolated hamster heart. Biochem. J. 268, 47-54.

Hatch, G.M., O, K., & Choy, P.C. (1989). Regulation of phosphatidylcholine metabolism in mammalian hearts. Biochem. Cell Biol. 67, 67-77.

Hatch, G.M., Angel, A., & Cao, S.G. (1994). Newly synthesized phosphatidylglycerol is preferentially utilized for cardiolipin biosynthesis. Can. Fed. Biol. Soc. Abstract.

Hatch, G.M. (1994). Cardiolipin biosynthesis in the isolated heart. Biochem. J. 297, 201- 208.

Heinecke, J.W., Rosen, H., & Chait, A. (1984). Iron and copper promote modification of low density lipoprotein by human arterial smooth muscle cells in culture. J. Clin. Invest. 74, 1890-1894.

Henriksen, T., Mahoney, E.M., & Steinberg, D. (1981). Enhanced macrophage degradation of low density lipoprotein previously incubated with cultured endothelial cells: recognition by receptor for acetylated low density lipoproteins. Proc. Natl. Acad. Sci. USA, 78, 6499-6503.

Hostetler, K.Y., van den Bosch, H., & van Deenen, L.L.M. (1971). Biosynthesis of cardiolipin in liver mitochondria. Biochim. Biophys. Acta 239, 113-119.

Hostetler, K.Y. (1982). Polyglycerophospholipids: phosphatidylglycerol, diphosphatidylglycerol and bis(monoacylglycero).phosphate. In: Phospholipids. (Hawthorne, J.N., & Ansell, G.B. eds.), pp. 215-261. Elsevier, Amsterdam.

Hostetler, K.Y. (1991). Effect of thyroxine on the activity of mitochondrial cardiolipin synthase in rat liver. Biochim. Biophys. Acta 1086, 139-140.

Hrelia, S., Biagi, P.L., Lamers, J.M., & Bordoni, A. (1992). Fatty acid composition of phosphoinositides in cultured cardiomyocytes: effects of docosahexaenoic acid and alpha 1-adrenoceptor stimulation. Cardioscience 3, 91-95.

Huterer, S.J., & Wherrett, J.R. (1989). Formation of bis(monoacylglycero).phosphate by a macrophage transacylase. Biochim. Biophys. Acta 1001, 68-75.

Huterer, S.J., & Wherrett, J.R. (1990). Formation of acylphosphatidylglycerol by a lysosomal phosphatidylcholine:bis(monoacylglycero).phosphate acyl transferase. Biochem. Cell Biol. 68, 366-372.

Innes, S.M., & Clandinin, M.T. (1981). Dynamic modulation of mitochondrial membrane physical properties and ATPase activity by diet lipid. Biochem. J. 198, 167-175.

Inoue, N., Hirata, K., Yamada, M., Hamamori, Y., Matsuda, Y., Akita, H., & Yokoyama, M. (1992). Lysophosphatidylcholine inhibits bradykinin-induced phosphoinositide hydrolysis and calcium transients in cultured bovine aortic endothelial cells. Circ. Res. 71, 1410-1421.

Ishidate, K., & Nakazawa, Y. (1992). Choline/ethanolamine kinase from rat kidney. Methods in Enzymol. 209, 121-134.

Kanfer, J.N. (1992). Serine-ethanolamine base-exchange enzyme from rat brain. Methods in Enzymol. 209, 341-348.

Kanoh, H., & Ohno, K. (1976). Solubilization and purification of rat liver microsomal 1,2-diacylglycerol: CDP-choline cholinephosphotransferase and 1,2-diacylglycerol: CDP-ethanolamine ethanolaminephosphotransferase. Eur. J. Biochem. 66, 201-210.

Khuu Thi Dinh, K.L., Demarne, Y., Nicholas, C., & Lhuillery, C. (1990). Effect of dietary fat on phospholipid class distribution and fatty acid composition of rat fat cell plasma membrane. Lipids 25, 278-283.

Kugiyama, K., Kerns, S.A., Morrisett, J.D, Roberts, R., & Henry, P.D. (1990). Impairment of endothelium-dependent arterial relaxation by lysolecithin in modified low-density lipoproteins. Nature 344, 160-162.

Kugiyama, K., Ohgushi, M., Sugiyama, S., Murohara, T., Fukunaga, K., Miyamoto, E., & Yasue, H. (1992). Lysophosphatidylcholine inhibits surface receptor-mediated intracellular signals in endothelial cells by a pathway involving protein kinase C activation. Circ. Res. 71, 1422-1428.

Lands, W.E.M. (1960). Metabolism of glycerolipids. II. The enzymatic acylation of lysolecithin. J. Biol. Chem. 253, 2233-2237.

Lands, W.E.M., & Hart, P. (1965). Metabolism of glycerophospholipids. VI. Specificities of acyl coenzyme A: phospholipid acyltransferase. J. Biol. Chem. 240, 1905-1912.

Leff, A.R., & Schumacker, P.T. (1993). Repriratory Physiology: Basics and Applications. W.B. Saunders Co., Philadelphia.

Liu, S., Tardi, P.G., Choy, P.C., & Man, R.Y.K. (1993). Effects of selenium supplement on the de novo biosynthesis of glycerolipids in the isolated rat heart. Biochim. Biophys. Acta. 1170, 307-313.

Lopaschuck, G. (1989). Alterations in myocardial fatty acid metabolism contribute to ischemic injury in the diabetic. Can. J. Cardiol. 5, 315-320.

Low, M.G. (1989). The glycosyl-phosphatidylinositol anchor of membrane proteins. Biochim. Biophys. Acta 988, 427-454.

Matsuzawa, Y., & Hostetler, K.Y. (1981). Studies on the drug-induced lipidosis: Subcellular location of phospholipid and cholesterol in the liver of rats treated with chloroquine or 4,4'-bis(diethylaminoethoxy).-diethyldiphenylethane. J. Lipid. Res. 21, 202-214.

McMaster, C.R., & Choy, P.C. (1992). Serine regulates phosphatidylethanolamine biosynthesis in the hamster heart. J. Biol. Chem. 267, 14586-14591.

Mendenhall, C.L., Bradford, R.H., & Furman, R.H. (1969). Effects of ethanol on glycerolipid metabolism in rat liver. Biochim. Biophys. Acta 187, 501-509.

Mok, A.Y.P., McDougall, G.E., & McMurray, W.C. (1992). CDP-diacylglycerol synthesis in rat liver mitochondria. FEBS Letters 312, 236-240.

Morel, D.W., DiCorlerleto, P.E., & Çhisolm, G.M. (1984). Endothelial and smooth muscle cells alter low density lipoprotein in vitro by free radical oxidation. Arteriosclerosis 4, 357-364.

Mukherjee, J.J., Jay, F.T., & Choy, P.C. (1993). Purification, characterization and modulation of a microsomal carboxylesterase in rat liver for the hydrolysis of acyl-CoA. Biochem. J. 295, 81-86.

O, K., Hatch, G.M., & Choy, P.C. (1988). Effect of diethyl ether on phosphatidylcholine biosynthesis in hamster organs. Lipids 23, 656-659.

O, K., Siow, Y.L., & Choy, P.C. (1989). Hamster liver cholinephosphotransferase and ethanolaminephosphotransferase are separate enzymes. Biochem. Cell Biol. 67, 680-686.

O, K., & Choy, P.C. (1990). Solubilization and partial purification of cholinephosphotransferase in hamster tissues. Lipids 25, 122-124.

O, K., & Choy, P.C. (1993). Effect of fasting on phosphatidylcholine biosynthesis in hamster liver: Regulation of cholinephosphotransferase activity by endogenous argininosuccinate. Biochem. J. 289, 727-733.

Paradies, G., & Ruggiero, R.M. (1988). Effect of hyperthyroidism on the transport of pyruvate in rat heart mitochondria. Biochim. Biophys. Acta 935, 79-86.

Paradies, G., & Ruggiero, R.M. (1990). Enhanced activity of the tricarboxylate carrier and modification of lipids in hepatic mitochondria from hyperthyroid rats. Arch. Biochem. Biophys. 278, 425-430.

Parthasarathy, S., Printz, D.J., Boyd, D., Joy, L., & Steinberg, D. (1986). Macrophage oxidation of low density lipoprotein generates a modified form recognized by the scavenger receptor. Arteriosclerosis 6, 505-510.

Parathasarathy, S., Steinbrecher, U.P., Barnett, J., Witztum, J.L., & Steinberg, D. (1985). Essential role of phospholipase A2 activity in endothelial cell-induced modification of low density lipoprotein. Proc. Natl. Acad. Sci. USA 82, 3000-3004.

Parathasarathy, S., Wieland, E., & Steinberg, D. (1989). A role for endothelial cell lipoxygenase in the oxidative modification of low density lipoprotein. Proc. Natl. Acad. Sci. USA 86, 1046-1050.

Parathasarathy, S., & Barnett, J. (1990). Phospholipase A2 activity of low density lipoprotein: Evidence for an intrinsic phospholipase A2 activity of apoprotein B-100. Proc. Natl. Acad. Sci. USA 87, 9741-9745.

Pelassy, C., Mary, D., & Aussel, C. (1991). Effects of the serine analogues isoserine and serinol on interleukin-2 synthesis and phospholipid metabolism in a human T cell line Jurkat. J. Lipid Mediat. 3, 79-89.

Pierce, G.N., & Dhalla, N.S. (1983). Sarcolemmal Na+-K+ATPase activity in the diabetic heart. Am. J. Physiol. 245, C241-C247.

Pierce, G.N., Kutryk, M.J.B., & Dhalla, N.S. (1983). Alterations in the Ca+2-binding by and composition of the cardiac sarcolemmal membrane in chronic diabetes. Proc. Natl. Acad. Sci., USA 80, 5412-5416.

Poorthuis, B.J., & Hostetler, K.Y. (1976). Studies on the subcellular localization and properties of bis(monoacylglycero).phosphate biosynthesis in rat liver. J. Biol. Chem. 251, 4596-4602.

Porter, T., & Kent, C. (1992). Choline/ethanolamine kinase from rat liver. Methods in Enzymol. 209, 134-146.

Post, .M., Batenburg, J.J., & van Golde, L.M.G. (1980). Effects of cortisol and thyroxine on phosphatidylcholine and phosphatidylglycerol synthesis by adult rat lung alveolar type II cells in primary culture. Biochim. Biophys. Acta 618, 308-317.

Post, M, Floros, J., & Smith, H.T. (1984). Inhibition of lung maturation by monoclonal antibodies against fibroblast-pneumonocyte factor. Nature 308, 284-286.

Post, M. Barsoumian, A., & Smith, B.T. (1986). The cellular mechanism of glucocorticoid acceleration of fetal lung maturation: Fibroblast pneumonocyte factor stimulates choline phosphate cytidylyltransferase activity. J. Biol. Chem. 261, 2179-2184.

Pritchard, P.H., Chiang, P.K., Cantoni, G.L., & Vance, D.E. (1982). Inhibition of phosphatidylethanolamine N-methylation by 3-deazaadenosine stimulates the synthesis of phosphatidylcholine via the CDP-choline pathway. J. Biol. Chem. 257, 6362-6367.

Rhee, S.G., Suh, P.-G., Ryu, S.-H., & Lee, S.Y. (1989). Studies of inositol phospholipid-specific phospholipase C. Science 244, 546-550.

Saito, T., Wolf, A., Menon, N.K., Saeed, M., Alves, C., & Bing, R.J. (1988). Lysolecithins as endothelium-dependent vascular smooth muscle relaxants that differ from endothelium-derived relaxing factor (nitric oxide). Proc. Natl. Acad. Sci. USA 85, 8246-8250.

Sanjanwala, M., Sun, G.Y., Cutrera, M.A., & MacQuarrie, R. (1988). Acylation of lysophosphatidylcholine in bovine heart muscle microsomes: Purification and kinetic properties of acyl-CoA: 1-acyl-sn-glycero-3-phosphocholine-O-acyltransferase. Arch. Biochem. Biophys. 265, 476-483.

Schlame, M., Brody, S., & Hostetler, K. Y. (1993). Mitochondrial cardiolipin in diverse eukaryotes: comparison of biosynthetic reactions and molecular acyl species. Eur. J. Biochem. 212, 727-735.

Schlame, M., & Hostetler, K.Y. (1991). Solubilization, purification, and characterization of cardiolipin synthase from rat liver mitochondria. J. Biol. Chem. 266, 22398-22403.

Schlame, M., & Rustow, B. (1990). Lysocardiolipin formation and reacylation in isolated rat liver mitochondria. Biochem. J. 272, 589-595.

Shier, W.T. (1977). Inhibition of acyl coenzyme A: Lysolecithin acyltransferase by local anesthetics, detergents and inhibitors of cyclic nucleotide phosphodiesterases. Biochem. Biophys. Res. Comm. 75, 186-193.

Skrbic, T.R., & Cumings, J.N. (1970). Fatty acids of lecithin in subcellular fractions during maturation of brain. J. Neurochem. 17, 85-90.

Snodgrass, P.J. (1981). Biochemical aspects of urea cycle disorders. Pediatrics 68, 273-283

Sohal, P.S., & Cornell, R.M. (1990). Sphingosine inhibits the activity of rat liver CTP:phosphocholine cytidylyltransferase expressed in COS cells. J. Biol. Chem. 265, 11746-11750.

Steinberg, D., Parthasarathy, S., Carew, T.E., Khoo, J.C., & Witztum, J.L. (1989). Beyond cholesterol, modifications of low-density lipoprotein that increase its atherogenicity. New Eng. J. Med. 320, 915-924.

Steinbrecher, U.P., Parthasarathy, S., Leake, D.S., Witztum, J.L., & Steinberg, D. (1984). Modification of low density lipoprotein by endothelial cells involves lipid peroxidation and degradation of low density lipoprotein phospholipids. Proc. Natl. Acad. Sci. USA, 83, 3883-3887.

Steinbrecher, U.P., & Pritchard, P.H. (1989). Hydrolysis of phosphatidylcholine during LDL oxidation is mediated by platelet-activating factor acetylhydrolase. J. Lipid Res. 30, 305-315.

Stremler K.E., Stafforini, D.A., Prescott, S.M., Zimmerman, G.A., & McInctyre, T.M. (1989). An oxidized derivative of phosphatidylcholine is a substrate for the platelet-activating factor acetylhydrolase from human plasma. J. Biol. Chem 264, 5331-5334.

Sundler, R. (1973). Biosynthesis of rat liver phosphatidylethanolamines from intraportally injected ethanolamine. Biochim. Biophys. Acta 306, 218-226.

Sundler, R. (1975). Ethanolamine cytidylyltransferase. J. Biol. Chem. 250, 8585-8590.

Takenawa, T., & Egawa, K. (1977). CDP-diglyceride: Inositol transferase from rat liver. J. Biol. Chem. 252, 5419-5423.

Tardi, P.G., Man, R.Y.K., McMaster, C.R., & Choy, P.C. (1990). The effect of methyl lidocaine on lysophospholipid metabolism in hamster heart. Biochem. Cell Biol. 68, 745-750.

Tardi, P.G., Man, R.Y.K., & Choy, P.C. (1992). The effect of methyl-lidocaine one the de novo biosynthesis of phospholipids in the isolated hamster heart. Biochem. J. 285, 161-166.

Tijburg, L.B.M., Vermeulen, P.S., & van Golde, L.M.G. (1992). Ethanolamine phosphate cytidylyltransferase. Methods in Enzymol. 209, 258-263.

Vance, D.E. (Ed.) (1989). CTP:cholinephosphate cytidylyltransferase, in Phosphatidylcholine Metabolism. CRC Press, Boca Raton, FL.

Vance, D.E. (1991). Phospholipid metabolism and cell signalling in eucaryotes. In: Biochemistry of Lipids, Lipoproteins and Membranes (Vance, D.E., & Vance, J.E., eds.), pp. 205-240. Elsevier Science Publishers, Amsterdam.

Vigo, C. Paddon, H.B., Millard, F.C., Pritchard, P.H., & Vance, D.E. (1981). Diethylstilbestrol treatment modulates the enzymatic activities of phosphatidylcholine biosynthesis in rooster liver. Biochim. Biophys. Acta 665, 546-550.

Vigo, C., & Vance, D.E. (1981). Effect of diethylstilbestrol on phosphatidylcholine biosynthesis and choline metabolism in the liver of roosters. Biochem. J. 200, 321-326.

Voelker, D.R. (1984). Phosphatidylserine functions as the major precursor of phosphatidylethanolamine in cultured BHK-21 cells. Proc. Natl. Acad. Sci. USA 81, 2669-2673.

Weinhold, P.A., Feldman, D.A., Quade, M.M., Miller, J.C., & Brooks, R.L. (1981). Evidence for a regulatory role of CTP:choline phosphate cytidylyltransferase in the synthesis of phosphatidylcholine in fetal lung following premature birth. Biochim. Biophys. Acta. 665, 134-144.

Weinhold, P.A., Quade, M.M., Brozowski, T.B., & Feldman, D.A. (1980). Increased synthesis of phosphatidylcholine by rat lung following birth. Biochim. Biophys. Acta. 617, 76-84.

Weinhold, P.A., & Feldman, D.A. (1992). Choline phosphate cytidylyltransferase. Methods in Enzymol. 209, 248-258.

Weinhold, P.A., Rounsifer, M.E., & Feldman, D.A. (1986). The purification and characterization of CTP:phosphorylcholine cytidylyltransferase from rat liver. J. Biol. Chem. 261, 5104-5110.

Wheeldon, L.W., Schumert, Z., & Turner, D.A. (1965). Lipid composition of heart muscle homogenate. J. Lipid Res. 6, 481-489.

White, D.A. (1973). The phospholipid composition in mammalian tissues. In: Form and Function of Phospholipids (Ansell, G.B., Hawthorne, J.N., & Dawson, R.M.C., eds.), pp. 441-482. Elsevier Scientific Publishing, Amsterdam.

Yamaoka, S., Urade, R., & Kito, M. (1990). Cardiolipin molecular species in rat heart mitochondria are sensitive to essential fatty acid-deficient dietary lipids. J. Nutr. 120, 415-421.

Yao, Z., Jamil, H., & Vance, D.E. (1990). Choline deficiency causes translocation of CTP:phosphocholine cytidylyltransferase from cytosol to endoplasmic reticulum in rat liver. J. Biol. Chem. 265, 4326-4331.

Yashiro, K., Kameyama, Y., Mizuno, M., Hayashi, S., Sakashita, Y., & Yokota, Y. (1989). Phospholipid metabolism in rat submandibular gland. Positional distribution of fatty acids in phosphatidylcholine and microsomal lysophospholipid acyltransferase systems concerning proliferation. Biochim. Biophys. Acta 1005, 56-64.

Yokoyama, M., Hirata, K., Miyake, R., Akita, H., Ishikawa, Y., & Fukuzaki, H. (1990). Lysophosphatidylcholine: Essential role in the inhibition of endothelium-dependent vasorelaxation by oxidized low density lipoprotein. Biochem. Biophys. Res. Comm. 168, 301-308.

Zelinski, T.A., Savard, J.D., Man, R.Y.K., & Choy, P.C. (1980). Phosphatidylcholine biosynthesis in isolated hamster hearts. J. Biol. Chem. 255, 11423-11428.

Zelinski, T.A., & Choy, P.C. (1982). Choline regulates phosphatidylethanolamine biosynthesis in the isolated hamster heart. J. Biol. Chem. 257, 13201-13204.

THE ROLE OF PHOSPHOLIPIDS IN CELL FUNCTION

William Dowhan

Advances in Lipobiology
Volume 2, pages 79-107.

ISBN 0-7623-0205-4

ABSTRACT

Phospholipids play multiple roles in cells in forming the permeability barrier of the cell membrane and intracellular organelles, in providing the supporting matrix and surface for many catalytic processes, in actively participating in signal transduction in response to both external and internal stimuli, and in providing precursors for signaling processes and macromolecular synthesis. Since phospholipids do not have inherent catalytic activity, function has usually been determined indirectly by studying effects on enzymatic activities analyzed *in vitro*. Since most phospholipids appear to be essential, alteration of phospholipid levels *in vivo* by applying molecular genetic approaches has been difficult to use for uncovering new functions for phospholipids or for studying the role of phospholipids in the *in vivo* state. Isolation of several mutants in *Escherichia coli* phospholipid metabolism which exhibit dramatic alterations in phospholipid metabolism while maintaining viability has made possible an investigation of phospholipid requirements for many essential and nonessential processes. This review will summarize many of the known roles for phospholipids and focus on the use of mutants in phospholipid metabolism to increase our understanding of the many functions of phospholipids. Although much of the review will focus on prokaryotic systems, similar approaches are now possible in eukaryotic systems, particularly yeast, which will be discussed.

I. INTRODUCTION

It is widely recognized that phospholipids play multiple roles in cell function. Their primary role is to define the permeability barrier of cells and organelles by forming a phospholipid bilayer with their hydrophobic domain oriented inward and shielded from water and their hydrophilic domain exposed and interacting with the aqueous environment (Tanford, 1973). This bilayer serves as the matrix and support for a vast array of proteins involved in important functions of the cell such as energy transduction, signal transduction, solute transport, DNA replication, protein targeting, cell-cell recognition, and so on. Phospholipids do not play a static role in these processes but are active participants which influence the properties of the proteins associated with the membrane and serve as precursors to important cellular components. Since phospholipids have no inherent catalytic activity, a major problem has been to develop approaches to systematically investigate the specific role phospholipids play in each cellular function. The majority of evidence concerning the role of phospholipids has come secondary to the study of some function in an *in vitro* reconstituted system. This approach does yield important clues as to function but can be fraught with artifact and may lead to conclusions which have no relationship to the *in vivo* state. Therefore, *in vivo* approaches are needed to validate the observations made *in vitro*. A classical approach to this end has been to employ genetic methods to isolate mutants in phospholipid synthesis and function. However, a complication in genetically dissecting phospholipid function *in vivo* is the pleiotrophic role these molecules play in determining both cell integrity and cell function. Elimination of or drastic alteration in the level of a particular phospholipid might be expected to affect many cellular processes simultaneously or worse yet compromise cell integrity before affecting cell function. Fortunately, recent results utilizing a series of mutants in phospholipid metabolism (Dowhan, 1992a) in *Escherichia coli* have demonstrated that meaningful data can be obtained using a combined molecular genetic and biochemical approach to both validate functions initially postulated on the basis of *in vitro* data and uncover new functions previously undocumented. The successful application of these molecular genetic approaches to *E. coli* should be generally applicable to more complicated biological systems such as yeast and even somatic cells. In addition, the basic principles underlying the function of phospholipids in *E. coli* should be relevant to phospholipid function in more complex organisms. This review will outline the approaches now possible in *E. coli*, summarize the some of the basic *in vivo* functions of phospholipids now documented in *E. coli*, and finally, suggest how these approaches might be applied to more complex systems.

II. PHOSPHOLIPID SYNTHESIS

E. coli contains two major classes of phospholipids (Raetz and Dowhan, 1990), one as esterified derivatives of a glucosamine phosphate dissaccharide backbone (lipid A-based) and the other as esterified derivatives of *sn*-glycerol-3-phosphate (referred to hereafter as phospholipids, see Figure 1).

Phospholipids make up (Figure 2) both leaflets of the inner membrane of *E. coli* and the inner leaflet of the outer membrane of *E. coli*; the lipid A derivatives form the outer leaflet of the outer membrane as well as the hydrophobic anchor of the lipopolysaccharide-O-antigen structures which extend from the surface of Gram-negative bacteria (Nikaido and Vaara, 1987).

In *E. coli* the major phospholipid classes (Raetz and Dowhan, 1990) are the zwitterionic phospholipid phosphatidylethanolamine (70-80%) and the anionic phospholipids phosphatidylglycerol (20-25%) and cardiolipin (5-10%). The remainder of the phospholipids shown in Figure 1 make up less than 5% of the total. To date no definite role other than intermediates to the synthesis of the major lipids has been attributed to these minor phospholipids. The varied fatty acid moieties and the different head groups of the three major phospholipids define the physical chemical properties of the membrane bilayer which acts as both the permeability barrier of the cell and the supporting matrix for many membrane associated processes. In addition the individual components of phosphatidylethanolamine and phosphatidylglycerol serve as precursors to other important macromolecules in the cell. These same multiple roles for phospholipids occur in eukaryotic cells so that the basic principles derived from studies in *E. coli* should be applicable to understanding the role of phospholipids in all organisms.

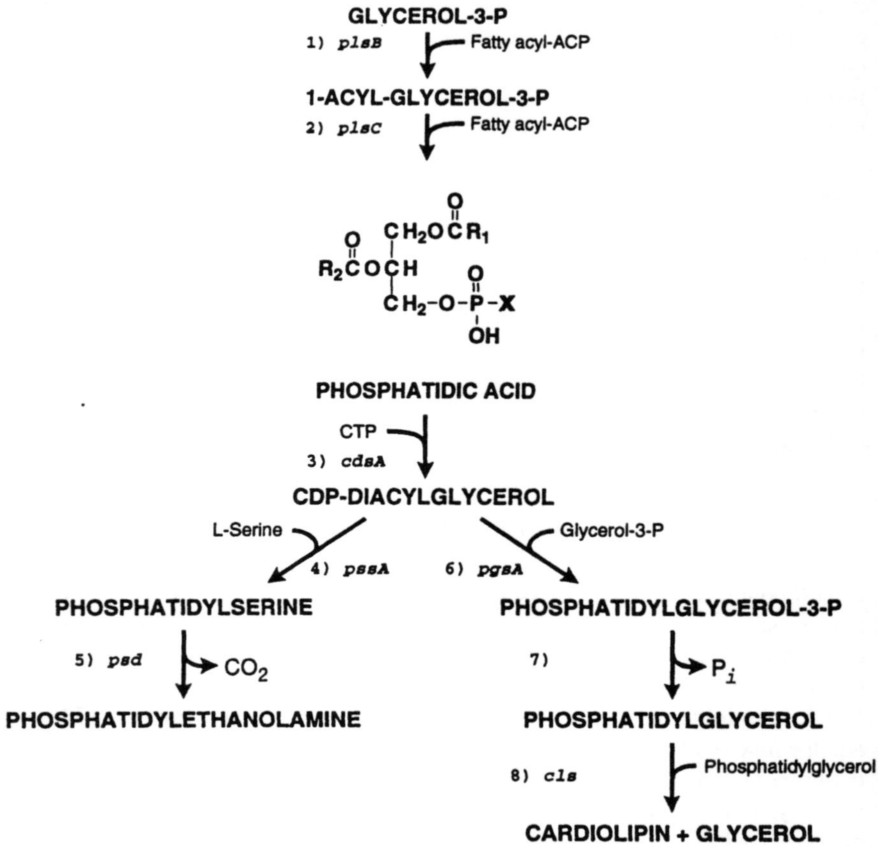

Figure 1. Pathway of phospholipid biosynthesis in *E. coli* and the associated genes. The moiety "X" attached to phosphatidic acid (Step 2) is defined for the product of each enzymatic reaction, respectively, and the charge of the head groups at physiological pH are indicated for the major phospholipids. The name of each gene is listed with the respective step catalyzed by the following enzymatic activities: (1) *sn*-Glycerol-3-phosphate acyltransferase; (2) 1-Acyl-*sn*-glycerol-3-phosphate acyltransferase (X = OH), (3) CDP-diacylglycerol synthase (X = CMP), (4) Phosphatidylserine synthase (X = serine), (5) Phosphatidylserine decarboxylase (X = ethanolamine; one positive and one negative charge), (6) Phosphatidylglycerophosphate synthase (X = *sn*-glycerol-3-phosphate), (7) Phosphatidylglycerophosphate phosphatase (X = glycerol; one negative charge); 8) Cardiolipin synthase (X = phosphatidylglycerol; two negative charges).

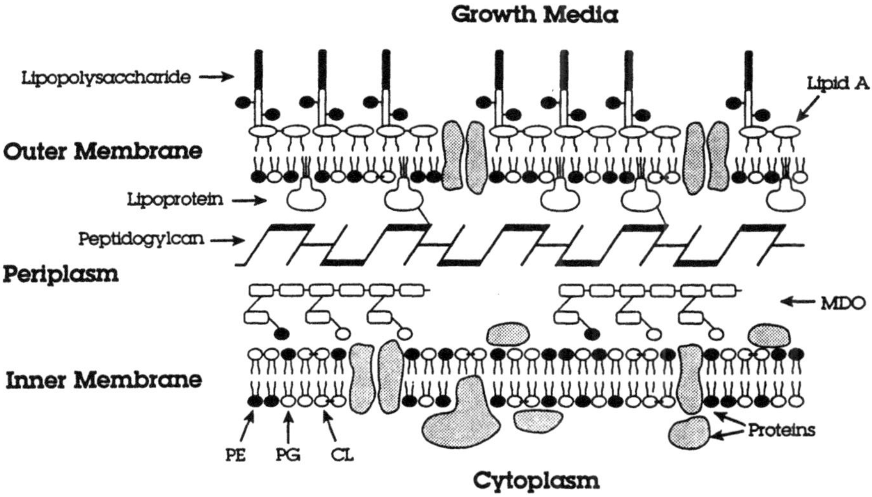

Figure 2. Model of the cell envelope of *E. coli*. The circles depict the head groups of phospholipids with the two phosphatidic acid moieties connected by a glycerol residue depicted for cardiolipin (CL). The filled circles connected to MDO (membrane derived oligosaccharide) and the core of lipopolysaccharide (open rectangle) are derived from the head group of phosphatidylethanolamine (PE). The open circles connected to MDO are derived from the head group of phosphatidylglycerol (PG). The closed rectangles of liposaccharide indicate the position of the O-antigens of *Enterobacteriaceae*. The acylated disaccharide structure of lipid A is shown by the connected ovals. The amino acid-sugar containing crosslinked structure of peptidoglycan is shown covalently attached through the carboxyl end of some but not all of the major outer membrane lipoprotein molecules which are modified at their amino termini with a diacylglycerol residue from PG and fatty acids from phospholipids.

The factors which determine the rate of total phospholipid synthesis and the total amount of phospholipid are still not well understood, but the committed step to the synthesis of phospholipids (Step 1, Figure 1) must be an important determinant. Although there is significant turnover and remodeling of the fatty acids once the major phospholipid classes are formed, primary determinants of the fatty acid composition of the phospholipids are the specificities of the enzymes which catalyze Steps 1 and 2 and the composition of the fatty acyl-acyl carrier protein pool (Vanden Boom and Cronan, 1989). Much of the turnover of the fatty acids of preformed phospholipids is due to the transfer of the acyl moieties for posttranslation modifications primarily of the major and minor lipoproteins of *E. coli*. (Jackowski et al., 1992; Jackowski and Rock, 1986). The factors which determine the amounts of the final three major phospholipids and particularly the balance between zwitterionic and anionic phospholipids are not well understood. The relative rates at which metabolites flow through Step 4 (the committed step to zwitterionic phospholipid synthesis) and Step 6 (the committed step to anionic phospholipid biosynthesis) coupled with the high but variable rate of utilization of particularly the polar headgroup of phosphatidylglycerol for modification of other macromolecules must be important determinants in this balance (Jackson et al., 1986); the polar head group of phosphatidylethanolamine does turnover but at a much lower rate than the polar head group of phosphatidylglycerol (Jackson and Kennedy, 1983; Miller and Kennedy, 1987; Schulman and Kennedy, 1977). The balance between phosphatidylglycerol and cardiolipin (determined largely by Step 8) remains constant until cells enter the late log phase of growth where the level of cardiolipin increases at the expense of phosphatidylglycerol (Tunaitis and Cronan, 1973).

The enzymes which catalyze Steps 1-8 in Figure 1 are all associated with the inner membrane of *E. coli* as integral membrane proteins except for phosphatidylserine synthase (Step 4) (Dowhan, 1992b). The synthase appears to be a cytoplasmic protein which peripherally associates with the membrane surface during catalysis. Therefore, the regulation and activity of this enzyme, like many protein kinase C isozymes (Bell and Burns, 1991) and the CTP: phosphocholine cytidylyltransferase of eukaryotic cells (Cornell, 1991; Johnson et al., 1992), may be highly sensitive to the surface properties of membranes in addition to the properties of the individual lipid molecules with which it interacts. In yeast and Gram-positive bacteria this enzyme is tightly membrane associated and has different mechanistic (Raetz et al., 1987) and physical properties (Carman and Bae, 1992; Dutt and Dowhan, 1981) than the *E. coli* enzyme (Dowhan, 1992b).

In eukaryotic cells phospholipid metabolism (Bishop and Bell, 1988; Carman and Henry, 1989) and function are more complicated due to compartmentalization of pathways, possibly multiple forms or dual localization for biosynthetic activities, and additional lipids which must be synthesized. The synthesis of cardiolipin and the majority of the phosphatidylglycerol is confined to the mitochondria (Schlame and Haldar, 1993; Zinser et al., 1991), and except for cardiolipin synthase, which in eukaryotic cells transfers a phosphatidic acid moiety from CDP-diacylglycerol to a molecule of phosphatidylglycerol (Schlame and Hostetler, 1991; Tamai and Greenberg, 1990), proceeds by reactions (Bishop and Bell, 1988; Carman and Henry, 1989) very similar to those reported in *E. coli* (Raetz and Dowhan, 1990). Phosphatidylserine is synthesized in the endoplasmic reticulum by a membrane bound synthase in yeast and by headgroup exchange between serine and either phosphatidylethanolamine or phosphatidylcholine in somatic cells (Kuge et al., 1986). In all eukaryotic cells the major phosphatidylserine decarboxylase (Step 5, Figure 1) is localized to the inner mitochondrial membrane (Kuge et al., 1991; Zinser et al., 1991) which suggests a requirement for translocation of phosphatidylserine and phosphatidylethanolamine between the endoplasmic reticulum and the mitochondria (Ardail et al., 1991; Hovius et al., 1992; Simbeni et al., 1993; Vance, 1991). However, recent evidence from yeast indicates there may be two phosphatidylserine decarboxylases (Clancey et al., 1993; Trotter et al., 1993) with the location of the second minor activity unknown. Eukaryotic cells have additional pathways not found in *E. coli* for the synthesis of phosphatidylethanolamine and phosphatidylcholine. In addition eukaryotic cells must synthesize phosphatidylinositol which is accomplished in the endoplasmic reticulum by a reaction similar to Step 4 in *E. coli* (substitution of inositol for serine).

Since the phosphatidylserine synthase and the phosphatidylglycerol phosphate synthase (Steps 4 and 6, Figure 1) catalyze the committed steps to the synthesis of the major zwitterionic and anionic phospholipids, respectively, of *E. coli*, this branch point has been the target of numerous studies aimed at understanding the role of phospholipid composition and specific phospholipids in *E. coli*. The basic approach outlined below has been to design specific mutants in the *pssA* and *pgsA* genes to both uncover previously unknown roles for phospholipids and to verify the *in vivo* significance of roles for phospholipids previously documented *in vitro*. These approaches make possible the coupling of the physiological properties of these mutants with *in vitro* biochemical studies on specific processes.

III. FUNCTIONS OF ANIONIC PHOSPHOLIPIDS

A. Overview of Functions

Phosphatidylglycerol is metabolically active (Figures 1 and 3) as evidenced by at least 40% of the glycerol headgroup turning over per generation (Kanfer and Kennedy, 1963). Phosphatidylglycerol serves as the immediate precursor to cardiolipin releasing free glycerol in the condensation of two phosphatidylglycerol molecules (Hirschberg and Kennedy, 1972); cardiolipin levels rise in late log phase due to an increase in *cls* gene expression (Heber and Tropp, 1991) and cardiolipin synthase activity (Hiraoka et al., 1993), but there appears to be little turnover of cardiolipin. The precise requirement for cardiolipin is not clear since mutants in the *cls* gene with cardiolipin levels reduced from 5% to less than 0.1% of total phospholipid appear to have few discernible changes in their properties (Nishijima et al., 1988) except a reduced viability after prolonged incubation in stationary phase (Hiraoka et al., 1993). However, as will be discussed in more detail later, cells lacking phosphatidylethanolamine have an absolute requirement for cardiolipin (DeChavigny et al., 1991). The lack of a strong dependence of wild type *E. coli* on cardiolipin is particularly interesting since *in vitro* experiments have indicated a strong dependence of the mitochondrial energy transducing systems of eukaryotic cells on cardiolipin (Hayer-Hartl et al., 1992; Robinson, 1993). This discrepancy between prokaryotic and eukaryotic cells may be due to the interchangeability of phosphatidylglycerol and cardiolipin for many functions in *E. coli* (as will be pointed out below) and the fact that cardiolipin and not phosphatidylglycerol is the major anionic phospholipid in mitochondria (Ardail et al., 1990;

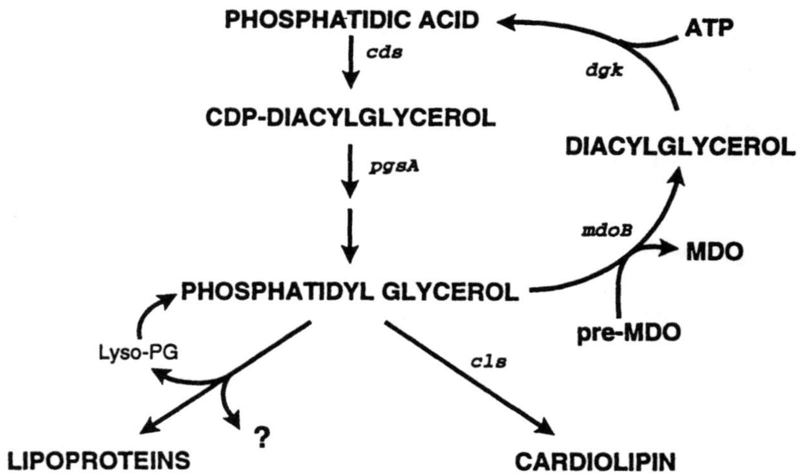

Figure 3. Turnover of phosphatidylglycerol and the diacylglycerol cycle of *E. coli*. Some of the genes which encode the enzymes for the indicated steps are shown. The product of posttranslational acylation of the lipoproteins is shown, but the lipid after transfer of diacylglycerol to the lipoproteins is not known.

Zinser et al., 1991). Therefore, it is particularly important to generate and characterize mutants in both phosphatidylglycerol and cardiolipin biosynthesis in eukaryotic cells in order to answer basic questions about the importance of these lipids for mitochondrial membrane structure and function in general and for understanding the involvement of cardiolipin in the basic mechanism of energy transduction. Preliminary studies on a mutant in somatic cells defective in mitochondrial phosphatidylglycerol synthesis (and therefore cardiolipin biosynthesis) shows a perturbation in mitochondrial function (Ohtsuka et al., 1993), but further studies will be required to separate specific from global effects of this mutation.

A major turnover route for phosphatidylglycerol is in the formation of periplasmically localized membrane derived oligosaccharide (MDO), the level of which is inversely proportional to the osmolarity of the growth medium (Kennedy, 1987). In this event the *sn*-glycerol-1-phosphate headgroup of phosphatidylglycerol is transferred by a membrane-associated enzyme (Jackson and Kennedy, 1983) to a small branched glucose-based oligosaccharide which is also decorated with succinyl and ethanolamine phosphate (derived from phosphatidylethanolamine (Miller and Kennedy, 1987)) residues leaving diacylglycerol as the lipid product; diacylglycerol is subsequently recycled (Figure 3) by the diacylglycerol kinase and re-enters the phospholipid biosynthetic pathway at phosphatidic acid (Walsh and Bell, 1992). The above lipid-dependent modification of MDO occurs on the periplasmic side of the inner membrane as evidenced by modification of arbutin (an acceptor-analog of MDO) which cannot penetrate the inner membrane (Bohin and Kennedy, 1984). The initial work on the diacylglycerol kinase in *E. coli* paved the way for many of the investigations of the level and role of the phosphatidylinositol-diacylglycerol cycle in signal transduction mediated by protein kinase C isozymes in animal cells (Bell and Burns, 1991). It is not clear whether the anionic phospholipid (phosphatidylglycerol)-diacylglycerol cycle of *E. coli* (Figure 3) (Jackson et al., 1984) has an associated signaling role (Raetz and Dowhan, 1990). This pathway appears to be one of the mechanisms available to *E. coli* to respond to changes in osmolarity of the growth medium. In low osmolarity growth medium the amount of MDO in the periplasmic space (Figure 2) increases which stimulates the rate of turnover of the ionic headgroup of phosphatidylglycerol by the above cycle without changing the steady state phospholipid composition. Growth of *E. coli* in the presence of a substitute MDO precursor molecule, arbutin, can increase the rate of turnover of phosphatidylglycerol 7-fold without appreciably changing the phospholipid composition of the membrane (Jackson et al., 1986). This result emphasizes the degree of regulation of the charge balance of the membrane phospholipid pool and the excess capacity for the synthesis of phosphatidylglycerol. Since *E. coli* has multiple mechanisms for dealing with changes in osmolarity of the growth medium (Ingraham, 1987), elimination of the synthesis of MDO is not lethal, but elimination of the diacylglycerol kinase is lethal in the presence of arbutin probably due to the high accumulation of diacylglycerol in the membrane (Jackson et al., 1984). Overproduction of either the *pssA* or *pgsA* gene product by 10- to 20-fold also does not influence the phospholipid charge balance of the membrane (Dowhan, 1992b; Dowhan, 1992c) which also emphasizes the tight regulation of the phospholipid composition of this organism. Short of introducing mutations, no conditions have been found to significantly vary the balance of zwitterionic and anionic phospholipids in the *E. coli* membrane.

Phosphatidylglycerol also serves as a donor of glycerol in the posttranslational modification of a number of lipoproteins of *E. coli* (Wu, 1987), one of which is a major outer membrane protein (*lpp* gene product) responsible for anchoring the peptidoglycan (Figure 2) to the inner surface of the outer membrane (Wu et al., 1983). In this posttranslational event the diacylglycerol moiety is added in thioether linkage to a cysteine of the prolipoprotein which after removal of the leader sequence becomes the amino terminus of the protein (Sankaran and Wu, 1994). The free amino group of cysteine becomes acylated by transacylation from existing phospholipid, contributing to much of remodeling of the acyl chains of phospholipids (Jackowski and Rock, 1986). Since phosphatidylethanolamine is the major phospholipid, *in vivo* experiments done in wild type cells skewed the results suggesting that this phospholipid was the sole donor of acyl groups in these reactions. *In vitro* studies have shown that all the major phospholipids of *E. coli* can serve as the acyl donor (Gupta and Wu, 1991). Subsequent studies employing a mutant of *E. coli* completely lacking phosphatidylethanolamine strongly suggest that the acyl donor is not restricted to phosphatidylethanolamine but can include phosphatidylglycerol (Gupta et al., 1991). Although the *lpp* gene product serves an important role in maintaining structural integrity of the outer envelope, this gene product is not absolutely essential for cell viability. Mutants lacking this gene product are hypersensitive to external assault and antibiotics, and tend to leak their periplasmic components to the growth medium (Yem and Wu, 1978).

Since the turnover products of phosphatidylglycerol (cardiolipin, MDO and the lipoproteins) appear not to be absolutely essential for viability, the question arose as to whether phosphatidylglycerol itself was essential. The question became more relevant with the isolation of viable mutants in the *pgsA* gene with very low levels of gene product activity and levels of phosphatidylglycerol and cardiolipin 10-fold lower than found in wild type cells (Miyazaki et al., 1985). However, generation of a null allele of the *pgsA* gene was subsequently found to be lethal (Heacock and Dowhan, 1987), and to date no conditions of growth or no suppressers of the null allele of the *pgsA* gene have been found. Of particular interest was the finding that mutant alleles such as *pgsA3* which encode a gene product with attenuated activity could not support growth of otherwise wild type cells but could support growth of cells carrying a mutation in the above mentioned *lpp* gene (Asai et al., 1989); therefore *lpp* mutations act as suppressers of *pgsA* mutations which show a reduced but not complete loss of phosphatidylglycerol synthesis. The mechanism for this suppression is not known, but one could speculate that elimination in the utilization of newly synthesized phosphatidylglycerol for lipoprotein modification on the inner surface of the inner membrane in a *pgsA* mutant might be sufficient to remove the limitation to growth imposed by this mutation; at the limiting level of phosphatidylglycerol in a *pgsA3* mutant the number of lipoprotein molecules required to be synthesized per generation (Nikaido and Vaara, 1987) is on the same order as the steady state level of phosphatidylglycerol molecules which are distributed over 3 phospholipid monolayers (Figure 2). These experiments established that some level of phosphatidylglycerol is essential (*lpp* mutations do not suppress the null allele (Dowhan, 1991)) but still left open the question of the requirement for phosphatidylglycerol in the null allele. Furthermore, the ability to suppress "leaky mutants" in the *pgsA* locus by a mutation in a process which utilizes phosphatidylglycerol suggested that additional suppresser mutations could be isolated in functions also requiring phosphatidylglycerol thus providing an independent avenue for uncovering additional functions for anionic phospholipids.

B. Perturbation of Anionic Phospholipid Levels

To systematically survey functions in which anionic phospholipids might be involved, a strain was developed in which the steady state level of these phospholipids can be regulated by artificially controlling the expression from the *pgsA* gene in response to the composition of the growth medium (Dowhan, 1992a; Dowhan, 1992c; Heacock and Dowhan, 1989). In this strain the normal chromosomal *pgsA* gene is inactivated by insertion of a drug marker, and a second single copy of the gene is integrated within the *lac* operon so that the expression of this copy is under the direct regulation of the *lac* promoter. The level of *pgsA* gene product (which catalyzes Step 6), the level of phosphatidylglycerol and cardiolipin, and the rate of cell growth are directly related to the level in the growth medium of the gratuitous inducer of the *lac* operon, IPTG (isopropyl-β-thiogalactoside). Cells arrest their growth when the level of gene product (about 2% of wild type activity) can no longer sustain sufficient synthesis of these two anionic phospholipids (reduced to 2-3% of total phospholipid) to maintain growth. However, cell arrest is fully reversible upon addition of IPTG to the growth medium indicating that the low level of anionic phospholipid synthesis still allowed by the fully repressed *lac* operon is bacteriostatic and not bacteriocidal. Therefore, these anionic phospholipids are not specifically required at high levels to maintain cell membrane integrity, structure or charge density, but at reduced levels can become limiting for other cell processes; a mutation in the *lpp* gene is also a suppresser of the IPTG requirement relieving the restraint on the cell growth rate but not the reduction in anionic phospholipid levels (Kusters et al., 1991). The precursors (mainly phosphatidic acid and CDP-diacylglycerol) are also anionic and accumulate to levels as high as 7%, but they cannot substitute for phosphatidylglycerol function in alleviating the growth arrest phenotype and appear not to be the cause of growth arrest since introduction of the *lpp* mutation allows growth without reducing the levels of the precursors (Kusters et al., 1991). This IPTG-dependent strain has been effectively used to uncover a role for anionic phospholipids in the process of translocation of outer membrane proteins across the inner membrane, to establish an involvement of anionic phospholipids in the initiation of DNA replication, and to establish a requirement for anionic phospholipids in the insertion of colicin toxins entering from the exterior of cells into the inner membrane bilayer.

C. Protein Translocation Across Membranes

Initially, it was observed that in mutants with reduced anionic phospholipid content, the preprotein precursors to several outer membrane proteins began to accumulate in the cytoplasm in significant amounts (de Vrije et al., 1988). More detailed investigation (Lill et al., 1990) indicated that the rate of translocation of these preproteins across the inner membrane was dependent on the level of anionic phospholipid suggesting that the optimal functioning of the translocation machinery may require significant levels of anionic phospholipids. Further biochemical dissection using an *in vitro* reconstituted inverted inner membrane vesicle system to study protein translocation showed that the degree to which such membranes supported translocation of purified precursors added to the outside (analogous to the cytoplasmic side in whole cells) of the vesicles was dependent on the level of phosphatidylglycerol and cardiolipin in the target membranes isolated from cells grown at different IPTG concentrations (Lill et al., 1990). One of the important interactions in this process is formation of a membrane associated complex between the cytoplasmic SecA protein, the membrane associated SecY and SecE proteins and the preprotein substrate. Anionic phospholipids appear to provide the appropriate binding surface for the cytoplasmic SecA protein through first an ionic interaction followed by a partial insertion of the SecA protein into the phospholipid bilayer (Oliver, 1993; Ulbrandt et al., 1992). These events result in a conformational change which apparently result in the induction of latent ATPase activity of the SecA protein. This binding and activation has been demonstrated *in vitro* using isolated membrane vesicles with increasing amounts of phosphatidylglycerol (Lill et al., 1990). Membranes isolated from cells grown in the absence of IPTG are inert for protein translocation in vitro but can be made competent by reconstituting them with anionic phospholipids added after the isolation of the membrane vesicles (Hendrich and Wickner, 1991). There appears to be no specificity with respect to the anionic head group for the *in vitro* reconstitution since phosphatidylethanol or phosphatidylinositol (neither of which are found in *E. coli*) are as effective as phosphatidylglycerol or cardiolipin (Kusters et al., 1991). This may explain why protein translocation is only slowed and not stopped in the absence of IPTG. Although the phosphatidylglycerol plus cardiolipin level is reduced 5- to 10-fold, the total anionic phospholipid pool (including phosphatidic acid) is only reduced about 2- to 3-fold. Since the leader sequences of the preprotein forms of translocated proteins are rich in positive charges, association between the leader sequences and anionic phospholipids may also be important in the protein translocation process. Similar experiments as to those outlined above have shown the same dependence on anionic phospholipids for both the *in vivo* and *in vitro* translocation of the M13 procoat protein (Kusters et al., 1994); insertion of procoat into the membrane is independent of the SecA protein but dependent on a cationic leader peptide (Gallusser and Kuhn, 1990). This dual role for anionic phospholipids in interacting with the leader peptide and the SecA protein may explain the added stability of the SecA protein-precursor-membrane complex over SecA-membrane association alone (Lill et al., 1990).

The combined *in vivo* and *in vitro* experiments made possible by this novel mutant in anionic phospholipid metabolism have clearly established a central role for anionic phospholipids in organization of the membrane associated complex necessary for translocation of proteins across the inner membrane of *E. coli*. A similar example of activation of a cytoplasmic protein through membrane association exists in eukaryotic cells for the activation of protein kinase C isozymes via association with phosphatidylserine, diacylglycerol and calcium in a membrane bound complex (Bell and Burns, 1991). The organization of cytoplasmic and membrane components via a common affinity for a minor phospholipid component certainly should serve as a working model for the development of experimental approaches in more complex systems where such genetic manipulation may not as yet be feasible for *in vivo* verification.

D. Colicin Insertion into Membranes

Bacterial toxins like the colicins also must insert into the membrane bilayer, but from the exterior of the inner membrane, in order to function in killing cells (Bourdineaud et al., 1990; Pattus et al., 1990). Colicin A and N form voltage-gated channels in the inner membrane of *E. coli* and in model phospholipid vesicles which selectively allow efflux of K^+ ions resulting in dissipation of the electrochemical gradient across the membrane leading to cell death. In model membranes this insertion has been shown to be absolutely dependent on anionic phospholipids. *In vivo* the complete process requires outer membrane receptor proteins and a series of cell envelope proteins responsible for delivering the colicin to the outer surface of the inner membrane (Lazdunski et al., 1988). The *pgsA* mutant in which the level of anionic phospholipids can be controlled by IPTG in the growth medium was used to determine whether assembly and function of colicin A *in vivo* also required anionic phospholipids (van der Goot et al., 1993). The lag time for assembly of colicin A (i.e., appearance of efflux) was inversely proportional to the anionic phospholipid content of the membrane and the rate of efflux was proportional to the level of anionic phospholipids. These results suggest a profound effect of the anionic phospholipid content of the membrane on the assembly and/or function of colicin A as one might predict from *in vitro* experiments. Unknown was whether the effect was on the receptor and delivery systems of the cell envelope or on the insertion process. Colicin N uses the same cellular machinery as colicin A for binding and delivery to the inner membrane (Wilmsen et al., 1990), but its insertion into the membrane may be less dependent on anionic phospholipids because it is a more basic protein. Colicin N assembly and function were found to be independent of anionic phospholipid content in the above mutant suggesting that the low amount of anionic phospholipids did not compromise the binding and delivery machinery but was below the threshold for efficient membrane insertion and function of colicin A but not colicin N (van der Goot et al., 1993). These results give further support for the importance of anionic phospholipids in the *in vivo* assembly of at least colicin A into the inner membrane bilayer.

E. Initiation of DNA Replication

Another example of potential involvement of anionic phospholipids is the initiation of DNA replication in *E. coli*. Historically, there has been considerable evidence suggesting a role of either the membrane surface or phospholipids in DNA replication (Kornberg and Baker, 1992). Association of the genome with the membrane has been observed at several points on the chromosome (Ryter, 1968) including the site of the origin (*oriC*) of replication (Craine and Rupert, 1978; Norris, 1990). Cessation of phospholipid synthesis using a *plsB* mutant (Step 1, Figure 1) results in a coordinate cessation in all macromolecular synthesis with specific inability to initiation a new round of DNA synthesis (McIntyre et al., 1977;

Pierucci and Rickert, 1985). Mutants unable to synthesize unsaturated fatty acids arrest at the point of initiation of DNA replication (Fralick and Lark, 1973) suggesting a relationship between membrane fluidity and DNA replication. The DnaA protein (primarily responsible for specific initiation of DNA replication at *oriC*) copurifies with phospholipid (Sekimizu et al., 1988) and *in vitro* requires anionic phospholipids for function (Crooke et al., 1992; Sekimizu and Kornberg, 1988; Yung and Kornberg, 1988). This protein binds to a specific site at *oriC* in both the ATP and ADP forms; however, the latter form cannot proceed to the remaining steps of initiation and does not readily exchange ADP for ATP; ATP is slowly hydrolyzed to ADP during the initiation process resulting in an incompetent complex. This complex can be reactivated *in vitro* by the presence of anionic phospholipids (either naturally occurring in *E. coli* or from other sources) containing unsaturated fatty acids. Activation occurs by facilitating the rapid exchange of ATP for ADP to regenerate the active DnaA protein.

In vivo support for this proposed involvement of anionic phospholipids has been obtained using the above IPTG-dependent mutant in anionic phospholipid biosynthesis (Xia and Dowhan, 1995). The rationale used was as follows. If anionic phospholipids are limiting for DnaA protein-dependent initiation at *oriC* in these mutants in the absence of IPTG, then mutations which suppress the need for DnaA protein-dependent initiation should also be suppressers of the requirement for IPTG for growth. Mutations in the *rnh* gene (encodes RNAse H) suppress the need for both DnaA protein and *oriC* (Horiuchi et al., 1984; Ogawa et al., 1984; von Meyenburg et al., 1987) allowing an alternate mode of initiation of DNA replication. Normally, the short RNA primers necessary for initiation are degraded by RNAse H at non-specific *oriK* sites around the chromosome but not at the *oriC* site protected by the DnaA protein; in *rnh* mutants the process of constitutive stable DNA replication (cSDR) can be initiated at the *oriK* sites independent of DnaA protein but the by-pass process is dependent on the RecA protein (Kogoma et al., 1985); one of the normal roles of RecA protein is to stabilize the open DNA duplex during recombination which may be its role in this alternate mode of initiation. As predicted by the above hypothesis, an *rnh* mutation suppresses the growth dependence on IPTG of the anionic phospholipid mutant without affecting the basal level of anionic phospholipid synthesis in the absence of IPTG (Xia and Dowhan, 1994). This suppression is dependent on a functional RecA protein further supporting induction of cSDR as the mechanism of suppression. However, *E. coli* carrying a *lexA*(def) mutation does not require a functional RecA protein for cSDR (Torrey and Kogoma, 1987), and it was found that the dependence on IPTG could also be suppressed in a *recA* mutant by introducing a *lexA*(def) mutation (Xia and Dowhan, 1995). This genetic evidence for the involvement of anionic phospholipids in DnaA protein-dependent initiation was further supported by the observation that anionic phospholipid mutants (which were also *rnh* mutants so they could grow in the absence of IPTG) could not maintain plasmids dependent on the *oriC* locus for initiation, but could maintain plasmids not dependent on the DnaA protein for initiation of replication (Xia and Dowhan, 1995). These results provide strong evidence from *in vivo* experiments in support of the *in vitro* observation implicating anionic phospholipids in DnaA-dependent initiation of DNA replication. However, supporting initiation of DNA replication is not the only essential role for anionic phospholipids since an *rnh* mutation cannot by-pass the lethal effect of a null allele of the *pgsA* gene.

Anionic phospholipids were found to inhibit the activity of eukaryotic nuclear DNA polymerases α, β, and mitochondrial DNA polymerase γ suggesting a possible role for phospholipids in nuclear and/or mitochondrial DNA replication (Yoshida et al., 1989). Inhibition occurred only in the absence of DNA template which was also observed for DnaA protein in the absence of ATP or ADP (Hwang et al., 1990); in *E. coli* this inhibition could be reversed (Hwang et al., 1990) by either phospholipases or DnaK protein (a heat shock/chaperone protein). Cardiolipin inhibited all the polymerases while phosphatidylinositol selectively inhibited polymerase γ. Phosphatidic acid showed some effect on polymerase α and strongly inhibited polymerase γ. Mutants in phosphatidylinositol synthesis in yeast have been isolated which demonstrate the absolute requirement for phosphatidylinositol (Nikawa et al., 1987), but DNA replication has not been investigated in these mutants. The isolation of mutants in the synthesis of cardiolipin in the mitochondria of yeast would be useful for extending studies of the effects of anionic phospholipids on mitochondrial DNA synthesis *in vivo*. Extrapolation of the results from *E. coli* would suggest that a search in eukaryotic cells for factors which modulate the effects of phospholipids on these polymerase activities would be a reasonable undertaking.

IV. FUNCTIONS OF PHOSPHATIDYLETHANOLAMINE

A. Overview of Functions

Due to the high phosphatidylethanolamine content of both the inner leaflet of the outer membrane (close to 90%) and both leaflets of the inner membrane (about 60%) (Raetz et al., 1979), this phospholipid must play an important structural role in the membrane. Phosphatidylethanolamine also serves as a precursor to the ethanolamine phosphate residues decorating both MDO (Miller and Kennedy, 1987) and probably the carbohydrate portion of the polysaccharides (Hasin and Kennedy, 1982) extending from the lipid A moieties of the outer membrane (Figure 2). Due to its high content in the membrane, this phospholipid is a major source of fatty acid used for modification of the major and minor lipoproteins (Jackowski and Rock, 1986) of *E. coli*. However, none of these functions appear to be essential individually or as a whole for *E. coli* survival since mutants have been isolated (DeChavigny et al., 1991) lacking phosphatidylethanolamine (less than 0.01% of the total phospholipid content).

Mutants lacking phosphatidylethanolamine were constructed by first making a null allele of the *pssA* gene (encoding the enzyme for Step 4, Figure 1) in a strain carrying a functional copy of the gene on a plasmid which was temperature sensitive for replication (DeChavigny et al., 1991). When this mutant is grown at the restrictive temperature for plasmid replication, daughter cells do not inherit a functional copy of the *pssA* gene and therefore lose the ability to make phosphatidylethanolamine. Cell growth arrests when the phosphatidylethanolamine content declines to about 35% of total phospholipid as had previously been seen with other mutants which were temperature sensitive in the *pssA* locus directly; the growth phenotype of these earlier mutants could be suppressed by addition of mmolar levels of Mg^{2+} ion to the growth medium (Ohta and Shibuya, 1977; Raetz et al., 1979), but since the gene product was still partially functional, the level of phosphatidylethanolamine remained at 35%. Surprisingly, addition of either Ca^{2+}, Mg^{2+} or Sr^{2+} to the growth medium (in decreasing order of effectiveness on a mmolar basis) of a strain carrying the null allele also suppressed the growth arrest phenotype following loss of the plasmid copy of the gene even though no phosphatidylethanolamine was now made (DeChavigny et al., 1991).

As noted earlier, the involvement of phosphatidylethanolamine in modification of MDO and the lipoproteins is not essential for cell viability. Whatever critical role phosphatidylethanolamine may play inside the cell can apparently be served by the normal high intracellular Mg^{2+} content of *E. coli* (Chang et al., 1986). Since both Ca^{2+} and Sr^{2+} are actively excluded from the interior of *E. coli* (Gangola and Rosen, 1987), the major suppressive effect of high divalent cation concentration appears to be a role normally fulfilled by phosphatidylethanolamine outside of the inner membrane. The only source of counterions exterior to the inner membrane to damp the high negative charge density on the membranes would be the growth medium. Since removal of divalent cations results in cell lysis, one role of phosphatidylethanolamine must be in maintaining structural integrity of the cell envelope and the outer membrane in particular primarily through its contribution of charged but neutral head groups. One consequence of the *pssA* null mutant would be the removal of phosphatidylethanolamine as the donor of the ethanolamine residues which decorate the carbohydrate portions of the outer membrane lipopolysaccharides (Hasin and Kennedy, 1982) and provide the only positive charge to these highly negative structures. Proper maintenance of outer membrane integrity does rely on low levels of calcium ion (Nikaido and Vaara, 1987), but may require much higher divalent metal ion concentrations in the absence of ethanolamine modification. A detailed analysis of lipopolysaccharide structure in this mutant has not been carried out, but it certainly exists in the mutant (Schnaitman, personnel communication). This is somewhat surprising since *in vitro* experiments had shown (Hinckley and Müller, 1972; Müller et al., 1972) a strong dependence on phosphatidylethanolamine for function of the sugar transferases responsible for synthesis of the complex carbohydrate portion of this structure.

B. Nonbilayer Forming Lipids

Another potentially important structural role of phosphatidylethanolamine may be the ability of this phospholipid to assume a non-bilayer structure, namely existing in the inverted hexagonal (H_{II}) phase (Cullis and de Kruijff, 1979; Lindblom and Rilfors, 1989) within a normal bilayer (Figure 4).

The potential for phospholipids to contribute bilayer discontinuity to the membrane may be important for insertion of proteins into the membrane, regulated passage of molecules through the membrane, membrane growth, membrane fusion, and membrane budding. Mutants lacking phosphatidylethanolamine maintain the normal protein to phospholipid ratio in their membranes and make up for the absence of zwitterionic phospholipid by increasing the amount of phosphatidylglycerol and cardiolipin while maintaining a normal fatty acid composition (DeChavigny et al., 1991); cardiolipin, but not phosphatidylglycerol, can assume the H_{II} phase dependent on the presence of specific divalent metal ions (Cullis and de Kruijff, 1979; Lindblom and Rilfors, 1989). The level of cardiolipin in the *pssA* null mutant is sensitive to the type of divalent metal ion in the growth medium which is not the case for wild type cells. When grown in Ca^{2+}, the cardiolipin level is about 25-30% of total phospholipid but reaches 40-45% when grown in Mg^{2+} or Sr^{2+}. The above three cations, but not Ba^{2+} or Na^{+}, also support growth in relation to the ability of these cations to induce a bilayer to H_{II} phase transition (Ca^{2+} > $Mg2^{+}$ > Sr^{2+}) as a function of temperature for either the total phospholipids extracted (Rietveld et al., 1993) from this mutant or for the cardiolipin alone (Killian

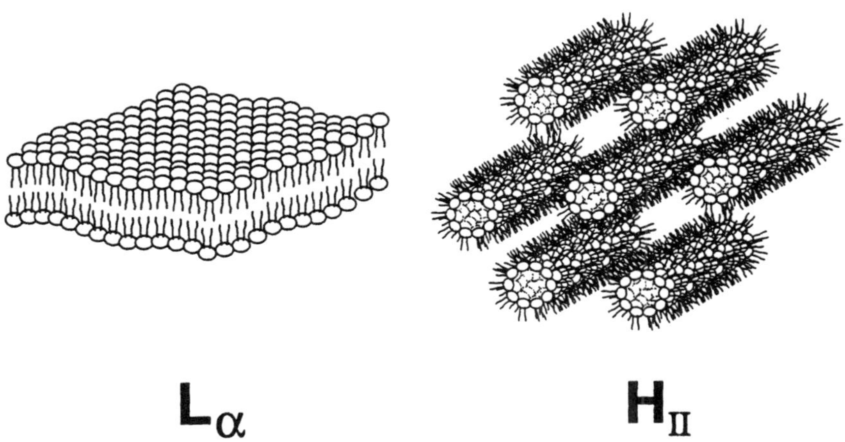

***Figure 4**.* Phospholipids such as PE and CL can exist in either a liquid crystalline phase (Lα) or an inverted hexagonal phase (H_{II}) depending on temperature and solvent conditions.

et al., 1994) from this mutant. Phospholipids extracted from cells grown at 37°C and in the optimal divalent metal ion concentrations, when suspended in the same divalent metal ion, exhibit a transition from bilayer to H_{II} phase beginning at about 45°C and extending over 20 to 30°C higher; these transitions are very similar to that for phospholipids extracted from wild type cells and analyzed in the absence of divalent metal ion. However, phospholipids extracted from cells grown in Ca^{2+} and then monitored for transition in the presence of Mg^{2+} exhibit a transition beginning 20°C higher than the transition induced by Ca2+; these membranes have a considerably reduced cardiolipin content, and Ca^{2+} is a much stronger inducer of this transition for cardiolipin than is Mg^{2+} (Cullis and de Kruijff, 1979). Conversely, phospholipids from cells grown in Mg^{2+} and analyzed in Ca^{2+} show a transition 20°C lower than the transition induced by Mg^{2+} and well below the temperature at which the culture was grown. With Sr^{2+} the ability to induce a transition is very dependent on the cation concentration with 12 mM inducing a transition at a lower temperature than either 10 or 15 mM; 20 mM Sr^{2+} neither supports growth of the *pssA* null mutant nor induces a transition to a nonbilayer phase for phospholipids extracted from cells grown in 15 mM Sr^{2+}. Therefore, the physical properties of the phospholipids extracted from a mutant lacking phosphatidylethanolamine and the growth properties of the mutants appear to be directly related to the cardiolipin content (Rietveld et al., 1993) and its physical properties (Killian et al., 1994).

In the absence of phosphatidylethanolamine the cell appears to regulate the level of cardiolipin in response to the potential for non-bilayer formation in response to the growth medium; normally, such an adjustment is not necessary since cardiolipin levels are low in wild type cells, and the non-bilayer-forming potential of phosphatidylethanolamine is not a function of the presence of divalent cations. Maintenance of a proper balance between bilayer and non-bilayer forming phospholipids is evident in several organisms (Goldfine et al., 1987; Wieslander et al., 1980), and the fact that *E. coli* has a mechanism to make such an adjustment in the absence of phosphatidylethanolamine suggests an important role for lipid polymorphism in normal cell function. These conclusions are further strengthened by the inability to eliminate cardiolipin (via introduction of a *cls* mutation) in strains defective (DeChavigny et al., 1991) in the *pssA* gene. Cardiolipin plus divalent metal ion appears to fulfill a critical structural role in the absence of phosphatidylethanolamine and should be considered in analyzing the importance of cardiolipin (Hoch, 1992) for the function of several components of the energy transduction system of mitochondria where cardiolipin is found in high concentrations. In addition the importance of non-bilayer structures to membrane function and the presence of phosphatidylethanolamine in all membranes of eukaryotic cells must be considered.

C. Lactose Transport

How the presence of phosphatidylethanolamine affects particular cellular functions has been addressed using mutants lacking this phospholipid. The ability to purify and reconstitute solute transport systems made possible the study *in vitro* of the possible influence of membrane phospholipid composition on the properties of transporters which move solutes across membranes. The need for phosphatidylethanolamine has been implicated in the function of these systems. The best studied example of such systems is the lactose permease (Kaback, 1992; Kaback et al., 1993) encoded by the *lac* operon of *E. coli*. Like most transporters of this type the permease is a polytopic protein with 12 membrane spanning domains connected by hydrophilic loops exposed on both the cytoplasmic and periplasmic side of the inner membrane. The transporter can function in an energy independent manner to facilitate equilibration of lactose or its analogs across the membrane, or in an energy dependent manner to actively cotransport a proton and substrate utilizing the energy of the proton electrochemical gradient ($\Delta\mu H^+$) to accumulate substrate. Upon reconstitution of the purified permease into proteoliposomes of defined phospholipid composition, a near absolute dependence on phosphatidylethanolamine was noted for reconstitution of energy dependent active transport of substrate but not for facilitated equilibration of substrate across the artificial phospholipid bilayer (Chen and Wilson, 1984; Page et al., 1988). Phosphatidylserine behaved much as did phosphatidylethanolamine and successive N-methylated derivatives of phosphatidylethanolamine were progressively less effective with phosphatidylcholine being completely ineffective in supporting active transport. These results strongly implicate a requirement for an ionizable amine, rather than a non-bilayer forming phospholipid, in either the assembly and/or function of this transporter.

However, studies of reconstituted membrane systems *in vitro* are fraught with potential difficulties. The differences observed between proteoliposomes of various phospholipid composition might be due to variability in the amount of the transport protein incorporated, in the magnitude and stability of the imposed $\Delta\mu H^+$, and in the internal volume of the proteoliposomes. The native conformation and properties of extracted and reconstituted proteins can be irreversibly altered. Finally, the influence of dynamic metabolism on regulation and function cannot be tested in reconstituted systems. Mutants lacking phosphatidylethanolamine afford a system in which to test *in vivo* the importance of this phospholipid in the assembly and function of specific processes like solute transport.

The *lac* permease is fully expressed and is membrane associated in cells lacking phosphatidylethanolamine (Bogdanov and Dowhan, 1995); however, the properties of the permease parallel those reported earlier in proteoliposomes lacking phosphatidylethanolamine. Mutant cells lacking phosphatidylethanolamine rapidly facilitate the uptake of metabolizable substrate which does not require energy, but under conditions where accumulation of substrate against a concentration gradient is measured (i.e., utilization of $\Delta\mu H^+$), the rate of uptake is reduced 20- to 50-fold over the rate in wild type cells. These results suggest that the lactose permease cannot efficiently couple active transport to energy metabolism in membranes lacking phosphatidylethanolamine as was noted *in vitro*. An unexpected observation was that the low K_m for transport normally associated with accumulation of substrate against a concentration gradient remained the same in the mutant both in the presence and absence of energy poisons (Bogdanov and Dowhan, 1995) while in the wild type this low K_m was significantly increased, as expected, in the presence of energy poisons to a high K_m normally associated with energy independent facilitated transport (Ghazi and Schechter, 1981).

The molecular basis for this lack of coupling is not understood, but preliminary evidence suggests that the permease is misassembled in cell membranes lacking phosphatidylethanolamine. Utilization of permease derivatives with specifically placed proteolytic sites (Bogdanov and Dowhan, 1994) indicates that some of the membrane spanning domains of the permease are assembled in an inverted orientation when compared to the same derivatives expressed in wild type cells. Therefore, the signals which are important in directing the assembly of some integral membrane proteins must include recognition of the phospholipid component of the membrane as was also noted for proteins which are translocated across the membrane (Kusters et al., 1994; Kusters et al., 1991; Lill et al., 1990). The *pssA* null mutant should prove useful for probing the role of zwitterionic phospholipids in the function of other transport systems and the assembly of polytopic membrane proteins.

D. Energy Transduction

Most aspects of energy transduction are normal in cells lacking phosphatidylethanolamine. Both whole cells (Bogdanov and Dowhan, 1995) and isolated inverted membrane vesicles (Mileykovskaya and Dowhan, 1993) can generate and maintain a membrane potential ($\Delta\mu H^+$) comparable to that of wild type cells. ATP hydrolysis and oxidation of succinate and lactate by inverted membrane vesicles isolated from mutant cells results in a proton gradient across the membrane comparable to that of membranes isolated from wild type cells. However, not all aspects of energy transduction are insensitive to phospholipid composition (Mileykovskaya and Dowhan, 1993). Oxidation of NADH, which occurs at a reduced rate in mutant membrane preparations, can only support a $\Delta\mu H^+$ of 20% that of wild type membranes. This reduced efficiency is not due to a reduction in the level of NADH dehydrogenase, as measured directly by using artificial electron acceptors, but appears to be due to a reduction in the interaction of the NADH dehydrogenase with its natural electron acceptor, ubiquinone 8. The level of this acceptor is also not reduced in cells lacking phosphatidylethanolamine. Artificial ubiquinones with much shorter hydrophobic side chains (and thus not as deeply imbedded in the membrane bilayer) restore the rate of oxidation of NADH and the magnitude of the membrane potential to wild type levels. During purification of the dehydrogenase, phospholipid and ubiquinone 8 remain associated with the protein indicative of a high affinity of the protein for these components (Jaworowski et al., 1981). Apparently some subtle change in the organization of the dehydrogenase and its surrounding ubiquinone-containing phospholipid environment is affected by the lack of phosphatidylethanolamine in the membrane resulting in a change in the apparent rate of electron transfer. Further biochemical studies are required to define at the molecular level the basis for this alteration in the function of the dehydrogenase.

A more serious defect in this mutant, which may account for a lower yield of cells in stationary overnight cultures, is a lack of cytochrome *d* (high oxygen affinity). Normally, cytochrome *d* levels increase dramatically as oxygen tension decreases (Iuchi et al., 1990) to compensate for the low oxygen affinity of cytochrome *o* present at all stages of growth; these two cytochromes act as the terminal donors of electrons to O_2. In this mutant the spectrum of cytochrome *d* is nearly absent and does not increase in late log phase (Mileykovskaya and Dowhan, 1993). Preliminary data suggest that the failure to induce cytochrome d levels in the late log phase of growth lies at the level of gene transcription (Mileykovskaya and Dowhan, personal observation) and therefore may be related to a more general problem in the recognition of extracellular signals in mutants lacking phosphatidylethanolamine. In addition cells lacking phosphatidylethanolamine fail to show a normal motility and chemotaxic response. Not only do they lack flagella necessary for motility, but they do not induce the *flaD* master operon (Shi et al., 1993), in response to stimuli, which is responsible for induction of a host of genes related to motility and chemotaxis (Macnab, 1987a; Macnab, 1987b). This lack of response may be related to a global lack of response by stressed cells or to more specific defects in the membrane-related signal transduction machinery of the cell. Again this mutant affords a vehicle to probe more fully questions of the importance of phospholipids in membrane signal transduction.

V. CONCLUDING REMARKS

The role of phospholipid composition and specific phospholipids in the synthesis, assembly and function of membrane components clearly requires further investigation. Obviously, not all integral membrane proteins of *E. coli* are misassembled in the absence of phosphatidylethanolamine as the lactose permease appears to be. However, this result coupled with the requirement for anionic phospholipids for protein translocation emphasizes the need for more studies on the role phospholipids play in assembly of membrane proteins. Little is known about the requirements for proper assembly of integral membrane proteins, as opposed to proteins which are translocated across membranes in secretory processes. Recent evidence suggests that insertion of integral membrane proteins occurs by translocation of individual domains across the membrane by both SecA protein-dependent and SecA protein-independent mechanisms depending on the charge and length of the segments (Andersson and von Heijne, 1993). Statistical analysis of and experimental work on the distribution of positive charges within the periplasmic and cytoplasmic hydrophilic loops of integral membrane proteins has revealed a strong bias for net positive charge ("positive inside rule") in the cytoplasmic loops and either neutral or negative charge in the periplasmic loops (von Heijne, 1992); similar predictions can be made about the cytoplasmic loops of eukaryotic proteins (Sipos and von Heijne, 1993). Is this bias due only to the positive outward sign of the membrane potential which would favor retention of positive domains on the cytoplasmic side or due to the interaction of the positive charges of the proteins with anionic phospholipids on the cytoplasmic side of the membrane? How does elimination of all the phosphatidylethanolamine affect the predictions of "positive inside-rule?" Are there additional protein-phospholipid interactions which determine the orientation of membrane spanning domains? The above effects of the lack of phosphatidylethanolamine on the assembly and function of lactose permease and the role of anionic phospholipids in protein translocation point to specific roles for the ionic groups of phospholipids in membrane protein assembly and/or function which must be accounted for before a complete understanding of membrane protein assembly can be attained.

Not all functions requiring phospholipids are affected to the same extent or at the same threshold level of a particular phospholipid. Those functions which are affected at lower levels of a particular phospholipid may have higher affinities for or be more specific for that phospholipid. Protein translocation is affected at a higher anionic phospholipid content than initiation of DNA synthesis which appears not to be growth limiting in a cell with a mutation in the major outer membrane lipoprotein while protein translocation is still compromised. Colicin A appears to be more sensitive to the anionic phospholipid content of membranes than colicin N which is a more basic protein. There must also be additional requirements for anionic phospholipids since the suppressers of the anionic phospholipid requirement for initiation of DNA replication and protein translocation are not suppressers of a null mutation in anionic phospholipid synthesis. This apparent hierarchy in the requirement for a particular phospholipid has made possible the genetic dissection of these multiple roles.

Due to the similarities between *E. coli* and eukaryotic cells with respect to various aspects of phospholipid synthesis, modern molecular genetic techniques can be exploited utilizing the existing mutants in *E. coli* phospholipid metabolism to extend our understanding of the function of these molecules in eukaryotic cells. Complementation of a mutation in the gene encoding phosphatidylserine decarboxylase of *E. coli* was used to clone the analogous gene (*PSD1*) in yeast (Clancey et al., 1993). The yeast enzyme is encoded by a nuclear gene which must be translocated as the nascent polypeptide across the outer mitochondrial membrane, after synthesis in the cytoplasm of yeast, followed by assembly into the inner mitochondrial membrane while the *E. coli* enzyme is synthesized in the cytoplasm and directly assembled into the inner cytoplasmic membrane (Dowhan and Li, 1992). The decarboxylases from *E. coli* (Dowhan and Li, 1992), yeast (Clancey et al., 1993; Trotter et al., 1993) and CHO cells (Nishijima et al., 1988) show remarkable homology including an unusual covalently bound pyruvate prosthetic group which in *E. coli* is generated by a posttranslational cleavage of a proenzyme form at an internal serine resulting in formation of the pyruvate prosthetic group located at the N-terminus of one of the two resulting subunits of the enzyme (Dowhan and Li, 1992). Given the above complexity in the synthesis and assembly of the decarboxylases, it is encouraging that cloning of this yeast gene related to phospholipid metabolism was possible by complementing of an *E. coli* mutant. Another example of functional expression of a yeast enzyme of phospholipid metabolism in *E. coli* is the yeast phosphatidylinositol synthase which will synthesize its phospholipid product in *E. coli* cells supplied with inositol (Nikawa et al., 1988). These results point out the utility of existing mutants in *E. coli* phospholipid biosynthesis for cloning important additional genes responsible for yeast and somatic cell phospholipid metabolism. Although several genes encoding phospholipid biosynthetic enzymes in eukaryotic cells have been cloned (Cui et al., 1993; Kalmar et al., 1990; Nikoloff and Henry, 1991), the *PSD1* gene is the only example of a cloned gene whose product is directly involved in mitochondrial phospholipid metabolism. The availability of additional genes responsible for mitochondrial phospholipid metabolism should make possible the construction of mutants similar to those outlined above for *E. coli*. Such mutants should be very useful in expanding our understanding of the role of phospholipids, particularly cardiolipin, in mitochondrial function.

REFERENCES

Andersson, H., and von Heijne, G. (1993). *Sec* dependent and *sec* independent assembly of *E. coli* inner membrane proteins: the topological rules depend on chain length. EMBO J. 12, 683-691.

Ardail, D., Privat, J.-P., Egret-Charlier, M., Levrat, C., Lerme, F., and Louisot, P. (1990). Mitochondrial contact sites: Lipid composition and dynamics. J. Biol. Chem. 265, 18797-18802.

Ardail, D., Privat, J.-P., Lerme, F., & Louisot, P. (1991). Involvement of contact sites in phosphatidylserine import into liver mitochondria. J. Biol. Chem. 266, 7978-7981.

Asai, Y., Katayose, Y., Hikita, C., Ohta, A., & Shibuya, I. (1989). Suppression of the lethal effect of acidic-phospholipid deficiency by defective formation of the major outer membrane lipoprotein in *Escherichia coli*. J. Bacteriol. 171, 6867-6869.

Bell, R. M., & Burns, D. J. (1991). Lipid activation of protein kinase C. J. Biol. Chem. 266, 4661-4664.

Bishop, W. R., & Bell, R. M. (1988). Assembly of phospholipids into cellular membranes: Biosynthesis, transmembrane movement and intracellular translocation. Ann. Rev. Cell Biol. 4, 579-610.

Bogdanov, M., & Dowhan, W. (1995). Phosphatidylethanolamine is required for *in vivo* function of the membrane-associated lactose permease of *Escherichia coli*. 8, ABS 1110.

Bogdanov, M., & Dowhan, W. (1994). Phosphatidylethanolamine (PE) is required for the correct membrane assembly of the *E. coli* lactose permease. FASEB J. J. Biol. Chem. 270, 732-739.

Bohin, J.-P., & Kennedy, E. P. (1984). Regulation of the synthesis of membrane-derived oligosaccharides in *E. coli*. J. Biol. Chem. 259, 9390-9393.

Bourdineaud, J. P., Boulanger, P., Lazdunski, C., & Letellier, L. (1990). *In vivo* properties of colicin A: Channel activity is voltage dependent but translocation may be voltage independent. Proc. Natl. Acad. Sci. USA 87, 1037-1041.

Carman, G. M., & Bae, L. M. (1992). Phosphatidylserine synthase from yeast. Methods Enzymol. 209, 298-305.

Carman, G. M., & Henry, S. A. (1989). Phospholipid biosynthesis in yeast. Ann. Rev. Biochem. 58, 635-669.

Chang, C.-F., Shuman, H., & Somlyo, A. P. (1986). Electron probe analysis, X-ray mapping and electron energy-loss spectroscopy of calcium, magnesium, and monovalent ions in log-phase and in dividing *Escherichia coli* B cells. J. Bacteriol. 167, 935-939.

Chen, C.-C., & Wilson, T. H. (1984). The phospholipid requirement for activity of the lactose carrier of *Escherichia coli*. J. Biol. Chem. 259, 10150-10158.

Clancey, C. J., Chang, S.-C., & Dowhan, W. (1993). Cloning of the gene (*PSD1*) encoding phosphatidylserine decarboxylase from *Saccharomyces cerevisiae* by complementation of an *Escherichia coli* mutant. J. Biol. Chem. 268, 24580-24590.

Cornell, R. B. (1991). Regulation of CTP:phosphocholine cytidylyltransferase by lipids. 1. Negative surface charge dependence for activation. Biochemistry 30, 5873-5880.

Craine, B. L., & Rupert, C. S. (1978). Identification of a biochemically unique DNA-membrane interaction involving the *E. coli* origin of replication. J. Bacteriol. 134, 193-199.

Crooke, E., Castuma, C. E., & Kornberg, A. (1992). The chromosome origin of *Escherichia coli* stabilizes DnaA protein during rejuvenation by phospholipids. J. Biol. Chem. 267, 16779-16782.

Cui, Z., Vance, J. E., Chen, M. H., Voelker, D. R., & Vance, D. E. (1993). Cloning and expression of a novel phosphatidylethanolamine N-methyltransferase. J. Biol. Chem. 268, 16655-16663.

Cullis, P. R., & de Kruijff, B. (1979). Lipid polymorphism and the functional roles of lipids in biological membranes. Biochim. Biophys. Acta 559, 399-420.

de Vrije, T., de Swart, R. L., Dowhan, W., Tommassen, J., & de Kruijff, B. (1988). Phosphatidylglycerol is involved in protein translocation across *Escherichia coli* inner membranes. Nature 334, 173-175.

DeChavigny, A., Heacock, P. N., & Dowhan, W. (1991). Phosphatidylethanolamine may not be essential for the viability of *Escherichia coli*. J. Biol. Chem. 266, 5323-5332.

Dowhan, W. (1991). Role of phospholipids in cell function. pp. 11-32. In: NATO ASI Series: Dynamics of Membrane Assembly (Op den Kamp, J.A.F., ed.), pp, 11-32. Springer-Verlag, Corsica, France.

Dowhan, W. (1992a). Strategies for generating and utilizing phospholipid synthesis mutants in *Escherichia coli*. Methods Enzymol. 209, 7-20.

Dowhan, W. (1992b). Phosphatidylserine synthase from *Escherichia coli*. Methods Enzymol. 209, 287-298.

Dowhan, W. (1992c). Phosphatidylglycerophosphate synthase from *Escherichia coli*. Methods Enzymol. 209, 313-321.

Dowhan, W., & Li, Q. X. (1992). Phosphatidylserine decarboxylase from *Escherichia coli*. Methods Enzymol 209, 348-59.

Dutt, A., & Dowhan, W. (1981). Intracellular distribution of phospholipid biosynthetic enzymes in Gram-positive bacteria: Characterization of a membrane-associated phosphatidylserine synthase. J. Bacteriol. 147, 535-542.

Fralick, J. A., & Lark, K. G. (1973). Evidence for the involvement of unsaturated fatty acids in initiating chromosome replication in *Escherichia coli*. J. Mol. Biol. 80, 459-475.

Gallusser, A., & Kuhn, A. (1990). Initial steps in protein membrane insertion. Bacteriophage M13 procoat protein binds to the membrane surface by electrostatic interaction. EMBO J. 9, 2723-2729.

Gangola, P., & Rosen, B. P. (1987). Maintenance of intracellular calcium levels in *Escherichia coli*. J. Biol. Chem. 262, 12570-12574.

Ghazi, A., & Schechter, E. (1981). Lactose transport in *E. coli* cells: Dependence of kinetic parameters on transmembrane electric potential difference. Biochim. Biophys. Acta 644, 305-315.

Goldfine, H., Johnston, N. C., Mattai, J., & Shipley, G. G. (1987). Regulation of bilayer stability in *Clostrium butyricum*: Studies on the polymorphic phase behavior of ether lipids. Biochemistry 26, 2814-2822.

Gupta, S. D., Dowhan, W., & Wu, H. C. (1991). Phosphatidylethanolamine is not essential for the N-acylation of apolipoprotein in *Escherichia coli*. J. Biol. Chem. 266, 9983-9986.

Gupta, S. D., & Wu, H. C. (1991). Indentification and subcellular localization of apolipoprotein N-acyltransferase of *E. coli*. FEMS Microbiol. Lett. 78, 37-42.

Hasin, M., & Kennedy, E. P. (1982). Role of phosphatidylethanolamine in the biosynthesis of pyrophosphoethanolamine residues in the LPS of *Escherichia coli*. J. Biol. Chem. 257, 12475-12477.

Hayer-Hartl, M., Schagger, H., von Jagow, G., & Beyer, K. (1992). Interaction of phospholipids with the mitochondrial cytochrome-*c* reductase studied by spin-label ESR and NMR spectroscopy. Eur. J. Biochem. 290, 423-430.

Heacock, P. N., & Dowhan, W. (1987). Construction of a lethal mutation in the synthesis of the major acidic phospholipids of *Escherichia coli*. J. Biol. Chem. 262, 13044-13049.

Heacock, P. N., & Dowhan, W. (1989). Alterations of the phospholipid composition of *Escherichia coli* through genetic manipulation. J . Biol. Chem. 264, 14972-14977.

Heber, S., & Tropp, B. E. (1991). Genetic regulation of cardiolipin synthase in *Escherichia coli*. Biochim Biophys Acta 1129, 1-12.

Hendrich, L. P., & Wickner, W. (1991). SecA protein needs both acidic phospholipids and SecY/E protein for functional high-affinity binding to the *Escherichia coli* plasma membrane. J. Biol. Chem. 266, 24596-24600.

Hinckley, A., & Müller, E. (1972). Reassembly of a membrane-bound multienzyme system. I. Formation of a particle containing phosphatidylethanolamine, lipopolysaccharide, and two glucosyltransferases. J. Biol. Chem. 247, 2623-2628.

Hiraoka, S., Matsuzaki, H., & Shibuya, I. (1993). Active increase in cardiolipin synthesis in the stationary growth phase and its physiological significance in *Escherichia coli*. Febs Lett 336, 221-224.

Hirschberg, C. B., & Kennedy, E. P. (1972). Mechanism of the enzymatic synthesis of cardiolipin in *Escherichia coli*. Proc. Natl. Acad. Sci. USA 69, 648-651.

Hoch, F. L. (1992). Cardiolipins and biomembrane function. Biochim Biophys Acta 1113, 71-133.

Horiuchi, T., Maki, H., & Sekiguchi, M. (1984). RNase H-defective mutants of *Escherichia coli*: A possible discriminatory role of RNase H in initiation of DNA replication. Mol. Gen. Genet. 195, 17-22.

Hovius, R., Faber, B., Brigot, B., Nicolay, K., & de Kruijff, B. (1992). On the mechanism of the mitochondrial decarboxylation of phosphatidylserine. J. Biol. Chem. 267, 16790-16795.

Hwang, D. S., Crooke, E., & Kornberg, A. (1990). Aggregated DnaA protein is dissociated and activated for DNA replication by phospholipase or DnaK protein. J. Biol. Chem. 265, 19244-19248.

Ingraham, I. (1987). Effect of temperature, pH, water activity, and pressure on growth. In: *Escherichia coli* and *Salmonella typhimurium*: Cellular and Molecular Biology. Vol. 2 (Neidhardt, F.C., Ingraham, J.L., Low, K.B., Magasanik, B., and Schaechter, M., eds.), pp. 1543-1554. American Society for Microbiology, Washington, DC.

Iuchi, S., Chepuri, V., Fu, H. A., Gennis, R. B., & Lin, E. C. (1990). Requirement for terminal cytochromes in generation of the aerobic signal for the *arc* regulatory system in *Escherichia coli*: Study utilizing deletions and *lac* fusions of *cyo* and *cyd*. J. Bacteriol. 172, 6020-6025.

Jackowski, S., Hsu, L., & Rock, C. O. (1992). 2-Acylglycerophosphoethanolamine acyltransferase/acyl-[acyl-carrier- protein] synthetase from *Escherichia coli*. Methods Enzymol. 209, 111-117.

Jackowski, S., & Rock, C. O. (1986). Transfer of fatty acid from the 1-position of phosphatidylethanolamine to the major outer membrane lipoprotein of *Eshcerichia coli*. J. Biol. Chem. 261, 11328-11333.

Jackson, B. J., Bohin, J.-P., & Kennedy, E. P. (1984). Biosynthesis of membrane-derived oligosaccharides: Characterization of *mdoB* mutants defective in phosphoglycerol transferase I activity. J. Bacteriol. 160, 976-981.

Jackson, B. J., Gennity, J. M., & Kennedy, E. P. (1986). Regulation of the balanced synthesis of membrane phospholipids: Experimental test of models for regulation in *Escherichia coli*. J. Biol. Chem. 261, 13464-13468.

Jackson, B. J., & Kennedy, E. P. (1983). The biosynthesis of membrane-derived oligosaccharides: A membrane-bound phosphoglycerol transferase. J. Biol. Chem. 258, 2394-2398.

Jaworowski, A., Mayo, G., Shaw, D. C., Campbell, H. D., & Young, I. G. (1981). Characterization of the respiratory NADH dehydrogenase of *Escherichia coli* and reconstitution of NADH oxidase in *ndh* mutant membrane vesicles. Biochemistry 20, 3621-3628.

Johnson, J. E., Kalmar, G. B., Sohal, P. S., Walkey, C. J., Yamashita, S., & Cornell, R. B. (1992). Comparison of the lipid regulation of yeast and rat CTP: phosphocholine cytidylyltransferase expressed in COS cells. Biochem J. 285, 815-820.

Kaback, H. R. (1992). The lactose permease of *Escherichia coli*: A paradigm for membrane transport proteins. Biochim. Biophys. Acta 1101, 210-213.

Kaback, H. R., Jung, K., Jung, H., Wu, J., Prive, G. G., & Zen, K. (1993). What's new with lactose permease. J. Bioenerg. and Biomembr. 25, 627-636.

Kalmar, G. B., Kay, R. J., Lachance, A., & Aebersold, R. (1990). Cloning and expression of rat liver CTP:phosphocholine cytidylyltransferase: An amphipathic protein that controls phosphatidylcholine synthesis. Proc. Natl. Acad. Sci. USA 87, 6029-6033.

Kanfer, J., & Kennedy, E. P. (1963). Metabolism and function of bacterial lipids. J. Biol. Chem. 238, 2919-2922.

Kennedy, E. P. (1987). Membrane-derived oligosaccharides. In: *Escherichia coli* and *Salmonella typhimurium*: Cellular and Molecular Biology. Vol. 1 (Neidhardt, F.C., Ingraham, J.L., Low, K.B., Magasanik, B., and Schaechter, M., eds.), pp.636-679. American Society for Microbiology, Washington, DC.

Killian, J. A., Koorengevel, M. C., Bouwstra, J. A., Gooris, G., Dowhan, W., & de Kruijff, B. (1994). Effect of divalent cations on lipid organization of cardiolipin isolated from *E. coli* strain AH930. Biochim. Biophys. Acta 1189, 225-232.

Kogoma, T., Skarstad, K., Boye, E., von Meyenburg, K., & Steen, H. B. (1985). RecA protein acts at the initiation of stable DNA replication in *rnh* mutants of *Escherichia coli* K-12. J. Bacteriol. 163, 439-444.

Kornberg, A., & Baker, T. A. (1992). DNA Replication. W. H. Freeman and Co.

Kuge, O., Nishijima, M., & Akamatsu, Y. (1986). Phosphatidylserine biosynthetic in Chinese hamster ovary cells. J. Biol. Chem. 261, 5790-5794.

Kuge, O., Nishijima, M., & Akamatsu, Y. (1991). A cloned gene encoding phosphatidylserine decarboxylase complements the phosphatidylserine biosynthetic defect of a Chinese hamster ovary mutant. J. Biol. Chem. 266, 6370-6376.

Kusters, R., Breukink, E., Gallusser, A., Kuhn, A., & de Kruijff, B. (1994). A dual role for phosphatidylglycerol in protein translocation across the *Escherichia coli* inner membrane. J. Biol. Chem. 269, 1560-1563.

Kusters, R., Dowhan, W., & de Kruijff, B. (1991). Negatively charged phospholipids restore prePhoE translocation across phosphatidylglycerol depleted *Escherichia coli* membranes. J. Biol. Chem. 266, 8659-8662.

Lazdunski, C., Baty, D., Geli, V., Cavard, D., Morlon, J., Lloubès, R., Howard, S. P., Knibiehler, M., Chartier, M., Varenne, S., Frenette, M., Dasseux, J.-L., & Pattus, F. (1988). The membrane channel-forming colicin A: Synthesis, secretion, structure, action and immunity. Biochim. Biophys. Acta 947, 445-464.

Lill, R., Dowhan, W., & Wickner, W. (1990). The ATPase activity of SecA is regulated by acidic phospholipids, SecY, and the leader and mature domains of precursor proteins. Cell 60, 271-280.

Lindblom, G., & Rilfors, L. (1989). Cubic phases and isotropic structures formed by membrane lipids—possible biological relevance. Biochim. Biophys. Acta 988, 221-256.

Macnab, R. M. (1987a). Motility and chemotaxis. In: *Escherichia coli* and *Salmonella typhimurium*: Cellular and Molecular Biology. Vol. 2 (Neidhardt, F.C., Ingraham, J.L., Low, K.B., Magasanik, B., and Schaechter, M., eds.), pp. 732-759. American Society for Microbiology, Washington, DC.

Macnab, R. M. (1987b). Flagella. In: *Escherichia coli* and *Salmonella typhimurium*: Cellular and Molecular Biology. Vol. 1 (Neidhardt, F.C., Ingraham, J.L., Low, K.B., Magasanik, B., and Schaechter, M., eds.), pp. 70-83. American Society for Microbiology, Washington, DC.

McIntyre, T. M., Chamberlain, B. K., Webster, R. E., & Bell, R. M. (1977). Mutants of *Escherichia coli* defective in membrane phospholipid synthesis. Effects of cessation and reinitiation of phospholipid synthesis on macromolecular synthesis and phospholipid turnover. J. Biol. Chem. 255, 4487-4493.

Mileykovskaya, E. I., & Dowhan, W. (1993). Alterations in the electron transfer chain in mutant strains of *Escherichia coli* lacking phosphatidylethanolamine. J. Biol. Chem. 268, 24824-24831.

Miller, K. J., & Kennedy, E. P. (1987). Transfer of phosphoethanolamine residues from phosphatidylethanolamine to the membrane-derived oligosaccharides of *Escherichia coli*. J. Bacteriol. 169, 682-686.

Miyazaki, C., Kuroda, M., Ohta, A., & Shibuya, I. (1985). Genetic manipulaiton of membrane phospholipid composition in *Escherichia coli: pgsA* mutants defective in phosphatidylglycerol synthesis. Proc. Natl. Acad. Sci. USA 82, 7530-7534.

Müller, E., Hinckley, A., & Rothfield, L. (1972). Studies of phospholipid-requiring bacterial enzymes. III. Purification and properties of uridine diphosphate glucose: lipopolysaccharide glucosyltransferase. J. Biol. Chem. 247, 2614-2622.

Nikaido, H., & Vaara, M. (1987). Outer membrane. In: *Escherichia coli* and *Salmonella typhimurium*: Cellular and Molecular Biology. Vol. 1 (Neidhardt, F.C., Ingraham, J.L., Low, K.B., Magasanik, B., and Schaechter, M., eds.), pp. 7-22. American Society for Microbiology, Washington, DC.

Nikawa, J.-I., Kodaki, T., & Yamashita, S. (1987). Primary structure and disruption of the phosphatidylinositol synthase gene *of Saccharomyces cerevisiae*. J. Biol. Chem. 262, 4876-4881.

Nikawa, J.-I., Kodaki, T., & Yamashita, S. (1988). Expression of the S. *cerevisiae PIS* gene and synthesis of phosphatidylinositol in *E. coli*. J. Bacteriol. 170, 4727-4731.

Nikoloff, D. M., & Henry, S. A. (1991). Genetic analysis of yeast phospholipid biosynthesis. Ann. Rev. Genet. 25, 559-583.

Nishijima, S., Asami, Y., Uetake, N., Yamagoe, S., Ohta, A., & Shibuya, I. (1988). Disruption of the *Escherichia coli* cls gene responsible for cardiolipin synthesis. J. Bacteriol. 170, 775-780.

Norris, V. (1990). DNA replication in *Escherichia coli* is initiated by membrane detachment of *oriC*. A model. J. Mol. Biol. 215, 67-71.

Ogawa, T., Pickett, G. G., Kogoma, T., & Kornberg, A. (1984). RNase H confers specificity in the *dnaA*-dependent initiation of replication at the unique origin of the *Escherichia coli* chromosome *in vivo* and *in vitro*. Proc. Natl. Acad. Sci. USA 81, 1040-1044.

Ohta, A., & Shibuya, I. (1977). Membrane phospholipid synthesis and phenotypic correlation of an *Escherichia coli* pss mutant. J. Bacteriol. 132, 434-443.

Ohtsuka, T., Nishijima, M., Suzuki, K., & Akamatsu, Y. (1993). Mitochondrial dysfunction of a cultured Chinese hamster ovary cell mutant deficient in cardiolipin. J. Biol. Chem. 268, 22914-22919.

Oliver, D. B. (1993). SecA protein: autoregulated ATPase catalyzing preprotein insertion and translocation across the *Escherichia coli* inner membrane. Mol. Microbiol. 7, 159-165.

Page, M. G. P., Rosenbusch, J. P., & Yamato, I. (1988). The effects of pH on proton sugar symport activity of lactose permease purified from *E. coli*. J. Biol. Chem. 263, 15897-15905.

Pattus, F., Massotte, D., Wilmsen, H. U., Lakey, J., Tsernoglou, D., Tucker, A., & Parker, M. W. (1990). Colicins: Prokaryotic killer-pores. Experientia 46, 180-192.

Pierucci, O., & Rickert, M. (1985). Duplication of *Escherichia coli* during inhibition of net phospholipid synthesis. J. Bacteriol. 162, 374-382.

Raetz, C. R., Carman, G. M., Dowhan, W., Jiang, R. T., Waszkuc, W., Loffredo, W., & Tsai, M. D. (1987). Phospholipids chiral at phosphorus. Steric course of the reactions catalyzed by phosphatidylserine synthase from *Escherichia coli* and yeast. Biochemistry 26, 4022-4027.

Raetz, C. R., & Dowhan, W. (1990). Biosynthesis and function of phospholipids in Escherichia coli. J. Biol. Chem. 265, 1235-1238.

Raetz, C. R. H., Kantor, G. D., Nishijima, M., & Newman, K. F. (1979). Cardiolipin accumulation in the inner and outer membranes of *Escherichia coli* mutants defective in phosphatidylserine synthetase. J. Bacteriol. 139, 544-551.

Rietveld, A. G., Killian, J. A., Dowhan, W., & de Kruijff, B. (1993). Polymorphic regulation of membrane phospholipid composition in *Escherichia coli*. J. Biol. Chem. 268, 12427-12433.

Robinson, N. (1993). Functional binding of cardiolipin to ctyochrome *c* oxidase. Bioenerg. and Biomembr. 25, 153-165.

Ryter, A. (1968). Association of the nucleus and the membrane of bacteria: A morphological study. Bacteriol. Rev. 32, 39-52.

Sankaran, K., & Wu, H. C. (1994). Lipid modification of bacterial prolipoprotein: Transfer of diacylglycerol moiety from phosphatidylclycerol. J. Biol. Chem. 269, 19701-19706.

Schlame, M., & Haldar, D. (1993). Cardiolipin is synthesized on the matrix side of the inner membrane of rat liver mitochondria. J. Biol. Chem. 268, 74-79.

Schlame, M., & Hostetler, K. Y. (1991). Solubilization, purification, and characterization of cardiolipin synthase from rat liver mitochondria: Demonstration of its phospholipid requirement. J. Biol. Chem. 266, 22398-22403.

Schulman, H., & Kennedy, E. P. (1977). Relation of turnover of membrane phospholipids to synthesis of membrane-derived oligosaccharides of *Escherichia coli*. J. Biol. Chem. 252, 4250-4255.

Sekimizu, K., & Kornberg, A. (1988). Cardiolipin activation of DnaA protein, the initiation protein of replication in *Escherichia coli*. J. Biol. Chem. 263, 7131-7135.

Sekimizu, K., Yung, B. Y., & Kornberg, A. (1988). The DnaA protein of *Escherichia coli*. Abundance, improved purification, and membrane binding. J. Biol. Chem. 263, 7136-7140.

Shi, W., Bogdanov, M., Dowhan, W., & Zusman, D. R. (1993). The *pss* and *psd* genes are required for mobility and chemotaxis in *Escherichia coli*. J. Bacteriol. 175, 7711-7714.

Simbeni, R., Tangemann, K., Schmidt, M., Ceolotto, C., Paltauf, F., & Daum, G. (1993). Import of phosphatidylserine into isolated yeast mitochondria. Biochim. Biophys. Acta 1145, 1-7.

Sipos, L., & von Heijne, G. (1993). Predicting the topology of eukaryotic membrane proteins. Eur. J. Biochem. 213, 1333-1340.

Tamai, K. T., & Greenberg, M. L. (1990). Biochemical characterization and regulation of cardiolipin synthase in *Saccharomyces cerevisiae*. Biochim. Biophys. Acta 1046, 214-222.

Tanford, C. (1973). The Hydrophobic Effect: Formation of Micelles and Biological Membranes. J. Wiley and Sons.

Torrey, T., & Kogoma, T. (1987). Genetic analysis of constitutive stable DNA replication in rnh mutants of *Escherichia coli*. Mol. Gen. Genet. 208, 420-427.

Trotter, P. J., Pedretti, J., & Voelker, D. R. (1993). Phosphatidylserine decarboxylase from *Saccharomyces cerevisiae*: Isolation of mutants, cloning of the gene and creation of a null allele. J. Biol. Chem. 268, 21416-21424.

Tunaitis, E., & Cronan, J. E., Jr. (1973). Characterization of the cardiolipin synthetase activity of *Escherichia coli* cell envelopes. Arch. Biochem. Biophys. 155, 420-427.

Ulbrandt, N. D., London, E., & Oliver, D. B. (1992). Deep membrane penetration of SecA involves a partial unfolding event promoted by high temperature and anionic lipids. FASEB J. 6, ABS 481.

van der Goot, F. G., Didat, N., Pattus, F., Dowhan, W., & Letellier, L. (1993). Role of acidic lipids in the translocation and channel activity of colicins A and N in *Escherichia coli* cells. Eur. J. Biochem. 213, 217-221.

Vance, J. E. (1991). Newly made phosphatidylserine and phosphatidylethanolamine are preferentially translocated between rat liver mitochondria and endoplasmic reticulum. J. Biol. Chem. 266, 89-97.

Vanden Boom, T., & Cronan, J. E., Jr. (1989). Genetics and regulation of bacterial lipid metabolism. Ann. Rev. Microbiol. 43, 317-343.

von Heijne, G. (1992). Membrane protein structure prediction. Hydrophobicity analysis and the positive-inside rule. J. Mol. Biol. 225, 487-494.

von Meyenburg, K., Boye, E., Skarstad, K., Koppes, L., & Kogoma, T. (1987). Mode of initiation of constitutive stable DNA replication in RNase H-defective mutants of *Escherichia coli* K-12. J. Bacteriol. 169, 2650-2658.

Walsh, J. P., & Bell, R. M. (1992). Diacylglycerol kinase *from Escherichia coli*. Methods Enzymol. 209, 153-162.

Wieslander, A., Christiansson, A., Rilfors, L., & Lindblom, G. (1980). Lipid bilayer stability in membranes. Regulation of lipid composition in *Acholeplasma laidlawi* as governed by molecular shape. Biochemistry 19, 3650-3656.

Wilmsen, H. U., Pugsley, A. P., & Pattus, F. (1990). Colicin N forms voltage- and pH-dependent channels in planar lipid bilayer membranes. Eur. Biophys. J. 18, 149-158.

Wu, H. C. (1987). Posttranslational modification and processing of membrane proteins in bacteria. In: Bacterial Outer Membrane as Model Systems. (Inouye, M., ed.), pp. 37-71. Academic Press.

Wu, H. C., Tokunaga, M., Tokunaga, H., Hayashi, S., & Giam, C.-Z. (1983). Posttranslational modification and processing of membrane lipoproteins in bacteria. J. Cell Biochem. 22, 161-171.

Xia, W., & Dowhan, W. (1995). *In vivo* evidence for the involvement of anionic phospholipids in the initiation of DNA replication in *E. coli*. Proc. Natl. Acad. Sci. USA 92, 783-787.

Yem, D. W., & Wu, H. C. (1978). Physiological characterization of an *E. coli* mutant atlered in the structure of murein lipoprotein. J. Bacteriol. 133, 1419-1426.

Yoshida, S., Tamiya, K. K., & Kojima, K. (1989). Interaction of DNA polymerases with phospholipids. Biochim. Biophys. Acta 1007, 61-66.

Yung, B. Y.-M., & Kornberg, A. (1988). Membrane attachment activates DnaA protein, the initiation protein of chromosome replication in *Escherichia coli*. Proc. Natl. Acad. Sci. USA 85, 7202-7205.

Zinser, E., Sperka, G. C. D., Fasch, E. V., Kohlwein, S. D., Paltauf, F., & Daum, G. (1991). Phospholipid synthesis and lipid composition of subcellular membranes in the unicellular eukaryote *Saccharomyces cerevisiae*. J. Bacteriol. 173, 2026-2034.

STRUCTURE, BIOSYNTHESIS, PHYSICAL PROPERTIES, AND FUNCTIONS OF THE POLAR LIPIDS OF *CLOSTRIDIUM*

Howard Goldfine

Advances in Lipobiology
Volume 2, pages 109-142.

ISBN 0-7623-0205-4

ABSTRACT

Clostridium is a very ancient and phylogenetically deep group of Gram-positive, spore-forming bacteria. Although the polar lipids of relatively few species have been studied, they have already revealed important and often novel information concerning structure, biosynthesis, and functions of these major membrane components. The polar lipids of most of the species that have been examined are characterized by relatively high concentrations of 1-(1′-alkenyl) 2-acyl glycerophospholipids (plasmalogens). In addition, some species have unique glycerol acetals and phosphatidylglycerol acetals of the major plasmalogens. Physical studies on these ether lipids have provided information on their conformation and phase behavior. These studies have also revealed the presence of regulatory mechanisms in clostridia that control both membrane fluidity and lipid polymorphism.

I. INTRODUCTION

As a consequence of wide variability in peptidoglycan, protein, and macroamphiphile compositions, the cell envelopes of prokaryotes are highly diverse. Membrane amphipathic lipids also vary widely among the major groups of prokaryotes and achieving an understanding of their individual functions presents a major challenge and an opportunity to lipid biochemists. From this viewpoint, the study of the structures, physical properties, and functions of clostridial lipids has provided insights into biochemical problems of wider scope and significance. Clostridia, like other Gram-positive bacteria, have a single cell membrane. Since there are no known lipid signaling pathways in prokaryotes we may assume for the present that the functions of their membrane lipids are confined to a small subset of potential functions for these compounds. Among these are: (1) To serve as the non-covalently bonded monomers of a semi- permeable membrane, (2) to provide a suitable matrix for the activities of membrane proteins, and (3) to serve as anchors for large covalently-linked structures which extend from the cell membrane, for example, lipoteichoic acids. In addition, some minor lipids have been identified as precursors of other lipids, while others have as yet undiscovered functions. Lastly, some major lipids also serve as precursors of other lipids, a process which will be discussed in greater detail in this review.

Recent studies on the phylogeny of the genus *Clostridium* using primarily 16S rRNA and some 5S rRNA sequences have revealed that these prokaryotes represent a very ancient and phylogenetically deep grouping which is likely to be divided into several new genera at some future time. Indeed, the basic description that holds this group together: Gram-positive, obligately anaerobic, non-sulphate-reducing, spore-forming rods (Cato et al., 1986), covers a wide diversity of biochemical, physiological, and metabolic characteristics. Some clostridia are saccharolytic, others are proteolytic, some are neither, and some are both (Cato and Stackebrandt, 1989). They are also widely diverse in cell wall (Schleifer and Stackebrandt, 1983) and fatty acid compositions (O'Leary and Wilkinson, 1988). It should not, therefore, be surprising that examination of only a few species has revealed major differences in their polar lipid compositions.

A recent study concluded that the rRNA homology group I of *Clostridium* (Johnson and Francis, 1975) is a phylogenetically distinct cluster which encompasses the type species *Clostridium butyricum*, and several other species, among which are *Clostridium beijerinckii, and Clostridium acetobutylicum* (Lawson et al., 1993). These three species are the best documented clostridia in terms of their polar lipid compositions, whereas the lipids of most of the other approximately 100 species remain largely unknown, thus representing a major gap in our knowledge of prokaryotic membrane lipids (O'Leary and Wilkinson, 1988).

II. POLAR LIPIDS OF *CLOSTRIDIUM*

A characteristic feature of the lipids of the genus *Clostridium* and of a number of other obligately anaerobic organisms is the presence of plasmalogens, which differ in having a 1-(1'-alkenyl) chain in place of the 1-acyl chain of the more common diacyl glycerolipids. Not all clostridia have plasmalogens, but among those that do they are accompanied by the corresponding diacyl phosphoglycerides or diacyl glycosylglycerides and the proportions of the plasmalogen and diacyl types vary from lipid-to-lipid and among species. When I refer to a given lipid I will use the modifier (diradyl) to include both types, and the abbreviations PX and PlaX to refer specifically to the diacyl or plasmalogen forms of phosphatidyl-X. Thus in addition to the major diacyl phosphoglycerides: phosphatidylethanolamine (PE), phosphatidyl-N-monomethylethanolamine (PME), phosphatidylglycerol (PG) and bisphosphatidylglycerol (cardiolipin) (CL), the corresponding plasmalogens, PlaE, PlaME, PlaG, and PlaCL (in which one or both halves of the molecule may be alkenylacyl) have also been found. Quantitative compositional data have recently been reviewed (O'Leary and Wilkinson, 1988; Goldfine, 1993). As in higher organisms, the alk- 1-enyl chains of bacterial plasmalogens are linked at the *sn*-1 oxygen of the glycerophosphate backbone (Hagen and Goldfine, 1967; Oulevey et al., 1986).

Along with PE, PG, CL and the corresponding plasmalogen forms of these lipids, a glycerol acetal of PlaE (GAPlaE) (1-(1'- glyceroalky) 2-acyl *sn*-glycero-3-phosphoethanolamine), is a major lipid of *C. butyricum* (Matsumoto et al., 1971) (Figure

1A), and the homologous glycerol acetal of PlaME (GAPlaME) is a major lipid of *C. beijerinckii* in which PME and PlaME largely replace the PE and PlaE of *C. butyricum* (Khuller and Goldfine, 1974; Johnston and Goldfine, 1983). Glycerol acetals of PlaE have been found in other clostridia including *C. acetobutylicum* (Johnston and Goldfine, 1983). Phosphatidylglycerol acetals of PlaE (PGAPlaE) (Figure 1B) (Johnston and Goldfine, 1988), of PlaG (PGAPlaG) (Figure 1C) and of PlaCL (PGAPlaCL) (Figure 1D) have recently been described (Johnston et al., 1994; Johnston and Goldfine, 1994). PGAPlaE and PGAPlaCL are normally minor lipids of *C. butyricum*. The latter is more prominent in *C. innocuum*, from which it has been isolated and characterized (Johnston et al., 1994; Johnston & Goldfine, 1994). PGAPlaG is present in trace amounts in both species. Other minor phospholipids of clostridia include phosphatidylserine (PS) and plasmenylserine (PlaS), which are intermediates in PE and PlaE biosynthesis in *C. butyricum*, *O*-acyl-PG (Koga and Goldfine, 1984), and phosphatidic acid (Goldfine and Hagen, 1972). *O*-aminoacyl esters of PG were identified as major phospholipids in *Clostridium perfringens* (MacFarlane, 1962).

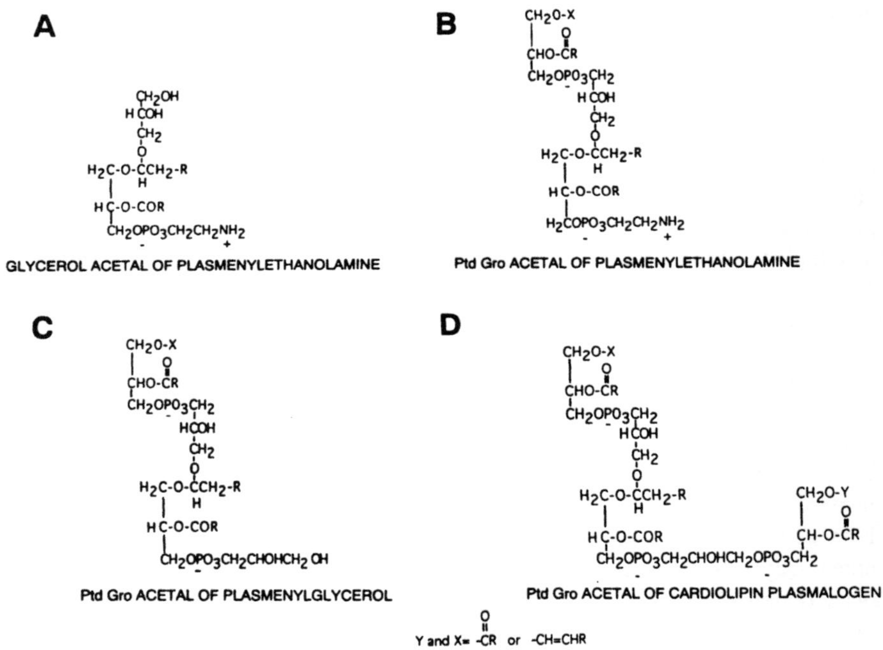

Figure 1. Structures of acetal ether lipids from clostridia. A. Glycerol acetal of plasmenylethanolamine. B. Phosphatidylglycerol acetal of plasmenylethanolamine. C. Phosphatidylglycerol acetal of plasmenylglycerol. D. Phosphatidylglycerol acetal of cardiolipin plasmalogen.

Clostridia, like other Gram-positive bacteria, can have significant levels of glycosyldiradylglycerols which are present in both diacyl and plasmalogen forms. *C. acetobutylicum* strain DSM 1731 grown under phosphate limitation has a galactosyl diradylglycerol and a β-glucosyl- galactosyl diradylglycerol. All of the glycolipid in this strain was reported to be of the alkenylacyl type (Oulevey et al., 1986). Mono and diglycosyldiradylglycerols were also found in *C. acetobutylicum* ATCC 4259 grown in media with phosphate supplementation (Johnston and Goldfine, 1992). In *C. innocuum* ATCC 14501 the major glycosyldiradylglycerols are d-Glc*p*(α1-3)radyl$_2$Gro and D-Gal*p*(α1-2)D-Glc*p*(α1-3)radyl$_2$Gro. Both glycolipids have some 1-O-[alk-1-enyl]-2-O- acyl species, in addition to diacyl species (Johnston et al., 1994). *C. acetobutylicum* 4259 (Johnston and Goldfine, 1992) and *C. innocuum* (Johnston et al., 1994) have significant amounts of phosphoglycolipids and two of those in the latter species have been identified as 2′-amino-1′,3′-dihydroxypropane-3′-P-6-D-Galp(α1-2)d-Glc*p*(α1-3)radyl$_2$Gro and a derivative of this lipid containing an acyl chain esterified to *O*-6 of the glucopyranosyl ring. These represent approximately 50 and 15 mol% of the total lipids, respectively (Johnston et al., 1994; Fischer et al, 1994).

III. LIPID BIOSYNTHESIS IN *CLOSTRIDIUM*

A. Synthesis of Diacylphosphoglycerides

The biosynthesis of the acyl forms of the polar lipids of *C. butyricum* and *C. beijerinckii* has recently been reviewed (Goldfine, 1993). The pathways involved have been elucidated by studies of the individual enzymes, and have been supported by isotopic labeling experiments with growing cultures (Baumann et al., 1965; Hagen and Goldfine, 1967; Koga and Goldfine, 1984; MacDonald and Goldfine, 1990). These pathways are the same as those elucidated for *E. coli* (Cronan and Rock, 1987; Pieringer, 1989), in which *sn*-glycerol-3-P is acylated in two distinct steps to form phosphatidic acid which is then reacted with CTP to form CDP-diacylglycerol. Unlike *E. coli*, the glycerophosphateacyltransferase is specific for acyl derivatives of the acyl carrier protein and does not use acyl-CoAs as acyl donors (Ailhaud et al., 1967; Goldfine et al., 1967; Goldfine and Ailhaud, 1971). The pathway to the amino-containing phospholipids and that leading to acidic phospholipids then diverge by reaction of CDP- diacylglycerol with L-serine to form PS or with *sn*-glycerol-3-P to form phosphatidylglycerolphosphate (PGP), respectively (Silber et al., 1980). PS is decarboxylated to yield PE (Verma and Goldfine, 1985), and in *C. beijerinckii* ATCC 6015 and in other strains of this species (Johnston and Goldfine, 1983), PE is N-methylated to form PME (Baumann et al., 1965). This reaction has not been demonstrated *in vitro*. (Strain ATCC 6015 was formerly designated *C. butyricum*, but is now classified as *C. beijerinckii* (Johnston and Goldfine, 1983), and will be so designated in the remainder of this review). As in *E. coli*, a PGP phosphatase removes the terminal phosphate to form PG, which can condense with another molecule of PG to form CL (Walton and Goldfine, 1987; Morii and Goldfine, 1990). The essentiality of these reactions for phospholipid biosynthesis in clostridia has not been confirmed by genetic experiments.

B. Anaerobic Biosynthesis of Plasmalogens

An oxygen-dependent pathway for plasmalogen synthesis in eukaryotic cells was elucidated beginning in the late 1960s by work in several laboratories, notably those of Snyder, Hajra, and Friedberg (Snyder, 1972; Hajra, 1983; Paltauf, 1983a). In this pathway long- chain alcohols, formed by reduction of long-chain fatty acyl-CoAs, displace the acyl chain of dihydroxyacetone phosphate (acyl-DHAP) to form alkyl-DHAP. After reduction of the keto group and acylation to form 1-alkyl-2-acyl *sn*-glycerol-3-P, this intermediate is converted to 1- alkyl-2-acyl-phosphoglycerides in reactions that parallel those for the formation of PE and PC. Lastly, the 1-0-alkyl chain is desaturated in the presence of O_2, NAD(P)H, and cytochrome b_5 reductase by an alkylacyl glycerylphosphorylethanolamine desaturase found in peroxisomes (Snyder, 1991).

The more ancient biosynthetic pathway to plasmalogens in anaerobes differs fundamentally from that in eukaryotic cells. Foremost among these differences is the absence from the environment of oxygen and the absence of the cellular cytochromes required for the desaturase reaction. In this regard there are strong parallels with the synthesis of unsaturated fatty acids in eukaryotes and prokaryotes. In the former, a desaturase similar to the alkylacyl glycerylphosphorylethanolamine desaturase removes two hydrogens from central carbons of a long-chain fatty acid. In most bacteria, the double bond is introduced during the elongation of medium-chain fatty acids, a process that does not require molecular oxygen (Bloch et al., 1961). Moreover, DHAP has been ruled out as an intermediate in anaerobic bacterial and rumen protozoal plasmalogen synthesis by experiments in which [2-^{3}H]glycerol was fed to *C. beijerinckii* (Hill and Lands, 1970), *Megasphaera elsdenii, Veillonella parvula* and to two anaerobic rumen protozoa (Prins and Van Golde, 1976). These experiments showed enrichment of ^{3}H relative to either ^{14}C-glycerol or to ^{3}H from 1,3 labeled glycerol in the glycerol moieties of diacyl- and alkenylacylphospholipids. An obligatory role for DHAP would have resulted in the loss of ^{3}H at C-2 of glycerol.

Studies on the incorporation of long-chain alcohols into clostridial plasmalogens have yielded mixed, but largely negative results (Goldfine and Hagen, 1972; Paltauf, 1983a). and experiments with alkyl glycerols (Goldfine and Hagen, 1972; Paltauf, 1983a) or alkyl- glycerylphosphorylethanolamine (Paltauf, 1983a) have likewise shown no incorporation into plasmalogens by clostridia. In contrast, long-chain fatty acids fed to *C. beijerinckii* ATCC 6015 were readily incorporated into the 1-alkenyl chains of plasmalogens (Baumann et al., 1965), as were the shorter chain precursors of long-chain fatty acids (Baumann et al., 1965; Hagen and

Goldfine, 1967). When carboxyl-labeled octanoic or decanoic acids were fed they were incorporated into saturated and unsaturated alkenyl and acyl chains and the specific activities of both types of chains were similar. As is the case for acyl chains, long-chain saturated precursors were predominantly incorporated into alkenyl chains without a mid-chain double bond (Baumann et al., 1965). These data indicated that the alkenyl chains shared similar biosynthetic pathways with the acyl chains. Furthermore, the monounsaturated aldehydes derived from the alkenyl chains had the same mixtures of positional isomers as the corresponding acyl chains in similar ratios (Goldfine and Panos, 1971). Thus it was reasonable to conclude that the cellular alkenyl chains are derived from cellular fatty acids. Similar conclusions were reached in kinetic studies of ^{14}C-acetate incorporation into the neutral lipids and phospholipids of *C. beijerinckii* (Hagen and Goldfine, 1967). In these studies, neutral alkenylether lipids did not appear to be precursors of the alkenylacyl glycerophospholipids, but a small pool of neutral precursor(s) could not be ruled out in these isotopic labeling experiments. Consistent with a potential role for long-chain aldehydes as precursors of the phospholipid alkenyl chains, was the important finding that 15% of ^{3}H from 1-^{3}H-1-^{14}C-palmitaldehyde was incorporated into the polar plasmalogens of *C. beijerinckii.* Thus, a long chain aldehyde can be incorporated into the alkenyl chain without undergoing prior oxidation to a fatty acid (Hagen and Goldfine, 1967). A long-chain acyl-CoA reductase activity was subsequently demonstrated in cell extracts of *C. beijerinckii* (Day et al., 1970) and partially purified (Day and Goldfine, 1978). Crude extracts were also able to reduce the long-chain aldehyde formed to the corresponding alcohol (Day et al., 1970; Day and Goldfine, 1978), but given the failure to obtain incorporation of alcohols in whole cells, this is a less likely intermediate. It should also be noted that one of the two deuterons of [2,2-$^2H_2$]palmitate or [2,2-$^2H_2$]oleate was incorporated into the sn-1 alkenyl chain of the plasmalogens of *C. butyricum*, which eliminates mechanisms that would lead to labilization of these protons (Malthaner et al., 1987b).

The kinetics of plasmalogen formation in clostridia and in other anaerobic bacteria have been studied by pulse and pulse-chase experiments with $^{32}P_i$. In all these experiments incorporation of P_i into diacyl phosphatides preceded incorporation into the corresponding plasmalogens. This was shown for *C. beijerinckii* ATCC 6015, which has PME and PlaME in addition to PE and PlaE (Baumann et al., 1965), and for *C. butyricum* IFO 3852 (Koga and Goldfine, 1984). In the latter species, experiments were also done with cells treated with hydroxylamine, an inhibitor of PS decarboxylase, which resulted in the cellular accumulation of radioactive PS and PlaS in a ratio ~15:1. When the inhibitor was removed, radioactivity was observed to flow into PE and more slowly into PlaE. Given the size and rate of labeling of the latter pool, it is not likely to have arisen from the small pool of labeled PlaS accumulated during hydroxylamine inhibition. Similar experiments were performed with *Megasphaera elsdenii*, a Gram-negative, anaerobic rumen bacterium. This species has a simple and markedly different phospholipid composition from those of clostridia. The principal lipids are: PlaE, 54%; PE, 8%; PlaS 26% and PS 10%. When growing *M. elsdenii* were pulsed with $^{32}P_i$, radioactivity was observed to accumulate first in PS and PlaS, with the latter lagging slightly behind the former. Label in both PE and PlaE appeared after an approximately 15 min lag. When PS-decarboxylase and growth were inhibited by addition of increasing concentrations of hydroxylamine, $^{32}P_i$ was observed to accumulate in increasing amounts in PlaS at the expense of PlaE, suggesting a precursor-product relationship between these two lipids (Prins et al., 1974).

Pulse and pulse-chase experiments with isotopic phosphate were also reported for *Selenomonas ruminantium*, another anaerobic, Gram-negative rumen bacterium containing PlaE, 45%; PE, 27%; PS, 17.5% and PlaS, 6.5%. In this organism PlaS does not predominate over PS, as it does in *M. elsdenii*. After growth in the presence of $^{32}P_i$, followed by a chase, radioactivity in PE and PS declined rapidly, but that in PlaE and PlaS remained relatively constant (Watanabe et al., 1984). However, when the side-chains of the lipids were labeled in a pulse-chase experiment with [^{14}C]caproic acid, radioactivity in PE and PS declined, but not as rapidly as in the $^{32}P_i$ pulse-chase, and radioactivity increased in PlaE while that in PlaS remained at a constant low level. The authors concluded that PlaE was probably formed from PlaS by decarboxylation. From the results with [^{14}C]caproate and [2-^{3}H]glycerol pulse-chase experiments, these workers argued for the presence of a large neutral lipid pool of 1-O- alk-1′-enyl-2-acylglycerols. This conclusion was supported by experiments with the fatty acid synthesis inhibitor, cerulenin. Koga and Goldfine (1984) reconciled their findings with *C. butyricum* with those obtained with *M. elsdenii* by consideration of the respective lipid pool sizes. The same arguments apply to *S. ruminantium*. In both Gram-negative species, there are large pools of PS and PlaS. If PlaS is derived from PS, possibly by way of a neutral lipid pool, and if both are decarboxylated at approximately equal rates, then the bulk of PlaE could come from PlaS, as proposed. This, however, is not likely to be the case in *C. butyricum*, since the kinetics of labeling of PlaE are not consistent with the very small pool of PS and PlaS.

In *E. coli* and in Gram-positive bacteria, phospholipid biosynthetic intermediates such as PA, CDP-diacylglycerol, PGP, and PS are present in very low amounts and have only been recognized in studies with cell extracts, in partially inhibited cell extracts or whole cells, or by the study of biosynthetic mutants (Raetz, 1986). Thus, intermediates of plasmalogen synthesis in obligately anaerobic bacteria are likely to be present in low amounts and to exhibit high turnover rates. The recently discovered PG acetal family of ether lipids in clostridia meets these criteria. PGAPlaE (Fig. 1B) was discovered in studies of *C. butyricum* grown in the presence of petroselinic acid (*cis*-6-18:1), under conditions of fatty acid auxotrophy (Goldfine et al., 1987b). In these cells, PGAPlaE represented 15 to 25% of total lipid phosphorus compared to cells that synthesize their own fatty acids which have about 5% of this lipid. Subsequent studies on the biosynthesis of PGAPlaE showed that like PlaE, the PlaE moiety of PGAPlaE was labeled slightly more slowly than PE after a pulse with $^{32}P_i$, and its specific activity was always lower than that of PE. The specific activity of the PlaE moiety of PGAPlaE was slightly higher or equal to that of PlaE at 10 min after the pulse, but lower thereafter. Pulse-chase experiments with $^{32}P_i$ in the absence and presence of hydroxylamine gave data consistent with the derivation of PlaE from PGAPlaE, but both could have been derived from a common precursor, possible a subset of the diacyl PE molecules, which turns over rapidly (MacDonald and Goldfine, 1990).

It is evident that there remain many unknown features of the anaerobic pathway to plasmalogens. Foremost is the mechanism of formation of the alkenyl ether bonds of plasmalogens and the ether bonds of the glycerol or PG acetals of plasmalogens. The very slow kinetics of labelling of GAPlaE in *C. butyricum* (Koga and Goldfine, 1984; MacDonald and Goldfine, 1990) argues strongly against a precursor role in plasmalogen formation for this lipid. A potential biosynthetic connection between the PG acetals of the plasmalogens and the plasmalogens themselves is seductive, but this idea has not been adequately tested. In addition, plasmalogens are formed anaerobically in Gram-negative bacteria in which neither PG nor acetal lipids have been found. To postulate a third mechanism for plasmalogen synthesis at this time is problematic but it must be considered a possibility in future work.

IV. PHYSICAL PROPERTIES OF CLOSTRIDIAL ETHER LIPIDS

Plasmalogens are not universally distributed in nature. Aerobic and facultatively anaerobic bacteria, fungi, and plant tissues, with few exceptions are either devoid of these lipids or have very low levels (Horrocks and Sharma, 1982). Saturated ether lipids with long-chain polyisopranoids joined to the *sn*-2 and *sn*-3 oxygens of glycerol are signature lipids of archaebacteria, which do not have the corresponding alkenyl ether lipids. The major groups of organisms with plasmalogens are anaerobic bacteria, certain groups of protozoa, and most tissues of higher eukaryotes (Horrocks and Sharma, 1982). The evolutionary gap between bacteria and eukaryotes suggests that plasmalogens have evolved at different times and very possibly to fulfill different as well as common purposes. The polar head group and hydrocarbon chain compositions of these lipids influence their physical behavior and even their chemical properties, and these must be considered in any discussion of their potential roles in the cell.

A. Packing Properties

Monolayer studies have shown that the 1-alk-1′-enyl ether linkage permits closer packing of choline plasmalogens than the corresponding diacyl lipids. The cross sectional area of the alkylacyl lipid was found to be slightly larger than that of the plasmalogen, but considerably less than that of the diacyl compound. However, the three classes of the glycerophosphoethanolamine lipids did not differ in packing densities in monolayers, presumably because their polar head group interactions were dominant (Smaby et al., 1983). Most of the available studies on the permeability of artificial membranes containing plasmalogens have been done with the choline-containing species, which are present in relatively few bacteria (Horrocks and Sharma, 1982). Liposomes made with PlaC were slightly more permeable to glucose and Rb+, and considerably more permeable to Cl- than liposomes made with PC (Paltauf, 1983b). (Conflicting results were recently reported by Chen and Gross, 1994). In considering the functions of plasmalogens in bacteria, the lack of data on the permeability of membranes composed of PlaE mixed with other lipids such as PG or PS is unfortunate.

B. Conformation and Motion

In diacylphosphoglycerides the *sn*-2 acyl chain is bent so that the first and second carbon atoms lie essentially parallel to the plane of the bilayer and the more distal carbons are essentially perpendicular to the bilayer plane as first shown for a dilauroyl PE (Hitchcock et al., 1974). This leads to inequivalence of the two protons on carbon-2 of the chain which can be detected by quadrupole deuterium NMR (Seelig and Seelig, 1980) (Table 1). The *sn*-1 chain of diacyl phosphoglycerides does not have this bend. Deuterium NMR experiments on (diradyl) PlaE or (diradyl) PlaME labeled biosynthetically at the *sn*-2 acyl chain with [2,2-$^2H_2$]palmitic acid in *C. butyricum* or *C. beijerinckii*, respectively (Malthaner et al., 1987b) and on PlaE labeled semisynthetically with deuterium on the same chain at C-2 (Malthaner et al., 1987a) revealed a quadrupole splitting of 23 to 25 kH for the plasmalogens (Table 1). The inequivalence of the two deuterons on the *sn*-2 acyl chains of diacylphosphoglycerides is absent and the quadrupolar splitting approaches that found typically for the *sn*-1 chain of diacyl lipids, suggesting that the proximal portion of the *sn*-2 chain of these plasmalogens is perpendicular to the bilayer surface. When deuterium was present at carbons 3 and 4, the *sn*-2 chains of (diradyl) PE produced only single quadrupole splittings indicating that these protons were very similar or identical for both the alkenylacyl and diacyl lipids (Malthaner et al., 1987b). When labeled biosynthetically with [2,2-$^2H_2$] oleic acid, which was predominantly incorporated into the *sn*-1 alkenyl and acyl chains of (diradyl) PE, quadrupole splittings of 8.1 and 26.0 kH were observed (Table 1). The smaller splitting, which was assigned to the alkenyl chain deuteron, suggested that the 1,2-double bond is aligned essentially parallel with the long axis of the hydrocarbon chain. Similar results were obtained with a mixed lipid fraction from *C. butyricum* containing (diradyl) PG and (tetraradyl) CL in which the alkenyl chains labeled at C-2 gave a quadrupolar splitting of 6.7 kH (Malthaner et al., 1987b). The combined results of studies with PlaE led to the conclusion that both chains are perpendicular to the bilayer plane at all segments and the corollary that the sn-2 chain of the plasmalogen is more fully extended than it is in PE. Semisynthetic labeling of the

Table 1. Quadrupolar Splittings (ΔvQ) Detected by Deuterium Magnetic Resonance of the C-2 Segment Of The *sn*-1 and *sn*-2 Chains of Plasmalogens Compared with Those of Diacyl Phosphoglycerides[a]

Phospholipid	*sn-1*	*sn-1(1)*	*sn-2(2)*	*Reference*
Diacyl-PE	25.0	13.0	19.0	Malthaner et al., 1987b
PlaE	8.1	25.0		Malthaner et al., 1987b
PlaE	N.D.[b]	23.2		Malthaner et al., 1987a
PlaME	N.D.	25.5		Malthaner et al., 1987b
PlaC	N.D.	13.2		Malthaner et al., 1987a
POPC	25.8	10.8	16.4	Malthaner et al., 1987b

Notes: [a] All measured at 30°C
[b] Not determined

C-2 segment of PlaC with deuterium produced quadrupolar NMR splittings of 15.1 and 19.3 kH at 0°C and a single splitting of 13.2 kH at 30°C (Malthaner et al., 1987a). Similar studies with PlaC resulted in more distinct splittings at 41°C, which were not as large as those obtained with similarly labeled PC (Pak et al., 1987). The reduction, but not elimination of the inequivalence of the two deuterons suggested either greater motional freedom or less bending of the *sn*-2 chain of PlaC than that of PC. In more recent studies employing truncated driven nuclear Overhauser enhancement, which provided estimates of internuclear distances, it was concluded that the carbon atoms of the proximal portion of the *sn*-2 chain of PlaC are nearly coplanar, in register with, (i.e., the carbon atoms of the proximal portion of each chain are directly adjacent to those of the neighboring chain), and parallel to the *sn*-1 aliphatic chain (Han and Gross, 1990). The proposed conformational model is compatible with that derived from deuterium NMR studies of PlaE (Malthaner et al., 1987a,b) and PlaC (Pak et al., 1987), but the results with PlaE suggest greater equivalence of the two C-2 protons.

Synthetic phospholipid bilayers and biological membranes including those of *E. coli* (Gally et al., 1979), and *A. laidlawii* (Stockton et al., 1977), display a characteristic hydrocarbon chain order profile which has been studied by a variety of physical techniques including the measurements of NMR quadrupolar splittings observed with deuterium-labeled acyl chains (Seelig and Seelig, 1980). These studies have shown a region of almost constant deuterium order parameter [S_{mol}] extending from the C-4 segment to the C-9 segment. [S_{mol}] then decreases towards the liquid like ends of the chains. These experiments were done with either synthetic diacyl phosphoglycerides or membranes containing diacyl phosphoglycerides, as is the case for *E. coli*, or mixtures of diacyl phosphoglycerides and glycosyl diacylglycerols, as is the case for *A. laidlawii*. Similar studies were carried out with the phospholipids, membranes, and whole cells of *C. butyricum* grown with palmitic acid labeled with deuterium at each of five different carbons from C-4 to C-13 (Johnston et al., 1987). The order profiles derived from the quadrupolar splittings for the phospholipid liposomes, cells and isolated membranes were similar to those obtained with diacyl lipids, with nearly constant [S_{mol}] from C-4 to C-9 and a large decrease in the order parameter at C-12 and C- 13. The [S_{mol}] values of the hydrocarbon chains in the whole cells and membranes were generally somewhat lower than those of the liposomal phospholipids (Figure 2), which is considered to result from a disordering effect of the non-lipid constituents of the cell membranes (Seelig and Seelig, 1980). The absolute [S_{mol}] value for the C-7 segment of the liposomes prepared from *C. butyricum* phospholipids, 0.47, was similar to that obtained for the C-6/C-7 segment of POPC, ~0.40, a surprising result considering the complexity of the *C. butyricum* lipids with their more heterogeneous chain compositions and three major phospholipid classes: diacyl, alkenylacyl and the glycerol acetal of PlaE.

C. Thermotropic Behavior of Ether Lipids

Gel to Liquid Crystalline Phase Transitions

Complete chemical synthesis of plasmalogens with defined alkenyl and acyl chains and the natural configuration has presented a severe test for organic chemists. Therefore, the initial approach to defining the gel to liquid crystalline (G → L) phase transition temperature of a plasmalogen was to isolate PlaE from *C. butyricum* grown under conditions of fatty acid auxotrophy with elaidic acid, which supported growth well and was incorporated into both hydrocarbon chains. The isolated PlaE had 99% 18:1 alkenyl chains and 92% 18:1 acyl chains. The near absence of cyclopropane chains indicated that the 18:1 chains were predominantly trans, as shown for *C. beijerinckii* (Khuller and Goldfine, 1975). Elaidate-enriched PlaE gave an endothermic transition at 33°C on heating (Tm), and an exothermic transition at 27°C on cooling compared with the reported values of 38° and 35°C, respectively for dielaidoyl PE (Table 2) (Goldfine et al., 1981). The thermotropic properties of a semisynthetic PlaE with 84% saturated sn-1 alkenyl chains, and *cis*-9-18:1 acyl chains were compared to those of diacyl PE with similar *sn*-1 acyl chains and the same *sn*-2 acyl chains. The G → L transition temperatures for PlaE and PE were 26° and 30°C, respectively. An alkylacyl PE with similar chain compositions gave a G → L transition temperature of 29.5°C (Table 2) (Lohner et al., 1984). Thus, the presence of the sn-1 alk-1-enyl chain in PlaE lowers the G → L transition temperature 4 to 6°C compared to PE.

The phase behavior of GAPlaE proved to be of greater interest. When this lipid was highly enriched in *trans*-9-18:1 (elaidic) chains, it produced a G → L transition at 34°C; however, on cooling, the L → G transition displayed supercooling with a Tm of 13°C. A similarly large hysteresis was observed with oleate-enriched GAPlaE, which melted at 12°C and formed a gel phase at -6°C (Table 2) (Goldfine et al., 1981). Slightly higher Tm values were obtained with another preparation of oleate-enriched GAPlaE (Goldfine et al., 1987a).

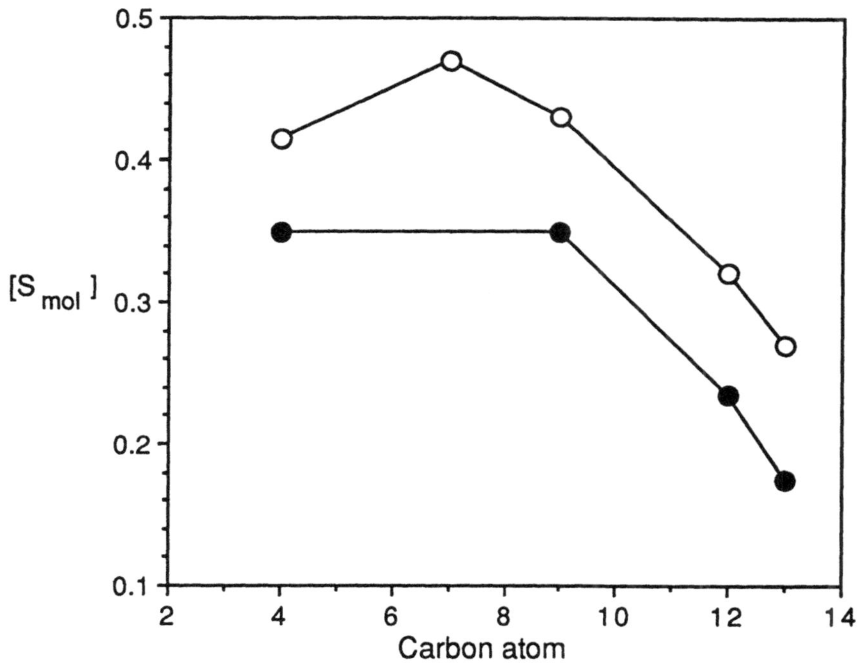

Figure 2. Variation of the deuterium order parameter [S_{mol}] with the acyl chain segment position for *C. butyricum.* ○, Lipid dispersion; ●, membrane preparation. Measuring temperature, 37°C (Johnston et al., 1987).

Lamellar to Non-Lamellar Transitions

A significant fraction of the lipids found in biological membranes readily form non-lamellar phases such as the reversed hexagonal (H_{II}) and cubic phases, at or near physiological conditions of temperature, pH, and salt concentration. Lipids that form the H_{II} phase generally have the shape of a truncated cone, in which the polar head group lies at the apex, because their acyl chains subtend a larger cross-sectional area than their polar head groups. Bilayer-preferring lipids, on the other hand tend to have a cylindrical shape (Israelachvili et al., 1980; Cullis and De Kruijff, 1979). Among the lipids of prokaryotes that can spontaneously form non-lamellar phases are: PE, monoglycosyldiacylglycerols, and cardiolipin (Seddon, 1990; Lindblom and Rilfors, 1989). All of these lipids are found in clostridia and other anaerobes.

The first indication that alkenyl ether lipids are more prone to form non-lamellar phases was the observation that (diradyl) PE (77% PlaE) isolated from bovine brain

Table 2. Gel to Liquid Crystalline Phase Transition Temperatures of Plasmalogens and Related Lipids

Phospholipid	*Heating (Tm)°C*[a]	*Cooling (Tm)°C*	*ΔH kCal/mol*	*Reference*
DEPE[b]	38	34.5, 36.5	4-7	van Dijck et al., 1976; Jackson & Sturtevant, 1977; Yang et al., 1979
DEPlaE[b]	33	27	5.7	Goldfine et al., 1981
DEGAPlaE[b]	34	13	12	Goldfine et al., 1981
DOPE	-13	~-23	11.2	van Dijck et al., 1976
DOGAPlaE[c]	12, 16	-6, -11.2 -2.2	9	Goldfine et al., 1981; Goldfine et al., 1987a
PE[d]	30		6.4	Lohner et al., 1984
PlaE[d]	26		5.2	Lohner et al., 1984
(Alkylacyl)PE[d]	29.5		6.2	Lohner et al., 1984

Notes: [a] Temperature at exothermic or endothermic peak maximum.
[b] Contained 10% < saturated acyl chains.
[c] ~7% 19:cyclopropane alkyl chains.
[d] *sn*-1 chains were 52-56% 18:0, 24.5-28% 16:0, 10% unsaturated; *sn*-2 (acyl) chains were *cis*-9-18:1.

gave an L → H transition at 18°C, a lower temperature than expected based on its fatty acyl and alkenyl chain composition (Boggs et al., 1981). Subsequent studies with the semisynthetic ethanolamine phospholipids described above demonstrated convincingly that the presence of the alk-1-enyl bond stabilizes the H_{II} phase. The L → H transition temperature for semisynthetic PlaE was 30°C compared to 68°C for the diacyl form and 53° for the alkylacyl form (Table 3) (Lohner et al., 1984).Similarly, (diradyl) PE (50 to 60% PlaE) isolated from *C. butyricum*, which had ~80% 18:1 plus 19:cyclopropane sn-1 alkenyl chains and ~78% 16:0 acyl chains, appeared to undergo a L → transition at 30 to 40°C as evidenced by an appropriate change in the deuterium quadrupolar splitting (Table 2) (Malthaner et al., 1987b). This value is close to that of the semisynthetic PlaE, with the reversed saturated/unsaturated chain distribution.

V. FUNCTIONS OF ETHER LIPIDS IN ANAEROBIC BACTERIA

A. Regulation of Membrane Fluidity

Beginning in the early 1960s with the landmark paper of Marr and Ingraham on the effect of growth temperature on the fatty acid composition of *E. coli* (Marr and Ingraham, 1962), the emphasis of much research on the membranes of prokaryotes centered on the regulation of the polar lipid hydrocarbon chain composition, which

Table 3. Lamellar to Reversed Hexagonal (HII) Phase Transition Temperatures of Plasmalogens and Related Lipids

Phospholipid	T_{BH}, °C	*References*
DOPE	10	(Cullis & Hope, 1985)
DEPE	60-63	Cullis & Hope, 1985
PE[a]	~68	Lohner et al., 1984
PlaE[a]	~30	Lohner et al., 1984
(Alkylacyl)PE[a]	~53	Lohner et al., 1984
PE/PlaE (46% PlaE)[b]	~0	Goldfine et al., 1987a
DOGAPlaE[c]	> 50	Goldfine et al., 1987a
PE/PlaE (~55% PlaE)[d]	30-40	Malthaner et al., 1987b

Notes: [a] See Table 2, note d for hydrocarbon chain compositions.
[b] Highly enriched in 18:1 and 19:cyclopropane alkenyl and acyl chains.
[c] Alkyl chains ~8% 19:cyclopropane.
[d] ~80% 18:1 and 19:cyclopropane alkenyl chains and 73-82% 16:0 acyl chains.

was shown to be necessary for modulating membrane fluidity or viscosity. These studies revolved around the central concept that a fluid or at least a partially fluid lipid bilayer is essential for lipid function and cell growth (Melchior, 1982). When the ambient temperature was decreased below Tm for the bulk lipid L → G transition, cells lost solute transport and other membrane functions, membranes could become leaky, and growth ceased. Bacteria were observed to prevent these undesirable consequences by synthesizing or incorporating increasing amounts of unsaturated fatty acids into membrane lipids when exposed to lower ambient temperatures. In some genera such as *Bacillus,* which have predominantly branched acyl chains, acyl-CoA desaturases were seen to be induced accompanied by a shift of biosynthesis from *iso-* to *anteiso-*branched chains , which also served to enhance membrane fluidity, at lower growth temperatures (Fulco, 1983; Kaneda, 1991).

Studies on anaerobic bacteria rich in plasmalogens demonstrated the participation of these lipids in regulating membrane lipid G → L transitions. The changes seen were different for each species studied, but the general effects were similar. At lower growth temperatures, *C. beijerinckii* had more unsaturated acyl chains in the plasmalogen-rich (diradyl) PE plus PME fraction and in total phospholipids. There was little overall effect of growth temperature on the degree of unsaturation of the ether-linked chains. Both acyl and ether-linked chains tended to be somewhat shorter at lower growth temperatures (Khuller and Goldfine, 1974; Goldfine et al., 1977).

The pattern of acyl, but not alkenyl chain response to growth temperature was even more pronounced in this species when cells were grown with suboptimal biotin in media supplemented with exogenous fatty acids. At 37°C growth on oleate resulted in 39% and 81% 18:1 plus 19:cyclopropane acyl and ether-linked chains, respectively. When grown at 25°C under the same conditions, the acyl chains were 72% 18:1 plus 19:cyclopropane, with only a small increase in ether-linked chain unsaturation. Thus growth at lower temperatures results in increasing acyl chain unsaturation independent of the source of the chains (Khuller and Goldfine, 1975). This has the overall effect of increasing the population of phospholipids with two unsaturated chains since the plasmalogens and their glycerol acetals generally have predominantly unsaturated alkenyl and alkyl chains (Khuller and Goldfine, 1975; Malthaner et al., 1987b). Studies on the effects of growth temperature on the acyl chains of *C. acetobutylicum* ATCC 39057 did not show changes in the degree of unsaturation, but did show an increase in 16:1 plus 17:cyclopropane with a concomitant decrease in 18:1 plus 19:cyclopropane acyl chains at 25°C compared to 38°C, resulting in shorter chains, and presumably lower melting phospholipids. No data were presented on alkenyl chain composition (Lepage et al., 1987). In another study on *C. acetobutylicum* ATCC 824, growth at lower temperatures also resulted in decreased acyl chain length. Alkenyl chains were not examined (Baer et al., 1987).

Changes in growth temperature for *C. beijerinckii* also led to changes in the lipid class composition which can be summarized as follows. Growth at 25° as opposed to 37°C resulted in increased levels of the total polyglycerolphospholipids (PG + CL), a decrease in the total PE + PME fraction in one study (Khuller and Goldfine, 1974) with little change in a second study (Goldfine et al., 1977), and a decrease in GAPlaE + GAPlaME (Goldfine et al., 1977). These changes were viewed as a means to increase the amounts of lower melting phospholipids, principally PG, at the expense of higher melting phospholipids, principally PE and PME. However, as will be discussed later, they will also affect lipid packing.

It is useful to compare these results with those obtained with Gram-negative species rich in ether lipids. In *Veillonella parvula* alkenyl chains were more unsaturated at 25°C than at 37°C, largely as a result of a decrease in 15:0 and an increase in 17:1 alkenyl chains at the lower temperature. The acyl chains were >70% unsaturated at both temperatures (Johnston and Goldfine, 1982). The phospholipids of *Megasphaera elsdenii* were highly enriched in monounsaturated and cyclopropane acyl and alkenyl chains at all growth temperatures between 27 and 42.5°C, but the lowest degree of unsaturation was observed at 37°C. At 42.5° there was a significant increase in unsaturated chains compared to 37°C. The (diradyl) PE/(diradyl)PS ratio of the polar lipids progressively increased at lower growth temperatures (Johnston and Goldfine, 1982).

These changes in lipid composition were shown to produce small changes in the presumed onset of the L → G transition of the bulk phospholipids of *C. beijerinckii* as determined with the ESR probe 2,2,6,6-tetramethylpiperidine-1-oxyl (TEMPO) and the fluorescent probe, 1-phenyl-6-phenylhexatriene (DPH). The beginning of the transition from the liquid crystalline to gel phase appeared to occur some 4 to 5°C lower for the lipids from *C. beijerinckii* grown at 25°C. When grown at 37°, a small fraction of the bulk lipids appeared to be ordered at 37°C, whereas the lipids from cells grown at 25° appeared to be completely melted at that temperature (Goldfine et al., 1977). For *V. parvula*, the changes in polar lipid composition for cells grown at lower temperatures resulted in 4 to 6°C decreases in both Tm and the completion of the G → L transition observed by DSC (Johnston and Goldfine, 1982). These modest changes were consistent with the rather narrow range of growth temperatures available for these organisms. In the case of *M. elsdenii*, no discernable G → L transitions were observed in membranes by freeze fracture electron microscopy (Verkley et al., 1975) or in the hydrated phospholipids by DSC (Johnston and Goldfine, 1982). Presumably the membrane lipids of this rumen bacterium were completely melted within the range of growth temperatures studied.

B. Regulation of Lipid Polymorphism

Numerous experiments have been done to explore the abilities of bacteria to adapt to the constraints imposed by either natural or induced fatty acid auxotrophy. The species studied include: *E. coli* fatty acid auxotrophic mutants (Silbert, 1975), *C. beijerinckii* (Khuller and Goldfine, 1975; Goldfine et al., 1977), and *C. butyricum* (Goldfine et al., 1981), grown in the absence of biotin, and the natural fatty acid auxotrophs, *A. laidlawii* (Wieslander and Rilfors, 1977; Wieslander et al., 1980; McElhaney, 1984; Rilfors et al., 1984; Silvius et al., 1980), and *Butyrivibrio* S2 (Hazlewood et al., 1980). For all of these organisms except *E. coli*, extensive incorporation of exogenous fatty acids resulted in significant changes in polar lipid composition, some of which were difficult to explain in terms of their potential effects on membrane fluidity. In an important shift away from the fluidity paradigm, Lindblom, Rilfors, Wieslander, and their colleagues (Wieslander et al., 1980; Wieslander et al., 1981) proposed that the changes in polar lipid class composition observed in *A. laidlawii* A strains could be better explained as a means of regulating the tendency of amphipathic lipids to form non-lamellar phases. As described above, certain membrane lipids by virtue of an effectively small polar head group and relatively large apolar chain cross-sectional area, tend to form the H_{II} phase at or close to physiological temperatures. Factors that promote this tendency include increased temperature, hydrocarbon chain unsaturation or increased chain length. According to lipid packing theory, the increased lipid preference for non-lamellar phases with for example a higher proportion of unsaturated chains can be balanced by the addition of lipids that stabilize the lamellar phase. Thus in *A. laidlawii*, the increased tendency of monoglucosyldiacylglycerol (MGDG) to form the H_{II} and cubic phases when the organism is fed *cis*-unsaturated fatty acids, is counterbalanced by a decrease in the ratio of MGDG to diglucosyldiacylglycerol (DGDG), which stabilizes the lamellar phase.

Although not as intensively studied as the regulation of membrane fluidity because of the difficult nature of the experiments and the paucity of experimental systems, this concept has derived considerable support from subsequent work with clostridia, which will be described below, *E. coli* (Rietveld et al., 1993; Killian et al., 1994; Rietveld et al., 1994), and eukaryotic cells including yeast (Tung et al., 1991), and rat liver (Jamil et al., 1993). Changes in polar head group composition in all of these systems can be explained by the need to maintain the bilayer arrangement while retaining a proportion of lipids that prefer non- lamellar phases so that the membrane is not very far from a lamellar to non-lamellar transition in thermodynamic terms. Monolayers of lipids that promote the formation of non-lamellar phases have a tendency to curl, which can be quantitatively expressed as an intrinsic radius of curvature, R_0. The greater the tendency to curl the smaller the value of R_0. It has been proposed that cells adjust the mix of lipids to produce intermediate values of R_0, which results in their membranes being close to bilayer instability. This in turn may be optimal for the functions of certain membrane proteins and for membrane processes such as fusion, endocytosis, and the translocation of largely hydrophilic molecules across the bilayer barrier (Gruner, 1985; Hui and Sen, 1989; De Kruijff, 1987).

Effects of Incorporation of Exogenous Fatty Acids

As described above, growth of C. beijerinckii (Khuller and Goldfine, 1975) and *C. butyricum* (Goldfine et al., 1981) in biotin- depleted media supplemented with fatty acids, which effectively blocked endogenous fatty acid synthesis, resulted in extensive incorporation of the fed fatty acids and large changes in the polar head group composition. Most notable were increases in the proportion of GAPlaE (Figure 1A) and decreases in (diradyl) PE when *C. butyricum* was fed oleic acid (*cis*-9-18:1), with much smaller changes in the GAPlaE/diradyl PE ratio when cells were fed elaidic acid (trans- 9-18:1). In *C. beijerinckii*, N-methylethanolamine is the predominant lipid head group base, and GAPlaME increased relative to (diradyl) PME in oleate-grown cells. Attempts to explain these results in terms of regulation of membrane fluidity and maintenance of G ↔ L transitions within a narrow range were unsatisfactory (Goldfine et al., 1977; Goldfine et al., 1981; Goldfine, 1984).

In order to explore this phenomenon in greater depth, *C. butyricum* was grown on mixtures of a saturated fatty acid (palmitic acid = P) and a monounsaturated fatty acid (oleic acid = O). The phospholipids were shown to have largely unsaturated ether-linked chains and ~70% saturated acyl chains upon growth in media with P/O = 8:2, in accord with the high selectivity for unsaturated *sn*-1 ether-linked chains and saturated *sn*-2 acyl chains in this organism. Growth was not seen with palmitic acid alone. As the proportion of oleic acid in the medium was increased to P/O ≤ 4:6, the acyl chains were still ≥ 60% saturated. At these lipid saturated/unsaturated chain ratios, which resemble those in cells grown with biotin, the ratio of GAPlaE/(diradyl) PE was close to 0.7. At medium P/O ≤ 2:8, the lipid acyl chains became increasingly unsaturated and the ratio of GAPlaE/(diradyl) PE increased to approximately 2.0, which was the ratio of these two lipids found in cells grown on 100% oleic acid (Figure 3) (Johnston and Goldfine, 1985). Thus, the

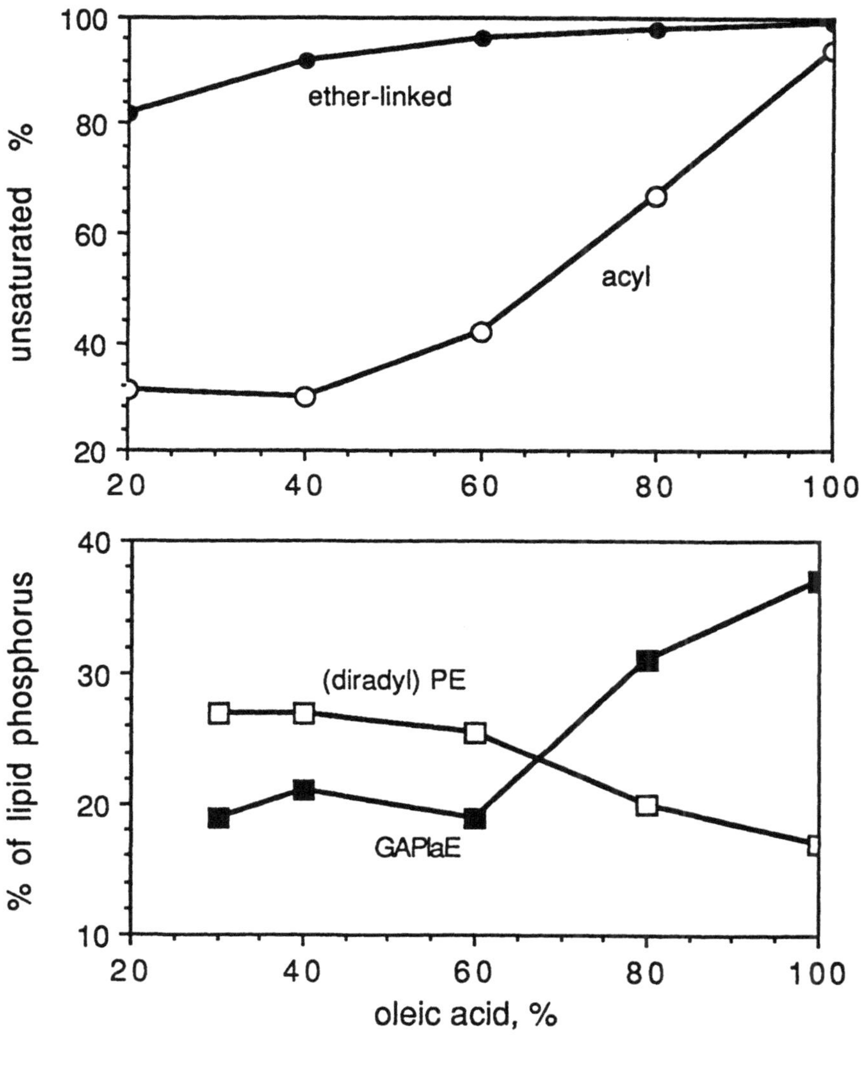

Figure 3. Upper panel: phospholipid acyl and ether-linked chain compositions (% unsaturated + cyclopropane chains) of *C. butyricum* grown on mixtures of oleic and palmitic acids in the absence of biotin. The % oleic acid in the medium is given on the x-axis. Lower panel: (diradyl) PE and GAPlaE compositions of the cells grown on mixtures of oleic and palmitic acids (Johnston and Goldfine, 1985).

GAPlaE/(diradyl) PE ratio was strongly influenced by the increasing presence of phospholipids with two unsaturated chains per glycerol. Earlier studies had shown that PE and GAPlaE were predominantly located in the outer leaflet of the plasma membrane of *C. butyricum* grown with biotin (Goldfine et al., 1982), and this strong asymmetry was maintained in cells grown with fatty acid supplementation in the absence of biotin (Johnston and Goldfine, 1985). Although the evidence is indirect, PG and CL are probably largely confined to the plasma membrane inner leaflet. On fatty acid supplementation, increasing medium oleate resulted in decreased PG, relatively unchanged CL and only a small change in the sum of these acidic phospholipids (Johnston and Goldfine, 1985). The near constancy of the sum of the acidic phospholipids may serve to preserve membrane charge density, a parameter that has been shown to be carefully regulated in *A. laidlawii* strain A (Rilfors et al., 1993).

The changes in polar lipid composition observed on fatty acid supplementation can be readily rationalized on consideration of phospholipid polymorphic phase behavior. (Diradyl) PE from oleate- grown cells undergoes a G → L transition at -1.9°C and an L → H transition at or near 0°C. Addition of one part GAPlaE to two parts (diradyl) PE from oleate-enriched cells elevated the lamellar to non- lamellar phase transition to 30°C, but a 1:1 mixture of these lipids formed a stable lamellar phase up to 46°C. Similarly, addition of oleate-enriched GAPlaE to DOPE (1:1) produced a lamellar phase at 35°C compared to the L → H transition temperature of 5 to 10°C for DOPE alone (Table 3), and a 2:1 ratio of these lipids produced a lamellar phase up to 46°C (Goldfine et al., 1987a). Thus, *C. butyricum* cells produce the lipid ratio required to stabilize the lamellar arrangement of the principal plasma membrane outer leaflet lipids up to ~10°C above their growth temperature.

As described above, a simple lipid packing concept can be used to explain the polymorphic phase behavior of these membrane phospholipids. As mixtures of polar lipids become increasingly populated with species having two unsaturated chains per glycerol, at a constant growth temperature, the area swept out by the chains increases and the average chain-length decreases, as a result of the kink introduced by the double bond. The individual lipids increasingly favor the shape of a truncated cone, which decreases the intrinsic radius of curvature and increases the tendency to form non-lamellar phases (Israelachvili et al., 1980; Cullis and De Kruijff, 1979; Gruner, 1985). The switch from (diradyl) PE to GAPlaE serves to increase the effective polar head group cross-sectional area (Goldfine and Johnston, 1980) giving the lipids a more cylindrical shape, and stabilizing the bilayer arrangement. Oleate-enriched GAPlaE like DOPC or DODGlcDG, forms a lamellar phase at temperatures > 50°C (Table 3).

Feeding *C. butyricum* terminally branched fatty acids, such as *anteiso*-C_{17} or *iso*-C_{14}, which was partly elongated by the cells to *iso*-C_{16}, resulted in membranes which had GAPlaE/diradyl PE ratios (r) lower than those seen in cells grown with *trans*-9-18:1 (r = 0.82) or with palmitic/oleic 4:6 (r = 0.7). The lowest ratio (r = 0.44) was observed in cells containing large amounts of iso- branched fatty acids

(Goldfine et al., 1987b). This observation was consistent with the known behavior of PE with iso-branched acyl chains which forms a lamellar phase up to 98°C (Silvius et al., 1985; Lewis et al., 1989). As discussed above, the presence of PlaE would lower T_{BH}, possibly by as much as 40°C depending on the ratio of PlaE to PE. Even so, less GAPlaE is needed to form a lamellar phase at the growth temperature of these cells, 37°C than for cells grown with unsaturated fatty acids. For cells grown on *anteiso*-C_{17}, r = 0.64, consistent with the observation that lipids with *anteiso* acyl chains tend to have a lower T_{BH} than lipids with similar *iso* chains (Rilfors et al., 1982; Lewis et al., 1989). When cells were grown with various *cis*-unsaturated fatty acids or a C_{19}-cyclopropane acid, r ranged from 1.8 for *cis*-6-18:1 to 3.1 for *cis*-11-18:1 (Goldfine et al., 1987b).

Growth of *C. acetobutylicum* in biotin-free media supplemented with fatty acids also led to significant changes in polar lipid composition which accompanied the induced changes in lipid acyl and alkenyl chain compositions. In addition to phospholipids identical to those found *in C. butyricum* , this species, has relatively large amounts of diradyl glycosyl glycerols, which in strain DSM 1731 are mono-galactosyl diradylglycerol (MGDG) and a β-glucosyl galactosyl diradylglycerol (DGDG) (Oulevey et al., 1986). Strain ATCC 4259 also has large amounts of unidentified phosphoglycolipids (Johnston and Goldfine, 1992). When grown on mixtures of palmitic and oleic acids, the pattern of incorporation was similar to that observed in *C. butyricum*, i.e., largely unsaturated ether-linked chains and predominantly saturated acyl chains except when oleic acid was increased to ≥ 80%. Increasing unsaturation of the lipid hydrocarbon chains resulted in increased GAPlaE/(diradyl) PE, increased (diradyl) PG/(tetraradyl) CL, and increased (diradyl) DGDG/MGDG ratios. However at 100% oleic acid the rise in the DGDG/MGDG ratio reversed, which was accompanied by an overall decrease in membrane glycolipid relative to phospholipid (Johnston and Goldfine, 1992). The effects of each of these lipid compositional changes on phase behavior can be predicted to result in stabilization of the bilayer arrangement of the membrane lipids, but no physical studies have been carried out with these mixtures of lipids.

Butyrivibrio S2, an obligately anaerobic rumen bacterium, has a unique set of plasmalogens in which a galactosyl alkenylacyl glycerol is cross-linked to a butanoylated PlaG by a C_{32} dicarboxylic acid (diabolic acid) which has a vicinal dimethyl substitution at the center of the chain (Clarke et al., 1980; Hazlewood et al., 1980). Thus the intact lipid can be thought of as a dimer of a plasmalogenic glycosyldiglyceride and PlaG in which the two glycerols are cross- linked by a diabolic acid as their common *sn*-2 chains. Other variations have been found including one in which two molecules of *sn*-1- alkenylglycero-3-phospho-*sn*-1'-glycerol butyroyl ester are joined through a diabolic acid. This organism grows in an environment rich in saturated long-chain fatty acids and *trans*-18:1 fatty acids produced by hydrogenation of plant polyunsaturated fatty acids. It incorporates various fatty acids into its complex mixture of polar lipids resulting in considerable alterations in their ratios. These changes were viewed as a means to maintain a sufficiently fluid membrane with the large amounts of saturated and *trans*-unsaturated chains present in the cellular lipids. However, effects on lipid polymorphism were not studied (Hauser et al., 1985).

Effects of Growth Temperature

As noted above, lowering the growth temperature of *C. beijerinckii* resulted in a decrease in the GAPlaE/(diradyl) PE + PME ratio (Goldfine et al., 1977) and an increase in total polyglycerolphospholipids (PG + CL). Since the ratio of PG to CL was not measured, only the effects on the ratio of the glycerol acetals to (diradyl) PE + PME can be considered. A decrease in this ratio at lower growth temperatures coupled with the increased levels of unsaturated chains in the (diradyl) PE + PME fraction will result in larger amounts of lipids that have the shape of a truncated cone and smaller amounts of the bilayer-preferring glycerol acetals of PlaE and PlaME, thus shifting the equilibrium towards the non-lamellar phase and counteracting the effects of temperature, which shifts the equilibrium towards the lamellar phase.

Effects of Loss of Plasmalogens

M. elsdenii, a Gram-negative, obligately anaerobic bacterium of the rumen, normally has a plasmalogen-rich lipid composition with the major lipids PE and PS having 87% and 72% plasmalogen respectively. These lipids display no G → L transitions over a wide- temperature range (Johnston and Goldfine, 1982), and are Lα at 30°C. Above 30°C both ^{31}P NMR and X-ray diffraction of the polar lipids revealed the gradual formation of an H_{II} phase with increasing temperature, which was not complete at 80°C. In contrast, phospholipids from a plasmalogen-deficient strain of this organism (Kaufman et al., 1988), appeared to form a relatively stable lamellar phase (Kaufman et al., 1990) despite the presence of more unsaturated acyl chains than the alkenyl chains they displaced (Kaufman et al., 1988). These observations were consistent with the finding that PlaE destabilizes the lamellar phase more than PE of similar chain compositions (Lohner et al., 1984).

Effects of Solvents and Alcohols Incorporated into the Cell Membrane

Many compounds in the environment are hydrophobic and small hydrophobic molecules that penetrate the cell envelope of bacteria will readily partition into the cell membrane. These compounds can have varying effects depending on their interactions with membrane lipids. In the case of n-alcohols, compounds shorter than C_8 disorder natural and model membrane lipids and lower T_m for the G → L transition, while longer chain alcohols have differing effects on motion depending on the techniques used to measure the order parameter (Wieslander et al., 1986). ^{2}H-NMR studies show an ordering effect of *n*-octanol on dimyristoyl PC membranes, whereas longer chains up to C_{14} had little effect (Westerman et al., 1988). With respect to the transition from a lamellar to a non-lamellar phase, ethanol stabilizes the lamellar arrangement for egg PE, *n*-butanol has only a small stabilizing effect, whereas the *n*-alcohols hexanol, octanol, decanol, and dodecanol lower T_{BH} for egg PE (Hornby and Cullis, 1981). Cyclohexane has the same effect on DPPC (McDanield et al., 1982).

If cells are grown with fixed hydrocarbon chains, according to the hypothesis that the cell strives to control lamellar to non-lamellar transitions in the cell membrane, one would expect the addition of *n*- hexanol, *n*-octanol, or cyclohexane to the growth medium of bacteria to cause an elevation of bilayer-stabilizing lipids relative to lipids that promote the formation of non-lamellar phases. The addition of n- butanol should have little effect. These predictions were fulfilled in experiments with *A. laidlawii* (Wieslander et al., 1986) and *C. butyricum* (MacDonald and Goldfine, 1991). As shown in Figure 4, increasing concentrations of cyclohexane and n-octanol resulted in marked elevations of the GAPlaE/diradyl PE ratio in *C. butyricum* grown in biotin-free media with elaidic acid, whereas *n*-butanol had no effect. Ethanol, had opposite effects in the two species. As expected, the addition of ethanol to cultures of *A. laidlawii* grown with oleic acid decreased the ratio of bilayer-stabilizing DGlcDG to bilayer- destabilizing MGlcDG, but in *C. butyricum* grown with elaidic acid the corresponding GAPlaE/(diradyl) PE ratio was elevated. It is possible that the disparate effects on the ratios of these polar lipids exerted by growth with ethanol can be explained by the fact the two organisms were grown with different fatty acids, but it is more likely to be the result of divergent interactions with the very different lipids of these organisms. When *C. butyricum* was grown in biotin-free media with oleic acid, which results in a high GAPlaE/(diradyl) PE ratio, the addition of *n*-alcohols and cyclohexane did not result in further changes in this ratio (MacDonald and Goldfine, 1991).

In general, these observations with *A. laidlawii* and *C. butyricum* could not be explained by the effects of the added compounds on G $\leftrightarrow$ L transitions or on lipid hydrocarbon chain order (Wieslander et al., 1986; MacDonald and Goldfine, 1991). For example, the replacement of diradyl PE by GAPlaE in elaidate-grown C. butyricum would have little effect on the G $\rightarrow$ L transition of the mixture, but it would suppress the L $\rightarrow$ G transition (Table 2). In these experiments the cells were grown at 37°C,. and the addition of cyclohexane or octanol should Lower T_m for the L $\rightarrow$ G transition. Additional elaidate- enriched GAPlaE would move the L $\rightarrow$ G transition in the same direction. On the other hand, it would move the L $\rightarrow$ H_{II} transition in the opposite direction from that exerted by these solvents.

Studies on *C. acetobutylicum*, which carries out an acetone/butanol fermentation, have also shown that the GAPlaE/diradyl PE ratio increases and the unsaturated/saturated acyl chain ratio decreases, as solvents accumulate in the culture fluid (Lepage et al., 1987). Similar changes were seen in the acyl chain composition along with shortening of the fatty acyl chains on addition of *n*-hexanol or *n*-octanol. Ethanol and butanol produced smaller changes in acyl chain composition. Thus, when cells are permitted to adjust both hydrocarbon chain and lipid class compositions in response to foreign molecules that alter membrane fluidity and perturb the lamellar/non-lamellar equilibrium, changes are observed that counteract these effects.

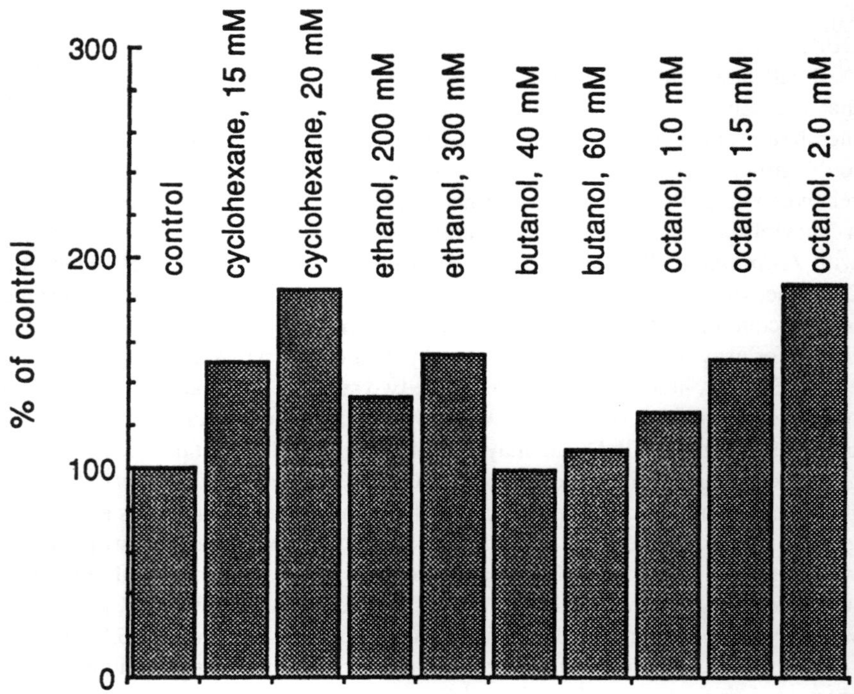

Figure 4. Effects of cyclohexane and n-alcohols on the GAPlaE/(diradyl) PE ratio of *C. butyricum* grown with elaidic acid in the absence of biotin in the presence of the indicated concentrations of these compounds. The ratio of GAPlaE/(diradyl) PE ratio of the control cells was 0.87 (MacDonald and Goldfine, 1991).

How is the Membrane Lipid Composition Regulated?

The regulation of bacterial polar lipid head groups as a means to control the balance between lipids that tend to form a lamellar phase and those that prefer one of the non-lamellar arrangements, has now been well documented. At this time, we can only speculate on the cellular mechanisms involved. The simplest hypothesis, is that the responsible enzymes, which are integral membrane proteins, respond to changes in the intrinsic radius of curvature, r_0 (Gruner, 1989). Accordingly, the activity of the glucosyl transferase responsible for conversion of MGlcDG to DGlcDG in *A. laidlawii* would be positively regulated by a decrease in r_0, thus increasing the DGlcDG/MGlcDG ratio, and negatively regulated by an increase in r_0 (Dahlqvist et al., 1992). In *M. elsdenii*, the PS decarboxylase which produces PE, would be negatively regulated by a decrease in r_0, thus increasing the PS/PE ratio, and positively regulated by an increase in r_0. These effects have recently been postulated to result from responses of these enzymes to changes in the lateral pressure exerted by the lipids (Gruner, 1989; Rilfors et al., 1993). Earlier models in which protein conformation, substrate affinity or substrate accessibility to the enzyme active site were affected by the transient formation of micro- domains of non-bilayer structures (Goldfine, 1985) are less plausible since there is little evidence for such structures in normal cell membranes. Nonetheless, this important membrane property is unlikely to be controlled by a unitary mechanism, consequently the activities and kinetics of the enzymes involved could be affected by yet unknown cellular parameters which are in turn affected by the packing properties of the membrane lipids (Rilfors et al., 1993). Lastly, the synthesis of the enzymes involved may be controlled at either the translational or transcriptional levels. We are at an early stage of our understanding of these regulatory systems. It will be necessary to study the enzymes involved *in situ* and after isolation and reconstitution. It will also be essential ultimately to study the regulation of their synthesis. Only then can a satisfactory picture possibly emerge.

VI. CONCLUSIONS

With a limited number of major (≥ 10% of total) amphipathic lipids in any given species, it would seem economical to have each lipid fulfill several functions. As described above, the hydrocarbon chain composition of each lipid class can be modulated by the cell in response to changes in temperature, in response to the presence of foreign molecules which partition into the membrane, or when the cell is constrained to use exogenous fatty acids as a result of natural or induced fatty acid auxotrophy. Changes in hydrocarbon chain composition will alter L ↔ G phase transitions, the motions of chains in the fluid phase, and other indicators of translational or rotational lipid motion which are generally thought of as membrane fluidity (Melchior, 1982; Russel, 1989).

Among bacterial lipids, PE, and more so PlaE, MGDG, and cardiolipin have strong tendencies to form the H_{II} and other non-lamellar phases, whereas PS, DGDG and PG usually form stable lamellar phases and when added to lipids that prefer non-lamellar arrangements, they serve to stabilize the bilayer. Thus, the presence of at least two major lipid species, as is the case in some of the Gram-negative anaerobes such as *M. elsdenii, V. parvula, Anaerovibrio lipolytica* and *S. ruminantium* (Wilkinson, 1988), appears to be adequate for the control of both membrane fluidity and lipid polymorphism. In *M. elsdenii* membrane fluidity is ensured by having large proportions of unsaturated and cyclopropane chains at all growth temperatures (Verkley et al., 1975; Johnston and Goldfine, 1982). The lamellar/non-lamellar equilibrium is dependent on the nature of the lipid chains, the PE/PS ratio and the plasmalogen content (Kaufman et al., 1990). Since (diradyl) PE is formed from (diradyl) PS by decarboxylation, the cell has to exert tight control over the extent of decarboxylation to maintain sufficient PS to ensure the continued presence of the bilayer (Silber et al., 1981; Watanabe et al., 1984).

As the complexity of the polar lipid composition increases, the potential for modulating these parameters expands. An example of increasing compositional complexity can be seen in clostridia that produce acetic and butyric acids as major products of fermentation, including *C. butyricum, C. beijerinckii,* and *C. acetobutylicum. C. butyricum* has four major polar lipids: (diradyl) PE, (diradyl) PG, GAPlaE, and (tetraradyl) CL. PE is usually between 50 to 80% PlaE, while PG and CL are about one-third plasmalogen (Goldfine et al., 1982; Johnston and Goldfine, 1983; Malthaner et al., 1987b). *C. beijerinckii* lipids have almost the same composition, but the predominant nitrogenous lipid base is N-methylethanolamine (Baumann et al., 1965; Goldfine et al., 1977; Johnston and Goldfine, 1983). Some strains of *C. acetobutylicum* have in addition to the lipids listed *for C. butyricum*, monoglycosyl- and diglycosyl-diradylglycerols, and strain ATCC 4259 has in addition large amounts of phosphoglycolipids of unknown structures (Oulevey et al., 1986; Johnston and Goldfine, 1992; Lepage et al., 1987). An expansion of the polar lipid repertoire may provide each organism with the means to resist distinctive environmental challenges.

The major end-products of sugar fermentation by *C. butyricum* are acetic and butyric acids, *C. beijerinckii* produces large amounts of butanol, isopropanol and some ethanol, and *C. acetobutylicum* produces butanol, acetone and some ethanol (Rogers, 1984). The latter two organisms switch from an acidogenic fermentation during which acetic and butyric acids are the major products, to a solventogenic fermentation, leading to the production of butanol and isopropanol or acetone. This switch usually occurs after the exponential growth phase in response to changes in the environment such as lower pH or phosphate limitation (Rogers, 1984).

Thus, initially during growth on adequate supplies of substrate, each organism may experience similar environments, but at later stages solvents are produced which can inhibit energy-producing pathways of fermentation, leading to growth inhibition (Linden and Moreira, 1983). The accretion of membrane lipid components through the addition of biosynthetic pathways or the regulation of existing pathways, may represent co-evolution with increasing versatility of solvent-generating pathways. The addition of an N-methyl group to PE, which differentiates *C. beijerinckii* from *C. butyricum*, raises T_{BH} of the lipid by 50°C (Tate et al., 1991), which may serve to provide greater stability to the bilayer in the presence of solvents. As shown for *C. acetobutylicum*, the regulation of lipid composition in response to a shift to high concentrations of a monounsaturated fatty acid in the cellular lipids involves complex changes in both glycolipids and phospholipids, with a greater dominance of phospholipids in terms of membrane concentration at the highest unsaturated lipid chain content (Johnston and Goldfine, 1992). At this time, only changes in phospholipids have been studied during the acetone-butanol fermentation of this organism, since the strain studied did not appear to have major glycolipids (Lepage et al., 1987). As in cells exposed to increased lipid unsaturation

(Johnston and Goldfine, 1992), an increase in GAPlaE relative to (diradyl) PE was observed in response to solvent accumulation during the solventogenic phase of growth.

The notion that increasing lipid complexity in the butyric acid- producing clostridia is related to the need to adjust to the high concentrations of membrane-perturbing solvents produced, can be tested genetically by disabling specific lipid biosynthetic pathways. For example, one could test the effects of eliminating one or both of the glycosyltransferases that convert diradylglycerol to galactosyl diradylglycerol and glucosyl galactosyl diradylglycerol in *C. acetobutylicum* on the ability of the cell to respond to a solvent challenge. It is known that there is wide variation in the size of the genome of *C. acetobutylicum* (Wilkinson and Young, 1993), and, as indicated above, strains appear to differ in their contents of glycolipids and phosphoglycolipids (Johnston and Goldfine, 1992; N.C. Johnston and H. Goldfine, unpublished). An examination of the abilities of these strains to produce solvents and to adapt to a variety of solvent challenges could be of potential value, but it is important to be aware of the possibility that these wild-type strains may have developed alternative strategies to resist the toxic effects of solvents.

In thinking about the special roles that plasmalogens play in anaerobes, it may be useful to consider why they did not survive as important lipids in facultative and in obligately aerobic prokaryotes. It is possible that their presence in unicellular organisms is incompatible with molecular oxygen. However, Raetz and colleagues have presented evidence that one potential role of plasmalogens in animal tissues is to protect against photosensitized oxygen-dependent killing (Zoeller et al., 1988; Morand et al., 1988). They postulate that plasmalogens are decomposed by disruption of the alk-1′-enyl bond in the presence of a photosensitizer and molecular oxygen, but the damage is repaired, thus protecting the cell from more severe damage. It would seem contradictory to postulate that plasmalogens in anaerobes are detrimental to survival in an aerobic environment. However, that is just what we have observed in studies on a plasmalogen-deficient strain of *M. elsdenii* (Kaufman et al., 1988). The wild-type, plasmalogen-rich strain was killed *more rapidly* in oxygenated media than the deficient strain (B.R. Phillips and H. Goldfine, unpublished). This was not a completely satisfactory test because the plasmalogen-deficient strain arose from a series of step- wise changes rather than from a single mutation. Indeed it appeared that the loss of plasmalogens led to better survival of the cells in the stationary phase for long periods of time (Kaufman et al., 1988). There are several significant differences between the bacterial and animal cell experiments, including the use of a photosensitizer with the mammalian cells as opposed to oxygen with the bacteria, and the postulated ability of the animal cell to repair the oxygen-mediated disruption of the alk-1′-enyl bond, which prokaryotes may not have developed.

Prokaryotic organisms have virtues aside from the simplicity of their structures and their relatively small and readily manipulated genomes. The study of bacterial membrane lipids over the past 35 years has revealed a very wide diversity of polar lipid structures and of the way these lipids are combined in different genera and species (Goldfine, 1982; Ratledge and Wilkinson, 1988). From archaebacteria with their *sn*-2,3-diphytanylglycerol diethers and dibiphytanyl diglycerol tetraethers (Kates, 1992; Koga et al., 1993), to acholeplasma with their mixtures of anionic phospholipids and glycosyl diacylglycerols (Rilfors et al., 1993), to enteric bacteria with their relatively simple mixtures of PE, PG, and CL (Cronan and Rock, 1987), and the anaerobic bacteria discussed in this review, each group of organisms has presented a challenge in achieving an understanding of how their individual lipid arrays have evolved to meet the internal requirements of the cell and the need to protect the cell against external environmental stresses. As illustrated in the reviews cited above, studies in depth of each of these groups of organisms has been rewarding in providing insights into the structures and functions of membrane lipids. They have also provided information and insights which will help the cell biologist trying to make sense of the bewildering complexities of eukaryotic subcellular membrane lipids and the regulation of their compositions.

ACKNOWLEDGMENTS

I wish to acknowledge with gratitude the long-term support of the National Institute of Allergy and Infectious Diseases (AI-05079 and AI-08903) for the work on bacterial lipids carried out in my laboratory. The efforts of the many students, and colleagues who contributed to this work is also gratefully acknowledged.

ABBREVIATIONS

CL, cardiolipin (bisphosphatidylglycerol); DEPE, dielaidoylphosphatidylethanolamine;
DHAP, dihyroxyacetonephosphate; DOPE, dioleoylphosphatidylethanolamine
DPPC, dipalmitoylphosphatidylcholine; DGDG, diglycosyldiacylglycerol
DGlcDG, diglucosyldiacylglycerol; DODGlcDG, dioleoyldiglucosyldiacylglycerol
GAPlaE, glycerol acetal of plasmenylethanolamine
GAPlaME, glycerol acetal of plasmenyl-N-methylethanolamine
MGDG, monoglycosyldiacylglycerol; MGlcDG, monoglucosydiacylglycerol
PE, phosphatidylethanolamine; PG, phosphatidylglycerol
PGA, phosphatidylglycerol acetal; PME, phosphatidyl-N-methylethanolamine
PS, phosphatidylserine; PlaE, plasmenylethanolamine
PlaG, plasmenylglycerol; PlaME, plasmenyl-N-methylethanolamine
PlaS, plasmenylserine; POPC, palmitoyl, oleoyl phosphatidylcholine
$G \rightarrow L$, gel to liquid crystalline; $L \rightarrow H$, lamellar to reversed hexagonal
T_{BH}, lamellar to reversed hexagonal phase transition temperature
T_m, temperature at peak maximum of a phase transition

REFERENCES

Ailhaud, G.P., Vagelos, P.R., & Goldfine, H. (1967). Involvement of acyl carrier protein in acylation of glycerol 3-phosphate in *Clostridium butyricum*. I. Purification of *Clostridium butyricum* acyl carrier protein and synthesis of long-chain derivatives of acyl carrier protein. J. Biol. Chem. 242, 4459-4465.

Baer, S.H., Blaschek, H.P., & Smith, T.L. (1987). Effect of butanol challenge and temperature on lipid composition and membrane fluidity of butanol-tolerant *Clostridium acetobutylicum*. Appl. Environ. Microbiol. 53, 2854-2861.

Baumann, N.A., Hagen, P-O., & Goldfine, H. (1965). Phospholipids of *Clostridium butyricum*: Studies on plasmalogen composition and biosynthesis. J. Biol. Chem. 240, 1559-1567.

Bloch, K., Baronowsky, P., Goldfine, H., Lennarz, W.J., Light, R., Norris, A.T., & Scheuerbrandt, G. (1961). Biosynthesis and metabolism of unsaturated fatty acids. Federation Proc. 20, 921-927.

Boggs, J.M., Stamp, D., Hughes, D.W., & Deber, C.M. (1981). Influence of ether linkage on the lamellar to hexagonal phase transition of ethanolamine phospholipids. Biochemistry 20, 5728-5735.

Cato, E.P., George, W.L., & Finegold, S.M. (1986). Genus Clostridium Prazmowski 1880, 23[AL] In: Bergey's Manual of Systematic Bacteriology, Vol. 2 (Sneath, P.H.A., Mair, N.S., Sharpe, M.E., & Holt, J.G., eds.), pp. 1141-1200. Williams and Wilkins, Baltimore.

Cato, E.P. & Stackebrandt, E. (1989). Taxonomy and Phylogeny In: Clostridia (Minton, N.P., & Clarke, D.J., eds.), pp. 1-26. Plenum Press, New York.

Chen, X., & Gross, R.W. (1994). Phospholipid subclass-specific alternations in the kinetics of ion transport across biologic membranes. Biochemistry 33, 13769-13774.

Clarke, N.G., Hazlewood, G.P., & Dawson, R.M.C. (1980). Structure of diabolic acid-containing phospholipids isolated from *Butyrivibrio* Sp. Biochem. J. 191, 561-569.

Cronan, J.E., Jr. & Rock, C.O. (1987). Biosynthesis of membrane lipids In: *Escherichia coli* and *Salmonella typhimurium*. Cellular and Molecular Biology Vol. 1 (Neidhardt, F.C., Ingraham, J.L., Low, K.B., Magasanik, B., Schaechter, M., & Umbarger, H.E., eds.), pp. 474-503. American Society for Microbiology, Washington,D.C.

Cullis, P.R., & De Kruijff, B. (1979). Lipid polymorphism and the functional roles of lipids in biological membranes. Biochim. Biophys. Acta 559, 399-420.

Cullis, P.R., & Hope, M.J. (1985). Physical properties and functional roles of lipids in membranes In: Biochemistry of Lipids and Membranes (Vance, D.E., & Vance,J.E., eds.), pp. 25-72. Benjamin/Cummins, Menlo Park, CA.

Dahlqvist, A., Andersson, S., & Wieslander, Å. (1992). The enzymatic synthesis of membrane glucolipids in *Acholeplasma laidlawii*. Biochim. Biophys. Acta 1105, 131-140.

Day, J.I.E., Goldfine, H., & Hagen, P-O. (1970). Enzymatic reduction of long chain acyl-CoA to fatty aldehyde and alcohol by extracts of *Clostridium butyricum*. Biochim. Biophys. Acta 218, 179-182.

Day, J.I.E., & Goldfine, H. (1978). Partial purification and properties of acyl-CoA reductase from *Clostridium butyricum*. Arch. Biochem. Biophys. 190, 322-331.

De Kruijff, B. (1987). Polymorphic regulation of membrane lipid composition. Nature 329, 587-588.

Fischer, W., Hartmann, R., Peter-Katalinic, J., & Egge, H. (1994). (*S*)-2-amino-1,3-propanediol-3-phosphate-carrying diradylglyceroglycolipids-Novelmajor membrane lipids of *Clostridium innocuum*. Eur. J. Biochem. 223, 879-892.

Fulco, A.J. (1983). Fatty acid metabolism in bacteria. Prog. Lipid Res. 22, 133-160.

Gally, H.U., Pluschke, G., Overath, P., & Seelig, J. (1979). Structure of *Escherichia coli* membranes. Phospholipid conformation in model membranes and cells as studied by dueterium magnetic resonance. Biochemistry 18, 5605-5610.

Goldfine, H. (1982). Lipids of procaryotes-structure and distribution. Curr. Topics Membr. Transp. 17, 1-43.

Goldfine, H. (1984). The control of membrane fluidity in plasmalogen-containing anaerobic bacteria. Biomembranes 12, 349-377.

Goldfine, H. (1985). Modulation of polar lipid composition by aliphatic chain unsaturation in bacteria. Curr. Topics Cell. Reg. 26, 163-174.

Goldfine, H. (1993). Phospholipid biosynthetic enzymes of butyric acid-producing clostuidia In: Genetics and Molecular Biology of Anaerobes (Sebald,M., ed.), pp. 349-357. Springer Verlag, N.Y.

Goldfine, H., Ailhaud, G.P., & Vagelos, P.R. (1967). Involvement of acyl carrier protein in acylation of glycerol 3-phosphate in *Clostridum butyricum* II. Evidence for the participation of acyl thioesters of acyl carrier protein. J. Biol. Chem. 242, 4466-4475.

Goldfine, H., & Ailhaud, G.P. (1971). Fatty acyl-acyl carrier protein and fatty acyl-CoA in the biosynthesis of phosphatidic acid in *Clostridium butyricum*. Biochem. Biophys. Res. Commun. 45, 1127-1133.

Goldfine, H., & Hagen, P-O. (1972). Bacterial plasmalogens In: Ether Lipids: Chemistry and Biology (Snyder,F., ed.), pp. 329-350. Academic Press, N.Y.

Goldfine, H., & Johnston, N.C. (1980). Regulation of membrane fluidity in anaerobes In: Membrane Fluidity: Biophysical Techniques and Cellular Regulation (Kates, M., & Kuksis, A., eds.), pp. 365-380. Humana Press, Clifton, N.J.

Goldfine, H., Johnston, N.C., & Phillips, M.C. (1981). Phase behavior of ether lipids from *Clostridium butyricum*. Biochemistry 20, 2908-2916.

Goldfine, H., Johnston, N.C., & Bishop, D.G. (1982). Ether phospholipid asymmetry in *Clostridium butyricum*. Biochem. Biophys. Res. Commun. 108, 1502-1507.

Goldfine, H., Johnston, N.C., Mattai, J., & Shipley, G.G. (1987a). The regulation of bilayer stability in *Clostridium butyricum*: Studies on the polymorphic phase behavior of the ether lipids. Biochemistry 26, 2814-2822.

Goldfine, H., Khuller, G.K., Borie, R.P., Silverman, B., Selick, H., Johnston, N.C., Vanderkooi, J.M., & Horwitz, A.F. (1977). Effects of growth temperature and supplementation with exogenous fatty acids on some physical properties of *Clostridium butyricum* phospholipids. Biochim. Biophys. Acta 488, 341-352.

Goldfine, H., & Panos, C. (1971). Phospholipids of *Clostridium butyricum* IV. Analysis of the positional isomers of monounsaturated and cyclopropane fatty acids and alk-1'-enyl ethers by capillary column chromatography. J. Lipid Res. 12, 214-220.

Goldfine, H., Rosenthal, J.J.C., & Johnston, N.C. (1987b). Lipid shape as a determinant of lipid composition in *Clostridium butyricum*. The effects of incorporation of various fatty acids on the ratios of the major ether lipids. Biochim. Biophys. Acta 904, 283-289.

Gruner, S.M. (1985). Intrinsic curvature hypothesis for biomembrane lipid composition: A role for nonbilayer lipids. Proc. Natl. Acad. Sci. USA 82, 3665-3669.

Gruner, S.M. (1989). Stability of lyotropic phases with curved interfaces. J. Phys. Chem. 93, 7562-7570.

Hagen, P-O., & Goldfine, H. (1967). Phospholipids of *Clostridium butyricum* III. Further studies on the origin of the aldehyde chains of plasmalogens. J. Biol. Chem. 242, 5700-5708.

Hajra, A.K. (1983). Biosynthesis of the *O*-alkyl ether lipids In: Ether Lipids. Biochemical and Biomedical Aspects (Mangold, H.K., & Paltauf, F., eds.), pp. 85-106. Academic Press, N.Y.

Han, X., & Gross, R.W. (1990). Plasmenylcholine and phosphatidylcholine membrane bilayers possess distinct conformational motifs. Biochemistry 29, 4992-4996.

Hauser, H., Hazlewood, G.P., & Dawson, R.M.C. (1985). Characterization of membrane lipids of a general fatty acid auxotrophic bacterium by electron spin resonance spectroscopy and differential scanning calorimetry. Biochemistry 24, 5247-5253.

Hazlewood, G.P., Dawson, R.M.C., & Hauser, H. (1980). The question of membrane fludity in an anaerobic general fatty acid auxotroph In: Membrane fluidity: Biophysical techniques and cellular regulation (Kates, M., & Kuksis, A., eds.),pp. 191-202. Humana Press, Clifton, N.J.

Hill, E.E. & Lands, W.E.M. (1970). Formation of acyl and alkenyl glycerol derivatives in *Clostridium butyricum*. Biochim. Biophys. Acta 202, 209-211.

Hitchcock, P.B., Mason, R., Thomas, K.M., & Shipley, G.G. (1974). Structural chemistry of 1,2-Dilauroyl-DL-phosphatidylethanolamine: Molecular conformation and intermolecular packing of phospholipids. Proc. Natl. Acad. Sci. USA 71, 3036-3040.

Hornby, A.P., & Cullis, P.R. (1981). Influences of local and neutral anaesthetics on the polymorphic phase preferences of egg yolk phosphatidylethanolamine. Biochim. Biophys. Acta 647, 285-292.

Horrocks, L.A., & Sharma, M. (1982). Plasmalogens and *O*-alkyl glycero phospholipids In: Phospholipids (Hawthorne, J.N., & Ansell, G.B., eds.), pp. 51-93. Elsevier Biomedical, Amsterdam.

Hui, S.-W. & Sen, A. (1989). Effects of lipid packing on polymorphic phase behavior and membrane properties. Proc. Natl. Acad. Sci. USA 86, 5825-5829.

Israelachvili, J.N., Marcelja, S., & Horn, R.G. (1980). Physical principles of membrane organization. Q. Rev. Biophys. 13, 121-200.

Jackson, M.B. & Sturtevant, J.M. (1977). Studies of the lipid phase transitions of *Escherichia coli* by high sensitivity differential scanning calorimetry. J. Biol. Chem. 252, 4749-4751s.

Jamil, H., Hatch, G.M., & Vance, D.E. (1993). Evidence that binding of CTP:phosphocholine cytidylyltransferase to membranes in rat hepatocytes is modulated by the ratio of bilayer- to non-bilayer-forming lipids. Biochem. J. 291, 419-427.

Johnson, J.L., & Francis, B.S. (1975). Taxonomy of the Clostridia: Ribosomal ribonucleic acid homologies among the species. J. Gen. Microbiol. 88, 229-244.

Johnston, N.C., & Goldfine, H. (1982). Effects of growth temperature on fatty acid and alk-1-enyl group compositions of *Veillonella parvula* and *Megasphaera elsdenii* phospholipids. J. Bacteriol. 149, 567-575.

Johnston, N.C., & Goldfine, H. (1983). Lipid composition in the classification of the butyric acid-producing clostridia. J. Gen. Microbiol. 129, 1075-1081.

Johnston, N.C., & Goldfine, H. (1985). Phospholipid aliphatic chain composition modulates lipid class composition, but not lipid asymmetry in *Clostridium butyricum*. Biochim. Biophys. Acta 813, 10-18.

Johnston, N.C., & Goldfine, H. (1988). Isolation and characterization of a novel four chain phospholipid, the phosphatidylglycerol acetal of plasmenylethanolamine from *Clostridium butyricum*. Biochim. Biophys. Acta 961, 1-12.

Johnston, N.C., & Goldfine, H. (1992). Replacement of the aliphatic chains of *Clostridium acetobutylicum* by exogenous fatty acids: Regulation of phospholipid and glycolipid composition. J. Bacteriol. 174, 1848-1853.

Johnston, N.C., & Goldfine, H. (1994). Isolation and characterization of new phosphatidylglycerol acetals of plasmalogens—A family of ether lipids in clostridia. Eur. J. Biochem. 223, 957-963.

Johnston, N.C., Goldfine, H., Malthaner, M., & Seelig, J. (1987). ^{2}H-NMR studies on ether lipid-rich bacterial membranes: Deuterium order profile of *Clostridium butyricum*. Biochim. Biophys. Acta 899, 302-306.

Johnston, N.C., Goldfine, H., & Fischer, W. (1994). Novel polar lipid composition of *Clostridium innocuum* as the basis for an assessemnt of its taxonomic status. Microbiology 140, 105-111.

Kaneda, T. (1991). Iso- and anteiso-fatty acids in bacteria: Biosynthesis, function, and taxonomic significance. Microbiol. Rev. 55, 288-302.

Kates, M. (1992). Archaebacterial lipids: structure, biosynthesis and function. Biochem. Soc. Trans. 58, 51-72.

Kaufman, A.E., Goldfine, H., Narayan, O., & Gruner, S.M. (1990). Physical studies on the membranes and lipids of plasmalogen-deficient *Megasphaera elsdenii*. Chem. Phys. Lipids 55, 41-48.

Kaufman, A.E., Verma, J.N., & Goldfine, H. (1988). Disappearance of plasmalogen-containing phospholipids in *Megasphaera elsdenii*. J. Bacteriol. 170, 2770-2774.

Khuller, G.K., & Goldfine, H. (1974). Phospholipids of *Clostridium butyricum* V. Effects of growth temperature on fatty acid, alk-1-enyl group, and phospholipid composition. J. Lipid Res. 15, 500-507.

Khuller, G.K., & Goldfine, H. (1975). Replacement of acyl and alk-1-enyl groups in *Clostridium butyricum* phospholipids by exogenous fatty acids. Biochemistry 14, 3642-3647.

Killian, J.A., Koorengevel, M.C., Bouwstra, J.A., Gooris, G., Dowhan, W., & De Kruijff, B. (1994). Effect of divalent cations on lipid organization of cardiolipin isolated from *Escherichia coli* strain AH930. Biochim. Biophys. Acta Bio-Membr. 1189, 225-232.

Koga, Y., & Goldfine, H. (1984). Biosynthesis of phospholipids in *Clostridium butyricum*: The kinetics of synthesis of plasmalogens and the glycerol acetal of ethanolamine plasmalogen. J. Bacteriol. 159, 597-604.

Koga, Y., Nishihara, M., Morii, H., & Akagawa-Matsushita, M. (1993). Ether polar lipids of methanogenic bacteria: Structures, comparative aspects, and biosynthesis. Microbiol. Rev. 57, 164-182.

Lawson, P.A., Llop-Perez, P., Hutson, R.A., Hippe, H., & Collins, M.D. (1993). Towards a phylogeny of the clostridia based on 16S rRNA sequences. FEMS Microbiol. Lett. 113, 87-92.

Lepage, C., Fayolle, F., Hermann, M., & Vandecasteele, J.-P. (1987). Changes in membrane lipid composition of *Clostridium acetobutylicum* during acetone-butanol fermentation: Effects of solvents, growth temperature and pH. J. Gen. Microbiol. 133, 103-110.

Lewis, R.N.A.H., Mannock, D.A., McElhaney, R.N., Turner, D.C., & Gruner, S.M. (1989). Effect of fatty acyl chain length and structure on the lamellar gel to liquid-crystalline and lamellar to reversed hexagonal phase transitions of aqueous phosphatidylethanolamine dispersions. Biochemistry 28, 541-548.

Lindblom, G., & Rilfors, L. (1989). Cubic phases and isotropic structures formed by membrane lipids -possible biological relevance. Biochim. Biophys. Acta 988, 221-256.

Linden, J.C., & Moreira, A. (1983). In: Basic biology of new developments in biotechnology (Hollaender, A., Laskin, A.I., & Rogers, P., eds.),pp. 377-403. Plenum Press, NY.

Lohner, K., Hermetter, A., & Paltauf, F. (1984). Phase behavior of ethanolamine plasmalogen. Chem. Phys. Lipids 34, 163-170.

MacDonald, D.L., & Goldfine, H. (1990). Phosphatidylglycerol acetal of plasmenylethanolamine as an intermediate in ether lipid formation in *Clostridium butyricum*. Biochem. Cell Biol. 68, 225-230.

MacDonald, D.L., & Goldfine, H. (1991). Effects of solvents and alcohols on the polar lipid composition of *Clostridium butyricum* under conditions of controlled lipid chain composition. Appl. Environ. Microbiol. 57, 3517-3521.

Macfarlane, M.G. (1962). Characterization of lipoamino-acids as *O*-amino-acid esters of phosphatidylglycerol. Nature 196, 136-138.

Malthaner, M., Hermetter, A., Paltauf, F., & Seelig, J. (1987a). Structure and dynamics of plasmalogen model membranes containing cholesterol: A deuterium NMR study. Biochim. Biophys. Acta 900, 191-197.

Malthaner, M., Seelig, J., Johnston, N.C., & Goldfine, H. (1987b). Deuterium NMR studies on the plasmalogens and the glycerol acetals of plasmalogens of *Clostridium butyricum* and *Clostridium beijerinckii*. Biochemistry 26, 5826-5833.

Marr, A.G., & Ingraham, J.L. (1962). Effect of temperature on the composition of fatty acids in *Escherichia coli*. J. Bacteriol. 84, 1260-1267.

Matsumoto, M., Tamiya, K., & Koizumi, K. (1971). Studies on neutral lipids and a new type of aldehydogenic ethanolamine phospholipid in *Clostridium butyricum*. J. Biochem. 69, 617-620.

McDanield, R.V., Simon, S.A., McIntosh, T.J., & Borovyagin, V. (1982). Interactions of benzene with bilayers. Thermal and structural studies. Biochemistry 21, 4116-4126.

McElhaney, R.N. (1984). The structure and function of the *Acholeplasma laidlawii* plasma membrane. Biochim. Biophys. Acta 779, 1-42.

Melchior, D.L. (1982). Lipid phase transitions and regulation of membrane fluidity in prokaryotes. Curr. Top. Membr. Transp. 17, 263-316.

Morand, O.H., Zoeller, R.A., & Raetz, C.R.H. (1988). Disappearance of plasmalogens from membranes of animal cells subjected to photosensitized oxidation. J. Biol. Chem. 263, 11597-11606.

Morii, H., & Goldfine, H. (1990). Phosphatidyltransferase of *Clostridium butyricum*: Specificity for diacylphosphoglycerides. Biochim. Biophys. Acta 1044, 394-398.

O'Leary, W.M., & Wilkinson, S.G. (1988). Gram-positive bacteria In: Microbial Lipids, Volume 1 (Ratledge, C. & Wilkinson, S.G., eds.), pp. 117-201, Academic Press, London.

Oulevey, J., Bahl, H., & Thiele, O.W. (1986). Novel alk-1-enyl ether lipids isolated from *Clostridium acetobutylicum*. Arch. Microbiol. 144, 166-168.

Pak, J.H., Bork, V.P., Norberg, R.E., Creer, M.H., Wolf, R.A., & Gross, R.W. (1987). Disparate molecular dynamics of plasmenylcholine and phosphatidylcholine bilayers. Biochemistry 26, 4824-4830.

Paltauf, F. (1983a). Biosynthesis of 1-*O*-(1′-Alkenyl) glyeerolipids (plasmalogens) In: Ether Lipids. Biochemical and Biomedical Aspects (Mangold, H.K., & Paltauf, F., eds.), pp. 107-128, Academic Press, NY.

Paltauf, F. (1983b). Ether lipids in biological and model membranes In: Ether lipids. Biochemical and Biomedical Aspects (Mangold, H.K., & Paltauf, F., eds.), pp. 309-353, Academic Press, NY.

Pieringer, R.A. (1989). Biosynthesis of non-terpenoid lipids In: Microbial lipids, Vol. 2 (Ratledge,C. & Wilkinson,S.G., eds.),pp. 51-114, Academic Press, London.

Prins, R.A., Akkermans-Kruyswijk, J., Franklin-Klein, W., Lankhorst, A., & Van Golde, L.M.G. (1974). Metabolism of serine and ethanolamine plasmalogens in *Megasphaera elsdenii*. Biochim. Biophys. Acta 348, 361-369.

Prins, R.A., & Van Golde, L.M.G. (1976). Entrance of glycerol into plasmalogens of some strictly anaerobic bacteria and protozoa. FEBS Lett. 63, 107-111.

Raetz, C.R.H. (1986). Molecular genetics of membrane phospholipid synthesis. Ann. Rev. Genet. 20, 253-295.

Ratledge, C., & Wilkinson, S.G. (1988). Microbial Lipids, Vol. 1 London, Academic Press.

Rietveld, A.G., Chupin, V.V., Koorengevel, M.C., Wienk, H.L.J., Downhan, W., & De Kruijff, B. (1994). Regulation of Lipid Polymorphism is essential for the viability of phosphatidylethanolamine-deficient Escherichia coli cells. J. Biol. Chem. 269, 28670-28675.

Rietveld, A.G., Killian, J.A., Dowhan, W., & De Kruijff, B. (1993). Polymorphic regulation of membrane phospholipid composition in *Escherichia coli*. J. Biol. Chem. 268, 12427-12433.

Rilfors, L., Khan, A., Brentel, I., Wieslander, Å., & Lindblom, G. (1982). Cubic liquid crystalline phase with phosphatidyl-ethanolamine from *Bacillus megaterium* containing branched acyl chains. FEBS Lett. 149, 293-298.

Rilfors, L., Lindblom, G., Wieslander, Å., & Christiansson, A. (1984). Lipid bilayer stability in biological membranes. Biomembranes 12, 205-245.

Rilfors, L., Wieslander, Å., & Lindblom, G. (1993). Regulation and physicochemical properties of the polar lipids in *Acholeplasma laidlawii* In: Subcellular Biochemistry, Volume 20: Mycoplasma Cell Membranes (Rottem,S., & Kahane, I., eds.), pp. 109-166. Plenum Press, NY.

Rogers, P. (1984). Genetics and biochemistry of *Clostridium* relevant to development of fermentation processes. Adv. Appl. Microbiol. 31, 1-60.

Russel, N.J. (1989). Functions of lipids: Structural roles and membrane functions In: Microbial Lipids Vol. 2 (Ratledge, C., & Wilkinson, S.G., eds.), pp. 277-365. Academic Press, London.

Schleifer, K.H., & Stackebrandt, E. (1983). Molecular systematics of prokaryotes. Annu. Rev. Microbiol. 37, 143-187.

Seddon, J.M. (1990). Structure of the inverted hexagonal (H_{II}) phase, and non-lamellar phase transitions of lipids. Biochim. Biophys. Acta 1031, 1-69.

Seelig, J., & Seelig, A. (1980). Lipid conformation in model membranes and biological membranes. Q. Rev. Biophys. 13, 19-61.

Silber, P., Borie, R.P., & Goldfine, H. (1980). The enzymes of phospholipid synthesis in *Clostridium butyricum*. J. Lipid Res. 21, 1022-1031.

Silber, P., Borie, R.P., Mikowski, E.J., & Goldfine, H. (1981). Phospholipid biosynthesis in some anaerobic bacteria. J. Bacteriol. 147, 57-61.

Silbert, D.F. (1975). Genetic modification of membrane lipid. Ann. Rev. Biochem. 44, 315-339.

Silvius, J.R., Lyons, M., Yeagle, P.L., & O'Leary, T.J. (1985). Thermotropic properties of bilayers containing branched-chain phospholipids. Calorimetric, Raman and ^{31}P NMR studies. Biochemistry 24, 5388-5395.

Silvius, J.R., Mak, N., & McElhaney, R.N. (1980). Lipid and protein composition and thermotropic lipid phase transitions in fatty acid-homogeneous membranes of *Acholeplasma laidlawii*. Biochim. Biophys. Acta 597, 199-215.

Smaby, J.M., Hermetter, A., Schmid, P.C., Paltauf, F., & Brockman, H.L. (1983). Packing of ether and ester phospholipids in monolayers. Evidence for hydrogen-bonded water at the *sn*-1 acyl group of phosphatidylcholines. Biochemistry 22, 5808-5813.

Snyder, F. (1972). The enzymic pathways of ether-linked lipids and their precursons In: Ether Lipids. Chemistry and Biology (Snyder, F., ed.), pp. 121-156. Academic Press, NY.

Snyder, F. (1991). Metabolism, regulation, and function of ether-linked glycerolipids and their bioactive species In: Biochemistry of lipids, lipoproteins and membranes (Vance ,D.E., & Vance, J., eds.), pp. 241-267. Elsevier Science, Amsterdam.

Stockton, G.W., Johnson, K.G., Butler, K., Tulloch, A.P., Boulanger, Y., Smith, I.C.P., Davis, J.H., & Bloom, M. (1977). Deuterium NMR study of lipid organisation in *Acholeplasma laidlawii* membranes. Nature 269, 267-268.

Tate, M.W., Eikenberry, E.F., Turner, D.C., Shyamsunder, E., & Gruner, S.M. (1991). Nonbilayer phases of membrane lipids. Chem. Phys. Lipids 57, 147-164.

Tung, B.S., Unger, E.R., Levin, B., Brasitus, T.A., & Getz, G.S. (1991). Use of an unsaturated fatty acid auxotroph of *Saccharomyces cerevisiae* to modify the lipid composition and function of mitochondrial membranes. J. Lipid Res. 32, 1025-1038.

van Dijck, P.W.M., De Kruijff, B., van Deenen, L.L.M., de Gier, J., & Demel, R.A. (1976). The preference of cholesterol for phosphatidylcholine in mixed phosphatidylcholine-phosphatidylethanolamine bilayers. Biochim. Biophys. Acta 455, 576-587.

Verkley, A.J., Ververgaert, P.H.J.T., Prins, R.A., & Van Golde, L.M.G. (1975). Lipid phase transitions of the strictly anaerobic bacteria *Veillonella parvula* and *Anaerovibrio lipolytica*. J. Bacteriol. 124, 1522-1528.

Verma, J.N., & Goldfine, H. (1985). Phosphatidylserine decarboxylase from *Clostridium butyricum*. J. Lipid Res. 26, 610-616.

Walton, P.A., & Goldfine, H. (1987). Transphosphatidylation activity in *Clostridium butyricum* Evidence for a secondary pathway by which membrane phospholipids may be synthesized and modified. J. Biol. Chem. 262, 10355-10361.

Watanabe, T., Okuda, S., & Takahashi, H. (1984). Turn-over of phospholipids in *Selenomonas ruminantium*. J. Biochem. 95, 521-527.

Westerman, P.W., Pope, J.M., Phonphok, N., Doane, J.W., & Dubro, D.W. (1988). The interaction of *n*-alkanols with lipid bilayer membranes: A ^{2}H-NMR study. Biochim. Biophys. Acta 939, 64-78.

Wieslander, Å., Christiansson, A., Rilfors, L., & Lindblom, G. (1980). Lipid bilayer stability in membranes. Regulation of lipid composition in *Acholeplasma laidlawii* as governed by molecular shape. Biochemistry 19, 3650-3655.

Wieslander, Å., Christiansson, A., Rilfors, L., Khan, A., Johansson, L.B.-Å., & Lindblom, G. (1981). Lipid phase structure governs the regulation of lipid composition in membranes of *Acholeplasma laidlawii*. FEBS Lett. 124, 273-278.

Wieslander, Å. & Rilfors, L. (1977). Qualitative and quantitative variations of membrane lipid species in *Acholeplasma laidlawii* A. Biochim. Biophys. Acta 466, 336-346.

Wieslander, Å., Rilfors, L., & Lindblom, G. (1986). Metabolic changes of membrane lipid composition in *Acholeplasma laidlawii* by hydrocarbons, alcohols, and detergents: arguments for effects on lipid packing. Biochemistry 25, 7511-7517.

Wilkinson, S.G. (1988). Gram-negative bacteria In: Microbial Lipids Vol. 1 (Ratledge, C., & Wilkinson, S. G., eds.), pp. 299-488, Academic Press, London.

Wilkinson, S.R. & Young, M. (1993). Wide diversity of genome size among different strains of *Clostridium acetobutylicum*. J. Gen. Microbiol. 139, 1069-1076.

Yang, R.D., Patel, K.M., Pownall, H.J., Knapp, R.D., Sklar, L.A., Crawford, R.B., & Morrisett, J.D. (1979). Biophysical properties of a major membrane phospholipid, dielaidoylphosphatidylethanolamine, found in an *Escherichia coli* fatty acid auxotroph. J. Biol. Chem. 254, 8256-8262.

Zoeller, R.A., Morand, O.H., & Raetz, C.R.H. (1988). A possible role for plasmalogens in protecting animal cells against photosensitized killing. J. Biol. Chem. 263, 11590-11596.

THE SPHINGOMYELIN CYCLE: THE FLIP SIDE OF THE LIPID SIGNALING PARADIGM

Yusuf A. Hannun and Supriya Jayadev

Advances in Lipobiology
Volume 2, pages 143-166.

ISBN 0-7623-0205-4

I. INTRODUCTION

Sphingolipids, first described by Johann Thudichum in 1874 (Thudichum, 1962), are ubiquitous lipid molecules of cellular membranes. With their characteristic amphipathic structure, sphingolipids have long been considered as important structural components of cells; crucial for maintaining membrane integrity and fluidity (Pascher, 1976; Holleran et al.,1991). Various sphingolipids have also been associated with and have been defined as markers of biological and disease phenotypes (see Hannun and Bell, 1989 for review). Studies with the addition of glycolipids and glycolipid-directed antibodies have also illustrated the ability of glycosphingolipids to regulate a number of biological end-points, including: growth, cell-cell contact, adhesion and differentiation (Hakomori, 1980; Hakomori, 1981; Hakomori and Kannagi, 1983; Okada et al., 1984; Okada et al., 1985). It is only recently, however, that sphingolipid-derived molecules have been assigned a role in intracellular signal transduction.

Recent work has renewed interest in sphingolipids by demonstrating that sphingosine, the common backbone structure of all sphingolipids, acts as a potent endogenous inhibitor of protein kinase C (PKC) (Hannun et al., 1986). This was the first demonstration of sphingolipids as regulators of a signaling pathway. Thus, the discovery of sphingosine's inhibition of PKC provided the impetus for studying sphingolipids as potential endogenous biomodulators. With the subsequent discovery of the "sphingomyelin cycle," sphingolipids have become increasingly recognized as cell regulatory and signaling molecules.

The sphingomyelin (SM) cycle was first described in the human leukemia-derived HL60 cell line, where it was found that an inducer of monocytic differentiation, dihydroxyvitamin D_3, was able to stimulate SM hydrolysis (Okazaki et al., 1989). This SM hydrolysis occurred rapidly and exhibited many characteristics of a signaling event. Loss of SM was: (1) dose and time dependent, (2) transient, and (3) specific, occurring in response to a specific subset of inducers of HL60 differentiation (Kim et al., 1991). More importantly, ceramide, the lipid breakdown product of SM, has been shown to regulate many of the downstream effects of these hormones/cytokines on cell growth, differentiation, and apoptosis (Okazaki et al., 1990; Kim et al., 1991; Obeid et al., 1993; Jarvis et al., 1994).

The description of the SM cycle therefore brought the role of sphingolipid-derived molecules into the forefront as potentially critical signaling molecules of the cell. Over the past few years a great deal of work has focused on: (1) understanding the regulation of the SM cycle, (2) determining the components of the cycle, and (3) defining the targets and biological functions of the liberated second messenger, ceramide. The purpose of this chapter is to provide insight into what has been learned in these areas of study.

II. BACKGROUND: SPHINGOLIPID METABOLISM AND BIOLOGY

A. Structure

Sphingolipids are composed structurally of a hydrophilic headgroup conjugated to the lipophilic ceramide backbone (Figure 1). Ceramide, in turn, is composed of a long-chain alcohol (sphingosine) and an amide-linked fatty acid (Vance and Vance, 1985). Classically, the distinct headgroups of sphingolipids have defined the specific sphingolipid species. For example, sphingomyelin (Figure 1) contains a choline phosphate group at the 1 hydroxyl position of ceramide whereas cerebroside contains either a glucose or a galactose in ester linkage (Hannun, 1991). Other complex glycolipids and gangliosides are the products of more complex glucose-containing headgroups. Additionally, for each sphingolipid, heterogeneity of backbone structure has also been determined (Karlsson, 1970). This is the result of: (1) different fatty acyl groups (with varying chain length and/or hydroxylation status) in amide linkage, and (2) heterogeneity in the sphingoid backbone although C_{18} sphingosine is the predominant species in most mammalian species. With these differences in headgroup and backbone structure, the complexity of sphingolipids far exceeds the complexity of the glycerophospholipids.

B. Metabolism

The biosynthesis of sphingolipids is initiated by the rate limiting serine palmitoyl transferase (Figure 2). This enzyme catalyzes the condensation of serine and palmitoyl CoA resulting in the formation of 3-ketosphinganine (Williams et al., 1984; Braun and Snell, 1968; Brady and Koval, 1958; Krisnangkura and Sweeley, 1976). The 3-ketosphinganine then serves as a direct precursor for dihydrosphingosine which is then acylated to dihydroceramide (Braun et al., 1970; Braun and Snell, 1968; Stoffel et al., 1968; Merrill, Jr. and Jones, 1990; Ong and Brady, 1973). The sphingolipid double bond is probably introduced at this step resulting in the formation of ceramide in the Golgi apparatus (Merrill Jr. and Wang, 1986; Stoffel and Bister, 1974; Ong and Brady, 1973; Kolesnick, 1991). Ceramide then serves as a precursor for sphingomyelin, ceramide-phosphoethanolamine, ceramide-phosphate, and cerebroside (Hannun, 1991). The cerebrosides in turn serve as precursors for sulfatides and gangliosides. In effect, ceramide plays a central role in sphingolipid biosynthesis nearly identical to the role of diacylglycerol (DAG) in glycerol phospholipid biosynthesis (Hannun and Bell, 1993).

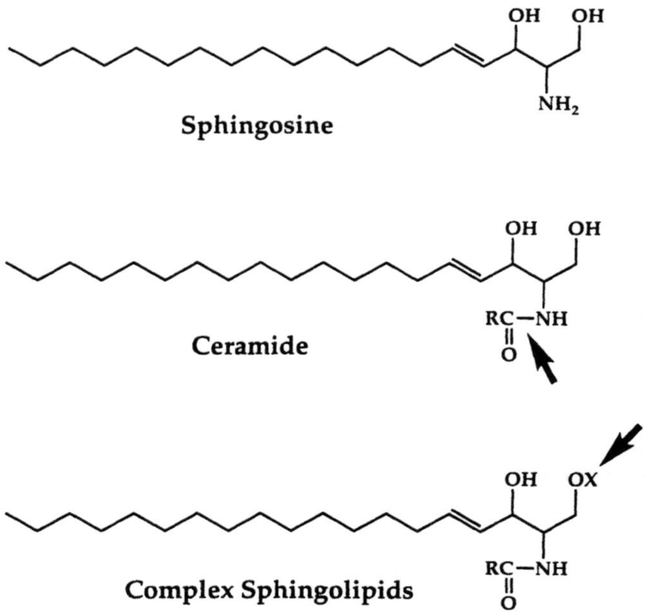

X =	Nomenclature
phosphorylcholine	**Sphingomyelin**
1-20 glycose units	**Neutral Sphingolipid**
glycose units + sulfate monoesters	**Sulfatide**
glycose units + sialic acid residues	**Ganglioside**

Figure 1. Sphingolipid nomenclature. The long-chain base sphingosine is usually composed of an 18 carbon chain containing two hydroxyl groups at carbons 1 and 3, an amine at carbon 2, and a trans double bond between carbons 4 and 5. Addition of an amide-linked fatty acid to sphingosine produces ceramide, the common backbone for more complex sphingolipids. In turn, complex sphingolipids are defined by their headgroup composition.

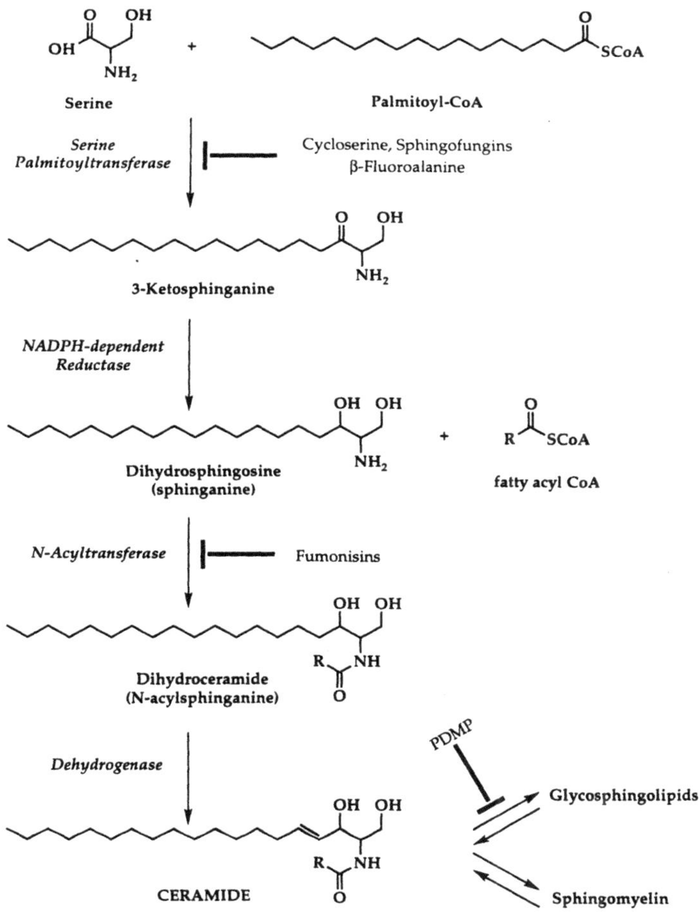

Figure 2. Sphingolipid biosynthesis and inhibitors of the synthetic pathway. The first committed reaction of sphingolipid biosythesis is the condensation of serine and palmitoyl-CoA. This initial enzymatic step, catalyzed by serine palmitoyl transferase, yields 3-ketosphinganine. The 3-ketosphinganine is rapidly reduced to dihydrosphingosine by a microsomal NADPH-dependent reductase. Formation of dihydrosphingosine is followed by addition of an amide-linked fatty acid. The enzyme involved, a microsomal acyl-CoA:long chain base N-acyltransferase, utilizes the various fatty acyl-CoAs with a relative order of efficiency which parallels their distribution ratios in sphingolipids. The product of this reaction, dihydroceramide, is then desaturated by a dehydrogenase to yield ceramide. This ceramide can then be utilized as the precursor for more complex sphingolipids. Various inhibitors have been described to interfere with this biosynthetic pathway, among them are: (1) cycloserine, sphingofungins and β-fluoroalanine that inhibit the serine palmitoyl transferase, (2) fumonisins that inhibit the N-acyltransferase, and (3) PDMP which inhibits the formation of cerebrosides.

The breakdown of complex sphingolipids proceeds through a stepwise pathway whereby distinct hydrolases cleave off various components of the sphingolipid headgroups resulting ultimately in the formation of ceramide (Spence, 1993; Chatterjee, 1993; Sandhoff and van Echten, 1993). Ceramide, in turn, can be further broken down to sphingosine through the action of poorly-characterized ceramidases (Hassler and Bell, 1993). Alternatively, ceramide, formed from sphingolipid catabolism, may be reincorporated into sphingolipid biosynthetic pathways.

A number of inhibitors have been discovered to modulate sphingolipid metabolism (Figure 2). Cycloserine and β-fluroalanine inhibit serine palmitoyl transferase and thus result in general loss of sphingolipids (Williams et al., 1987; Sundaram and Lev, 1984; Medlock and Merrill Jr., 1988). The sphingofungins, isolated as natural products with antifungal activity, also act by inhibiting serine palmitoyl transferase (Horn et al., 1992; Zweerink et al., 1992). Recently, the fumonisins have been shown to inhibit the acylation of dihydrosphingosine and sphingosine; resulting in accumulation of these two sphingoid bases and subsequent decreases in the levels of dihydroceramide, ceramide, and complex sphingolipids (Wang et al., 1991; Wang et al., 1992; Shier, 1992; Merrill et al., 1993; Kaneshiro et al., 1992). PDMP and related analogs have been developed as inhibitors of cerebroside synthase. Their action results in a decrease in the levels of cerebroside and complex sphingolipids with a concomitant increase in the levels of ceramide (Inokuchi et al., 1989; Rosenwald et al., 1992).

The availability of these inhibitors has provided important insight into potential roles for sphingolipids in cell growth regulation and viability, and promises to provide new insight into other functions of sphingolipids.

C. Biology

In addition to the wealth of insight on sphingolipid structure, a number of approaches have been undertaken in order to acquire insight into potential functions associated with individual sphingolipids. Thus, the species of sphingolipids present in different cells and tissue types and their changes during development, differentiation, and oncogenesis have been characterized. This has led to identification of unique sphingolipids and gangliosides as specific antigens, tumor markers and markers of specific cell lineages and distinct phases of development (for reviews, please refer to (Hakomori and Igarashi, 1993; Fishman et al., 1993; Lingwood, 1993; Fredman, 1993; Tettamanti and Riboni, 1993)). In cell biologic approaches, various sphingolipids have been added to cells in culture and changes in cell function and behavior have been examined. Such studies have resulted in the identification of a number of effects of specific sphingolipids on cell growth, differentiation, and cell-cell contact responses (Hakomori, 1981; Hakomori and Kannagi, 1983; Okada et al., 1984; Okada et al., 1985). Such findings have also been supported with studies using antibodies directed against individual glycolipids

(Nudelman et al., 1983; Thudichum, 1962). In addition, various sphingolipids have been implicated as either specific receptors for extracellular agents (such as choleratoxin) or as coreceptors (for various matrix proteins such as fibronectin) (Hannun and Bell, 1989; Hakomori and Kannagi, 1983). The mechanisms by which these glycolipids regulate cell function, however, have remained for the most part rather elusive. Taken together, these studies suggest that the complexity of sphingolipid structure may be matched by an equally complex role in the regulation of cell and tissue behavior.

III. THE SPHINGOMYELIN CYCLE

Since its initial description, the SM cycle has been found to exist in multiple cell systems ranging from monocytes (Obeid et al., 1993) and lymphocytes (Niculescu et al., 1993) to fibroblasts (Ballou et al., 1992) and glioma cells. The number of inducers shown to be capable of signaling through this cycle has also expanded. In HL60 cells, other inducers of monocytic differentiation - tumor necrosis factor α (TNFα) (Kim et al., 1991), γ-interferon (Kim et al., 1991) and brefeldin A[2]—were found to cause SM hydrolysis and ceramide generation. In other cell systems, interleukin-1 (Ballou et al., 1992; Mathias et al., 1993), dexamethasone (Ramachandran et al., 1990), terminal complement complexes (Niculescu et al., 1993), and NGF[1] have also been found to be agonists capable of signaling through this cascade. With this expanding list of agonists and cell lines being defined, the SM cycle has evolved into a more general mechanism of cell signaling.

A. Stages of the SM Cycle

The SM cycle consists of two phases (Figure 3): (1) the SM hydrolysis stage, and (2) the SM resynthesis stage. In the first stage, the interaction of agonist with receptor stimulates a neutral sphingomyelinase (SMase) (Okazaki et al., 1989), possibly through mobilization of arachidonic acid (AA). This SMase activity has been found to reside in the cytosolic fraction of cells (Okazaki et al., 1994). Upon activation, however, this enzyme must translocate to a target membrane where it catalyzes the cleavage of membrane SM. Much like the activity of phospholipase C (PLC), the SMase cleaves the phosphodiester bond linking the hydrophilic head group to the hydrophobic portion of the lipid. The consequence of this hydrolysis is the generation of choline-phosphate and ceramide. Ceramide in turn, transduces many of the downstream effects of the agonist (see below). What role, if any, the generated choline-phosphate plays in regulating agonist response remains to be determined.

Very little is understood about the mechanisms by which agonist-occupation of receptor couples to SMase. Recently, some insight has come from work with the TNFα - TNF receptor system. TNFα, like some of the other agonists known to couple to the SM signaling cascade, interacts with a unique family of receptors. These

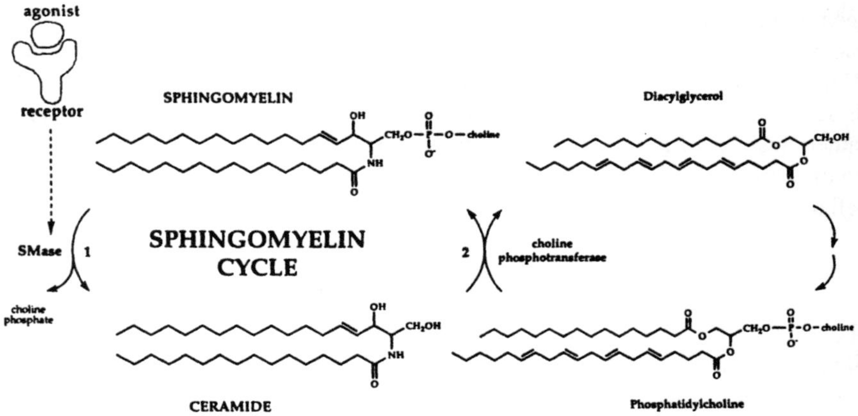

Figure 3. The sphingomyelin cycle. In the hydrolysis stage of the sphingomyelin cycle (#1 in figure), agonist occupation of receptor leads to the activation of a neutral, cytosolic sphingomyelinase (SMase). This SMase hydrolyzes sphingomyelin to generate choline-phosphate and the second messenger ceramide. In the regenerative phase (#2 in figure), the activity of a choline phosphotransferase leads to the resynthesis of sphingomyelin from ceramide. The choline phosphate headgroup is thought to come from phosphatidylcholine; thus, the generation of sphingomyelin is accompanied by concomitant increase in diacylglycerol levels.

receptors have no known intrinsic function and are not known to be coupled with proteins which direct their function. However, a number of signaling events have been variably associated with TNF receptor occupation, including: G-proteins, phosphorylation cascades, cAMP, oxygen radicals, phosphatases, phospholipase C, phospholipase D, and phospholipase A_2 (see (Vilcek and Lee, 1991) for review).

Recently, studies from our laboratory have begun to implicate the phospholipase A_2/AA pathway in the coupling of TNFα to SMase (Jayadev et al., 1994). AA has been shown to be rapidly generated upon TNFα stimulation in many cell systems. In HL-60 cells, the addition of TNFα results in rapid mobilization of arachidonic acid within 3-10 minutes; preceding the effects of TNFα on ceramide mobilization. Exogenous addition of AA was able to inhibit growth similar to the action of ceramide. Furthermore, free arachidonic acid, but not its methyl ester or alcohol analogs, could rapidly stimulate SM hydrolysis and ceramide generation in intact cells. Moreover, in an *in vitro* assay, AA was able to stimulate a neutral cytosolic SMase. These results suggested that the activation of PLA_2 by TNFa may couple to activation of SMase. In support of a role for PLA_2, the addition of melittin, a potent activator of PLA_2, resulted in rapid and dramatic SM hydrolysis and ceramide formation. However, unlike AA, melittin was unable to activate the neutral sphingomyelinase in the *in vitro* assay, suggesting that the effects of melittin occurred indirectly; probably through activation of PLA_2. These studies suggest one possible biochemical link between TNFα receptor activation and SMase activation. This is further supported by other studies that have implicated PLA_2 in the cytotoxic mechanism of action of TNFα (Suffys et al., 1987; Neale et al., 1988; Mutch et al., 1992; Hayakawa et al., 1993).

At this point, AA meets many (but not all) of the criteria required of a molecule which plays an intermediary role between TNFα and SMase (Jayadev et al., 1994): (1) AA stimulates both SM hydrolysis and ceramide generation. (2) AA stimulates a neutral, cytosolic SMase in an *in vitro* assay system. (3) The kinetics of AA release and AA-stimulated SM hydrolysis fit well with the observed kinetics of TNFα-stimulated activation of the SM cycle. (4) AA shows specificity. (5) AA is able to elicit the same biology as ceramide in terms of growth inhibition.

In contrast, Schutze and co-workers have suggested that DAG may function as a mediator between TNFα and the SM cycle (Schutze et al., 1991; Schutze et al., 1992). They found that upon activation of the p55 TNF receptor, a phosphatidylcholine-specific phospholipase C was rapidly activated. In their cell lysate system, exogenous addition of DAG stimulated the activity of an acidic SMase. However, in studies with intact cells, neither DAG nor phorbol esters (potent PKC activators) were capable of stimulating SM hydrolysis or ceramide generation (Kim et al., 1991; Jayadev et al., 1994). Thus, the relevance of DAG as an intermediary between TNFα and the SM cycle in an intact cell remains to be determined.

Almost nothing is known about what regulates the onset of the second stage of the SM cycle; namely, the resynthesis of SM. However, the consequence of this regenerative phase is the restoration of SM and ceramide to basal levels. SM is resynthesized by the transfer of the choline-phosphate head group from phosphatidylcholine to ceramide (see Figure 3) (Ullman and Radin, 1974; Diringer et al., 1972; Marggrag et al., 1981). In addition to the resynthesis of SM during this regenerative phase, diacylglycerol (DAG) is also formed. Thus, the removal of one second messenger molecule—ceramide—leads to the generation of another - DAG. This converse relationship between signal transduction molecules is intriguing since it may reflect an inverse relationship in their functions (see below).

IV. CERAMIDE BIOLOGY

The identification of ceramide as a sphingolipid-derived product whose generation is regulated by the action of cytokines raises the possibility that ceramide may participate in mediating, at least part of, the cellular activities of these cytokines. Along these lines, the role of ceramide as a mediator for TNFα has received particular attention. TNFα is a pleiotropic cytokine with multiple cellular activities including induction of differentiation, cell death, mitogenesis (in a few cell types), and regulation of multiple signaling pathways, transcription factors, and oncogenic products (see Vilcek and Lee, 1991 for review). In initial studies, it was shown that cell-permeable ceramides were able to induce monocytic differentiation of HL60 cells; thus mimicking the effects of both 1,25-dihydroxyvitamin D_3 and TNFα (Okazaki et al., 1990). Moreover, ceramide was shown to inhibit the growth of a number of leukemia cells and to exert cytotoxic effects.

Investigation of the cytotoxicity of ceramide resulted in the identification of the ability of ceramide to induce programmed cell death (apoptosis). Thus, exogenous ceramides induce the morphologic characteristics of apoptotic cell death and cause DNA fragmentation in the characteristic DNA ladder observed with apoptosis (Obeid et al., 1993). These effects were specific to ceramide in that closely related molecules such as dihydroceramide were inactive. The cytotoxicity of ceramide also mimicked the effects of TNFα in that it was seen primarily in cell lines known to respond to the cytotoxic/cytocidal activities of TNFα. Importantly, the ability of both TNFα and ceramide to exert cytotoxicity is countered by co-stimulation of cells with phorbol esters or diacylglycerols (Obeid et al., 1993). In these cell types activation of PKC seems to enhance viability and to counteract the effects of the ceramide pathway. Thus, ceramide is emerging as an important regulator of growth, differentiation, and apoptosis.

Many of the effects of ceramide may result from regulation of transcription factors. *C-myc*, a transcription factor associated with high proliferative capacity, has been shown to be down-regulated upon exposure of cells to ceramide (Kim et al., 1991). Others have suggested that ceramide may regulate another transcription factor, NFκB (Schutze et al., 1992; Bell, 1986). NFκB is rapidly activated in response to TNFα, and translocates to the nucleus where it participates in gene transcription through specific NFκB-binding DNA elements. Thus, ceramide may function, in part, through this pathway. However, the balance of emerging evidence suggests that ceramide is not sufficient to activate NFκB, and it is not determined whether it plays a necessary role in this pathway (Dbaibo et al., 1993).

V. CELLULAR TARGET(S) FOR CERAMIDE

If ceramide participates in a signal transduction pathway (in analogy with DAG), then an important cornerstone in establishing this role for ceramide is to identify a proximal target that is responsive to ceramide and serves to mediate at least some of its activities. (Analogous to PKC). Also, identification of such a target would be critical in defining the mechanism of action of ceramide. Therefore, effort has been directed at identifying the most proximal target for ceramide action. These studies have so far identified two candidate targets: a serine/threonine protein phosphatase and a serine/threonine protein kinase.

Investigation of the effects of ceramide on cellular protein phosphorylation led to the observation that ceramide may enhance protein dephosphorylation. Prompted by these findings, the effects of ceramide on protein phosphatases was examined *in vitro*. These results led to the identification of a ceramide activated protein phosphatase (CAPP). CAPP was initially isolated from crude cytosolic preparations of T9 glioma cells as a phosphatase activatable by low micromolar concentrations of ceramide (Dobrowsky and Hannun, 1992). It is a serine/threonine phosphatase of the PP2A class and can be potently inhibited by the fatty acid polyether, okadaic acid (Dobrowsky et al., 1993; Fishbein et al., 1993). Thus, an inhibitor of ceramide-biology is also emerging in the from of okadaic acid. In cellular systems tested (growth inhibition[3], apoptosis[4] (Obeid et al., 1993), and *c-myc* down regulation[3]) okadaic acid has been found to inhibit the effects of ceramide. Therefore, these pharmacologic studies are beginning to identify a role for a PP2A-like enzyme in the mechanism of action of ceramide. More direct evidence for a role for CAPP in mediating cellular activities of ceramide have come from biochemical studies on the specificity of activation of CAPP by ceramide analogs (see next section). These studies are beginning to disclose a very close analogy between the specificity of activation of CAPP *in vitro* by ceramide analogs and the ability of these analogs to induce apoptosis, growth suppression, and down regulation of the *c-myc* protooncogene. This specificity suggests that CAPP may function as the proximal mediator of the effects of ceramide.

In other studies, the possibility that ceramide may activate a protein kinase has been suggested. In initial studies, it was demonstrated that sphingosine induced phosphorylation of the EGF receptor at threonine 669 (Faucher et al., 1988; Davis et al., 1988). In these initial studies, C_2-ceramide failed to induce this phosphorylation.; however, in subsequent studies, both sphingosine and C_8-ceramide were shown to activate this kinase (Goldkorn et al., 1991; Mathias et al., 1991). This magnesium-dependent, proline-directed, serine/threonine kinase has been shown to be activated by ceramide and TNFα in intact and in cell free systems (Mathias et al., 1991).

However, very little is known about the function this ceramide-activated kinase plays in regulating agonist function. Moreover, with the recent partial purification of this 97kDa enzyme, it has become apparent that this activity is not a direct target of ceramide *in vitro* (Liu et al., 1994). This kinase may therefore emerge as a downstream target for ceramide.

At this time, the only known direct *in vitro* target for ceramide action is CAPP. The substrates on which this phosphatase works remain to be identified and future studies are being directed at defining these substrates as well as at increasing understanding of the mechanism of ceramide activation.

VI. STRUCTURAL SPECIFICITY

Initial studies into the biological activity of ceramide were performed by addition of exogenous ceramide to cells. Since naturally occurring ceramides are difficult to introduce into aqueous solutions, a number of short-chain analogs of ceramide were developed (see Figure 4). One of the first to be utilized, C_2-ceramide was shown to potentiate the effect of dihydroxyvitamin D_3 on HL60 cell differentiation

(Okazaki et al., 1990; Bielawska et al., 1992b). Addition of C_2-ceramide alone was also able to stimulate growth inhibition and stimulate differentiation. Furthermore, the potency of C_2-ceramide was shown to be a function of the available lipid to cell ratio as opposed to the absolute concentration of ceramide (Bielawska et al., 1992b). Thus, ceramide exhibited characteristics prototypic of a lipid bioeffector.

Studies with the water soluble ceramides illustrated that the effectiveness of ceramide was not the consequence of a detergent-like perturbation of membrane structure. Furthermore, work with the closely related molecules, sphingosine and N-ethylsphingosine, demonstrated the necessity for the amide-linked acyl chain to produce activity (Okazaki et al., 1990). Extension of the study of ceramide's specificity required the development of structural analogs with less bulky backbone structure. Towards this end a number of N-acyl-phenylaminoalacohol analogs of ceramide were developed (see figure 4) (Bielawska et al., 1992a). These structural analogs provided the opportunity to study the effects of a number of factors on ceramide specificity: (1) the importance of the free hydroxyl groups, (2) the minimum hydrophobicity required; and, most importantly, (3) stereospecificity. These analogs demonstrated that the hydroxyls at positions 1 and 3 of the sphingosine backbone had modest effects on function. In contrast, the presence and length of the amide-linked fatty acyl chains were critical for biologic activity, probably indicating that a minimum hydrophobicity was critical to bioactivity. The compounds also illustrated the importance of stereochemistry to ceramide function. The D enantiomers were found to be the active forms with the naturally occurring *erythro* isomer exhibiting the greatest potency in both growth inhibition and differentiation assays. These analogs begin to demonstrate the stereospecificity of ceramide, and with the advent of short-chain ceramide stereoisomers these studies of specificity can be further extended.

Finally, using dihydroceramide (see Figure 4), the importance of the trans 4,5 double bond has been assessed. Short-chain dihydroceramide analogs were developed for assays of biological activity. In all systems tested (including the yeast *S. cerevisiae* (Fishbein et al., 1993), human lymphocytes (Bielawska et al., 1993), and monocytes (Obeid et al., 1993)) dihydroceramide was unable to produce the biological effects of ceramide on growth inhibition, apoptosis, and *c-myc* down regulation (Bielawska et al., 1993). The lack of activity of dihydroceramide was not due to poor cellular uptake or more rapid metabolism compared to ceramide. In fact, the uptake of dihydroceramide when assessed using radiolabeled analogs was equivalent to the cellular uptake of ceramide (Bielawska et al., 1993). Moreover, both tritium-labeled ceramide and tritium-labeled dihydroceramide underwent minimal metabolic transformation in leukemia cells (Bielawska et al., 1993). Thus, the cellular levels of both ceramide and dihydroceramide were equivalent under the conditions used for the various biological assays. Therefore, this specificity indicates differences in the mechanism of action of these two lipids rather than differences in pharmacologic behavior. In corroboration of this, it was shown that ceramide but not dihydroceramide was able to activate CAPP *in vitro* (Do-

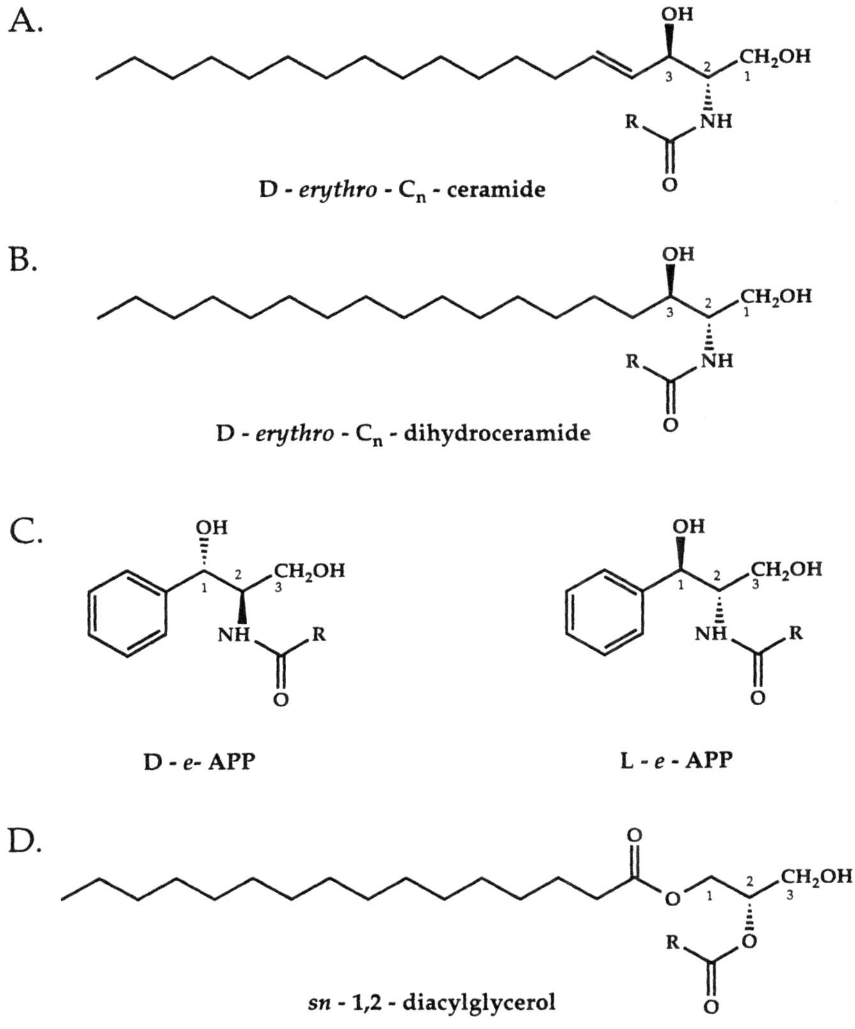

Figure 4. Structural comparison of ceramide and related molecules. (A) Structure of the naturally occurring D-erythro form of ceramide. the length and saturation of the fatty acid in amide linkage produces part of the structural diversity observed in sphingolipids. (B) Structure of dihydroceramide, which differs from ceramide only in the absence of a 4-5 trans double bond. (C) Structures of N-acyl-phenylaminoalcohol analogs of ceramide. These analogs have a less bulky backbone but preserve the free hydroxyls, the amide linked fatty acid and the stereospecificity of ceramide. (D) The structure of diacylglycerol.

browsky et al., 1993). These studies, therefore, further support a role for CAPP in ceramide-mediated biology (see above).

These studies also present an important insight into the sphingolipid double bond. This double bond emerges as critical for defining the functional activity of ceramide since both ceramide and dihydroceramide are known to exist in the cell. Moreover, dihydroceramide serves as a metabolic precursor to ceramide (Stoffel and Bister, 1974; Ong and Brady, 1973). At the biochemical level, it is not clear whether the double bond is introduced directly into dihydroceramide to form ceramide or whether dihydroceramide is incorporated first into more complex sphingolipids (such as sphingomyelin and cerebroside) prior to the introduction of the double bond. According to the latter scheme, the possibility emerges that ceramide may not function as a metabolic precursor in the biosynthesis of sphingolipids but may only function as a sphingolipid-breakdown product. This may distinguish a signaling role for ceramide from a purely metabolic function. More importantly, this may serve to prevent the exposure of cells to significant levels of ceramide in the resting condition. In addition, these metabolic steps appear to occur in the Golgi apparatus such that ceramide, if present as a metabolic precursor, is localized to the Golgi and is only exported outside the Golgi after processing into higher sphingolipids; again underscoring the requirement for cells to sequester ceramide and prevent exposure (perhaps by dissociating ceramide from its signaling targets).

VII. DISCUSSION

These studies are beginning to define a sphingomyelin/ceramide pathway of signal transduction and cell regulation. Insight into this pathway is still, however, at a preliminary stage and a number of key questions arise: (1) Is ceramide a second messenger? (2) Is ceramide a critical mediator of apoptosis? (3) What are the relative roles of ceramide and DAG in the regulation of growth and viability?

A. Ceramide as a Second Messenger

Ceramide, the critical signaling molecule of the SM cycle, meets many of the criteria required of a second messenger: (1) it is generated as a consequence of agonist stimulation, (2) its accumulation is transient, (3) its generation precedes the biologic response that it is thought to mediate, (4) it interacts with and specifically regulates the function of a target molecule, (5) addition of exogenous ceramide mimics the biologic effects of agonist, and (6) it exhibits specificity of action.

According to the above, ceramide meets the same criteria for a second messenger function as DAG or cAMP. However, unlike cAMP, both DAG and ceramide require additional investigation into their physiologic signaling function. Both are important intermediates in lipid metabolism, and they do not appear to have arisen (in an evolutionary sense) as specifically targeted for signal transduction in contra-distinction to the current understanding of cAMP as a specific metabolic product of signaling pathways. The case for the second messenger function of DAG has been strengthened immeasurably through the identification of phorbol esters as naturally-derived molecules (from plants) that mimic the action of endogenous DAGs. These studies suggest that plants have evolved their own (protective) mechanisms that operate by targeting a physiologic signal transduction pathway. Phorbol esters induce rapid inflammatory responses resulting in skin irritation and (probably avoidance behavior) thus offering protection to plants. An important expectation, therefore, is that a similarly important physiologic pathway mediated by ceramide may be targeted by other naturally occurring products. However, to date, a molecule analogous to PMA in the ceramide pathway has not been discovered.

In spite of the above accumulating evidence for a second messenger function of ceramide (and even for DAG), other scenarios (albeit more complex) can be envisioned where neither ceramide nor DAG function as signaling molecules. Thus, DAG and ceramide may be generated in response to physiologic agonists either as a byproduct of the "real" agonist (for example PA may be the real agonist in the phospholipase D pathway whereas DAG may be a metabolic product) or DAG and ceramide may serve as precursors to as yet unidentified signaling molecules (such as monoacylglycerol or phosphatidic acid in the case of DAG, or sphingosine and ceramide phosphate in the case of ceramide). Moreover, in the case of DAG, the ability to activate PKC may reflect the ability of DAG to mimic the action of a more physiologic endogenous activator. Accordingly, it has been difficult to prove conclusively that DAG, generated endogenously, activates PKC physiologically. For example, PIP2 has been shown to be an even more potent activator of PKC than DAG in vitro . Thus, a signaling function for DAG and ceramide is not established conclusively. Current data would argue strongly in support of a signaling function for DAG. The current data also provides almost as much support for a signaling function for ceramide. Unfortunately, providing conclusive evidence would require specific targeting of the signaling function of either DAG or ceramide through either genetic, biochemical, or pharmacologic approaches that do not appear to be realizable at the present.

At this point, two levels of criteria for the study of lipid second messengers may be considered. At a first level, identification of a physiologically-relevant putative lipid second messenger requires the following:

1. Demonstration that the endogenous generation of the candidate lipid occurs in response to physiologic stimuli and extracellular agents. (This is well-established for DAG and ceramide.)
2. Demonstration of biologic/cellular activity of this lipid. This is more feasible with cell-permeable analogs of the candidate lipids, but may be quite difficult with polar lipids (such as phosphatidic acid) that may not gain access to intracellular compartments. The cellular activity of the lipid second messenger should mimic, at least in part, the activity of the

extracellular agent inducing the formation of the candidate lipid. (This has been demonstrated repeatedly for ceramide and DAG).

3. Demonstration that the kinetics of generation of the lipid second messenger should precede the responses they are supposed to mediate.

The above criteria establish conditions sufficient (but not necessary) to assign a second messenger function to a lipid: i.e., inducer (A) → lipid second messenger (B) and lipid second messenger → cellular activity (C) (A → B, B → C and A → C).

These criteria, although sufficient to propose a second messenger function for candidate lipids, are not conclusive. Additional criteria should lend further support by evaluating the necessary conditions for the role of the lipid in mediating biologic responses; i.e., A → B → C. This is notoriously difficult to establish in mammalian cell systems especially if B is hypothesized as the only (or major) mediator of the effects of A on C. However, this second level of study can be approached and requires the following:

1. Definition of the lipid precursors and the metabolic pathways operating in the generation of these lipids. This should result in defining molecular mechanisms in generating the lipid second messenger. (For example, identification of PI-specific phospholipases C and their regulation by either direct tyrosine phosphorylation or through G proteins has established firmly the significance of signaling pathways activated by ligands resulting in formation of DAG.)
2. Establishing a direct molecular target for the action of the candidate lipid second messenger. This target should be directly regulated by the candidate lipid *in vitro*. Evidence should then be provided for activation of this target in cells. (In the case of DAG, the identification of PKC as a kinase activated by DAG *in vitro* and in cells has provided a major cornerstone in establishing a second messenger function. Similar evidence is accruing for ceramide and CAPP.)
3. Establishing structural specificity for the action of the candidate lipid second messenger, both in cellular studies and *in vitro*. The more rigorous the structural specificity the easier it is to accept the physiologic function of lipid second messengers. (For example, DAG demonstrates stereospecificity and a strict structural specificity. Similarly ceramide activities demonstrate a strict requirement for the 4-5 double bond).
4. Definition of specific roles for the candidate lipid and its target in the cellular activities of the extracellular agent hypothesized to act through the generation of this lipid. This may be obtained through the use of specific pharmacologic inhibitors and/or through molecular genetic manipulations.

Although the "technical" issue of establishing solid criteria for a lipid second messenger are more complex than for other molecules (such as cAMP and IP_3), these speculations and concerns should not detract from the overall advances in knowledge obtained through the study of DAG and PKC on the one hand and ceramide and CAPP on the other. In the case of the DAG/PKC pathway, important insight has been gained into the family of PKC isoenzymes, substrates, down stream targets, multiple biologic activities, and effects on significant transcription factors and gene regulatory events. In essence, these studies are beginning to delineate pathways of cell regulation. They also highlight the significance of lipid metabolism as an early cell regulatory/signaling event (irrespective of the identity of the "true" signaling molecule). Similarly, it is hoped that the studies on the SMase/ceramide/CAPP pathway should result in novel insight into pathways involved in growth regulation and signal transduction. These studies have already identified a novel sphingomyelinase, rapid turnover of sphingolipid metabolism, and the regulation of a protein phosphatase by ceramide. More importantly, these studies have allowed identification of a growth suppressor pathway activated by ceramide.

B. Ceramide- Versus DAG-Mediated Biology

With the definition of multiple cellular activities, ceramide is emerging as the sphingolipid counterpart of the glycerophospholipid signaling molecule, DAG (see Figure 5). However, ceramide appears to regulate opposing functions to DAG. For example, ceramide induces apoptosis in many cell types whereas DAG enhances viability. Moreover, DAG and PMA prevent the apoptotic activities of ceramide. In this context, it should be noted that DAG has been studied extensively (but not exclusively) as a lipid second messenger responsible for transducing mitogenic or growth stimulatory messages within cells. Ceramide on the other hand is becoming recognized as a growth inhibitory molecule. Even with regards to the cellular targets of these lipids, there is diversity of function. One of the major targets for DAG has been identified as PKC; thus, DAG works through the stimulation of protein phosphorylation. In contrast, the only direct target for ceramide so far identified is a phosphatase, suggesting that ceramide works through the inhibition of protein phosphorylation.

These comparisons between the opposing effects of DAG and ceramide raise tantalizing possibilities on the role of lipid-derived molecules in signal transduction and cell regulation. Both ceramide and DAG occupy central and critical roles in the metabolic pathways of sphingolipids and glycerophospholipids, respectively (see Figure 5). Yet, both molecules may have been recruited as specific second messengers/lipid mediators. On the one hand, these may reflect independent functions of these lipid molecules with their basic metabolic functions distinct from their signaling roles. Thus the enzymes involved in signaling function may be distinct from the enzymes involved with intermediary lipid metabolism. On the

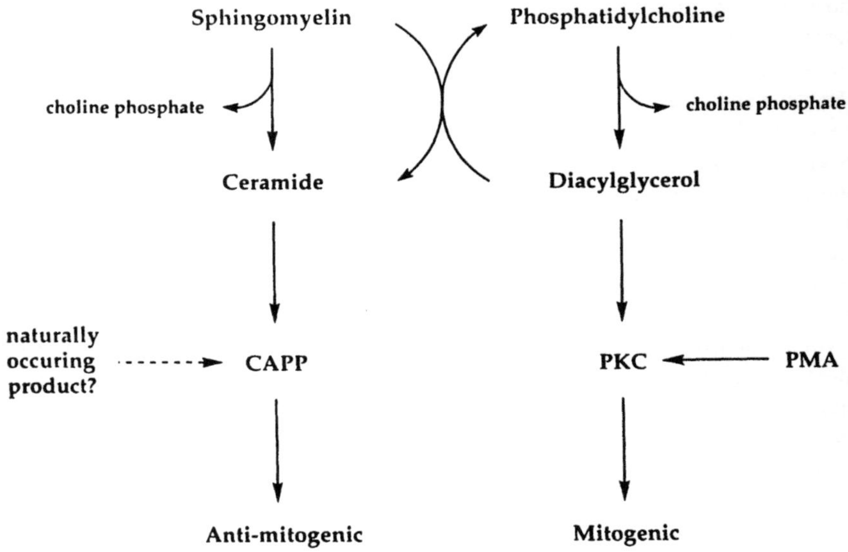

Figure 5. Comparison of ceramide and diacylglycerol signal transduction. Ceramide, generated from sphingomyelin, probably acts via ceramide activated protein phosphatase (CAPP) to transduce anti-mitogenic signals of cells. In contrast, diacylglycerol, generated from glycerophospholipid, acts via protein kinase C (PKC) to stimulate mitogenic responses. See text for details.

other hand, these two roles of ceramide and DAG may reflect an intrinsic coupling of lipid metabolism and cell regulation. According to this scenario, changes in sphingolipid or glycerolipid metabolism may result in changes in the levels of ceramide and DAG, respectively. These changes are then sensed by the cells through the interaction of these molecules with distinct effector molecules (such as CAPP and PKC). It is tempting to speculate that in the case of major life events of cells (such as decisions involving viability or apoptosis) the patterns of lipid metabolism may prove critical in deciding the fate of the cell. Moreover, and as a corollary, it may be that the relative levels of both DAG and ceramide are more critical for dictating cell viability or death as opposed to the absolute levels of either of these two lipid mediators.

These opposing functions of ceramide and DAG also suggest possible mechanistic interactions. If the predominant effects of DAG on viability are through PKC and the effects of ceramide on apoptosis are through CAPP, then the possibility emerges that phosphorylation of key substrates may play a decision making role in the regulation of viability or apoptosis. According to this hypothesis, induction of phosphorylation of these substrates by PKC enhances viability whereas dephosphorylation of these substrates by CAPP would drive cells towards apoptosis. Obviously, more complex scenarios could be envisioned with multiple down stream interactions of the DAG/PKC pathway with the ceramide/CAPP pathway. Nevertheless, this area of study promises important insight into the molecular mechanisms regulating cell viability and death.

VIII. CONCLUSIONS

The studies on the sphingomyelin cycle offer an initial and important insight into regulated metabolism of sphingolipids and on possible second messenger/cell regulatory roles for sphingolipid derived molecules as exemplified by ceramide. These studies have two major implications. First, they illustrate the potential role of regulated sphingolipid metabolism in the diverse processes of signal transduction and cell regulation. Accordingly, the SM cycle may serve as a prototype of sphingolipid signal transduction. Studies conducted over the last two to three years have already begun implicating sphingosine (Zhang et al., 1990a; Zhang et al., 1990b), sphingosine phosphate (Ghosh et al., 1990; Zhang et al., 1991; Olivera and Spiegel, 1993), and sphingosine phosphorylcholine (Desai and Spiegel, 1991; Desai et al., 1993) as specific regulators of diverse functions (such as calcium mobilization, thymidine uptake, and cell motility). Therefore, the study of sphingolipids in a signaling modality may promise the discovery of novel second messengers and the elucidation of important signal transduction pathways.

A second implication of these studies relates to the specific insight that the study of the SM/ceramide pathway is beginning to shed on cellular mechanisms involved in growth suppression and apoptosis. Unlike the study of mitogenesis where intensive investigation has identified multiple oncogenes, oncogene products, signal transduction pathways, and transcription factors involved in the mitogenic response, much less insight has been generated in the study of growth inhibitory mechanisms. Although conceptually growth inhibitory pathways had been suggested for decades, this field of biology received its first molecular link with the identification of Rb as a growth suppressor molecule whose deletion contributes to the pathogenesis of cancer (Goodrich and Lee, 1993). It is obvious now that a number of extracellular cytokines and agents may act to direct cells to either apoptosis or growth inhibition. In addition, a number of tumor suppressor genes in addition to Rb have been identified and implicated in the pathogenesis of cancer. However, the intracellular mechanisms regulating growth inhibition and tumor suppression, for the most part, have remained elusive. In this context, the identification of ceramide as a growth suppressor lipid promises important insight into these pathways by providing a signal transduction link that may help in unraveling these mechanisms.

NOTES

1. Activation of the sphingomyelin cycle through the low affinity neurotrophin receptor. R.T. Dobrowsky, M.H. Castellino, M.V. Chao, and Y.A. Hannun, (1994). Science 265, 1596-1599.

2. Brefeldin A has been found to be an inducer of HL-60 monocytic differentiation. C.M. Linardic and Y.A. Hannun, (1996). Cell growth and differentiation 7, 765-774.

3. R.M. Wolf, R.T. Dobrowsky, L.M. Obeid, and Y.A. Hannun, (1994). J. Biol. Chem. 269, 19605-19609.

4. L.M. Obeid, L.A. Karolak, and Y.A. Hannun, unpublished observations.

REFERENCES

Ballou, L.R., Chao, C.P., Holness, M.A., Barker, S.C., & Raghow, R. (1992). Interleukin-1-mediated PGE_2 production and sphingomyelin metabolism. Evidence for the regulation of cyclooxygenase gene expression by sphingosine and ceramide. J. Biol. Chem. 267, 20044-20050.

Bell, R.M. (1986). Protein kinase C activation by diacylglycerol second messengers. Cell 45, 631-632.

Bielawska, A., Crane, H.M., Liotta, D., Obeid, L.M., & Hannun, Y.A. (1993). Selectivity of ceramide-mediated biology. Lack of activity of *erythro*-dihydroceramide. J. Biol. Chem. 268, 26226-26232.

Bielawska, A., Linardic, C.M., & Hannun, Y.A. (1992a). Ceramide-mediated biology. Determination of structural and stereospecific requirements through the use of N-acyl-phenylaminoalcohol analogs. J. Biol. Chem. 267, 18493-18497.

Bielawska, A., Linardic, C.M., & Hannun, Y.H. (1992b). Modulation of cell growth and differentiation by ceramide. FEBS Letters 307, 211-214.

Brady, R.O., & Koval, G.J. (1958). The enzymatic synthesis of sphingosine. J. Biol. Chem. 233, 26-31.

Braun, P.E., Morell, P., & Radin, N.S. (1970). Synthesis of C_{18}- and C_{20}-dihydrosphingosines, ketodihydrosphingosines, and ceramides by microsomal preparations from mouse brain. J. Biol. Chem. 245, 335-341.

Braun, P.E., & Snell, E.E. (1968). Biosynthesis of sphingolipid bases. J. Biol. Chem. 243, 3775-3783.

Chatterjee, S. (1993). Neutral sphingomyelinase. Adv. Lipid Res. 26, 25-48.

Davis, R.J., Girones, N., & Faucher, M. (1988). Two alternative mechanisms control the interconversion of functional states of the epidermal growth factor receptor. J. Biol. Chem. 263, 5373-5379.

Dbaibo, G.S., Obeid, L.M., & Hannun, Y.A. (1993). Tumor necrosis factor-α (TNF-α) signal transduction through ceramide. Dissociation of growth inhibitory effects of TNF-α from activation of nuclear factor-κB. J. Biol. Chem. 268, 17762-17766.

Desai, N.N., Carlson, R.O., Mattie, M.E., Olivera, A., Buckley, N.E., Seki, T., Brooker, G., & Spiegel, S. (1993). Signaling pathways for sphingosylphosphorylcholine-mediated mitogenesis in Swiss 3T3 fibroblasts. J. Cell Biol. 121, 1385-1395.

Desai, N.N., & Spiegel, S. (1991). Sphingosylphosphorylcholine is a remarkably potent mitogen for a variety of cell lines. Biochem. Biophys. Res. Comm. 181, 361-366.

Diringer, H., Marggraf, W.D., Koch, M.A., & Anderer, F.A. (1972). Evidence for a new biosynthetic pathway of sphingomyelin in SV40 transformed mouse cells. Biochem. Biophys. Res. Comm. 47, 1345-1352.

Dobrowsky, R.T., & Hannun, Y.H. (1992). Ceramide stimulates a cytosolic protein phosphatase. J. Biol. Chem. 267, 5048-5051.

Dobrowsky, R.T., Kamibayashi, C., Mumby, M.C., & Hannun, Y.A. (1993). Ceramide activates heterotrimeric protein phosphatase 2A. J. Biol. Chem. 268, 15523-15530.

Faucher, M., Girones, N., Hannun, Y.A., Bell, R.M., & Davis, R. (1988). Regulation of the epidermal growth factor receptor phosphorylation state by sphingosine in A431 human epidermoid carcinoma cells. J. Biol. Chem. 263, 5319-5327.

Fishbein, J.D., Dobrowsky, R.T., Bielawska, A., Garrett, S., & Hannun, Y.A. (1993). Ceramide-mediated growth inhibition and CAPP are conserved in *Saccharomyces cerevisiae*. J. Biol. Chem. 268, 9255-9261.

Fishman, P.H., Pacuszka, T., & Orlandi, P.A. (1993). Gangliosides as receptors for bacterial enterotoxins. Adv. Lipid Res. 25, 165-187.

Fredman, P. (1993). Glycosphingolipid tumor antigens. Adv. Lipid Res. 25, 213-234.

Ghosh, T.K., Bian, J., & Gill, D.L. (1990). Intracellular calcium release mediated by sphingosine derivatives generated in cells. Science 248, 1653-1656.

Goldkorn, T., Dressler, K.A., Muindi, J., Radin, N.S., Mendelsohn, J., Menaldino, D., Liotta, D., & Kolesnick, R.N. (1991). Ceramide stimulates epidermal growth factor receptor phosphorylation in A431human epidermoid carcinoma cells. J. Biol. Chem. 266, 16092-16097.

Goodrich, D.W., & Lee, W. (1993). Molecular characterization of the retinoblastome susceptibility gene. Biochim. Biophys. Acta 1155, 43-61.

Hakomori, S. (1980). Glycolipid changes associated with oncogenesis and ontogenesis, and the inhibition of the process of transformation by monovalent anti-glycolipid antibodies. Adv. Pathobiol. 7, 270-281.

Hakomori, S. (1981). Glycosphingolipids in cellular interaction, differentiation, and oncogenesis. Ann. Rev. Biochem. 50, 733-764.

Hakomori, S., & Igarashi, Y. (1993). Gangliosides and glycosphingolipids as modulators of cell growth, adhesion, and transmembrane signaling. Adv. Lipid Res. 25, 147-162.

Hakomori, S., & Kannagi, R. (1983). Glycosphingolipids as tumor-associated and differentiation markers. J. Natl. Cancer Inst. 71, 231-251.

Hannun, Y.A. (1991). Encyclopedia of Human Biology. In: Dulbecco, R., eds.pp. 179-189, San Diego. Academic Press Inc.,

Hannun, Y.A., & Bell, R.M. (1989). Functions of sphingolipids and sphingolipid breakdown products in cellular regulation. Science 243, 500-507.

Hannun, Y.A., & Bell, R.M. (1993). The sphingomyelin cycle: A prototypic sphingolipid signaling pathway. Adv. Lipid Res. 25, 27-41.

Hannun, Y.A., Loomis, C.R., Merrill Jr., A.H., & Bell, R.M. (1986). Sphingosine inhibition of protein kinase C activity and of phorbol dibutyrate binding *in vitro* and in human platelets. J. Biol. Chem. 261, 12604-12609.

Hassler, D.F., & Bell, R.M. (1993). Ceramidases: Enzymology and metabolic roles. Adv. Lipid Res. 26, 49-57.

Hayakawa, M., Ishida, N., Takeuchi, K., Shibamoto, S., Hori, T., Oku, N., Ito, F., & Tsujimoto, M. (1993). Arachidonic acid-selective cytosolic phospholipase A_2 is crucial in the cytotoxic action of tumor necrosis factor. J. Biol. Chem. 268, 11290-11295.

Holleran, W.M., Feingold, K.R., Mao-Qiang, M., Gao, W.N., Lee, J.M., & Elias, P.M. (1991). Regulation of epidermal sphingolipid synthesis by permeability barrier function. J. Lipid Res. 32, 1151-1158.

Horn, W.S., Smith, J.L., Bills, G.F., Raghoobar, S.L., Helms, G.L., Kurtz, M.B., Marrinan, J.A., Frommer, B.R., Thornton, R.A., & Mandala, S.M. (1992). Sphingofungins E and F: Novel serine palmitoyltransferase inhibitors from *Paecilomyces variotii*. J. Antibiot. 45, 1692-1696.

Inokuchi, J., Momosaki, K., Shimeno, H., Nagamatsu, A., & Radin, N.S. (1989). Effects of D-threo-PDMP, an inhibitor of glucosylceramide synthetase, on expression of cell surface glycolipid antigen and binding to adhesive proteins by B16 melanoma cells. J. Cell. Physiol. 141, 573-583.

Jarvis, W.D., Kolesnick, R.N., Forani, F.A., Taylor, R.S., Gewirtz, D.A., & Grant, S. (1994). Induction of apoptotic DNA damage and cell death by activation of the sphingomyelin pathway. Proc. Nat. Acad. Sci. USA 91, 73-77.

Jayadev, S., Linardic, C.M., & Hannun, Y.A. (1994). Identification of arachidonic acid as a mediator of sphingomyelin hydrolysis in response to tumor necrosis factor a. J. Biol. Chem. 269, 5757.

Kaneshiro, T., Vesonder, R.F., & Peterson, R.E. (1992). Fumonisin-stimulated *N*-acetyldihydrosphingosine, *N*-acetylphytosphingosine, and phytosphingosine products of *Pichia (Hansenula) ciferri*, NRRL Y-1031. Curr. Microbiol. 24, 319-324.

Karlsson, K. (1970). On the chemistry and occurence of sphingolipid long-chain bases. Chem. Phys. Lipids 5, 6-43.

Kim, M., Linardic, C., Obeid, L., & Hannun, Y. (1991). Identification of sphingomyelin turnover as an effector mechanism for the action of tumor necrosis factor α and γ-interferon. J. Biol. Chem. 266, 484-489.

Kolesnick, R.N. (1991). Sphingomyelin and derivatives as cellular signals. Prog. Lipid Res. 30, 1-38.

Krisnangkura, K., & Sweeley, C.C. (1976). Studies on the mechanism of 3-ketosphinganine synthetase. J. Biol. Chem. 251, 1597-1602.

Lingwood, C.A. (1993). Berotoxins and their glycolipid receptors. Adv. Lipid Res. 25, 189-211.

Liu, J., Mathias, S., Yang, Z., & Kolesnick, R.N. (1994). Renaturation and tumor necrosis factor-a stimulation of a 97-kDa ceramide-activated protein kinase. J. Biol. Chem. 269, 3047-3052.

Marggrag, W.D., Anderer, F.A., & Kanfer, J.N. (1981). The formation of sphingomyelin from phosphatidylcholine in plasm membrane preparations from mouse fibroblasts. Biochim. Biophys. Acta 664, 61-73.

Mathias, S., Dressler, K.A., & Kolesnick, R.N. (1991). Characterization of a ceramide-activated protein kinase: Stimulation by tumor necrosis factor α. Proc. Nat. Acad. Sci. USA 88, 10009-10013.

Mathias, S., Younes, A., Kan, C., Orlow, I., Joseph, C., & Kolesnick, R.N. (1993). Activation of the sphingomyelin signaling pathway in intact EL4 cells in a cell-free system by IL-1β. Science 259, 519-522.

Medlock, K.A., & Merrill Jr., A.H. (1988). Inhibition of serine palmitoyltransferase in vitro and long-chain base biosynthesis in intact chinese hamster ovary cells by β-chloroalanine. Biochem. 27, 7079-7084.

Merrill Jr., A.H., & Jones, D.D. (1990). An update of the enzymology and regulation of sphingomyelin metabolism. Biochim. Biophys. Acta 1044, 1-12.

Merrill Jr., A.H., & Wang, E. (1986). Biosynthesis of long-chain (sphingoid) bases from serine by LM cells. J. Biol. Chem. 261, 3764-3769.

Merrill, A.H.,Jr., van Echten, G., Wang, E., & Sandhoff, K. (1993). Fumonisin B_1 inhibits sphingosine (sphinganine) *N*-acyltransferase and *de novo* sphingolipid biosynthesis in cultured neurons *in situ*. J. Biol. Chem. 268, 27299-27306.

Mutch, D.G., Powell, C.B., Kao, M., & Collins, J.L. (1992). Resistance to cytolysis by tumor necrosis factor α in malignant gynecological cell lines is associated with the expression of protein(s) that prevent the activation of phospholipase A_2 by tumor necrosis factor α. Cancer Res. 52, 866-872.

Neale, M.L., Fiera, R.A., & Matthews, N. (1988). Involvement of phospholipase A_2 activation in tumour cell killing by tumor necrosis factor. Immunology 64, 81-85.

Niculescu, F., Rus, H., Shin, S., Lang, T., & Shin, M.L. (1993). Generation of diacylglycerol and ceramide during homologous complement activation. J. Immun. 150, 214-224.

Nudelman, E., Kannagi, R., Hakomori, S., Parsons, M., Lipinski, M., Wiels, J., Fellous, M., & Tursz, T. (1983). A glycolipid antigen associated with Burkitt lymphoma defined by a monoclonal antibody. Science 220, 509-511.

Obeid, L.M., Linardic, C.M., Karolak, L.A., & Hannun, Y.A. (1993). Programmed cell death induced by ceramide. Science 259, 1769-1771.

Okada, Y., Matsuura, H., & Hakomori, S. (1985). Inhibition of tumor cell growth by aggregation of tumor-associated glycolipid antigen: A close functional association between gangliotriaosylceramide and transferrin receptor in mouse lymphoma L-5178Y. Cancer Res. 45, 2793-2801.

Okada, Y., Mugnai, G., Bremer, E.G., & Hakomori, S. (1984). Glycosphingolipids in detergent-insoluble substrate attachment matrix (DISAM) prepared from substrate attachment material (SAM). Their possible role in regulating cell adhesion. Exp. Cell Res. 155, 448-456.

Okazaki, T., Bell, R.M., & Hannun, Y.A. (1989). Sphingomyelin turnover induced by vitamin D_3 in HL-60 cells. J. Biol. Chem. 264, 19076-19080.

Okazaki, T., Bielawska, A., Bell, R.M., & Hannun, Y.A. (1990). Role of ceramide as a lipid mediator of 1α,25-dihydroxyvitamin D_3-induced HL-60 cell differentiation. J. Biol. Chem. 265, 15823-15831.

Okazaki, T., Bielawska, A., Domae, N., Bell, R.M., & Hannun, Y.A. (1994). Characteristics and partial purification of a novel cytosolic, magnesium-independent, neutral sphingomyelinase activated in the early signal transduction of 1a,25-dihydroxyvitamin D3-induced HL-60 cell differentiation. J. Biol. Chem. 269, 4070-4077.

Olivera, A., & Spiegel, S. (1993). Sphingosine-1-phosphate as second messenger in cell proliferation induced by PDGF and FCS mitogens. Nature 365, 557-560.

Ong, D.E., & Brady, R.N. (1973). *In vivo* studies on the introduction of the 4-*t*-double bond of the sphingenine moiety of rat brain ceramides. J. Biol. Chem. 248, 3884-3888.

Pascher, I. (1976). Molecular arrangements in sphingolipids. Comformation and hydrogen bonding of ceramide and their implication on membrane stability and permeability. Biochim. Biophys. Acta 455, 433-451.

Ramachandran, C.K., Murray, D.K., & Nelson, D.H. (1990). Daxamethasone increases neutral sphingomyelinase activity and sphingosine levels in 3T3-L1 fibroblasts. Biochem. Biophys. Res. Comm. 167, 607-613.

Rosenwald, A.G., Machamer, C.E., & Pagano, R.E. (1992). Effects of a sphingolipid synthesis inhibitor on membrane transport through the secretory pathway. Biochem. 31, 3581-3590.

Sandhoff, K., & van Echten, G. (1993). Ganglioside metabolism—Topology and regulation. Adv. Lipid Res. 26, 119-142.

Schutze, S., Berkovic, D., Tomsing, O., Unger, C., & Kronke, M. (1991). Tumor necrosis factor induces rapid production of 1'2'diacylglycerol by a phosphatidylcholine-specific phospholipase C. J. Exp. Med. 174, 975-988.

Schutze, S., Potthoff, K., Machleidt, T., Berkovic, D., Wiegmann, K., & Kronke, M. (1992). TNF activates NF-κB by phosphatidylcholine-specific phospholipase C-induced "acidic" sphingomyelin breakdown. Cell 71, 765-776.

Shier, W.T. (1992). Sphingosine analogs: An emerging new class of toxins that includes the fumonisins. J. Toxicol. Toxin Rev. 11, 241-257.

Spence, M.W. (1993). Sphingomyelinases. Adv. Lipid Res. 26, 3-23.

Stoffel, W., & Bister, K. (1974). Studies on the desaturation of sphinganine. Hoppe-Seyler's Z. Physiol. Chem. 355, 911-923.

Stoffel, W., LeKim, D., & Sticht, G. (1968). Stereospecificity of the NADPH-dependent reduction of 3-oxodihydrophingosine (2-amino-1-hydroxyoctadecane-3-one). Hoppe-Seyler's Z. Physiol. Chem. 349, 1637-1644.

Suffys, P., Beyaert, R., Van Roy, F., & Fiers, W. (1987). Reduced tumour necrosis factor-induced cytotoxicity by inhibitors of the arachidonic acid metabolism. Biochim. Biophys. Acta 149, 735-743.

Sundaram, K.S., & Lev, M. (1984). Comparative inhibition of bacterial and microsomal 3-ketodihydrosphingosine sythetase by L-cycloserine and other inhibitors. Antimicrob. Agents Chemother. 26, 211-213.

Tettamanti, G., & Riboni, L. (1993). Gangliosides and modulation of the function of neural cells. Adv. Lipid Res. 25, 235-267.

Thudichum, J.L.W. (1962). A Treatise on the Chemical Constitution of the Brain. Archon Books, Hamden, CT.

Ullman, M.D., & Radin, N.S. (1974). The enzymatic formation of sphingomyelin from ceramide and lecithin in mouse liver. J. Biol. Chem. 249, 1506-1512.

Vance, D.E., & Vance, J.E. (1985). Biochemistry of Lipids and Membranes. The Benjamin/Cummings Publishing Company, Inc., Menlo Park, CA.

Vilcek, J., & Lee, T.H. (1991). Tumor necrosis factor: New insights into the molecular mechanisms of its multiple actions. J. Biol. Chem. 266, 7313-7316.

Wang, E., Norred, W.P., Bacon, C.W., Riley, R.T., & Merrill, A.H.,Jr. (1991). Inhibition of sphingolipid biosynthesis by fumonisins. Implications for diseases associated with *Fusarium moniliforme*. J. Biol. Chem. 266, 14486-14490.

Wang, E., Ross, P.F., Wilson, T.M., Riley, R.T., & Merrill, A.H.,Jr. (1992). Increases in serum sphingosine and sphinganine and decreases in complex sphingolipids in ponies given feed containing fumonisins, mycotoxins produced by *Fusarium moniliforme*. J. Nutr. 122, 1706-1716.

Williams, R.D., Sgoutas, D.S., Zaatri, G.S., & Santoianni, R.A. (1987). Inhibition of serine palmitoyltransferase activity in rabbit aorta by L-cycloserine. J. Lipid Res. 28, 1478-1481.

Williams, R.D., Wang, E., & Merrill Jr., A.H. (1984). Enzymology of long-chain base synthesis by liver: Characterization of serine palmitoyltransferase in rat liver microsomes. Arch Biochem Biophys 228, 282-291.

Zhang, H., Buckley, N.E., Gibson, K., & Spiegel, S. (1990a). Sphingosine stimulates cellular proliferation via a protein kinase C-independent pathway. J. Biol. Chem. 265, 76-81.

Zhang, H., Desai, N.N., Murphey, J.M., & Spiegel, S. (1990b). Increases in phosphatidic acid levels accompany sphingosine-stimulated proliferation of quiescent Swiss 3T3 cells. J. Biol. Chem. 265, 21309-21316.

Zhang, H., Desai, N.N., Olivera, A., Seki, T., Brooker, G., & Spiegel, S. (1991). Sphingosine-1-phosphate, a novel lipid, invoved in cellular proliferation. J. Cell Biol. 114, 155-167.

Zweerink, M.M., Edison, A.M., Wells, G.B., Pinto, W., & Lester, R.L. (1992). Characterization of a novel, potent, and specific inhibitor of serine palmitoyltransferase. J. Biol. Chem. 267, 25032-25038.

ROLE OF PHOSPHOLIPID CATABOLISM IN HYPOXIC AND ISCHEMIC INJURY

Haichao Wang, D. Corinne Harrison-Shostak, Xue Feng Wang, Anna Liisa Nieminen, John J. Lemasters, and Brian Herman

Advances in Lipobiology
Volume 2, pages 167-194.

ISBN 0-7623-0205-4

ABSTRACT

The finding that hypoxic/ischemic injury leads to plasma membrane bleb formation provided the first evidence that the plasma membrane may serve as a site of expression of hypoxic/ischemic injury. Subsequent studies have focused on the mechanisms by which hypoxic/ischemic injury leads to plasma membrane damage. Numerous studies have demonstrated an association between alterations in plasma membrane phospholipid metabolism and myocardial, renal and hepatic hypoxic/ischemic injury, documented as a temporal correlation between phospholipid degradation and loss of cell viability. Structural and topographical alterations in plasma membrane phospholipid organization and order during hypoxic/ischemic injury have also been observed. Improved survival of hypoxic/ischemic cells and tissues in the presence of phospholipase inhibitors, anti-phospholipase antibodies and acidic intracellular pH (pH_i), has been taken as evidence of the involvement of pH-dependent phospholipases in hypoxic/ischemic injury. The recent discovery that: (1) A calcium-independent phospholipase A_2 selectively hydrolyzes the plasmalogen molecular species in ischemic myocardium sarcolemma, (2) Hypoxic/ischemic injury in hepatocytes leads to the expression of a group II 14 kDa phospholipase A_2, (3) Incubation of hepatocytes with antisense oligonucleotides directed against a group II 14 KDa phospholipase A_2 protects cells against hypoxic injury, and (4) acidotic pH protects cells against hypoxic injury and inhibits phospholipase A_2 activity, suggests that pH-dependent phospholipase A_2 activity is a critical regulator of hypoxic/ischemic injury.

I. INTRODUCTION

In hypoxic and ischemic injury, the primary cause of cell injury is oxygen deprivation. Typically, ischemia results from the interruption of the blood supply and is of clinical consequence in myocardial infarction, stroke and a number of other medical conditions. Oxygen uptake by the tissue itself leads to virtually absolute anoxia within seconds. Such anoxia inhibits mitochondrial oxidative phosphorylation completely, and the resultant decrease of ATP stimulates glycolysis. As a consequence of lactic acid accumulation, H^+ release after ATP hydrolysis and release of H^+ ions from intracellular organelles, pH_i decreases by as much as 2 units. Thus, two cardinal features of ischemia are anoxia and a 10-100 fold increase in intracellular H^+ concentration. Hypoxia differs from ischemia in that perfusion persists during hypoxia. Hypoxia results from respiratory failure, tissue hypoperfusion or a combination of the two. In contrast to ischemia, during hypoxia, some amount of oxygen constantly enters the tissue, although not enough to sustain normal mitochondrial function, and the decrease in pH_i is less marked. In both ischemia and hypoxia, recovery entails reoxygenation and return to physiological pH. The presence of oxygen during reperfusion has been suggested to be of major importance in reperfusion (reoxygenation) injury, but as will be discussed below, return of pH to physiological levels is a very important factor in precipitating lethal reperfusion injury (pH paradox).

II. ALTERATIONS IN THE PLASMA MEMBRANE DURING HYPOXIC/ISCHEMIC INJURY

Numerous studies suggest that defects in the cell membrane is an early feature of hypoxic/ischemic cell injury (Jennings et al., 1972; Chien et al., 1978, 1981; Farber et al., 1981; Bonventre et al., 1984; Buja et al., 1985; Gunn et al., 1985; Lemasters et al., 1987; Smith et al., 1981). These changes are manifested at both a gross and microscopic level. Faber and co-workers suggested that irreversible hypoxic/ischemic injury is characterized by dysfunction of cellular membranes (Chien et al., 1978, 1979; Farber et al., 1978, 1981,1984; Starke et al., 1986). These alterations in the plasma membrane are thought to underlie the dysfunction of membrane transport systems and enzymes, as well as uncoupling of membrane cytoskeletal interactions, and weakening of the plasma membrane permeability barrier which accompany hypoxic/ischemic injury (Chien et al., 1984, 1985).

A. Morphology

Plasma membrane blebbing, the development of bulb-like protrusions on the cell surface, is an early morphological alteration of cell injury. Formation of membrane blebs is an early consequence of hypoxic/ischemic stress in a wide variety of cells including hepatocytes (Herman et al., 1988), renal proximal tubule epithelium (Kreisberg et al., 1980), cardiac myocytes (Bond et al., 1991, 1993; Harper et al., 1993), and Erlich ascites tumor cells (Bonventre et al., 1985). Morphologically, blebs appear as bulb-like projections which initially originate from the tips of surface microvilli. They are phase-lucent and delimited by the plasma membrane when viewed by phase contrast microscopy. Three recognizable stages of bleb growth have been defined during hypoxic injury in hepatocytes (Figure 1). Stage I is characterized by formation of numerous small blebs. In Stage II, these small blebs undergo enlargement by a process of fusion and coalescence resulting in the formation of a few large terminal blebs. Near the end of Stage II, cells begin to swell, and one or more of terminal blebs undergoes lysis resulting in the loss of the plasma membrane permeability barrier. Loss of the plasma membrane permeability barrier initiates Stage III of injury and is coincident with the onset of irreversible injury and the onset of cell death. Stage I and II of injury are fully reversible upon reoxygenation (Herman et al., 1988).

Ultrastructural studies demonstrate that the bleb contents are somewhat variable (Lemasters et al., 1981,1983). Most blebs contain only amorphous material, rough or smooth endoplasmic reticulum, occasional glycogen rosettes and free ribosomes. Larger organelles such as mitochondria, lysosomes, Golgi apparatus and peroxisomes are not found within the blebs (Lemasters et al., 1987; Herman et al., 1988).

Bleb formation appears to occur as a result of breakdown of cytoskeleton-membrane interactions between microvillar core structures and the plasma membrane (Lemasters et al., 1983). Alterations of cytosolic free Ca^{2+}, cellular thiol status,

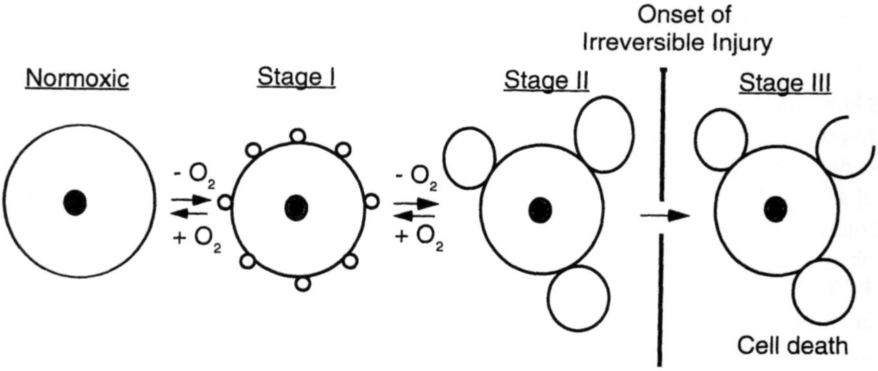

Figure 1. Progression of plasma membrane alterations during ischemic or hypoxic injury. Stage I, formation of numerous small membrane blebs; Stage II, formation of a few large blebs by coalescence and fusion of small blebs; Stage III, breakdown of the permeability barrier in large plasma membrane blebs. Stage I and II are reversible upon reoxygenation.

pH_i, cellular ATP and regional plasma membrane composition have all been postulated to cause bleb formation (Smith et al., 1984). Regional alterations in the lipid composition of the blebbed membrane, changes in membrane fluidity and creation of new membrane domains may result in local membrane weakness and play an important role in bleb formation (Florine-Casteel et al., 1991; Wang et al., 1993, 1994).

Blebbed membranes do not appear more permeable to charged compounds than healthy plasma membranes. Bleb formation during anoxia in isolated hepatocytes did not result in increased plasma membrane leakiness to fluorescent organic cations (extracellular propidium iodide, ethidium bromide and trypan blue) or anions (intracellular Fura-2 and BCECF) until shortly before the onset of cell death (Lemasters et al., 1987; Herman et al., 1988; Nicotera et al., 1989; and Gores et al., 1989). Moreover, hypoxic hepatocytes were still able to maintain cytosolic free Ca^{2+} concentrations constant during blebbing (Anderson et al., 1987; Lemasters et al., 1987; and Gores et al., 1989).

B. Phospholipid Organization

Cellular membranes are composed of proteins and lipids which serve as an interface between the cell and its external environment. Cellular membranes are highly selective permeability barriers containing specific molecular pumps and gates which regulate the transport of molecules into and out of cells and thus regulate the composition of the cytoplasm. The lipid components of the plasma membrane include phospholipids, glycolipids, and cholesterol. Phospholipids are the major component in all biological membranes, and are derived from either glycerol or sphingosine. Phospholipids derived from glycerol are called phosphoglycerides, and consist of a glycerol backbone, two fatty acid chains, and a phosphorylated alcohol of serine, ethanolamine, choline, glycerol, or inositol. Disruption of the lipid order during hypoxic/ischemic injury jeopardizes the normal structure and function of the cellular membrane, resulting in loss of cellular homeostasis and cell death.

There is increasing evidence that changes in the physical state of the plasma membrane are a major factor in the evolution of irreversible hypoxic/ischemic injury. Aggregation of intramembranous particles with the appearance of particle-free, presumably gel-phase regions has been observed in the plasma membrane of ischemic rat liver cells by freeze-fracture electron microscopy (Farber et al., 1978). Moreover, the fluidity of the plasma membrane is altered during anoxic and hypoxic/ischemic injury (Herman et al., 1988; Frederiks et al., 1984; Florine-Casteel et al., 1991; Sun et al., 1993; Wang et al., 1992, 1993, 1994). An increase in plasma membrane order parameter measured by ESR spectroscopy has been observed in rat hepatocytes after 4 hours of anoxia (Farber et al., 1981).

Detailed characterization of plasma membrane structure during hypoxic/ischemic injury is complicated by the difficulty of isolating plasma membrane fractions from injured cells (Farber et al., 1978; Frederiks et al., 1984). In studies employing isolated membrane preparations, it is often difficult to know whether the isolation procedure altered the native distribution of membrane phospholipids. The use of cell suspensions is complicated by the fact that the localization of the reporting probe in intact cells is not always known with certainty. Thus, to properly examine plasma membrane structure during the evolution of hypoxic/ischemic injury, new techniques that circumvent these potential pitfalls needed to be designed.

To better define alterations in plasma membrane structure as they relate to loss of plasma membrane permeability during hypoxic/ischemic injury in rat hepatocytes, we employed four fluorescent microscopic spectroscopic techniques, Digitized Fluorescence Polarization Microscopy (DFPM), Fluorescence Quenching Imaging (FQI) microscopy, Fluorescence Resonance Energy Transfer (FRET) microscopy, and Fluorescence Recovery after Photobleaching (FRAP), to detail alterations of plasma membrane structure during hypoxic/ischemic injury (Florine-Casteel et al., 1990; Wang et al., 1993, 1994). DFPM allows phospholipid structural order to be monitored in specific regions of the plasma membrane while FQI and FRET microscopy provide the ability to obtain two-dimensional information regarding phospholipid order and organization. FRAP can be used to quantitatively measure lipid or protein lateral diffusion during hypoxic injury. During hypoxic injury, DFPM demonstrated that the plasma membrane of cells becomes less fluid (Florine-Casteel et al., 1990). The lipid order parameter (S), which was 0.75 in normoxic cells, increased in a spatially heterogeneous manner to 0.95 in plasma membranes of hypoxic hepatocytes, and remained constant throughout the course of cell injury. This spatially heterogeneous change in the physical state of the plasma membrane might lead to the onset of irreversible injury via destabilization of the membrane or by the formation of weak points at the boundaries between gel- and fluid-phase domains, leading to loss of plasma membrane permeability barrier. For example, alterations in the cholesterol/phospholipid ratio and/or the double bond index of the constituent fatty acids has been shown to alter plasma membrane structure during hypoxic/ischemic injury (Figure 2). In fluid-phase liposomes, as well as a variety of biological membranes including rat liver plasma membranes, cholesterol has been shown to have an ordering effect that increases as the degree of phospholipid acyl chain unsaturation decreases (van Blitterswijk et al., 1987). Alternatively, hypoxic/ischemic injury may proceed by an indirect mechanism in which altered lipid structure leads to changes in membrane protein structure/function, resulting in altered enzyme activities and/or transport properties that could also lead to the eventual collapse of the plasma membrane permeability barrier.

Similar findings were obtained when FQI and FRET microscopy were employed. The data obtained using these two distinct types of fluorescent microscopic imaging approaches indicated that shortly after induction of hypoxic injury in hepatocytes, the plasma membrane fluidity increases. Subsequently, as injury progresses, a spatially and temporally heterogeneous change in both the organization and the degree of fluidity/rigidity of the plasma membrane of hypoxic cells occurs (Wang et al., 1993). In general, there is a transition from a fluid to a more rigid (gel) membrane. This transition occurred first in the cell body and subsequently in the plasma membranes blebs and interestingly, seemed to involve increased aggregation of phosphotidylethanolamine rather than phosphotidylcholine. These data are consistent with the hypothesis that hypoxic/ischemic injury leads to domain formation in the plasma membrane of cells, resulting in the eventual

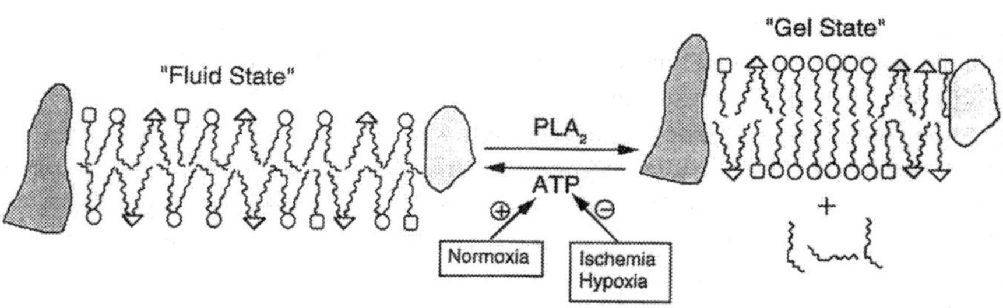

Figure 2. Hydrolysis of the membrane phospholipids at the sn-2 position by phospholipase A_2 leads to an increase in lipid order during ischemic or hypoxic injury. Hypoxic and ischemic injury leads to rapid depletion of ATP, preventing fatty acyl chain reacylation, resulting in segregation of phosphatidylethanolamine into regions of gel phase lipid. "square-shaped," cholesterol; "round-circle," phosphatidylethanolamine; and "triangle-shaped," phosphatidylcholine.

weakening of plasma membrane integrity barrier, loss of this barrier, and onset of cell death.

C. Membrane Composition

The development of membrane damage is related closely to the onset of irreversible injury *in vivo* and *in vitro*, suggesting that this may be the proximate cause of cell death during ischemia (Chien et al., 1981). The mechanism of plasmalemmal disruption in ischemia is unknown, although several investigators have suggested that degradation of sarcolemmal phospholipids may play a role (Jennings et al., 1981). Normally, plasma membrane phospholipid composition is maintained by a balance between phospholipid degradation and resynthesis. Degradation of membrane phospholipids is a two-step process involving phospholipases that remove one fatty acyl chain from the phospholipid followed by the removal of the second by a lysophospholipase. Activation of endogenous phospholipases and/or inhibition of reacylation of lysophospholipids would thus be the presumed bases for the accelerated degradation of membrane phospholipids in hypoxic/ischemic injury (Figure 2). Lysophosphatides are the principal end-products of endogenous phospholipid degradation during myocardial ischemia. These lysophospholipids may cause a variety of functional changes in the plasma membrane. Very small increases in lysophospholipids content or fatty acid content have been shown to have a profound impact on the permeability characteristics of cellular membranes (Chien et al., 1981; Katz et al., 1981) and on the electrophysiological properties of the cell membrane (Corr et al., 1981, 1982; Gross et al., 1982).

The possible importance of phospholipid catabolism in plasma membrane disruption is suggested by several observations. Numerous studies have demonstrated an association between alterations in phospholipid metabolism and membrane damage during myocardial cell injury (Chien et al., 1981, 1985; Van der Vusse et al., 1982; Buja et al., 1985; Gunn et al., 1985), and there is increasing evidence that progressive alterations in membrane phospholipids contribute to functional derangement in ischemic myocardium (Chien et al., 1979, 1981; Gunn et al., 1985; Jennings et al., 1985; Buja et al., 1985, 1988; Hazen et al., 1991). Accumulation of potentially damaging concentrations of metabolic intermediates (lysophospholipids, acyl-CoA esters, long-chain acyl carnitine) appears to be one of the causes of the accelerated membrane damage in ischemic cells (Farber et al., 1981). Consistent with earlier observations (Chien et al., 1979), phospholipid depletion was found to be a major event in the cellular injury induced by acute myocardial ischemia in rats. It has been reported that lysophospholipid production proceeds rapidly in ischemic rabbit hearts (Sobel et al., 1978). A close temporal correlation between myocardial phospholipid degradation and depletion of creatine kinase, a marker of irreversible cell injury and necrosis after coronary artery occlusion in control animals has also been reported (Chiariello et al., 1987). There is an association between rapid degradation of phospholipids and hepatocellular necrosis during *in vivo* hepatic ischemia (Chien et al., 1977, 1978). In both hepatic and myocardial ischemia, the time course of loss of phospholipids paralleled that of loss of reversibility of cell injury. Chien et al. (1981) reported a 10% decrease of total phospholipids, specifically phosphatidylethanolamine and phosphatidylcholine, in ischemic canine myocardium. Furthermore, a several fold increase in free fatty acids, including arachidonic acid, accompanies myocardial ischemia in dog (Chien et al., 1984). These phospholipid and fatty acid changes in ischemic myocardium were shown to be of sufficient magnitude to induce a significant Ca^{2+} permeability defect in isolated cardiac membranes. Inhibition of reacylation of lysophospholipids generated at a normal or an accelerated rate also very likely contributes to phospholipid degradation (Buja et al., 1988).

Some studies have suggested that phospholipid degradation is a more prominent factor in hepatic ischemic injury than in myocardial ischemic injury. As much as 15 to 20% loss of total tissue phospholipids during the first 30 min and a 40% loss during the 3 hr of hepatic ischemia has been reported (Chien et al., 1978). This loss of phospholipid content could be suppressed by chlorpromazine pretreatment, which also significantly decreased the extent of liver cell necrosis produced by 3 hr of ischemia (Chien et al., 1977, 1978). The much greater rate of phospholipid degradation in liver as compared with heart has also been observed in studies of phospholipase activity in homogenates of these tissue (Hostetler et al., 1980).

III. ROLE OF PHOSPHOLIPASES IN HYPOXIC/ISCHEMIC MEMBRANE INJURY

Biochemical studies of ischemic rat liver and ATP-depleted rat cardiac myocytes indicate a relationship between accelerated membrane phospholipid degradation and the progression of cell injury, with phospholipase inhibitors exhibiting a protective effect against the development of irreversible injury. Buja and colleagues (1985) have proposed that hypoxic/ischemic injury leads to an increase in Ca^{2+} which activates a Ca^{2+}-dependent phospholipase, resulting in the breakdown of cell membranes and production of free fatty acids and lysophospholipids that are toxic to cells. More recent studies demonstrating the involvement of Ca^{2+}-independent phospholipase A_2 in hypoxic/ischemic injury, the increased expression of certain phospholipase A_2s during hypoxic/ischemic injury and the protection of hepatocytes from hypoxic injury through the use of antisense oligonucleotides directed against a pH-dependent phospholipase A_2, strongly suggest that hypoxic/ischemic injury is mediated in large part through phospholipase A_2-mediated degradation of cellular membranes.

D. Phospholipase A_2 Activity

Phospholipases A_2 are a family of enzymes that hydrolyze the *sn*-2 ester of glycerolphospholipids, producing free fatty acids and lysophosphatides. They are classed as high (60-100 kDa) or low (12-15 kDa) molecular weight, with the low molecular weight class further subdivided into group I and group II PLA_2 (Gassama-Diagne et al., 1989; Cockcroft et al., 1991; Cordella-Miele et al., 1993; Hazen et al., 1990; Mayer et al., 1993). Both phospholipase A_1 and A_2 activities have been demonstrated in rat liver plasma membranes (Victoria et al., 1971; Nachbaur et al., 1972; Newkirk et al., 1973). Phospholipase activity has also been found in myocardial tissue in association with the plasma membrane (Franson et al., 1978), as well as mitochondria (Weglicki et al., 1971), microsomes (Weglicki et al., 1971), and lysosomes (Franson et al., 1972). These endogenous phospholipases are stimulated to a variable extent by increased intracellular calcium concentration (Franson et al., 1972, 1978; Weglicki et al., 1971, 1980). The predominant phospholipases in myocardial tissue, designated A_1 and A_2, hydrolyze diacyl phospholipids to form lysophosphatides and free fatty acids (Weglicki et al., 1980). Intracellular PLA_2 plays an important role in liberating arachidonic acid for eicosinoid biosynthesis (Van der Bosch, 1980) and its activity is very sensitive to pH with maximal activation occurring at slightly alkaline pH (Frei et al., 1979). Under normal physiological circumstances, lysophospholipids are catabolized by re-esterification by LPC:LPC transacylase or deacylation by lysophopholipase. However, under hypoxic/ischemic conditions, the transacylase and lysophospholipase enzymes are inhibited both by the accumulation of long chain acylcarnitines themselves (Gross et al., 1983) as well as by the decreased pH_i of hypoxic/ischemic injury (Gross et al., 1982). This may be partially responsible for the increased levels of lysophospholipids found during hypoxic/ischemic injury, as has been documented by numerous investigators (Corr et al., 1982, 1984; Steenbergen et al., 1984; Shaikh et al., 1981).

At least three different procedures have been used to measure the phospholipase A_2 activity: (1) arachidonic acid release (Harrison et al., 1991), (2) *in vitro* assays using a phospholipase substrate (e.g., plasmalogen substrate for Ca^{2+}-independent PLA_2), and (3) *In vivo* assays using an intramolecularly quenched pyrene-labeled phospholipid analogue as a substrate (Hazen et al., 1991; and Thuren et al., 1988). The first two assay methods suffer from the inability to definitively determine the site of phospholipase A_2 activity. Cellular sites of arachidonic acid release monitored using radioactive tracers cannot be visualized, and the use of isolated membrane fractions make interpretation difficult for reasons previously discussed. *In vivo* assays of phospholipase activity especially when combined with optical microscopy, obviate these problems. The phospholipid analog 1-palmitoyl-2,6(pyrene-1-yl)hexanoyl-sn-glycero-3-phospho-N-(trinitrophenyl) aminoethanol (PPHTE) in which pyrene fluorescence is intramolecularly quenched by the trinitrophenyl group, can be used as a substrate for phospholipase A_2. Upon phospholipase A_2 catalyzed hydrolysis of this molecule pyrene monomer fluorescence emission intensity increases as a result of the transfer of the pyrene fatty acid to the aqueous phase. This assays allow the detection of picomole levels of free fatty acid production (Thuren et al., 1988). However, this assay is not specific to PLA_2 as it is also sensitive to PLC and PLD. Recently, a new phospholipase A substrate, bis-BODIPY-C_{11}-PC (1,2-bis-(4,4-difluoro-5,7-dimethyl-4-bora-3a, 4a-diaza-s-indacene-3-undecanoyl)-sn-glycero-3-phosphocholine) (Molecular Probes, Inc.) has been synthesized and employed to measure phospholipase A activity. This phospholipase A substrate, bis-BODIPY-glycerophosphocholine, contains BODIPY fluorophores at the same position of each of the two fatty acyl chains, leading to self-quenching of the BODIPY FL fluorophores. Fatty acyl cleavage by phospholipase A_1 or A_2 relieves this self-quenching, resulting in increased fluorescence.

The biochemical basis for membrane injury may be related to accelerated phospholipid degradation by phospholipases in conjunction with inhibition of the reacylation of the resultant lysophosphatides due to ATP deficiency (Chien et al., 1981; Gunn et al., 1985: Figure 2). Accumulation of lysophosphatides as a result of the action of endogenous phospholipases will occur if the rate of diacyl phospholipid hydrolysis exceeds the combined rates of (1) hydrolysis of lysophospholipids by lysophospholipases, (2) ATP-dependent reacylation of lysophosphatides, and (3) removal of lysolipids from membranes by binding to albumin, which has a high affinity for these amphiphilic compounds (Fiehn et al., 1970). Most of the processes which would diminish lysophospholipid accumulation are impaired during ischemia (Corr et al., 1981, 1982), and significant lysophospholipid accumulation has been reported to occur in ischemic myocardium *in vivo* (Corr et al., 1982; Gross et al., 1982). Loss of arachidonic acid and a concomitant reduction in lipid fluidity, attributed to phospholipase A_2 activation, have been reported in calcium-treated rat hepatocyte plasma membranes (Storch et al., 1985). Release of arachidonic acid has also been implicated in irreversible injury of ATP-depleted rat cardiac myocytes (Jones et al., 1989). pH-dependent production of lysolipids during hypoxic/ischemic injury and improved cell survival in the presence of phospholipase inhibitors has been taken as evidence of the involvement of phospholipases in hypoxic/ischemic injury (Chien et al., 1977; Chiariello et al., 1987; Sen et al., 1988; Harrison et al., 1991; Hazen et al., 1991; Armstrong et al., 1991; Prasad et al., 1991; Sargent et al., 1992; Kikuchi-Yanoshita et al., 1993). Using confocal microscopy, we have recently been able to monitor phospholipase A activity in the plasma membrane of hypoxic hepatocytes at very high spatial resolution (Wang et al., unpublished data). Our findings indicate that phospholipase A activity is higher in cell body region compared to the blebs in hypoxic hepatocytes. In addition, phospholipase A activity increased substantially (up to 10-fold) during the early stages of the hypoxic/ischemic injury in the cell body region, while remaining relatively constant in the blebbed region of the membrane. Moreover, phospholipase A activity was found to be suppressed in hypoxic hepatocytes at acidic extracellular pH.

B. Inhibition of Phospholipase A_2 Activity

The findings of increased phospholipase activity, arachidonic acid production and lysolipid production following hypoxic/ischemic injury have led investigators to suggest a cause and effect relationship. However, although previous studies have documented a temporal correlation between arachidonic acid release and the onset or irreversible cell injury, they do not necessarily indicate a causal relationship between these two events. The availability of agents that would inhibit the release of arachidonic acid from ischemic cells allows for a more critical assessment of the role of phospholipid degradation in the development of irreversible injury, and might also provide insight into the biochemical mechanisms that are responsible for the liberation of arachidonic acid. A number of compounds with varying specificities have been employed as well the use of antibodies and antisense oligonucleotides directed against the various types of phospholipases (Table 1).

The phospholipase inhibitors mepacrine, quinacrine and chlorpromazine have been shown to reduce hypoxic/ischemic injury (Chien et al., 1979; Das et al., 1986; and Chiariello et al., 1990). Mepacrine completely prevented accumulation of fatty acids during ischemia and reperfusion (van Bilsen et al., 1990; Table 1), and reduced release of arachidonate and fatty acids from cell membranes during metabolic inhibition (Jones et al., 1989; and Armstrong et al., 1991). It also reduced morphologic changes, e.g., the size of ischemic-reperfusion infarcts, in ischemic rat (Chiariello et al., 1987) and dog hearts (Chiariello et al., 1990). In addition, Armstrong and Ganote (1991) reported that quinacrine delayed cell death in cultured cardiomyocytes subjected to metabolic inhibition (Table 1). In another study examining the cardioprotective activity of quinacrine in isolated globally ischemic rat hearts, quinacrine significantly improved reperfusion, contractile function and reduced lactate dehydrogenase release (Sargent et al., 1992). These data support the hypothesis that cell death after severe ATP depletion is mediated by membrane damage.

However, these reports should be interpreted with care as quinacrine is not a specific inhibitor of phospholipase A_2. There have been several reports that quinacrine can inhibit the Na^+/Ca^{++} exchanger (Stepherd et al., 1991) and phospholipase C activity (Otani et al., 1988; Shaikh et al., 1987; and Prasad et al., 1991). It also inhibits cyclooxygenase activity and prostaglandin production (Blackwell 1978). Furthermore, amphipathic drugs such as mepacrine and chlorpromazine can interact directly with membranes and alter their fluidity and permeability (Prasad et al., 1991).

Using a cultured myocardial cell model, a synthetic phospholipase inhibitor, U26,384, was found to inhibit the degradation of myocardial cell phospholipids during ischemia; specifically, it prevented the degradation of phosphatidylcholine and the release of arachidonic acid after ATP depletion (Sen et al., 1988; Table 1). Inhibition of phospholipid degradation by U26,384 prevented the development of sarcolemmal membrane defects and the release of creatine kinase from the cultured myocardial cells during ATP depletion (Sen et al., 1988). These results support the hypothesis that the development of sarcolemmal membrane injury and the associated loss of cell viability are causally related to progressive phospholipid degradation. One caviat of these studies is that the specificity of U266,384 for individual

Table 1. Effects of Inhibition of Phospholipase A2 Activity on Release of Fatty Acids/Arachidonic Acid and Cell Viability During Ischemic and Hypoxic Injury

PLA2 Inhibitors	Release of FA/AA	References	Cell Viability	References
Mepacrine*	Significantly preserved phospholipid Completely prevented accumulation of FA Delayed release of AA	Chiariello et al., 1987 van Bilson et al., 1990 Harrison et al., 1991	Significantly delayed cell death Delayed loss of cell viability Significantly reduced lactate dehydrogenase release, delayed cell death	Armstrong et al., 1991 Harrison et al., 1991 Sargent et al., 1992
Dibucaine	Delayed release of AA	Harrison et al., 1991	Delayed cell death	Harrison et al., 1991
U26,384**	Blocked the degradation of PC and release of AA	Sen et al., 1988	Prevented creatine kinase release	Sen et al., 1988
Bromoel lactone***	Inhibited FA release	Hazen et al., 1991	Not done.	
Acidotic pH (pH 6.5)	Delayed release of AA to a similar extent as mepacrine and dibucaine	Harrison et al., 1991	Delayed loss of cell viability to a similar extent as mepacrine and dibucaine	Harrison et al., 1991

Continued

Table 1. Continued

PLA2
Antibodies
Inhibited the degradation of PC and PI

Inhibited the degradation of PE
Prasad et al., 1991

Kikuchji-Yanoshita et al., 1993
Significantly reduced the release of both lactate dehydrogenase and creatine kinase
Prasad et al., 1991

PLA2
Antisense-Oligo

Delayed cell death
Wang et al., 1994, This publication.

Notes: "*" - Antimalarial drug; "**" - Substrate analogue; "***" - Mechanism-based inhibitor; AA - Arachidonic Acid; FA - Fatty Acid; PC - Phosphotidylcholine; PE - Phosphotidylethanolamine; PI - Phosphotidylinositol PLA2 and hypoxic/ischemic injury

phospholipase activities (phospholipase A versus phospholipase C) has not been determined. Attempts to use U26,384 and related analogs to directly assess this relationship in intact heart models have been hampered by the low solubility of the compound in aqueous solutions.

Oxidant chemicals causing thiol oxidation or alkylation, lipid peroxidation, and oxygen free radical formation are cytotoxic. During hypoxic injury, a low level of oxygen is present and may allow the production of oxygen free radicals. For many oxidant agents, disruption of mitochondrial oxidative phosphorylation may be a common mode of action (Nieminen et al., 1990). Oxidant chemicals frequently lead to mitochondrial depolarization and ATP depletion (Hunter et al., 1976; Masaki et al., 1989; Gunter et al., 1990; and Nieminen et al., 1990). Recently, cyclosporin A was shown to inhibit the mitochondrial permeability transition and to block the conductance of mitochondrial "megachannels" (Fournier et al., 1987; Crompton et al., 1988; and Broekemeier et al., 1989). Blockade by cyclosporin A alone persisted for only a short period of time. However, combining cyclosporin A and a phospholipase inhibitor produced a longer lasting inhibition of the permeability transition than cyclosporin A alone (Broekemeier et al., 1989). Furthermore, cyclosporin A in combination with trifluoperazine or mepacrine has been shown to be very effective in reducing iodoacetate-induced cell killing (Broekemeier et al., 1989). This marked protection by cyclosporin A in combination with inhibitors of phospholipase (trifluoperazine, mepacrine, and dibucaine) suggests that hypoxic injury may disrupt mitochondrial function by promoting a permeability transition of the inner mitochondrial membrane (Imberti et al., 1992), and that the mitochondria may be one site of action of phospholipase activity. This permeability transition would uncouple oxidative phosphorylation and inhibit aerobic ATP production. This interpretation is supported by experiments where hepatocytes were exposed to chemical hypoxia (iodoacetate plus cyanide), in the presence or absence of cyclosporin A and dibucaine; cyclosporin A plus dibucaine provided protection against lethal cell injury (Imberti et al., 1992).

The recent discovery of the involvement of a calcium-independent phospholipase A_2 in ischemic myocardium has led to the development of a mechanism-based inhibitor of this phospholipase. This mechanism-based inhibitor, the haloenol lactone, (*E*)-6-(bromomethylene)tetrahydro-3-(1-naphthalenyl)-2H-pyran-2-one (Bromoenol lactone, Compound 1), is a substrate for, covalently binds to, and irreversibly inhibits canine myocardial cytosolic calcium-independent phospholipase A_2 (Hazen et al., 1991). It demonstrates a > 1000-fold selectivity for the myocardial calcium-independent phospholipase over a number of calcium-dependent phospholipase A_2's. However, while this inhibitor has been shown to substantially reduce arachidonic acid release during ischemic injury in myocardium, to date, studies on the protective effect of this mechanism-based inhibitor on hypoxic/ischemic injury have not been reported (Table 1).

In myocardium, brain, and non-excitable cells such as hepatocytes (Bing et al., 1973; Bonventre et al., 1985; Gores et al., 1988; Gores et al., 1989; Katsura et al., 1991; Scholz et al., 1992), Erlich ascites tumor cells (Pentilla et al., 1974), and renal proximal tubule cells (Kreisberg et al., 1980), acidosis protects against lethal cell injury during hypoxic/ischemic injury. Even small reductions of pH_0 protect substantially against lethal hypoxic/ischemic, toxic and reperfusion injury to hepatocytes (Herman et al., 1990; Gores et al., 1990; Harrison et al., 1991; and Kawanishi et al., 1991). Moreover, we have documented that protection is exerted through intracellular acidification, rather than by alterations ATP depletion or to direct effects of acidic extracellular pH on the plasma membrane (Gores et al., 1988, 1989). In addition, numerous reports indicated that reperfusion injury to hepatocyte suspensions and cultures, perfused liver, sinusoidal endothelial cells, and cultured neonatal and adult rat cardiac myocytes involves a pH paradox (Caldwell-Kendel et al., 1989; Currin et al., 1991; Chacon et al., 1994; Harper et al., 1993; Bond et al., 1991, 1993, 1994; Lemasters et al., 1992, 1993). Thus, the degradative processes leading to hypoxic/ischemic/reperfusion injury are highly pH-dependent.

To determine whether hypoxic/ischemic injury in hepatocytes was due to activation of a pH-dependent phospholipase, both acidosis (pH 6.5) and phospholipase inhibitors were tested for their ability to suppress arachidonic acid release and delay the cell death during hypoxic injury. Acidosis was found to inhibit arachidonic acid release to a similar extent as the phospholipase inhibitors, mepacrine and dibucaine (Harrison et al., 1991; Table 1). Notably, both acidosis and the phospholipase inhibitors also delayed the onset of cell death to a similar extent as arachidonic acid release. These finding suggest that a pH-dependent phospholipase causes alterations in plasma membrane phospholipid composition after ATP-depletion which contribute to lethal cell injury in hepatocytes.

Phospholipase A_2 antibodies have recently been employed to examine the role of phospholipase A_2 in myocardial phospholipid degradation and cellular injury during reperfusion of ischemic myocardium (Prasad et al., 1991). Hearts perfused with anti phospholipase A_2 antibodies retained a significantly higher amount of phosphatidylcholine and phosphatidylinositol and produced corresponding lower amounts of lysophosphatidylcholine and nonesterified fatty acids after 30 min of reperfusion following 30 min of normothermic global ischemia compared with hearts perfused with nonimmune immunoglobulin G (Table 1). Measurements of phospholipase activity in subcellular organelles of anti phospholipase A_2 treated hearts showed decreased activity in the cytosol, mitochondria, and microsomes. These results support the earlier suggestion that myocardial phospholipase A_2 catalyzes the breakdown of membrane phospholipids during hypoxic/ischemic/reperfusion injury (Chien et al., 1979, 1981, 1985; Otani et al., 1981, 1989; Zalewski et al., 1988).

Similarly, almost 20% of the cellular phosphatidylethanolamine (PE) was found to be hydrolyzed upon incubation of heart homogenates for 1 hour under ischemic conditions, whereas negligible hydrolysis of PE was observed in controls (Kikuchi-Yanoshita et al., 1993). Incubation of ischemic heart homogenates with anti-rat type II phospholipase A_2 antibodies resulted in substantial suppression of phosphatidylethanolamine degradation, implicating involvement of the rat 14-kDa type II phospholipase A_2 in the degradation of phosphatidylethanolamine in ischemic heart. Addition of EDTA at high concentrations inhibited this reaction, indicating that the phospholipase A_2 activity observed in the present study may differ from the myocardium cytosolic Ca^{2+}-independent enzyme (Kikuchi-Yanoshita et al., 1993; Hazen et al., 1990). This is further supported by the finding that phosphotidylethanolamine was exclusively hydrolyzed in this study, whereas the Ca^{2+}-independent myocardial enzyme preferentially hydrolyzes plasmalogen.

C. Mechanisms of Phospholipase A_2 Activation

The mechanism(s) underlying the regulation of phospholipase activity during hypoxic/ischemic injury remain unknown. Evidence accumulated over the past twenty years has suggested the existence of phospholipases whose activity are both sensitive and insensitive to Ca^{2+}. In addition, other cellular proteins have recently been described which can modulate the activity of phospholipases.

An increase in free Ca^{2+} has been hypothesized to activate phospholipases, proteases, and endonucleases during hypoxic/ischemic injury leading to high levels of lysophospholipids, disruption of the cytoskeleton and other effects (Buja et al., 1985, 1988; and Cheung et al., 1986). However, biochemical studies of the Ca^{2+} sensitivities of these phospholipases suggest that relatively high levels of Ca^{2+} (e.g.,

10 to 100 μM) appear to be required to activate phospholipases. Thus, for Ca^{2+}-sensitive phospholipases to be involved in hypoxic/ischemic injury, increases in Ca^{2+} of this magnitude would have to occur.

To determine whether increases in cytosolic free Ca^{2+} occur during hypoxic injury, we have employed Multiparameter Digitized Video Microscopy (MDVM), which permits the quantitative estimation of free Ca^{2+} in single living cells labeled with probes whose fluorescence is responsive to specific cellular parameters of interest (i.e., Ca^{2+}) (Lemasters et al., 1987; Gores et al., 1988; and Gores et al., 1990). As previously discussed, cells formed surface blebs within 10 to 20 min after initiation of hypoxic injury and most cells lost viability within 1 hour. However, an increase of free Ca^{2+} was not required for bleb formation (Nieminen et al., 1988). Blebbing occurred early after the initiation of injury and cells eventually died without an increase in free Ca^{2+}. Just prior to the onset of cell death, free Ca^{2+} increased rapidly in high Ca^{2+} buffer (1.2 mM), but not in low Ca^{2+} buffer (< 1 μM), however, the rates of cell killing were the same.

Numerous other reports support these findings. In aequorin-loaded isolated cardiac cells or cultured kidney cells, a small (less than threefold) increase in free Ca^{2+} was observed following anoxic injury (Snowdowne et al., 1985, 1985, 1985). In isolated cardiac myocytes exposed to 30 min of anoxia and substrate deprivation, neither total, cytosolic nor mitochondrial Ca^{2+} increased (Cheung et al., 1982, 1984, 1986). In papillary muscle of the ferret, Guanieri (1987) found no change in free Ca^{2+} concentration with ischemic perfusion. Mandel and co-workers (1988), in renal tubular cells found that Ca^{2+} accumulated in the mitochondria during severe hypoxia, but not during anoxia. The Ca^{2+} accumulation was readily reversible with reoxygenation and did not appear to be related to the observed mitochondrial dysfunction. Other investigators, using metabolic inhibitors to mimic hypoxic/ischemic injury, found that cyanide exposure of red cells for 1 h depletes ATP, but has no effect on free Ca^{2+} (Simons, 1983). No increase in free Ca^{2+} was observed in isolated cardiac cells following metabolic inhibition (Cobbold et al., 1985). Thus, the majority of the evidence tends to dissociate an increase in free Ca^{2+} from cell injury, and argues that increases in free Ca^{2+} do not regulate the activation of phospholipase A_2. Our findings have also recently been confirmed by Farber and his colleagues (Pastorin, et.al., 1993).

As numerous reports exist documenting the involvement of phospholipase A_2s in hypoxic/ischemic injury, other pathways of activation have been identified. In myocardium, the majority of measurable phospholipase A_2 activity is Ca^{2+}-independent and selectively hydrolyzes plasmalogen substrate (Wolf et al., 1985; and Hazen et al., 1990, 1991). This phospholipase was identified as a Ca^{2+}-independent phospholipase A_2 based on the following properties: (1) concomitant production of lysoplasmenylcholine and *sn*-2 fatty acid from plasmenylcholine substrate, (2) Maximal enzymatic activity in the absence of Ca^{2+}, and, (3) a 16-fold higher maximum reaction velocity utilizing plasmenylcholine compared to phosphatidylcholine substrate at multiple surface concentrations (Hazen et al., 1991).

This Ca^{2+}-independent phospholipase A_2 has a pH optimum of 6.4, and catalyzed the specific cleavage of the sn-2 fatty acid from diacyl glycerolphospholipids. It also possessed selectivity for hydrolysis of phospholipids containing arachidonic acid at the sn-2 position in comparison to oleic or palmitic acid (Hazen et al., 1990).

To examine the potential role of this Ca^{2+}-independent phospholipases A_2 in mediating membrane dysfunction during early myocardial ischemia, the temporal course of alterations in phospholipase A_2 activity during global ischemia in Langendorf perfused rabbit hearts was studied (Ford et al., 1991). Membrane-associated Ca^{2+}-independent plasmalogen-selective phospholipase A_2 activity increased over 400% during 2 min of global ischemia, was nearly maximally activated (> 10-fold) after only 5 min of ischemia, and remained activated throughout the entire ischemic interval examined (2-60 min). Moreover, activation of membrane-associated phospholipase A_2 was essentially complete before electron microscopic evidence of cellular damage. Using inhibitors of protein and RNA synthesis (cyclohexamide and actinomycin D), it was demonstrated that the appearance of microsomal phospholipase A_2 activity did not require ischemia-induced transcription or translation (Hazen et al., 1991). These data suggests that a membrane-associated calcium-independent phospholipase A2 that selectively hydrolyzes plasmalogen molecular species is the likely enzymatic mediator of accelerated phospholipid catabolism during early myocardial ischemia (Hazen et al., 1991), and that ischemic injury initiates the activation of a latent integral membrane phospholipase A2 which is highly selective for plasmalogen.

Phospholipases may also be regulated by other cellular proteins. Recently, a mammalian protein similar to melittin has been identified, named the phospholipase A_2-activating protein (PLAP) (Clark et al., 1987, 1991). PLAP was initially purified from mammalian cells using glutaraldehyde-cross-linked antibodies against melittin (Clark et al., 1987). PLAP stimulates phospholipase A_2 activity with phosphatidylcholine as its specific substrate. This finding may indicate that PLAP regulates only a subset of phospholipase A_2 enzymes. Recently, the gene coding for PLAP was cloned and characterized. Treatment of smooth muscle and endothelial cells with leukotriene D4 was found to increase the steady-state level of PLAP, which coincided with an increased amount of PLAP, suggesting that PLAP protein synthesis may be regulated at the transcriptional level (Clark et al., 1991).

While PLAP may play a role in the propagation of a number of inflammatory disease processes, its role if any in hypoxic/ischemic injury is not known. To determine if PLAP might regulate phospholipase activity during hypoxic/ischemic injury, we analyzed the expression of PLAP mRNA in normoxic and hypoxic rat hepatocytes by Northern blotting. PLAP mRNA levels were substantially decreased in the hypoxic rat hepatocytes relative to that in normoxic hepatocytes (Figure 3). Thus, PLAP is not a likely regulator in the activation of phospholipase A_2 during hypoxic/ischemic injury.

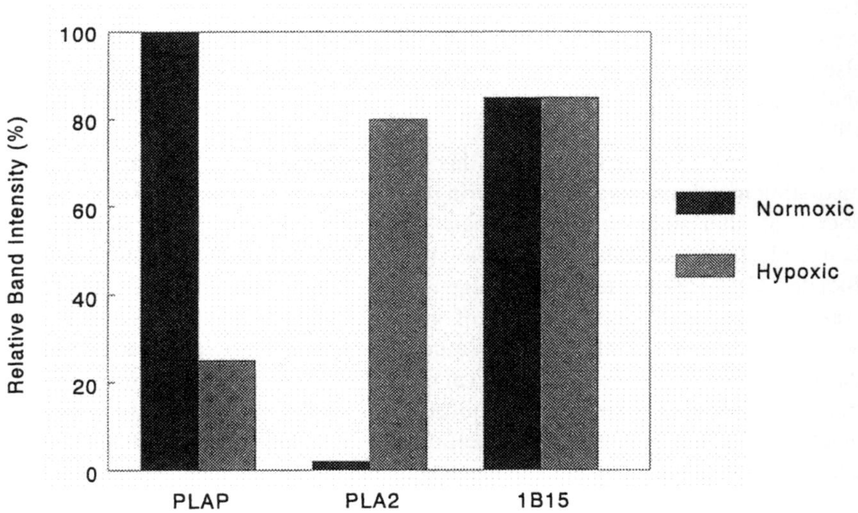

Figure 3. Relative mRNA levels of PLAP, PLA_2, and 1B15 in normoxic vs. hypoxic rat hepatocytes. Data are from densitomatric scans of Northern blots of equivalent amounts of normoxic and hypoxic (30 minutes) rat total RNA. The mRNA level of a house-keeping gene, IB15, remained unchanged in hypoxic rat hepatocytes (relative to normoxic hepatocytes), while the mRNA level of PLAP decreased during hypoxic injury and that of a Group II 14 kDa PLA_2 increased substantially.

D. Phospholipase A_2 Gene Expression

In addition to activation of preexisting phospholipases, hypoxic/ischemic injury may also stimulate the preferential expression of certain phospholipases. To examine this possibility, we selectively cloned genes that were preferentially expressed in the hypoxic hepatocytes using subtractive cloning (Wang et al., 1994, unpublished data). From the subtractive (hypoxic minus normoxic) cDNA library, several positive clones which were uniquely or preferentially expressed in hypoxic rat hepatocytes were obtained by differential screening. One of the positive clones that was preferentially expressed in hypoxic hepatocytes hybridized to a 45-mer oligonucleotide cDNA probe of the rat liver 14 kDa group II phospholipase A_2. Northern blot analysis of normoxic and hypoxic hepatocyte mRNA confirmed that the mRNA level of this message for this phospholipase was substantially increased in hypoxic hepatocytes (Figure 3). Moreover, the size of the expressed mRNA matched the size of the mRNA of the rat liver 14 kDa group II phospholipase A_2 enzyme. This phospholipase displays an alkaline pH optimum and suggests that in liver, a group II phospholipase A_2, or an group II PLA_2-isomer, is substantially up-regulated in hypoxic hepatocytes. More recent experiments employing antisense oligonucleotides directed against this phospholipase have shown complete protection against hypoxic injury in rat hepatocytes (Figure 4). This phospholipase may therefore play an important role in the membrane injury of hepatocytes during hypoxic/ischemic injury.

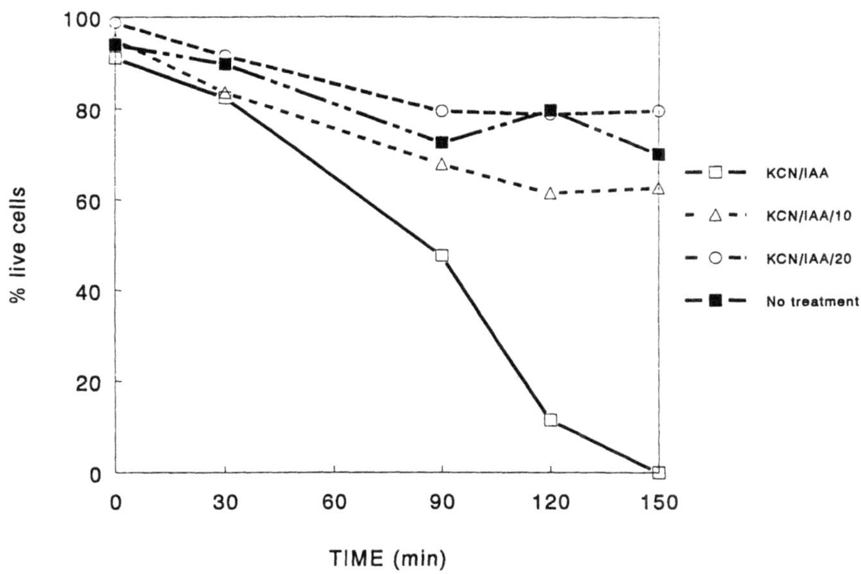

Figure 4. Inhibition of hypoxic injury by group II 14 kDa PLA_2-specific anti-sense oligonucleotides. Rat hepatocytes were cultured in serum-free medium in the presence or absence of antisense group II 14 KDa oligonucleotides (10 or 20 μM) for six hours before the initiation of hypoxic injury (KCN/IAA). LDH release was used as a measure of cell viability. Anti-sense DNA oligonucleotides dramatically inhibit hypoxic cell death.

IV. SUMMARY

Numerous studies have demonstrated an association between alterations in plasma membrane phospholipid metabolism and myocardial, renal and hepatic hypoxic/ischemic injury. Structural and topographical alterations in plasma membrane phospholipid organization and order during hypoxic/ischemic injury have also been observed, and improved survival of hypoxic/ischemic cells and tissues in the presence of phospholipase inhibitors, anti-phospholipase antibodies and acidotic pH_i. A Ca^{2+}-independent phospholipase A_2 which selectively hydrolyzes the plasmalogen molecular species in ischemic myocardial sarcolemma has been identified and the expression of a group II phospholipase A_2 has been found to be induced during hypoxic injury in hepatocytes. The prevention of hypoxic injury in

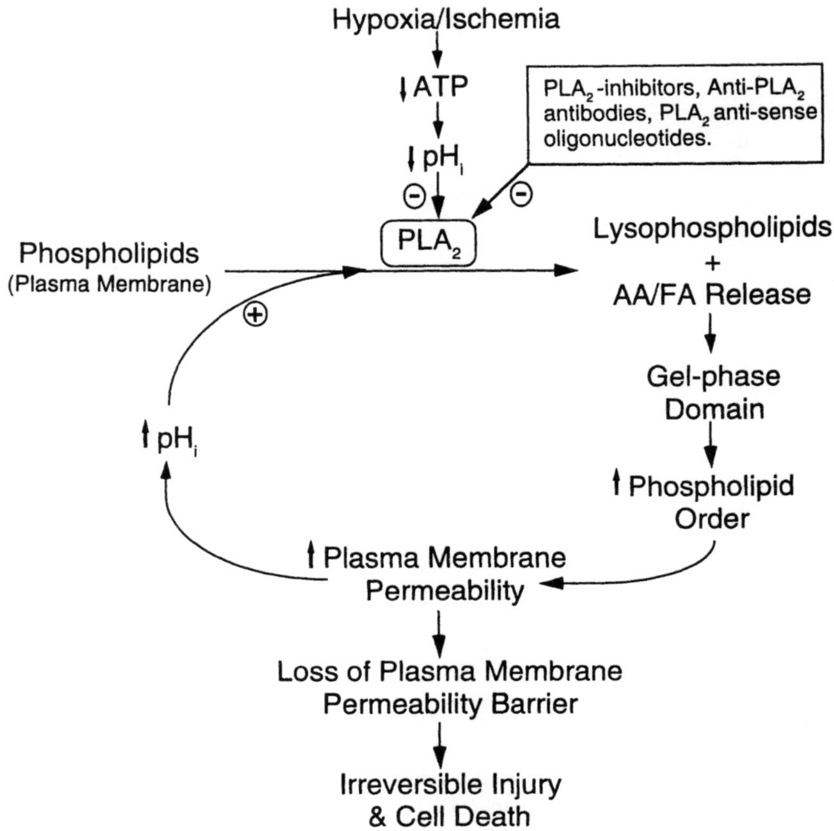

Figure 5. Mechanism of hypoxic and ischemic injury. Hypoxic/ischemic injury leads to ATP depletion and intracellular acidosis. The acidic pH_i partially inhibits phospholipid hydrolysis mediated by a pH-dependent phospholipase A_2. However, this phospholipase A_2 activity operates at a low level leading to an increase in the hydrolysis of plasma membrane phospholipids hydrolysis and increases in the phospholipid order, plasma membrane permeability, and intracellular pH_i. As pH_i rises, the enzymatic activity of the PLA_2 is accelerated, leading to further membrane damage, and eventually irreversible cell death. PLA_2-specific chemical inhibitors, antibodies, and anti-sense oligonucleotides can inhibit the release of arachidonic acids/fatty acids and delay and prevent cell death.

hepatocytes with antisense oligonucleotides directed against a group II 14 kDa phospholipase A_2 strongly suggests that this phospholipase A_2 activity is a critical regulator of hypoxic/ischemic injury. Based on these findings, we propose a model for the mechanism of hypoxic/ischemic injury (Figure 5). In a normoxic cell, plasma membrane phospholipid status is maintained as a balance between degradation and ATP-dependent synthesis. Hypoxic/ischemic injury results in ATP depletion and intracellular acidification which decreases PLA_2 activity. However, a low level of PLA_2 activity persists, leading to production of lysolipids and alteration in plasma membrane phospholipid organization. These alterations in plasma membrane phospholipids organization continue as PLA_2 activity continues in conjunction with an inhibition of reacylation because of lack of ATP. This eventually increase plasma membrane permeability to the extent that H^+ ions leak out of the cell, raising pH_i and further activating PLA_2. This continues until the plasma membrane permeability barrier in lost and irreversible injury and cell death occurs.

V. FUTURE PERSPECTIVES

The information presented in this chapter suggests that the plasma membrane is a key site of injury during hypoxia and ischemia. However, the specific phospholipase(s), the mechanism of activation and regulation of their activity during hypoxic/ischemic injury are still not understood. In addition, it is also not clear whether phospholipase activation alone is responsible for the lethal cell injury of hypoxia/ischemia, or whether other enzymatic activities and events also occur during hypoxic/ischemic injury which can act independently or in concert with phospholipase activation to cause cell damage. The identification and characterization of genes that are preferentially over- as well as underexpressed during hypoxic/ischemic injury should yield important information on the cellular functions which contribute to cell injury. The search for synthetic inhibitors of phospholipase activities should continue and be tested in both cultured cell and intact organ models to critically evaluate the role of phospholipid degradation in the development of irreversible hypoxic/ischemic injury. Other strategies, including the use of antisense oligonucleotides directed against genes whose expression is increased during hypoxic/ischemic injury and modulation of the pH of reperfusion solutions, should also be developed as a means to rescue injured tissue.

ACKNOWLEDGMENT

This work was supported by Grants AGO7218 and DK30874 from the National Institute of Health, the Gustavus and Louise Pfeiffer Research Foundation and Grant J-1433 from the Office of Naval Research.

REFERENCES

Anderson, B. S., Aw, T. Y., & Jones, D. P. (1987). Mitochondrial transmembrane potential and pH gradient during anoxia. Am. J. Physiol. 252, C349-C355.

Armstrong, S. C., & Ganote, C. E. (1991). Effects of the phospholipase inhibitor mepacrine on injury in ischemic and metabolically inhibited adult isolated myocytes. Am. J. Pathol. 138, 545-555.

Bing, O. H. L., Brooks, W. W., & Messer, J. V. (1973). Heart muscle viability following hypoxic/ischemic: Protective effect of acidosis. Science 180, 1297-1298.

Blackwell, G. J. (1978). Phospholipase A_2 and platelet activation. Adv. Prostaglandin Thromboxane Leukotriene Res. 3, 137-142.

Bond, J. M., Herman, B., & Lemasters, J. J. (1991). Protection by acidotic pH against anoxia/reoxygenation injury to rat neonatal cardiac myocytes. Biochem. Biophys. Res. Comm. 179, 798-803.

Bond, J. M., Chacon, E., Herman, B., & Lemasters, J. J. (1993). Intracellular pH and calcium homeostasis during the pH paradox of reperfusion injury to cultured neonatal rat cardiac myocytes. Am. J. Physiol. 34, C129-C137.

Bond, J. M., Harper, I. S., Chacon, E., Reece, J. M., Herman, B., & Lemasters, J. J. (1994). The pH paradox in the pathophysiology of reperfusion injury to rat neonatal cardiac myocytes. Ann. of the New York Acad. of Sci. 723, 25-37.

Bonventre, J. V. (1984). Cellular response to ischemia. In: Acute Renal Failure. (Soliz, K., &Whelton, A., eds.), pp. 195-218. Delcker, NY.

Bonventre, J. V., & Cheung, J. Y. (1985). Effects of metabolic acidosis on viability of cells exposed to anoxia. Am. J. Physiol. 249, C149-C159

Broekemeier, K. M., Dempsey, M. E., & Pfeiffer, D. R. (1989). Cyclosporin A is a potent inhibitor of the inner membrane permeability transition in liver mitochondria. J. Biol. Chem. 264, 7826-7830.

Buja, L. M., Hagler, H. K., Parsons, D., Chien, K., Reynolds, R. C., & Willerson, J. T. (1985). Alterations of ultrastructure and elemental composition in cultured neonatal rat cardiac myocytes after metabolic inhibition with iodoacetic acid. Lab. Invest. 53(4), 397-412.

Buja, L. M., Hagler, H. K., & Willerson, J. T. (1988). Altered calcium homeostasis in the pathogenesis of myocardial ischemic and hypoxic injury. Cell Calcium. 9, 205-217.

Caldwell-Kenkel, J. C., Currin, R. T., Tanaka, Y., Thurman, R. G., & Lemasters, J .J. (1989). Reperfusion injury to endothelial cells following cold ischemic storage of rat livers. Hepatology 10(3), 292-299.

Chacon. E., Reece, J. M., Nieminen, A.-L., Zahrebelski, G., Heman, B., & Lemasters, J .J. (1994). Distribution of electrical potential, pH, free Ca^{2+}, and volume inside cultured rabbit cardiac myocytes during chemical hypoxic/ischemic: a multiparameter digitized video microscopic study. Biophys. J. (in press).

Cheung, J. Y., Thompson, I. G., & Bonventre, J. V. (1982). Effects of extracellular calcium removal and anoxia on isolated rat myocytes. Am. J. Physiol. 241, C3-C8.

Cheung, J. V., & Swidler, M. (1984). Calcium dependency of prostaglandin E2 production in rat glomerular mesangial cells. Evidence that protein kinase C modulates the calcium-dependent activation of phospholiippase A_2. J. Clin. Invest. 82, 168-176.

Cheung, J. Y., Bonventre, J. V., Malis, C. D., & Leaf, A. (1986). Calcium and ischemic injury. N. Eng. J. Med. 314, 1670-1676. .

Chiariello, M., Ambrosio, G., Cappelli-Bigazzi, M., Nevola, E., Perone-Filardi, P., Marone, G., & Condorelli, M. (1987). Inhibition of ischemia-induced phospholipase activation by quinacrine protects jeopardized myocardium in rats with coronary artery occlusion. J. Pharmacol. Exp. Ther. 241, 560-568.

Chiariello, M., Ambrosio, G., Cappelli, B. M., Perrone, F. P., Tritto, I., Nenola, E., and Golino, P. (1990). Reduction in infarct size by the phospholipase inhibitor quinacrine in dogs with coronary artery occlusion. Am. Heart J. 120(4), 801-807.

Chien, K. R., Abrams, J. Pfau, R. G., & Farber, J. L. (1977). Prevention by chlorpromazine of ischemic liver cell death. Am. J. Pathol. 88, 539-558.

Chien, K. R., Abrams, J., Serroni, A., Martin, J. T., & Farber, J. L. (1978). Accelerated phospholipid degradation and associated membrane dysfunction in irreversible ischemic liver cell injury. J. Biol. Chem. 253, 4809-4817.

Chien, K. R., Pfau, R. D., & Farber, J. L. (1979). Ischemic myocardial cell injury: prevention by chlorpromazine of an accelerated phospholipid degradation and associated membrane dysfunction. Am. J. Pathol. 97, 505-529

Chien, K. R., Reeves, J. P., Buja, L. M ., Bontes, F. J., Parkey, R. W., & Willerson, J. T. (1981). Phospholipid alterations in canine ischemic myocardium. Temporal and topographical correlations with Tc-99-m-PPi accumulation and an *in vitro* sarcolemmal Ca^{2+} permeability defect. Circ. Res. 48, 711-719

Chien, K. R., Han, A., Sen, A., Buja, L. M., & Willerson, J. T. (1984). Accumulation of unesterified arachidonic acid in ischemic canine myocardium. Relationships to a phosphatidlycholine deacylation-reacylation cycle and the depletion of membrane phospholipids. Circ. Res. 54, 313-322.

Chien, K. R., Sen, A., Reynolds, R., Chang, A., Kim, Y., Gunn, M. D., Buja, L. M., & Willerson, J. T. (1985). Release of arachidonate from membrane phospholipids in cultured neonatal rat myocardial cells during adenosine triphosphate depletion. J. Clin. Invest. 75, 1770-1780.

Clark, M. A., Conway, T. M., Shorr, R. G. L., & Crooke, S. T. (1987). Identification and isolation of a mammalian protein which is antigenically and functionally related to the phospholipase A_2 stimulatory peptide melittin. J. Biol. Chem. 262 (9), 4402-4406.

Clark, M. A., Ozgur, L. E., Conway, T. M., Dispoto, J., Croole, S. T., & Bomalaski, J. S. (1991). Cloning of a phospholipase A_2-activating protein. Proc. Natl. Acad. Sci. 88, 5418-5422.

Cobbold, P. H., Bourne, P. K., & Cubbertson, K. S. R. (1985). Evidence from acquorin for injury of metabolically inhibited myocytes independently of free Ca^{2+}. Basic Res. Cardiol. 80(suppl 2), 155-158.

Cockcroft, S., Nielson, C. P., & Stutchfield, J. (1991). Is phospholipase A_2 activation regulated by G-proteins? Biochem. Soc. Trans. 19(2), 333-336.

Cordella-Miele, E., Miele, L., & Mukherjee, A .B. (1993). Identification of a specific region of low molecular weight phospholipases A_2 (residues 21-40) as a potential target for structure-based design of inhibitors of these enzymes. Proc. Natl. Acad. Sci. 90, 10290-10294.

Corr, P. B., Lee, B. I., & Sobel, B. E. (1981). Electrophysiological and biochemical derangements in ischemic myocardium: interactions involving the cell membrane. Acta. Med. Scand. 651(suppl), 59-68.

Corr, P. B., Gross, R. W., & Sobel, B. E. (1982). Arrhythmogenic amphiphilic lipids and the myocardial cell membrane. J. Mol. Cell. Cardiol. 14, 619-626.

Corr, P. B., Gross, R. W., & Sobel, B. E. (1984). Amphipathic metabolites and membrane dysfunction in ischemic myocardium. Circ. Res. 55, 135-154.

Crompton, M., Ellinger, H., & Costi, A. (1988). Inhibition by cyclosporin A of a Ca^{2+}-dependent pore in heart mitochondria activated by inorganic phosphate and oxidative stress. Biochem. J. 255, 357-360.

Currin, R.T., Gores, G. J., Thurman, R. G., & Lemasters, J. J. (1991) Protection by acidotic pH against anoxic cell killing in perfused rat liver: Evidence for a "pH paradox". FASEB J. 5, 207-210.

Das, D. K., Engleman, R. M., Rousou, J. A., Breyer, R. H., Otani, H., & Lemeshow, S. (1986). Role of membrane phospholipids in myocardial injury induced by ischemia and reperfusion. Am. J. Physiol. 251, H71-H79.

Farber, J. L., Martin, J. T., & Chien, K. R. (1978). Irreversible ischemic cell injury. Prevention by chlorpromazine of the aggregation of intramembranous particles of rat liver plasma membranes. Am. J. Pathol. 92, 713-732.

Farber, J. L., Chien, K. R., & Mittnacht, S., Jr. (1981). The pathogenesis of irreversible cell injury in ischemia. Am. J. Pathol. 102, 271-281.

Farber, J. L., & Young, E. E. (1981). Accelerated phospholipid degradation in anoxic rat hepatocytes. Arch. Biochem. Biophys. 211, 312-320.

Farber, J. L., & Gerson, R. J. (1984). Mechanisms of cell injury with hepatotoxic chemicals. Pharmacol. Rev. 36, 71S.

Fiehn, W., & Hasselbach, W. (1970). The effect of phospholipase A on the calcium transport and the role of unsaturated fatty acids in ATPase activity of sarcoplasmic vesicles. Eur. J. Biochem. 13, 510-518.

Florine-Casteel, K., Lemasters, J. J., & Herman, B. (1990). Phospholipid order in gel- and fluid- phase cell-size liposomes measured by digitized video fluorescence polarization microscopy. Biophys. J. 57, 1199-1215.

Florine-Casteel, K., Lemasters, J. J., & Herman, B. (1990). Digitized fluorescence polarization microscopy of DPH and related probes in cell-size vesicles composed of gel- or fluid- phase phospholipid. In: Optical Microscopy for Biology (Herman, B., & Jacobson, K., eds.), pp. 559-574. Alan R. Liss, Inc. NY.

Florine-Casteel, K., Lemasters, J. J., & Herman, B. (1991). Lipid order in hepatocyte plasma membrane blebs during ATP depletion measured by digitized video fluorescence polarization microscopy. FASEB J. 5, 2078-2084.

Ford, D. A., Hazen, S. L., Saffitz J. E., & Gross R. W. (1991). The rapid and reversible activation of calcium-independent plasmalogen-selective phospholipase A_2 during myocardial ischemia. J. Clin. Invest. 88, 331-335

Fournier, N., Ducet, G., & Crevat, A. (1987). Action of cyclosporine on mitochondrial calcium fluxes. J. Bioenerg. Biomembr. 19, 297-303.

Franson, R. C., Waite, M., & Weglichi, D. B. (1972). Phospholipase A activity of lysosomes of rat myocardial tissue. Biochemistry 11, 472-476.

Franson, R. C., Pang, D. C., Towle, D. W., & Weglichi, D. B. (1978). Phospholipase A activity of highly enriched preparations of cardiac sarcolemma from hamster and dog. J. Mol. Cell. Cardiol. 10, 921-930.

Frederiks, W. M., Myagkaya, G. L., v. Veen, H. A., & Vogels, I. M. C. (1984). Biochemical and ultrastructural changes in rat liver plasma membranes after temporary ischemia. Virchows Arch. B. (Cell Pathol.) 46, 269-282.

Frei, E. & Zahler, P. (1979). Phospholipase A_2 from sheep erythrocyte membranes Ca^{2+} dependence and localization. Biochem. Biophys. Acta. 550, 450-463.

Gassama-Diagne, A., Fauvel, J., & Chap, H. (1989). Purification of a new calcium-independent, high molecular weight phospholipase A_2/lysophospholipase (phospholipase B) from guinea pig intestinal brush-border membrane. J. Biol.Chem. 264(16), 9470-9475.

Gores, G. J., Nieminen, A.-L., Fleishman, K. E., Dawson, T. L., Herman, B., & Lemasters, J. J. (1988). Extracellular acidosis delays onset of cell death in ATP-depleted hepatocytes. Am. J. Physiol. 255, C315-C322.

Gores, G. J., Nieminen, A.-L., Wray, B. E., Herman, B., & Lemasters, J. J. (1989). Intracellular pH during 'chemical hypoxic/ischemic' in cultured rat hepatocytes: Protection by intracellular acidosis against the onset of cell death. J. Clin. Invest. 83, 386-396.

Gores, G. J., Flarsheim, C. E., Dawson, T. L., Nieminen, A.-L., Herman, B., & Lemasters, J. J. (1989). Swelling, reductive stress and cell death during chemical hypoxic/ischemic in hepatocytes. Am. J. Physiol. 257, C347-C354.

Gores, G. J., Herman, B., & Lemasters, J. J. (1990). Plasma membrane bleb formation and rupture: a common feature of hepatocellualr injury. Hepatology 11(4), 690-698.

Gross, R. W., & Sobel, B. E. (1982). Lysophosphatidylcholine metabolism in the rabbit heart. Characterization of metabolic pathways and partial purification of myocardial lysophospholipase-transacylase. J. Biol. Chem. 257, 6702-6708.

Gross, R. W., Drisdel, R. C., & Sobel, B. E. (1983). Rabbit myocardiac lysophospholipase-transacylase. J. Biol. Chem. 258, 15165-15172.

Gross, R. W. (1985). Identification of plasmalogen as the major phospholipid constituent of cardiac sarcoplasmic reticulum. Biochemistry 24, 1662-1668.

Guarnieri, T. (1987). Intracellular sodium-calcium dissociation in early contractile failure in hypoxic/ischemic ferret papillary muscles. J. Physiol. 388, 449-465.

Gunn, M. D., Sen, A., Chang, A., Willerson, J. T., Buja, L. M., & Chien, K. R. (1985). Mechanisms of accumulation of arachidonic acid in cultured myocardial cells during ATP depletion. Am. J. Physiol. 249, H1188-H1194.

Gunter, T. E., & Pfeiffer, D. R. (1990). Mechanisms by which mitochondria transport calcium. Am. J. Physiol. 258, C755-C786.

Harper, I. S., Bond, J. M., Chacon, E., Reece, J. M., Herman, B., & Lemasters, J. J. (1993) Inhibition of Na^+/H^+ exchange preserves viability, restores mechanical function, and prevents the pH paradox in reperfusion injury to rat neonatal myocytes. Basic Research in Cardiology, 88, 422-430.

Harrison. D. C., Lemasters, J. J., & Herman, B. (1991). A pH-dependent phospholipase A_2 activity contributes to loss of plasma membrane integrity during chemical hypoxic/ischemic in rat hepatocytes. Biochem. Biophys. Res. Comm. 174, 54-659.

Hazen, S. L., Stuppy, R. J., & Gross, R. W. (1990). Purification and characterization of canine myocardial cytosolic phospholipase A_2. J. Biol. Chem. 265(18), 10622-10630.

Hazen, S. L., Ford, D. A., & Gross, R. W. (1991). Activation of a membrane-associated phospholipase A_2 during rabbit myocardial ischemia which is highly selective for plasmalogen substrate. J. Biol. Chem. 266(9), 5629-5633.

Hazen, S. L., Zupan, L. A., Weiss, R. H., Getman, D. P., & Gross, R. W. (1991). Suicide inhibition of canine myocardial cytosolic calcium-independent phospholipase A2. J. Biol. Chem. 266(11), 7227-7232.

Hazen, S. L. & Gross, R. W. (1991). ATP-dependent regulation of rabbit myocardial cytosolic calcium-independent phospholipase A_2. J. Biol. Chem. 266(22), 14526-14534.

Hazen, S. L. & Gross, R. W. (1992). Identification and characterization of human myocardial phospholipase A_2 from transplant recipients suffering from end-stage ischemic heart disease. Circ. Res. 70, 486-495.

Herman, B., Nieminen, A.-L., Gores, G., & Lemasters, J. J. (1988). Loss of plasma membrane permeability barrier in relation to onset of irreversible injury in isolated rat hepatocytes. FASEB J. 2, 146-151.

Herman, B., Gores, G. J., Nieminen, A.-L., Kawanishi, T. A., Harman, B., & Lemasters, J. J. (1990) Calcium and pH in anoxic and toxic injury. Crit. Rev. Toxicol. 21, 127-148.

Hostetler, K. Y., & Hall, L. B. (1980). Phospholipase C activation of rat tissues. Biochem. Biophys. Res. Commun. 96, 388-393.

Hunter, D. R., Haworth, R. A., & Southard, J. H. (1976). Relationship between configuration, function, and permeability in calcium-treated mitochondria. J. Biol. Chem. 251, 5069-5077

Imberti, R., Nieminen, A.-L., Herman, B., & Lemasters, J. J. (1992). Synergism of cyclosporin A and phospholipase inhibitors in protection against lethal injury to rat hepatocytes from oxidant chemicals. Res. Comm. Chem. Path. Pharm. 78, 27-38.

Jennings, R. B., Hawkins, H. K., Lowe, K. E., Hill, M. L., Klotman, S., & Reimer, K. A. (1978). Relation between high energy phosphate and lethal injury in myocardial ischemia in the dog. Am. J. Pathol. 92, 187-214.

Jennings, R.B., & Reimer, K. A. (1981). Lethal myocardial ischemia injury. Am. J. Pathol. 102, 241-255.

Jennings, R. B., Reimer, K. A., & Steenbergen, C., (1985). Myocardia ischemia and reperfusion: Role of calcium. In: Control and Manipulation of Calcium Movement (Parratt, ed.), pp. 273-302. Raven Press, NY.

Jones, R. L., Miller, J. C., Hagler, H. K., Chien, K. R., Willerson, J. T., & Buja, L. M. (1989). Association between inhibition of arachidonic acid release and prevention of calcium loading during ATP depletion in cultured rat cardiac myocytes. Am. J. Pathol. 135, 541-556.

Katsura, K.-I., Ekholm, A., & Siesjö, B. K. (1991). Coupling among changes in energy metabolism, acid-base homeostasis, and ion fluxes in ischemia. Can. J. Physiol. Pharcol. 70, S170-S171.

Katz, A. M., & Messineo, F. C. (1981). Lipid-membrane interactions and the pathogenesis of ischemic damage in the myocardium. Circ. Res. 48, 1-17.

Kawanishi, T., Nieminen, A.-L., Herman, B., & Lemasters, J. J. (1991). Suppression of Ca^{2+} oscillations in cultured rat hepatocytes by chemical hypoxic/ischemic. J. Biol. Chem. 266(30), 20062-20069.

Kikuchi-Yanoshita, R., Yanoshita, R., Kudo, I., Arai, H., Takamura, T., Nomoto, K.-I., & Inoue, K. (1993). Preferential hydrolysis of phosphatidylethanolamine in rat ischemic heart homogenates during *in vitro* incubation. J. Biochem. 114, 33-38.

Kreisberg, J. I., Mills, J. W., Jarrell, J. A., Rabito, C. A., & Leaf, A. (1980). Protection of cultured renal tubular epithelial cells from anoxic cell swelling and cell death. Proc. Natl. Acad. Sci. 77(9), 5445-5447.

Lemasters, J. J., Ji, S., & Thurman, R. G. (1981). Centrilobular injury following hypoxia in isolated, perfused rat liver. Science 213, 661-663.

Lemasters, J. J., Stemkowski, C. J., Ji, S., & Thurman, R. G. (1983). Cell surface changes and enzyme release during hypoxia and reoxygenation in the isolated, perfused rat liver. J. Cell Biol. 97, 778-786.

Lemasters, J. J., DiGuiseppi, J., & Herman, B. (1987). Alterations in cytosolic calcium and mitochondrial membrane potential during toxic injury in cultured hepatocytes. Nature 325, 78-81.

Lemasters, J. J., Caldwell-Kenkel, J. C., Gao, W., Nieminen, A.-L., Herman, B., & Thurman, R. G. (1992). Hypoxic/ischemic, ischemic and reperfusion injury in the liver. In: Pathophysiology of Reperfusion Injury (Das, D.K., ed.), pp. 101-135. CRC Press, Boca Raton, FL.

Lemasters, J. J. & Thurman, R. G. (1993) Hypoxic/ischemic and reperfusion injury to liver. Progr. Liver Dis. 11, 85-114.

Masaki, N., Kyle, M. E., Serroni, A., & Farber, J. L. (1989). Mitochondrial damage as a mechanism of cell injury in the killing of cultured hepatocytes by tert-butyl hydroperoxide. Arch. Biochem. Biophys. 270, 672-680.

Mayer, R. J. & Marshall, L.A.. (1993). New insights on mammalian phospholipase A_2(s); comparison of arachidonolyl-selective and -nonselective enzymes. FASEB J. 7, 339-348.

Mandel, L. J., Takona, T., Soltoff, S. P., Jacobs, W. R., LeFurgey, A., & Ingram, P. (1988). Multiple roles of calcium in anoxic-induced injury in renal proximal tubules. In: Cell Calcium and the Control of Membrane Transport (Mandel, L. J., & Eaton, D. C., eds.), pp. 277-285. Rockefeller University Press,NY.

Nachbaur, J., Colbeau, A., & Vignais, P. M. (1972). Distribution of membrane-confined phospholipase A in the rat hepatocyte. Biochim. Biophys. Acta. 274, 426-446.

Newkirk, J. D., & Waite, M. (1973). Phospholipid hydrolysis by phospholipases A_1 and A_2 in plasma membranes and microsomes of rat liver. Biochim. Biophys. Acta. 298, 562-576.

Nicotera, P., Thor, H., & Orrenius, S. (1989). Cytosolic-free Ca^{2+} and cell killing hepatoma 1c1c7 cells exposed to chemical anoxia. FASEB J. 3, 59-64.

Nieminen, A-L., Gores, G. J., Wray, B. E., Tanaka, Y., Herman, B., & Lemasters, J. J. (1988). Calcium dependence of bleb formation and cell death in hepatocytes. Cell Calcium 9, 237-246.

Nieminen, A-L., Dawson, T. L., Gores, G. J., Kawanishi, T., Herman, B., & Lemasters, J. J. (1990). Protection by acidotic pH and fructose against lethal injury to rat hepatocytes from mitochondrial inhibition, ionophores and oxidant chemicals. Biochem. Biophys. Res. Commun. 167, 600-606.

Otani, H., Engelman, M., Breyer, R. H., Rousou, J. A., Lemeshow, S., & Das, D. K. (1981). Mepacrine, a phospholipase inhibitor. A potential tool for modifying reperfusion injury. J. Thorac. Cardiovasc. Surg. 92, 247-254.

Otani, H., & Das, D. K. (1988). Enhanced phosphodiesteratic breakdown and turnover of phosphoinositides during reperfusion of ischemic rat heart. Circ. Res. 63, 930-936.

Otani, H., Prasad, R. M., Jones, R. M. & Das, D. K. (1989). Mechanism of membrane phospholipid degradation in ischemic-reperfused rat hearts. Am. J. Physiol. 257, H252-H253.

Pastorino, J.G., Snyder, J.W., Serrani, A., Hoek, J.B., and Farber, J.L. (1993) Cyclosporin and carnitine prevent the anoxic cell death of cultured hepatocytes by inhibiting the mitochondrial permeability transition. J. Biol. Chem. 268, 13791-13798.

Pentilla, A., & Trump, B. F. (1974). Extracellular acidosis protects Ehrlich ascites tumor cells and rat renal cortex against anoxic injury. Science 185, 277-278.

Prasad, M. R., Popescu, L. M., Moraru, I. I., Liu, X., Maity, S., Engelman, R. M., & Das, D. K. (1991). Role of phospholipases A2 and C in myocardial ischemic reperfusion injury. Am. J. Physiol. 260, H877-H883.

Sargent, C. A., O. Vesterqvist, J. R. McCullough, M. L. Ogletree, & G. J. Grover. (1992). Effects of the phospholipase A2 inhibitors quinacrine and 7,7-dimethyleicosadienoic acid in isolated globally ischemic rat hearts. J. Pharmacol. Exp. Therapeut. 262(3), 1161-1167.

Scholz, W., Albus, U., Linz, W., Martorana, P., Lang, H. J., & Schölkens, B. A. (1992). Effects of Na^+/H^+ exchange inhibitors in cardiac ischemia. J. Mol. Cell. Cardiol. 24, 731-740.

Sen, A., Miller, J. C., Reynolds, R., Willerson, J. T., Buja, L. M., & Chien, K. R. (1988). Inhibition of the release of arachidonic acid prevents the development of sarcolemmal membrane defects in cultured rat myocardium cells during adenosine triphosphate depletion. J. Clin. Invest. 82, 1233-1238.

Shaikh, N. A., & Downar, E. (1981). Time course of changes in porcine myocardial phospholipid levels during ischemia. Circ. Res. 49, 316-325..

Shaikh, N. A., Downar, E., & Butany, J. (1987). Amidarone-an inhibitor of phospholipase C activity: a comparative study of inhibitory effets of amidarone, chloroquine, and chlorpromazine. Mol. Cell. Biochem. 7, 163-172.

Simons, T. J. B. (1983). The role of calcium in the regulation of sugar transport in the pigeon red blood cell. J. Physiol. 338, 501-525.

Smith, M. T., Thor, H., & Orreius, S. (1981). Toxic injury to isolated hepatocytes is not dependent on extracellular calcium. Science 213, 257-261.

Smith, M. T., Thor, H., Jewell, S. A., Bellomo, G., Sandy, M. S., & Orrenius, S. (1984). Free radical changes in the surface morphology of isolated hepatocytes. In: Free Radicals in Molecular Biology: Aging and Disease (Armstrong, D., Sohal, R. S., Cutler, R. G., eds.), pp. 33-340. Raven Press, NY.

Snowdowne, K. W. & Borle, A. B. (1985). Effects of low extracellular sodium on cytosolic ionized calcium: Na^+-Ca^{2+} exchange as a major calcium influx pathway in kidney cells. J. Biol. Chem. 260(28), 14998- 15507.

Snowdowne, K. W., Freudenrich, C. C., & Borle, A. B. (1985). The effects of anoxia on cystolic free calcium, calcium fluxes and cellular ATP levels in cultured kidney cells. J. Biol. Chem. 260(21),11619-11626.

Snowdowne, K. W., Ertel, R. J., & Borle, A. B. (1985). Measurement of cystolic calcium with aequorin in dispersed rat ventricular cells. J. Mol. Cell. Cardiol. 17, 233-241.

Sobel, B. E., Corr, P. B., Robinson, A. K., Goldstein, R. A., Witkowski, F. X., & Klein, M. S. (1978). Accumulation of lysophosphoglycerides with arrhythmogenic properties in ischemic myocardium. J. Clin. Invest. 62, 546-553.

Starke, P. E., Hock, J. B., & Farber, J. L. (1986). Calcium-dependent and calcium-independent mechanisms of irreversible cell injury in cultured hepatocytes. J. Biol. Chem. 261, 3006-3012.

Steenbergen, C., & Jennings, R. B. (1984). Relationship between lysophospholipids accumulation and plasma membrane injury during total *in vitro* ischemia. J. Mol. Cell. Cardiol. 16, 605-621.

Stepherd, N., Kavaler, F., & Spielman, W. (1991). Cadmium block of isometric contractions of isolated bullfrog atrial cells. Am. J. Physiol. 260, C249-C258.

Storch, J., & Schachter, D. (1985). Calcium alters the acyl chain composition and lipid fluidity of rat hepatocyte plasma membranes *in vitro*. Biochim. Biophys. Acta. 812, 473-484.

Sun, F. F., Fleming, W. E., & Taylor, B. M. (1993). Degradation of membrane phospholipids in the cultured human astroglial cell line UC-11MG during ATP depletion. Biochem. Pharmacol. 45(5), 1149-1155.

Thuren, T., Vitanen, J. A., Somerharju, P. T., & Kinnumen, P. K. J., (1988). Phospholipase A_2 assay using an intramolecularly quenched pyrene labeled phospholipid as substrate. Analyt. Biochem. 170, 248-255.

van Bilson, M., van der Vusse, G. J., Willemsen, P. H. M., Coumans, W. A., Roemen, T. H. M., & Reneman, R. S. (1990). Effects of nicotinic acid and mepacrine on fatty acid accumulation and myocardial damage during ischemia and reperfusion. J. Mol. Cell. Cardiol. 22, 155-163.

van Blitterswijk, W. J., van der Meer, B. W., & Hilkmann, H. (1987). Quantitative contributions of cholesterol and the individual classes of phospholipids and their degree of fatty acyl (un)saturation to membrane fluidity measured by fluorescence polarization. Biochemistry 26, 1746-1756.

Van der Bosch, H. (1980). Intracellular phospholipases A. Biochem. Biophys. Acta. 604, 191-246.

van der Vusse, G. I., Roeman, T. H. M., Prinzen, F. W., Coumans, W. A., & Reneman, R. S. (1982). Uptake and tissue content of fatty acids in dog myocardium under normoxic and ischemic conditions. Circ. Research 50, 538-546.

Van Schaik, R. H. N., N. M. Verhoeven, F. W. Neijs, A. J. Aarsman, & H. Van den Bosch. (1993). Cloning of the cDNA coding for 14 kDa group II phospholipase A_2 from rat liver. Biochim. Biophys. Acta. 1169, 1-11.

Victoria, E. J., van Golde, L. M. G., Hostetler, K. Y., Scherphof, G. L., & van Deenen, L. L. M. (1971). Some studies on the metabolism of phospholipids in plasma membranes form rat liver. Biochem. Biophys. Acta. 239, 443-457.

Wang, X. F., Kuo, S. C., Lemasters, J. J., & Herman, B. (1992). Measurements of plasma membrane architecture during hypoxic/ischemic using multiple fluorescent spectroscopic techniques. SPIE Proceedings. 1604, 309-318.

Wang, X. F., Lemasters, J. J., Kuo, S. C., & Herman, B. (1993). Multiple microscopic techniques for the measurement of plasma membrane lipid structure during hypoxic/ischemic. Opt. Eng. 32, 284-290.

Wang, X. F., Lemasters, J. J., & Herman, B. (1993). Plasma membrane architecture during hypoxic/ischemic injury in rat hepatocytes measured by fluorescence quenching and resonance energy transfer. J. BioImag. 1, 30-39.

Wang, X. F. Lemasters, J. J. & Herman, B. (1994). Plasma membrane phopholipase A_2 activity during hypoxic/ischemic injury in rat hepatocytes. Biophys. J. (in preparation).

Wang, X. F. Lemasters, J. J., & Herman, B. (1994). Plasma membrane lipid and protein lateral diffusion during hypoxic injury in rat hepatocytes. Biophys. J. (submitted).

Weglichi, W. B., Waite, B. M., Sisson, P., & Shohet, S. B. (1971). Myocardial phospholipse A of microsomal and mitochondrial fractions. Biochem. Biophys. Acta. 231, 512-519.

Weglichi, W. B., Owens, K., Kennett, F. F., Kessner, A., Harris, L., Wise, R. M., & Vahouny, G. V. (1980). Preparation and properties of highly enriched cardiac sarcolemma from isolated adult myocytes. J. Biol. Chem. 255, 305-3609.

Wolf, R. A., & Gross, R. W. (1985). Identification of neutral active phospholipase C which hydrolyzes choline glycerophospholipids and plasmalogen selective phospholipase A_2 in canine myocardium. J. Biol. Chem. 260, 7295-7303.

Zalewski, A., Goldberg, S., & Maroko, P. R. (1988). The effects of phospholipase A_2 inhibition on experimental infarct size, left ventricular size, left ventricular hemodynamics, and regional myocardial blood flow. J. Cardiol. 21, 247-257.

THE REGULATION OF CARNITINE ACYLTRANSFERASES AND THEIR ROLE IN CELLULAR METABOLISM

Janet H. Mar and Jeanie B. McMillin

Advances in Lipobiology
Volume 2, pages 195-225.

ISBN 0-7623-0205-4

I. INTRODUCTION

A. The Role of Carnitine in the Oxidation of Long Chain Fatty Acids

Greater than three decades ago, carnitine was demonstrated to be essential to the oxidation of long chain free fatty acids in a variety of tissue preparations (Fritz, 1959; Fritz and McEwen, 1959). The general site of action was shown to be at the level of long chain acyl-CoA. Since β-oxidation was catalytically stimulated by carnitine and since acylcarnitines increased mitochondrial respiration (Bremer, 1962a), it was concluded that carnitine stimulates the oxidation of fatty acids by forming acylcarnitines. The compound formed by mitochondria upon incubation with palmitoyl-CoA and carnitine was identified chemically as palmitoylcarnitine (Fritz and Yue, 1963). Subsequently, the enzyme responsible for this catalysis was partially separated from mitochondria (Norum, 1964) and the kinetic equilibrium of the reaction was calculated to be 0.45. The reaction kinetics for palmitoyl-CoA and carnitine was established in the presence and absence of detergent. Pronounced substrate inhibition of the enzyme, carnitine palmitoyltransferase (CPT), by palmitoyl-CoA was shown to be competitive with respect to carnitine and this inhibitory action was prevented by the presence of detergent (Bremer and Norum, 1967a; Bremer and Norum, 1967b). The mechanism of action of detergent under these conditions is most likely explained by the binding of detergent to hydrophobic inhibitory sites on CPT, and not by complex formation of the detergent with palmitoyl-CoA (Bremer and Norum, 1967b). At this early time, the importance of mitochondrial membrane integrity to the susceptibility of CPT to other synthetic inhibitors of the catalytic reaction was also recognized. Bremer demonstrated that only the l-isomers of acylcarnitines are metabolized by mitochondria (Bremer, 1962b). Subsequently, Fritz and Marquis (1965) found that the carnitine analogue, (+)palmitoylcarnitine, inhibited both β-oxidation and CPT activity in intact mitochondria. However, detergent solubilization of the mitochondrial membranes prevented this inhibitory interaction. These studies, carried out in the late 1960s, are forerunners of experiments published over twenty years later which have characterized further the nature of inhibitory sites on CPT as well as the early suggestion (Norum, 1964; Kopec and Fritz, 1971) that multiple proteins with CPT reactivity may be present in the cell.

The concept that carnitine palmitoyltransferase (CPT) exists on two separate sites on the mitochondrial membrane was first suggested by Fritz and Marquis (1965). The barrier to palmitoyl-CoA oxidation, but not to that of palmitoylcarnitine was identified as the mitochondrial inner membrane (CPT site "A" or site "I") where the direction of the reaction from the site of fatty acid activation proceeds according to:

$$\text{Palmitoyl-CoA} + \text{Carnitine} \rightarrow \text{Palmitoylcarnitine}$$

The generation of palmitoylcarnitine (permeable) from palmitoyl-CoA (impermeable) allows acyl groups to pass across to the matrix and the site of β-oxidation.

Likewise, there should be CPT activity located in or in proximity to the mitochondrial matrix which catalyzes the reverse reaction:

$$\text{Palmitoylcarnitine} + \text{CoA} \rightarrow \text{Palmitoyl-CoA} + \text{Carnitine}$$

Kopec and Fritz (1971) subsequently purified an enzyme from calf liver that they designated carnitine palmitoyltransferase II. This assignment was based on the finding that this preparation showed higher specificity towards long-chain acylcarnitines compared to long-chain acyl-CoAs which is consistent with activity located in the matrix at site "B" or "II" (Fritz and Marquis, 1965). The conclusion that both CPT-I (defined as outside the mitochondrial inner membrane matrix barrier) and CPT-II (matrix-associated) activities are located in the mitochondrial inner membrane comes chiefly from the laboratories of Yates and Garland (1966; 1970) as well as Hoppel and Tomec (1972). The former investigators demonstrated that in unbroken mitochondria, CPT-I (CPT-A) was non-latent and easily solubilized by sonication. As anticipated from the impermeability of the inner membrane to palmitoyl-CoA, only CPT-I was accessible to the inhibitory effects of bromostearoyl-CoA (Yates and Garland, 1970). Subsequent localization of CPT activity to liver mitochondrial inner membrane used the digitonin method for mitochondrial fractionation to separate outer membranes from inner membrane/matrix and soluble activities (Hoppel and Tomec, 1972). No CPT activity could be found either in the outer membrane fraction or as part of the intermembrane space. The ability of digitonin to solubilize CPT activity in a pattern distinct from the release of adenylate kinase suggested that part of the CPT activity is loosely bound to the outer aspect of the mitochondrial inner membrane (Hoppel and Tomec, 1972). A similar conclusion was subsequently reached by Brosnan, Kopec, and Fritz (1973) using phospholipase to fractionate beef liver mitochondria. The question of the mitochondrial membrane localization of these activities is especially relevant to more recent studies which address the molecular uniqueness of CPT-I and CPT-II activities, as well as the presence of specific membrane targetting sequences. Because of these considerations, the mitochondrial location of CPT activities remains an important issue, as will be discussed in section VII.

B. Role of CPT-I in the Control of β-Oxidation and Ketone Body Production

The discovery of the carnitine-dependent pathway for fatty acid oxidation is significant historically in that it has produced a more complete understanding of metabolic interplay and control of energy production in a variety of cell types. Elucidation of the protein chemistry and kinetic mechanisms of the carnitine palmitoyltransferase enzymes has been particularly important with respect to manipulating the system in the therapeutic management of diseases, including genetic deficiencies in the carnitine pathway, accumulation of carnitine esters with associated arrhthymogenesis in cardiac ischemia and the control of diabetic ketosis.

The finding that the (+) acylcarnitines are not β-oxidized (Bremer, 1962b) and that the unnatural isomer acts to inhibit CPT in intact mitochondria (Fritz and Marquis, 1965) allowed investigators to test specifically the role of CPT in the control of the rates of fatty acid oxidation and esterification as well as the relationship between fatty acid oxidation and gluconeogenesis. The ability of (+)decanoylcarnitine to block β-oxidation, ketone body production and gluconeogenesis from lactate in the perfused liver provided the initial proof that the rate of fatty acid oxidation (not the levels of plasma free fatty acids) and consequent changes in the mitochondrial matrix redox, ATP/ADP and products of β-oxidation are responsible for the rate of glucose production by the liver (Williamson et al., 1968). The lack of effect of (+)decanoylcarnitine on the β-oxidation of octanoate (and subsequent stimulation of gluconeogenesis) (Williamson et al., 1968) was also observed when the inhibitor was employed in a model of experimental diabetic ketoacidosis (McGarry and Foster, 1973). By comparing rates of ketone body formation from octanoic acid and octanoylcarnitine in livers from fed and fasted rats, McGarry and Foster provided the first physiological evidence that the regulation of β-oxidation appears to be imparted by the carnitine acyltransferase reaction (McGarry and Foster, 1974a). Inhibition of β-oxidation of oleic acid in the presence of the physiological isomer, (-)carnitine, but not of octanoic acid, by (+) octanoylcarnitine [an agent with less detergent activity than (+)-decanoylcarnitine (McGarry and Foster, 1974b)] further established mediation by the carnitine acyltransferases in control of ketogenesis. The action of (+) acylcarnitines to inhibit ketogenesis from fatty acid and carnitine was subsequently shown to involve the carnitine acylcarnitine translocase so that acylcarnitine uptake into the mitochondria in exchange for carnitine out is competitively blocked by these analogues (Pande and Parvin, 1976). However, a premonitory observation was made in which it was noted that the fed liver supports high rates of ketogenesis from short chain fatty acids whereas the carnitine-dependent oxidation of long-chain fatty acids is significantly reduced in these livers when compared to the fasted state (McGarry and Foster, 1974a).

The inducible effects of fasting and diabetes on the carnitine-dependent rates of β-oxidation suggested either direct effects on the expression of enzyme activity or the presence of an "activation" site on the enzyme which controls ketogenic potential(McGarry ey al., 1975). Acute augmentation of hepatic long-chain fatty acid oxidation by either anti-insulin serum or by glucagon, and the proportional acceleration in the oxidation of (-)- octanoylcarnitine suggested that the conversion of acylcarnitine into acyl-CoA was increased via activation of CPT-II (McGarry et al., 1975). Although it had been suggested that hormones which appear to activate ketogenesis, e.g., glucagon (Christianson, 1977), may do so as a result of inhibition of fatty acid esterification (McGarry and Foster, 1974a) as a result of down-regulation of glycerophosphate acylation, it has been demonstrated that primary control is imparted by CPT, as well as by acylcarnitine to free carnitine ratios (Bremer and Wojtczak, 1972: McGarry et al., 1973). Even so, the suggestion that CPT could become rate controlling for β-oxidation was controversial since direct measurement of CPT activity in rat liver mitochondria indicated that the measured activity of that enzyme *in vitro* "far exceeds" the capacity for β-oxidation (Van Tol and Hülsmann, 1969; Pande, 1971). Nonetheless, long-term adaptations in CPT activities have been demonstrated in liver homogenates from both fasted or fat-fed rats (Aas and Daae, 1971) and in liver mitochondria from diabetic rats (Cook, 1984). In the latter experiments, the changes in activity can be attributed to CPT-I rather than CPT-II. Increased activity of the outer transferase was also measured in intact liver mitochondria from rats fasted for 24 hours (Bremer, 1981). This aspect of CPT regulation will be considered in more detail later in the discussion. On the other hand, CPT-II activity has been shown to increase slightly but significantly (25%) after a 24-hour fast (Bremer, 1981), a finding that is consistant with the finding of increased ketogenesis from (-)-octanoylcarnitine (McGarry et al., 1975). Covalent modification of CPT-II by phosphorylation has been proposed to increase the activity of this protein (Harano et al., 1985); control of hepatocellular CPT-I activity by phosphorylation-dephosphorylation may also be regulated short-term in culture by various agonists (Guzmán and Castro, 1991). The apparent susceptibility of CPT-I and CPT-II to phosphorylation events requires further study to elucidate not only the molecular basis for this phenomenon, but also its physiological impact relative to the effect of malonyl-CoA on overall flux through β-oxidation.

II. CONTROL OF β-OXIDATION AND CPT-I ACTIVITY BY MALONYL-COA

A series of studies on the activation of ketogenesis by Exton et al. (1969), as well as those described by McGarry and Foster (above) established the association between increased CPT activity and conditions of carbohydrate and insulin deficiency. The suspicion that some component of carbohydrate metabolism plays an important role in the up-regulation of CPT activity was supported by the observation that liver glycogen content varies in a reciprocal fashion with the rates of hepatic β-oxidation (McGarry et al., 1975). Although increases in carnitine concentration stimulate overall rates of ketogenesis and β-oxidation in the hepatocyte (Christiansen et al., 1976), no stimulatory effect of carnitine occurs in the presence of high liver glycogen levels (Robles-Valdes et al., 1976). Therefore, the discovery of malonyl-CoA by McGarry, Mannaerts and Foster (1977) as the effector of fatty acid flux through β-oxidation represents an important breakthrough toward understanding the interplay between carbohydrate and lipid metabolism. Malonyl-CoA represents the initial step in the metabolic conversion of carbohydrate (i.e., acetyl-CoA) to the pathway of fatty acid biosynthesis. Its concomittent role in the control of β-oxidation at the level of mitochondrial CPT-I was a seminal finding which explains the sparing effect of carbohydrate on fatty acid oxidation, evidence for which was established in 1954 (Lossow and Chaikoff, 1954). The influence of hormones on fatty acid synthesis and oxidation was subsequently explained by direct effects on the tissue levels of malonyl-CoA (McGarry et al., 1978a). The site of action of glucagon to lower malonyl-CoA was attributed directly to the glucagon-mediated decrease in acetyl-CoA carboxylase activity (Cook et al., 1977). The assignment of CPT-I as the site of malonyl-CoA inhibition was supported by three important findings (McGarry et al., 1978b) which bear on our understanding of the CPT system, an issue which continues to be debated to this day. First, high concentrations of malonyl-CoA (20 μM) abolish the oxidation of palmitoyl-CoA, the physiological substrate for CPT-I, but has no effect on the oxidation of palmitoylcarnitine (which bypasses CPT-I for transesterification to CoA by CPT-II in the matrix). Secondly, two CPT activities, one sensitive and the second insensitive to malonyl-CoA, are measured in mechanically disrupted mitochondria. Finally, the release of a malonyl-CoA insensitive fraction by Tween-20 and the retention of malonyl-CoA sensitive activity in the pellet suggested a relationship between the membrane association of CPT-I and malonyl-CoA sensitivity. These basic issues have been expanded and refined by the laboratories of McGarry and others to support the concept that CPT-I and CPT-II are distinct proteins with differing regulatory properties (see below).

With the understanding that malonyl-CoA is the predominant regulator of flux through β-oxidation, subsequent studies on the sensitivity of CPT-I to malonyl-CoA began to reveal the plasticity of CPT-I to changing hormonal and dietary status. The first suggestion that factors other than alterations in the tissue concentrations of malonyl-CoA may play a significant role in the rates of ketosis during starvation and diabetes came from Cook et al. (1980) who demonstrated that the rates of ketogenesis in liver mitochondria from fed and 48-hour fasted rats responded differently to malonyl-CoA so that the K_I for malonyl-CoA was increased in starvation. Likewise, Bremer (1981) and Saggerson and Carpenter (1981a,1981b) found that inhibition of CPT-I by malonyl-CoA was decreased by fasting. In the former study (Bremer, 1981) the activity of CPT-I was almost doubled by a 24-hour fast, and total CPT activity (here total activity was the activity measured in the presence of detergent concentrations of (+) palmitoylcarnitine) increased by 25%. The observation that malonyl-CoA inhibition of CPT activity disappeared with detergent solubilization was confirmed in these studies (Bremer, 1981). Cook (1984) later reported cooperativity in the inhibition of CPT-I by malonyl-CoA in liver mitochondria from rats fasted for 72 hours. Consistant with previous reports, a 57% increase in CPT-I was observed with no effects on the K_m for oleoyl-CoA. Cooperative inhibition kinetics in the presence of malonyl-CoA also suggested that the kinetics of malonyl-CoA interaction with the enzyme is not classical competitive. In this regard, and as will be discussed later, there is no similar response of the cardiac mitochondrial CPT-I to changes in malonyl-CoA inhibition with fasting, and the presence of malonyl-CoA does not impart sigmoidicity to the inhibition kinetics of the cardiac enzyme (Cook et al., 1984).

Changes in liver CPT-I and in its malonyl-CoA sensitivity also occur in animal models of diabetic ketosis and are very similar to what is observed with starvation. Cook et al. (1984) demonstrated that the sensitivity of CPT-I to malonyl-CoA is diminished in streptozotocin-induced diabetes. In liver mitochondria from the 48-hour diabetic rat, the K_I for malonyl-CoA increases by ten-fold and insulin treatment for five days decreases the K_I toward control values (Gamble and Cook, 1985). It was proposed that insulin modulates the ketotic state by increasing the affinity of CPT-I for malonyl-CoA, thereby decreasing the rate of ketogenesis. Diabetes produces a 30% increase in CPT-I and a 40% increase in total activity (Gamble and Cook, 1985). Insulin treatment returns the activities of CPT-I and sensitivity to malonyl-CoA to control levels by 2-4 hours. Interestingly, control activity remains elevated at 30% two hours after insulin injection. In contrast, similar changes observed in CPT-I and malonyl-CoA inhibition with fasting do not respond to short-term insulin treatment (Gamble and Cook, 1985; Pénicaud et al., 1991). Later studies by Grantham and Zammitt (1988), however, found that rapid reversal of the ketotic state by insulin is not accompanied by recovery of CPT-I to its normal properties. Instead 24 hours of insulin treatment was required to restore the sensitivity of CPT-I to malonyl-CoA ; during this time the activity of the enzyme decreases only marginally. In line with the latter study, it was later demonstrated that a 24-hour insulin clamp is needed to reverse the effects of starvation on the activity of liver CPT-I, liver levels of malonyl-CoA, as well as the K_I for malonyl-CoA (Pénicaud et al., 1991). Therefore, the changes in the properties of CPT -I are slowly reversed by insulin, whereas the ketotic state is rapidly reversed, most likely due to the anti-lipolytic effect of insulin. The delayed change suggests an alteration in protein synthesis or a delayed response to some metabolite effector which acts to up-regulate the enzyme. Since glucagon is not decreased by the clamp, it was concluded that insulin is the major effector of CPT-I. Consistant with this interpretation, the same laboratory did not observe any short-acting (30 - 60 minutes) effects of either insulin or glucagon on the kinetic properties of CPT in an isolated liver cell preparation, even though care was taken to perserve the phosphorylation state of the cells during digitonin permeabilization (Boon and Zammit, 1988). This observation is in contrast to the studies by Harano et al. (1985) who reported that glucagon stimulates CPT through an increased affinity for palmitoyl-CoA, suggesting that hepatic fatty acid oxidation is regulated by glucagon-mediated increases in CPT phosphorylation through a c-AMP dependent protein kinase. However, since the protein which is phosphorylated is the 69 KD subunit of CPT-II, one may not be able to extrapolate these results to other studies which specifically examine hormonal effects on the malonyl-CoA sensitive CPT-I. Although phosphorylation of CPT-I could conceivably occur and result in the increased activity of CPT-I, others have attributed the ketogenic effects of glucagon to be due, in part, to a 67% increase in liver carnitine with no change in serum concentrations, suggesting that glucagon increases carnitine production in the liver (Parvin and Pande, 1979). Fasting and glucagon infusion were also observed to increase mitochondrial carnitine content. Since the carnitine acylcarnitine translocase facing the mitochondrial matrix is considered to be subsaturated with carnitine, an increase in mitochondrial carnitine as a result of glucagon treatment or under conditions of ketogenesis would explain the increased ability of acylcarnitines to be oxidized under these conditions, apart from any specific effect on CPT-I or CPT-II. Thus, although the consensus is that insulin has important implications for the control of mitochondrial CPT-I, the role of glucagon in this process is still unclear. It should be mentioned that hyperthyroidism (Stakkestad and Bremer, 1983) and high fat feeding (Pégovier et al., 1988) also increase the K_I of CPT-I for malonyl-CoA whereas hypothyroidism (Stakkestad and Bremer, 1983) and chronic alcohol feeding (Guzmán et al., 1987) decrease this value. The importance of the effects of diabetes and the nutritional state of the animal on the transcriptional and translational regulation of these proteins has been considered by Brady et al. (1992).

III. CPT REGULATION IN THE FETAL AND NEW-BORN ANIMAL

Immediately following birth, there is an apparent increased capacity of liver mitochondria for fatty acid oxidation and a resultant physiological hyperketonemia which continues during suckling (Girard et al., 1985). Coordinated changes in insulin (decrease) and glucagon (increase) accompany the approximate 5 to 8-fold increase in long chain fatty acid oxidation over the period extending from birth to 24 hours (Ferré et al., 1983; Herbin et al., 1987).

Plasma free fatty acids are also increased and hepatic carnitine concentrations are already high at birth due to its rapid placental transfer (Hahn et al., 1980). Consistant with increased fat oxidation and decreased lipogenesis at birth, malonyl-CoA levels decrease by 90% in hepatocyes isolated from newborn and 24-hour old rabbits (Herbin et al., 1987). However, these authors were able to dissociate partially the drop in malonyl-CoA concentrations from the 6-fold increase in oleate oxidation by inhibiting acetyl-CoA carboxylase. Although malonyl-CoA levels drop 10-fold, there is no associated stimulation of oleate oxidation (Herbin et al., 1987). Other investigators confirm the apparent absence of a reciprocal relationship between ketogenesis and lipogenesis in livers from the newborn (Ferré et al., 1983). This conclusion provides a rationale for the observation that there is a rapid induction of fatty acid oxidation in the newborn rat even when malonyl-CoA levels are high (Ferré et al., 1983). To provide a mechanistic explanation for these findings, Saggerson and Carpenter (1982) were the first to measure a 15- to 30-fold decrease in sensitivity to malonyl-CoA when CPT-I from the 24-hour old rabbit is compared to either the fetus or, in later experiments, to the newborn in work cited above (Herbin et al., 1987). CPT-I activity itself is rapidly elevated by 3 to 6-fold (Saggerson and Carpenter, 1982).

Despite the latter observation, these investigators concluded that the major influence in the increased rates of carnitine-dependent fatty acid oxidation is not malonyl-CoA or even enhanced CPT-I activity, but rather the impressive fall in sensitivity of this enzyme to malonyl-CoA. In contrast to CPT-I, little or no change in the activity of CPT-II is measured during the first 24 hours after birth (Saggerson and Carpenter, 1982). Two reports of significantly elevated expression of mRNA for CPT-II in both rat and mouse hearts at birth (Wu and Thornburg, 1993; Gelb, 1993) may reflect a developmental response to the initially depressed rates of palmitate oxidation in cardiac homogenates of newborn rats (Wittels and Bressler, 1965).

In adult rat liver, glucagon increases long chain fatty acid oxidation mainly by decreasing malonyl-CoA concentrations (McGarry and Foster, 1980), with no consistantly observed changes in CPT (Boon and Zammit, 1988). In contrast to the adult, glucagon may represent the signal for molecular events which produce a decrease in malonyl-CoA sensitivity of CPT-I in the liver of new-born rabbits (Girard et al., 1992). Prolonged exposure of cultured fetal hepatocytes to glucagon (24 to 48 hours) produces a time-dependent increase in the IC_{50} of CPT-I for malonyl-CoA (Prip-Buus et al.,1990). The change in this property of the enzyme is probably not due to either phosphorylation or dephosphorylation events (Harano et al. 1985) since there is a time delay for induction of the desensitization, nor is it likely that increased synthesis of new enzyme occurs, since total activity does not change. This observation is extremely relevant to the possibility of the presence of CPT-I isoforms, each with different IC_{50} values for malonyl-CoA which are individually characteristic of heart, muscle, brain and liver (see below). It is also possible that glucagon may be affecting the expression of a separate gene which encodes a malonyl-CoA binding protein that associates with CPT-I. Evidence for and against the existence of this protein and its role in the regulation of CPT-I activity will be discussed later. Alternately, glucagon may affect the microenvironment of CPT-I so that its aggregation and/ or membrane association is altered (Girard et al., 1992).

The additional role of dietary lipid on the response of CPT-I to malonyl-CoA has been established by Decaux et al. (1988) by studying the suckling to weaning transition. Rats weaned on a high fat diet maintain the ability to oxidize fatty acids at high rates with no reciprocally related changes in the rates of lipid synthesis in the liver. Most importantly, all of the kinetic characteristics of CPT-I, i.e., the decreased affinity for malonyl-CoA and increased enzyme activity, remain at the same levels seen during the suckling period (Decaux et al., 1988). Thus, changes in these parameters which occur upon weaning to a high carbohydrate diet appear to be linked to the nutrition of the animal rather than to its developmental stage. Indeed, the control by fatty acids of transcriptional expression of mitochondrial enzymes of β-oxidation (Gulick et al., 1993) may also be relevant to control of CPT-I. Initiation of medium chain acyl-CoA dehydrogenase transcription by fatty acids appears to be an autoregulatory phenomenon which involves activation of the nuclear peroxisomal proliferator-activated receptor (PPAR, an orphan receptor transcription factor) and retinoid X receptor at an upstream promoter element (Gulick et al., 1993). At the present time, there are no studies which have addressed the relevance of these specific mechanisms for CPT. In addition to fatty acids, it has been proposed that hypolipidemic agents induce peroxisomal proliferation with potent stimulation of peroxisomal enzymes of β-oxidation via PPAR (Issemann and Green, 1990). Of relevance to mitochondrial enzymes of β-oxidation, total mitochondrial CPT activity has been shown to increase significantly in the liver following treatment of rats with sulfur-substituted dicarboxylic acids (Berge et al., 1989). These authors propose that non-β-oxidizable fatty acids are much more potent than high levels of oxidizable fatty acids in inducing enzymes involved in fatty acid metabolism by proliferation of peroxisomes and mitochondria. Formation of acylcarnitines of hypolipidemic drugs may also inhibit CPT-I and mitochondrial β-oxidation with subsequent induction of peroxisomal enzymes (Foxworthy and Eacho, 1988; Gerondaes et al., 1988).

IV. OTHER CELLULAR PROTEINS WITH CARNITINE ACYLTRANSFERASE ACTIVITIES

Only two carnitine acyltransferases were known up to 1971, i.e., the carnitine acetyltransferase and the carnitine palmitoyltransferase. A medium-chain octanoyltransferase from calf liver was characterized in 1972 (Solberg, 1972). These results supported the initial observation of Kopec and Fritz (1971) of an activity specific for octanoylcarnitine in mitochondrial extracts from beef heart. Subsequently, Clarke and Bieber (1981a) reported that two proteins could account for all the activity of heart mitochondrial acyltransferases with medium chain acyl groups as substrate, i.e., carnitine acetyltransferase and carnitine palmitoyltransferase. A peak of medium chain length CPT eluted from a gel filtration column demonstrated that palmitoyltransferase activity is less than half of the activity for decanoyl-CoA. This finding has been confirmed by Mynatt et al. (1992) who found that acyl-CoA substrate specificity for heart outer mitochondrial membrane CPT-I was $C10 > C16 > C14 > C12 > C18 = C8$. Markwell et al. (1976), however, demonstrated the presence of extramitochondrial carnitine octanoyltransferase activity in peroxisomes and microsomes with little or no detectable palmitoyltransferase activity. An easily solubilized carnitine palmitoyltransferase from bovine liver mitochondria (Ramsay et al. 1987) was proposed to be distinct from CPT-II and therefore, likely represented mitochondrial CPT-I. The latter author subsequently attributed this activity to a peroxisomal origin due to its cross-reactivity with mouse peroxisomal COT (Ramsay, 1988). The maximum velocity of the peroxisomal protein is with palmitoyl-CoA (Ramsay, 1988). Derrick and Ramsay (1989) compared the overt mitochondrial and peroxisomal acyltransferases and found them both to be inhibitable by malonyl-CoA with overlapping chain-length specificities. Due to historical considerations, the soluble form of CPT (peroxisomal) has subsequently been referred to as COT, a term which reflects its wide-range substrate specificities. Since peroxisomes preferentially oxidize long- and very long-chain fatty acids, the purification of a peroxisomal acyltransferase suggested a carnitine-dependent pathway for fatty acid transfer across the peroxisomal membrane. Although some studies indicate that peroxisomes oxidize fatty acids in the absence of carnitine (Thomas et al., 1980; Appelkvist and Dallner, 1980), an ATP-dependent structural latency of these organelles has been reported which explains the high permeability of the peroxisomal membrane *in vitro* (Wolvetang et al., 1990). However, differential transport of palmitate, as palmitoyl-CoA, and lignoceric acid, as the free acid into rat liver peroxisomes has been attributed to a differing peroxisomal topology rather than to leaky membranes. Singh et al. (1992) have demonstrated that the active site for palmitoyl-CoA ligase is on the cytoplasmic surface whereas the active site of lignoceroyl-CoA ligase is on the luminal surface of the intact peroxisomal membrane. On the other hand, the presence of a medium-chain carnitine acyltransferase activity (COT) in peroxisomes has been proposed to be a specific pathway for the export of chain-shortened fatty acids to the cytosol (Bieber,1988). A functional explanation for these immunologically unique carnitine acyltransferases in peroxisomal β-oxidation (and also found in the microsomes, see below) is still under experimental evaluation. The hypothesis that peroxisomal COT may have the capacity to regulate peroxisomal β-oxidation is based on the apparent susceptibility of this enzyme to inhibition by TDGA-CoA (Skorin et al., 1992) and is consistant with the previous findings (Derrick and Ramsay, 1989) describing malonyl-CoA regulation of peroxisomal carnitine acyltransferase activity. The existence of a carnitine-dependent pathway for export of fatty acid from the peroxisome would be a pausible mechanism by which the cell conserves energy by bypassing the cycle of acyl-CoA hydrolysis in the cytoplasm and reactivation in the mitochondria (Ramsay and Arduini, 1993).

The second extramitochondrial location of carnitine acyltransferase activity is microsomal (Markwell et al., 1976) and the cellular content of this enzyme in the microsomes in the liver is quantitatively significant. The liver microsomal enzyme is a medium/long-chain transferase, designated as COT by Lilly et al. (1990). In addition, microsomal COT is inhibited by malonyl-CoA and etomoxiryl-CoA, although the I_{50} concentrations required are 200-fold higher than with mitochondrial CPT-I. These authors demonstrated that liver microsomal COT is antigenically different from either mitochondrial CPT-II or peroxisomal COT (Lilly et al., 1990). The enzyme has been purified and has a four-fold preference for decanoyl-CoA over palmitoyl-CoA (a similar order of substrate preference has been shown for mitochondrial CPT). The molecular weight of the purified protein is 50-54,000, again with no cross-reactivity with mitochondrial CPT or peroxisomal COT (Chung and Bieber,1993). The amino terminal sequence of this protein has a different primary structure from either CPT-I (Kolodziej and Zammit, 1993) or CPT-II (Woeltje et al., 1990b). The sequence, however, does have 100% homology with a phosphatidyl inositol-specific phospholipase Cα, although it lacks phospholipase activity. Furthermore, the protein disulfide oxidoreductase and thiol protease activities ascribed to a protein with this sequence have been separated from CPT (COT) in 900-fold enriched preparations (Murthy and Pande, 1993). It is of much interest that this protein has been shown to increase its levels of expression following starvation and heat stress (Lee, 1992). The same protein is apparently raised in activity by 100-fold in the brain by the cumulative actions of estrogen and leuteinizing hormone releasing hormone (Mobb et al., 1990). A similar protein has been also measured in microsomes from rat (Fogle and Bieber, 1978) and dog (McMillin et al., 1992) heart. This protein has been purified from dog cardiac sarcoplasmic reticulum and shows specificity for octanoyl-CoA compared to palmitoyl-CoA; it is also inhibited by etomoxiryl-CoA with 6-fold greater potency than liver COT (unpublished results, author's laboratory). The quantitative importance of this protein to the metabolism of the heart cell is unknown, since the membrane surface area of cardiac sarcoplasmic reticulum is less than 3.5% (Page and McAllister, 1973). Likewise, the presence of a malonyl-CoA sensitive carnitine palmitoyltransferase in the erythrocyte plasma membrane may represent another species in the CPT family (Ramsay et al., 1991). It has been suggested that this activity may act to "buffer acyl-CoA present in the red blood cell during turnover of membrane phospholipids" (Ramsay et al., 1991). It is also possible that its existence represents remnant activity derived from the immature erythrocyte; however, more work is needed to establish these interesting possibilities.

V. KINETIC PROPERTIES OF CARNITINE ACYLTRANSFERASE ACTIVITIES

A. Mitochondrial CPT-II

Separation and characterization of cellular carnitine medium and long-chain acyltransferases are complicated by their natural association with membranes of differing phospholipid composition and the necessity to employ detergents to solubilize and isolate individual proteins. The substrates, palmitoyl-CoA and palmitoylcarnitine, also form micelles which alters interpretation of kinetic parameters. Mitochondrial carnitine palmitoyltransferase II has been purified from liver (Kopec and Fritz, 1971; Miyazawa et al. 1983) and heart (Clarke and Bieber, 1981) and therefore, is the first protein to be studied extensively. The primary amino acid sequences of the heart and liver proteins are very similar as reflected by antibody cross-reactivity between tissues (Kolodziej et al., 1992) and to the nucleotide-sequence homology between the cloned CPT-II proteins from liver

(Woeltje et al., 1990a; Woeltje et al., 1990b) and heart (personal communication, Dr. Patrick Umeda). The pBKS-CPT-II.4 cDNA cloned from rat liver also hybridizes with a 2.4 kilobase sequence from neonatal rat heart total RNA (Hudson et al., 1993). The monomeric size of CPT-II is approximately 70,000 (Zammit et al., 1980), with a molecular weight of associated subunits estimated to vary in size from 150,000 (Kopec and Fritz, 1973) to 510,000, the latter estimated from molecular seiving columns (Fiol and Bieber, 1984).

CPT purified from calf liver mitochondria using an aqueous extraction medium demonstrates hyperbolic kinetics with K_m values of 31 μM for palmitoyl-CoA, 2 mM for carnitine, and 170 μM for palmitoylcarnitine (Norum, 1964). The substrate specificities for the non-ionic detergent solubilized enzyme from calf liver was highest for palmitoyl-CoA (Kopec and Fritz, 1971), but the pattern shifted to preference for medium chain fatty acids with Triton X-100 (Clarke and Bieber, 1981b). However, a range of K_m values was observed for the various chain-length carnitine esters, depending on the presence or absence of detergent as well as the monomeric versus micellar range of substrate concentrations (Clarke and Bieber, 1981b). This careful study helps to explain, in part, the variety of Michaelis constants reported for mitochondrial CPT, and stresses caution in the use of kinetic parameters to distinguish between potential isozymes of carnitine acyltransferase activities in the cell. An alternative approach to the study of CPT isoforms is the use of carnitine analogues to discriminate between the various activities (Murthy et al., 1990; Saeed et al.,1993).

Further work from the same laboratory reveals sigmoidal kinetics of purified beef heart mitochondrial CPT with palmitoyl-CoA as substrate, but not octanoyl-CoA (Fiol and Bieber, 1984). Dramatic effects of pH on the kinetic parameters suggested that this enzyme may have different catalytic properties depending upon its membrane location (cytosolic versus matrix face). The substrate cooperativity and absence of malonyl-CoA effects on the detergent-solubilized, purified enzyme indicate the following. When the enzyme is in its native state, malonyl-CoA has the ability to interact, not as a classical competitive inhibitor of CPT, but rather in an allosteric manner; however, this interaction is lost (regulator subunit?) upon enzyme solubilization (Fiol and Bieber, 1984). Evidence that CPT-I and CPT-II are the same protein will be discussed later and contrasted with the opposite view that they are, in fact, very different proteins.

B. Mitochondrial CPT-I

As might be anticipated from the discussion on purified CPT-II activity above, the kinetic analysis of CPT-I is complicated, not only by artifacts inherent in measuring membrane-associated proteins, but also by a plethora of data which point out the complex nature involved in regulation of CPT-I activity. Molecular weight estimates of CPT-I are based on irreversible binding of the inhibitor, tetradecylglycidic acid, to 90 or 86 kilodalton proteins in liver and muscle mitochondria, respectively. This difference in molecular weight from the value of 66 to 70,000 estimated for CPT-II form the basis for the suggestion that CPT-I and CPT-II are separate proteins (McGarry et al., 1989). On the other hand, the observation of sigmoidal behavior of purified mitochondrial CPT (CPT-II) is consistant with the sigmoidal kinetics present in malonyl-CoA-sensitive CPT-I activity measured in intact mitochondria (Saggerson and Carpenter, 1983; Gamble and Cook, 1985) and, therefore, with the idea that CPT-I and -II are the same enzyme.

In kinetic measurements on CPT-I in isolated mitochondria, bovine serum albumin is routinely added to prevent detergent-associated effects of the micellar substrate, palmitoyl-CoA, on the enzyme itself (would would lead to a decrease in malonyl-CoA sensitivity) and on the latency of CPT-II (as a result of membrane permeabilization). Cook and Gamble (1987) recognized the theoretical problem of acyl-CoA binding to albumin which could impart a sigmoidal appearance to CPT-I kinetics when the substrate is varied (Bartlette et al., 1985). When CPT-I kinetics are analyzed to reflect either mitochondrially bound palmitoyl-CoA (Pauly and McMillin, 1988) or free palmitoyl-CoA, as determined by fluorescent binding (Richards et al., 1991), the highly sigmoidal profile obtained in the presence of albumin is lost and the Hill coefficients reported in both studies are 1.2 and 1.04, respectively. Thus, the sigmoidal behavior is likely to be due to the artifact introduced by plotting total palmitoyl-CoA concentration versus activity where albumin reduces the availability of free substrate. It could also be argued that any intrinsic sigmoidicity in CPT-I kinetics would be obscured by the presence of albumin.

Studies on mitochondrial CPT-I by Mills et al. (1983) and Grantham and Zammit (1986) suggest the presence of a regulatory site on CPT-I, distinct from its active site. When the kinetics of mitochondria from liver, heart and skeletal muscle are followed over a pH range from 6.8 to 7.6, a decrease in the inhibition of the enzyme by malonyl-CoA is noted with no change in the K_m for palmitoyl-CoA. It was proposed that a competitive interaction between malonyl-CoA and palmitoyl-CoA takes place at a site distinct from the active site. The assignment of a low affinity malonyl-CoA binding site as a second, allosterically regulated acyl-CoA site is based on evidence that palmitoyl-CoA can only displace malonyl-CoA bound to this site but *not* to the high affinity inhibitor-binding site (Grantham and Zammit, 1986). These data are also consistant with the partial competitive inhibition kinetics of mitochondrial CPT-I. In this model, low affinity binding of malonyl-CoA to the enzyme could modify the high affinity malonyl-CoA interaction with the active site of CPT-I so that palmitoylcarnitine continues to be produced for β-oxidation even at high concentrations of malonyl-CoA in the cell (McMillin et al., 1994). The pattern of partial competitive kinetics has been attributed to mitochondrial isolation artifacts and exposure of malonyl-CoA insensitive CPT-II activity, particularly in the case of cardiac mitochondria where mechanical disruption of the tissue is required (Mynatt et al., 1992). It is interesting that CPT-I kinetics maintain a pattern of partial competitive inhibition both in isolated hepatocytes (Boon and Zammit,

1988) and in isolated neonatal cardiac myocytes (McMillin et al., 1994) where mitochondrial kinetics can be observed *in situ*, independent of preparational anomalies.

Carnitine has been reported to interact with the high affinity malonyl-CoA binding site on CPT-I (Bird and Saggerson, 1985). Evidence for this conclusion is based on the finding that both the K_m for carnitine and the inhibition by malonyl-CoA decrease as the assay pH is increased. Other investigators have observed an inverse relationship between the inhibitory potency of malonyl-CoA and the K_m for carnitine when mitochondria isolated from a variety of tissues are compared (Mills et al., 1983). Together with the finding that carnitine acts to modify the malonyl-CoA binding environment, facilitating reversible competition of malonyl-CoA with bromoacetyl-CoA (Edwards et al., 1985), these data support the hypothesis that the carnitine and high affinity malonyl-CoA binding sites are closely associated. Definitive assignment of the molecular topology and cooperativity between active and regulatory site interaction(s) awaits the availability of a purified, active preparation of CPT-I and genetic manipulation of the protein to probe potential binding domains of both substrates and regulators.

VI. ROLE OF CPT IN CARDIAC FATTY ACID METABOLISM

The existence of a cardiac-specific isoform of CPT-I is likely based on several lines of evidence, some of which have been discussed briefly in the previous section. First, as mentioned, the molecular weight of the TDGA-CoA-labeled protein from cardiac mitochondria is different from the molecular weight of the liver protein. Cardiac CPT-I also has been found to be immunologically distinct from the liver CPT-I (Kolodziej et al., 1992, Woeltje et al.,1990). Secondly, the K_I of the cardiac CPT-I for malonyl-CoA is greater than 20-fold lower (0.1 μM) compared to the K_I of the liver enzyme (2.7 μM) while the K_m for carnitine is 6-fold higher for heart compared to liver (McGarry et al., 1983). Importantly, in direct contrast to hepatic CPT-I, neither the activity of cardiac CPT-I nor its sensitivity to malonyl-CoA are subject to long-term regulation by fasting-feeding cycles (Cook, 1984; Mynatt et al., 1992; McGarry et al., 1983). However, similar to liver, cardiac tissue levels of malonyl-CoA are decreased by fasting (McGarry et al., 1983) and are acutely raised by the presence of insulin in isolated adult myocytes (Awan and Saggerson, 1993). Since the heart is not a lipogenic tissue and lacks fatty acid synthetase activity (Masoro and Porter, 1965; Wit-Peeters et al., 1970), it is presumed that the primary role of malonyl-CoA in the heart is to regulate β-oxidation. An important issue with respect to malonyl-CoA turnover in the heart is the relative absence of a known quantitatively important pathway for malonyl-CoA disposal which is a prerequisite to accomodate the observed fluctuations in tissue levels of malonyl-CoA with the feeding/fasting cycle. Awan and Saggerson (1993) speculate that the condensing enzyme associated with the microsomal fatty acid elongation in the heart may provide a significant "sink" for malonyl-CoA binding and turnover. Although the quantitative significance of this pathway remains to be established, it may represent one mechanism for controlling free levels of cytosolic malonyl-CoA and consequently, CPT-I activity. Nonetheless, variations in malonyl-CoA concentrations in the heart do occur and are reponsive to the physiological state of the organ. Increases in malonyl-CoA in the isolated, working rat heart are associated with decreased rates of β-oxidation (Saddick et al. 1993). In agreement with Awan and Saggerson, these authors conclude that synthesis of malonyl-CoA is directly related to heart CPT-I activity and regulation of β-oxidation. However, many questions are still unresolved, and these issues in part have been recently considered (Awan and Saggerson, 1993). Tissue concentrations of malonyl-CoA have been estimated at 3 to 9 μM which should effectively inhibit cardiac β-oxidation *if* (1) this measurement accurately reflects the cytosolic free concentration of malonyl-CoA and (2) the inhibition kinetics for cardiac CPT-I are simple competitive, which in all likelihood is not the case. It has been suggested that a large proportion of the cardiac malonyl-CoA is compartmentalized within the mitochondrial matrix, for example, or that the majority of malonyl-CoA is bound to malonyl-CoA binding proteins, e.g., the fatty acid elongation system in the microsomes, which serve as a low-affinity "sink" to bind and metabolize cytosolic malonyl-CoA. The low abundance of microsomal membranes in the heart, and the fact that the major protein within these microsomes is the Ca^{++}-ATPase are factors which must also be considered in assigning a potentially regulatory role for this pathway in the control of malonyl-CoA levels in the heart. One could also speculate that a large number of low affinity malonyl-CoA binding sites should still be in equilibrium with a high affinity site for malonyl-CoA on CPT-I. In this scenario, the high affinity site, i.e., CPT-I, should remain an effective competitor for inhibitor binding and consequently reflect the expected suppression of β-oxidation. Consistant with this anomaly is the lack of any known condition which produces total suppression of fatty acid oxidation in the heart. The finding of a positive correlation between malonyl-CoA levels and rates of fatty acid oxidation in the isolated, working rat heart (Saddick et al., 1993), however, provides strong support for the view that most of the cardiac malonyl-CoA is sequestered in a compartment to which CPT-I is not accessible. In the latter studies, β-oxidation is still ongoing at the highest malonyl-CoA concentrations attained, presumably reflecting levels which exceed the K_I of cardiac mitochondrial CPT-I for malonyl-CoA by many orders of magnitude. An alternate possibility is that malonyl-CoA binds at some low affinity site on CPT-I, with subsequent effects on catalysis consistant with partial inhibition kinetics. This type of interaction would not only increase the calculated K_I value, but also explain the inability to suppress completely β-oxidation in the heart. A similar mechanism has been proposed for liver mitochondria where exposure to malonyl-CoA and subsequent binding to the low-affinity site produces a time-dependent *increase* in the sensitivity of the enzyme for malonyl-CoA (Zammit et al., 1984). This increased affinity is attributed to a "malonyl-CoA-induced transition" from a form of CPT-I with low affinity for malonyl-CoA, to one with high affinity. Like partial inhibition kinetics, this observation requires a change in tertiary structure of CPT-I which affects its sensitivity to malonyl-CoA. However, because of the apparent kinetic and molecular differences between cardiac and liver CPT-I, the effects of malonyl-CoA binding on catalysis is likely dependent upon the predominant isoenzyme in the tissue under investigation.

Another issue of direct relevance to the control of β-oxidation in the heart is the regulation of the cardiac isoform of acetyl-CoA carboxylase (ACC)(Thampy, 1989; Bianchi et al., 1990) by hormones and metabolites. Bianchi et al. (1990) determined that the tissue levels of cardiac ACC (280,000 daltons) are not affected by the nutritional state. This is in contrast to the liver isoforms (both the 265,000 and 280,000 dalton proteins) which both decrease with fasting and increase in content with carbohydrate refeeding. Subsequent studies by Saddik et al (1993) suggest that the activity of cardiac ACC is regulated primarily by availability of its substrate, acetyl-CoA. This attractive proposition would provide a "messenger" role for acetyl-CoA, so that when the levels of this metabolite rise as a result of fatty acid oxidation, cardiac ACC becomes activated resulting in a down-regulation of flux through β-oxidation. The absence of any citrate dependence for cardiac ACC activity as reported by these authors, however, is in contrast to the published activation of the purified cardiac ACC by citrate (Thampy, 1989) and the report that malonyl-CoA levels can also be acutely regulated by insulin (increased malonyl-CoA) and epinephrine (decreased malonyl-CoA) (Awan and Saggerson, 1993). Although it is clear that the cardiac ACC undergoes phosphorylation (Thampy, 1989), the acute hormonal effects observed by Awan and Saggerson were speculated to result from tissue fluctuations in cytosolic palmitoyl-CoA levels. However, quantitative assessment of the the contribution of soluble acyl-CoA to total cardiac acyl-CoA concentrations reveals that acyl-CoA ester accumulation in the cytosol is limited by compartmentalization of 95 % of total tissue coenzyme A to the mitochondrial matrix (Idell-Wenger et al., 1978). The report of decreased tissue levels of malonyl-CoA in the heart with fasting may support either a role for phosphorylation or acetyl-CoA availability in the fasting response. Thus, metabolic control mechanisms characteristic of the liver may not directly reflect a similar environment in the heart.

VII. MOLECULAR CHARACTERIZATION OF CARNITINE ACYLTRANSFERASE(S)

For the past few years, investigators have focussed on the separate identities of the various cellular carnitine acyltransferases, including mitochondrial CPT-II and the three malonyl-CoA sensitive activities which are in contact with the cytosolic compartment, i.e., mitochondrial CPT-I, microsomal and peroxisomal COT. The greatest controversy developed concerning the relationship between the two mitochondrial CPT activities. The major "hard" piece of evidence is the demonstration that two separate and active CPT proteins can be purified from mitochondria. A protein demonstrating sensitivity to covalent inhibitors such as CoA esters of oxirane carboxylic acids has been purified in inactive form. An active protein designated CPT-II because of its insensitivity to malonyl-CoA has been purified and its kinetic mechanism exhaustively investigated. However, the finding that only one protein which has CPT activity can be purified from liver and heart has been taken by some investigators as evidence that CPT-I and CPT-II are identical proteins.

A. What is the Evidence that CPT-I and CPT-II are Different Proteins?

Much of the evidence related to the mitochondrial CPT system which has been established over the last ten years is not debated, but has been subject to interpretational differences. Since excellent reviews have recently appeared addressing both sides of this issue (McGarry et al., 1989; Saggerson et al., 1992), the data concerning this question will be presented briefly.

The first model of the mitochondrial CPT system is one where CPT -I and CPT-II are regarded as proteins with identical catalytic mechanism and structure (Figure 1). In this model, CPT-I is localized to the outer aspect of the inner mitochondrial membrane (other orientations are possible, see below), and CPT-II faces the matrix and the enzymes of β-oxidation. This orientation was suggested originally by Fritz and Yue (1963), experimentally verified by Hoppel and Tomec (1972) and recently confirmed by the latter laboratory using the French Press to fracture and separate the outer and inner membranes of liver mitochondria (Hoppel, 1991). In contrast to other recent studies (see below), the outer mitochondrial membrane fraction was found to be devoid of carnitine palmitoyltransferase activity.

Even if CPT-I and CPT-II share an identity, they still differ in that only CPT-I as measured in intact mitochondria, is inhibited by malonyl-CoA. The presence of a malonyl-CoA binding subunit associated with CPT only on the cytosolic-facing surface could explain the specificity of the inhibition (Figure 1). Dissociation of the malonyl-CoA binding protein from interaction with CPT has been explained as detergent disruption of the native quaternary structure; this concept is further supported by the conferral of cooperativity by malonyl-CoA on the reaction kinetics of CPT-I, suggesting oligomeric association is important to the heterotropic response. A malonyl-CoA binding protein has been identified as a 90,000 dalton protein which is covalently labeled by [H-3]tetradecylglycidyl-CoA (TDG-CoA). It is important to note that CPT-II is not labeled by TDG-CoA, consistant with the lack of inhibition of this protein by either TDG-CoA or malonyl-CoA (DeClercq et al., 1987). This information can be interpreted as either supporting the idea that (1) CPT-I activity represents an oligomeric association between CPT (II) and a different malonyl-CoA binding peptide, or (2) CPT-I is a different protein from CPT-II, which contains, within its primary sequence, a binding site for malonyl-

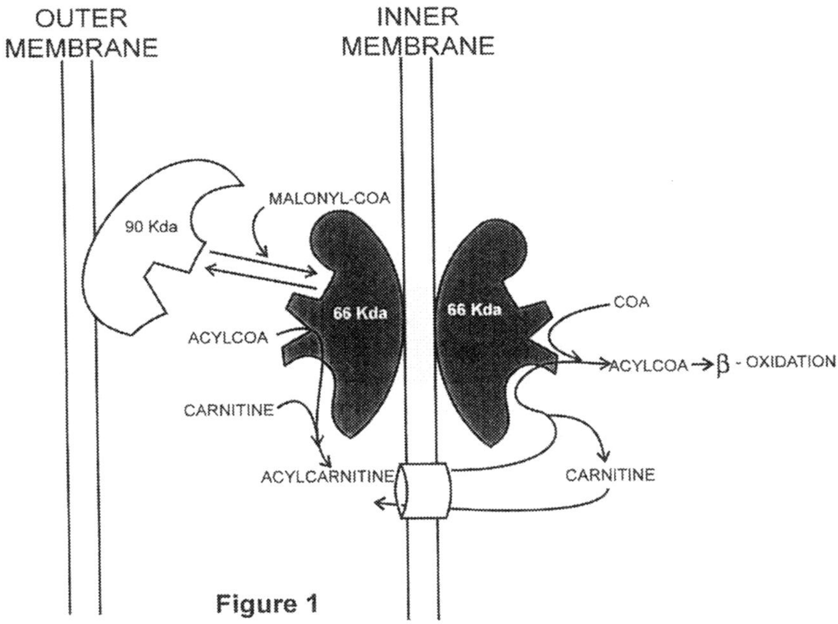

Figure 1.

CoA. To prove the former concept, it must be important to reconstitute the malonyl-CoA sensitive activity in a system where a protein corresponding to CPT-II is the only activity present which can catalyze palmitoylcarnitine formation. Ghadiminejad and Saggerson (1990) isolated a malonyl-CoA binding fraction with no CPT activity from rat liver mitochondrial outer membranes and combined this fraction with detergent-solubilized inner membranes containing malonyl-CoA-insensitive CPT activity. Under conditions where polyethyleneglycol was added to enhance catalytic and regulatory interaction, CPT-II was inhibited by 53 % in the presence of 100 μM malonyl-CoA. These data reemphasize the notion that the catalytic and malonyl-CoA binding entities are discreet. However, the localization by other investigators (Murthy and Pande, 1990) of malonyl-CoA sensitive CPT-I to the outer mitochondrial membrane may offer an alternative action for PEG in the activation of CPT-I catalysis. Chung et al (1992) presented data which demonstrate that malonyl-CoA sensitivity can also be conferred on purified CPT-II in the presence of phospholipids by the addition of immunoaffinity fractions containing malonyl-CoA binding protein. Woldegiorgis et al (1992) also showed that an 86 kD_a protein conferred malonyl-CoA sensitivity to malonyl-CoA-insensitive CPT-II. Recently, a malonyl-CoA-sensitive CPT activity from a rat heart mitochondrial particle was demonstrated to possess the catalytic properties of CPT-I and to be similar, if not identical, to the catalytic unit of CPT-II (Kerner et al., 1994). These investigators conclude that mitochondrial CPT can exist as two forms: one of which is sensitive to malonyl-CoA and one of which is not. According to this concept, catalytic activity is changed by association of the catalytic protein (CPT) with another, regulatory protein (Figure 1).

The concept that malonyl-CoA sensitive CPT-I represents a different isoform of mitochondrial CPT has been developed largely by the work of several independent laboratories. The interesting kinetic differences of substrate and inhibitor affinities between tissues led to the first indication that that CPT-I is different from CPT-II. In contrast, cross reactivity of anti-CPT-II antibody with CPT-II from a variety of tissue sources suggests that this protein is highly conserved across tissue and species (Finocchiaro et al., 1990). It has been argued that the differences observed between hepatic and muscle CPT, for example, could reflect a different membrane environment which has been shown to affect the sensitivity of CPT-I from purified outer mitochondrial membranes to malonyl-CoA (Kolodziej and Zammit, 1990). However, work from McGarry's laboratory has provided strong evidence that the 86,000-90,000 dalton malonyl-CoA binding protein is CPT-I. Competitive binding studies using malonyl-CoA and synthetic inhibitors of the CPT-I reaction revealed labeling of only a single protein from liver (90 kD_a) or from muscle (86 kDa) (DeClerq et al., 1987). Because this protein is covalently labeled with inhibitor, it is catalytically inactive; however, the labeling patterns reflect the CPT-I inactivation kinetics in vitro, so that these workers conclude the inhibitors are binding to the catalytic center of CPT-I and not to some separate binding protein. Antibodies raised to mitochondrial CPT-II do not react with CPT-I suggesting that they are immunologically distinct proteins (Woeltje et al., 1987). This conclusion was also reached by Kolodziej et al (1992) who also demonstrated that CPT-I in either heart, skeletal muscle or brown adipose tissue is only weakly reactive against the antibody to the liver [H-3] TDG-CoA labeled protein. The lack of molecular identity of these proteins could explain the markedly differing kinetic characteristics observed in the corresponding tissues.

Detergent solubilization of a malonyl-CoA insensitive CPT-II activity in active form which does not bind malonyl-CoA or TDG-CoA reflects that this is: (1) a malonyl-CoA resistant protein and (2) a protein which is stable and easily solubilized by detergent. Although membrane-bound malonyl-CoA-sensitive CPT demonstrates activity, this activity is destroyed in the presence of a powerful detergent, like Triton-X 100 (Woeltje et al., 1987). However, in later studies from Murthy and Pande (1990), malonyl-CoA sensitive CPT-I activity was successfully solubilized in the presence of octylglucoside-containing glycerol. This protein was separated from CPT-II and shown to be a single peptide from labeling of outer membrane preparations with [H-3] etomoxir (Murthy and Pande, 1990). Successful solubilization of CPT-I activity has subsequently been reported by Kolodziej et al. (1992). These authors conclude that CPT-I and CPT-II are separate proteins, in agreement with Woeltje et al (1987). The ability to obtain an active malonyl-CoA sensitive protein separate from CPT-II activity and the cloning and expression of CPT-I using a monoclonal antibody directed against the dinitrophenyl-derivatized etomoxir-CoA binding protein strengthens the case that CPT-I and CPT-II are distinct proteins (Esser et al., 1993a; Esser et al., 1993b). Further substantiation that the malonyl-CoA binding protein is CPT-I rests in the finding that there are regions of "strong identity" between the amino acid sequence of CPT-I and that of CPT-II, peroxisomal COT, and choline acetyltransferase (Esser et al., 1993b).

The molecular topology of CPT-I and the membrane with which it is associated is still speculative. Murthy and Pande (1990) have provided evidence that the malonyl-CoA binding site for CPT-I and the overt catalytic activity of CPT-I reside on opposite sides of the outer mitochondrial membrane in liver (Figure 2). Whereas a malonyl-CoA binding site is located on the outer surface (cytosol-facing) of the outer membrane, the catalytic region resides on the inner aspect of the outer membrane (facing the intermembrane space).

The proposed hydrophobic nature of this protein reveals only one potential peptide sequence which could serve as a membrane-spanning domain (Esser et al., 1993a). This region has been sequenced and its topology projected by Kolodziej and Zammitt (1993) based on predicted orientation of eukaryotic membrane-spanning proteins (Hartmann et al., 1989). The observed orientation of a cytosolic-facing malonyl-CoA binding peptide has been assigned to the N terminal cytosolic sequence (Figure 2). Whether this peptide acts as a regulatory acyl-CoA/low affinity malonyl-CoA binding site (a "transducer" region to report cytosolic concentrations of acyl-CoA and malonyl-CoA) or as a high affinity inhibitor site remains speculative; however, we have depicted this orientation with the high affinity site within the catalytic center of the protein (intermembrane facing) (Figure 2).

B. Genetic Deficiency in CPT

An inherited defect in fatty acid metabolism involving CPT deficiency was first described by DiMaurio and DiMaurio (1973). Since then numerous cases have been documented. Broadly, CPT deficiency may be divided into two forms: a muscular and a hepatic form. The muscular form is more common and usually occurs in adults and is characterized by exercise-induced myoglobinuria. Patients with the muscular form of CPT deficiency show a decrease in CPT-II activity in all tissues tested; however, CPT-I activity is unaffected (Demaugre et al. 1988; Tien et al., 1989). The more rare, hepatic form is early-onset, appearing in newborns and infants and is associated with hypoglycemia and low ketogenesis. In these patients, CPT-I, but not CPT-II, activity is depressed in liver and fibroblasts. Surprizingly, CPT-I activity in muscle is unaffected. This clear assignment of CPT enzyme deficiencies to different disease phenotypes has been taken by some investigators

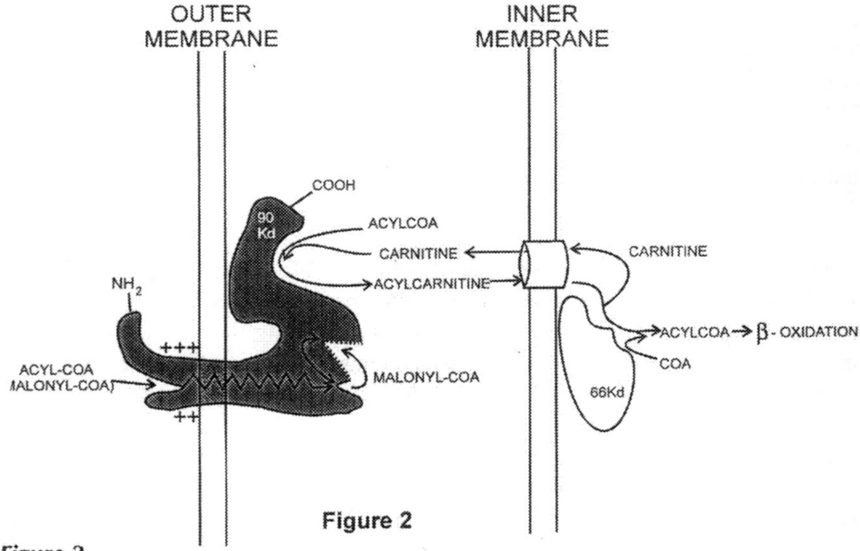

Figure 2.

as evidence for the presence of different CPT enzymes. However, this distinction is blurred by the finding that hepatic symptoms may be observed in some muscle forms of CPT deficiency (Taroni et al., 1993).

DiDonato's group has identified and characterized the molecular lesions in the CPT II protein which are associated with at least three forms of CPT deficiency (Taroni et al., 1992; Taroni et al., 1993). Two of these involve missense mutations resulting in non-conservative amino acid substitutions (serine to leucine and arginine to cysteine). CPT-II proteins with either of these mutations are unstable; consequently, patients carrying either one of these mutations have depressed CPT-II catalytic activity. Interestingly, CPT-I activity appears to be unaffected. The third CPT-II lesion is lethal and involves an eleven nucleotide duplication which results in early termination and generation of a truncated CPT-II protein. Cells of patients with this defect showed absent or severely reduced CPT activity. The location of the human CPT-II gene has been mapped to chromosome 1, region 1q12-1pter (Taroni et al., 1993; Minoletti et al., 1992). Thus, analysis of chromosomal lesions which may also be associated with the various CPT-II defects is now possible. Characterization of the molecular defects associated with the classical hepatic form of CPT deficiency is less advanced. However, the isolation of a cDNA encoding liver mitochondrial CPT-I by McGarry's group (Esser et al., 1993a) should allow for isolation of its human homologue and rapid molecular identification and characterization of the CPT-I lesions associated with the hepatic forms of CPT deficiency.

VIII. FUTURE PROSPECTUS

The application of molecular techniques has yielded cDNA clones encoding two proteins with CPT activities. Based on their sensitivities to malonyl-CoA, these two cDNAs most likely encode proteins representing CPT-I and CPT-II. The availability of the complete nucleotide and amino acid sequence of both enzymes should be of great value in designing experiments to probe the molecular identity and the structure/function relationships of the CPT enzymes. For example, by comparing the biochemical characteristics of wildtype and *in vitro* mutated CPT enzymes, one should gain insight into such questions as: where are the catalytic and inhibitor binding domains of CPT? What are the molecular constraints that affect the binding of substrates and inhibitors such as malonyl-CoA? How are CPT-I and CPT-II targeted to their respective mitochondrial locations and what are their topologies on the membrane? Mitochondrial proteins encoded by nuclear genes have been shown to contain mitochondria targeting sequences (Pfanner and Neupert, 1990). Analysis of the CPT-II protein sequence demonstrates that its N-terminus contains a consensus mitochondrial targeting sequence which is clipped after transport. A similar sequence has not been identified in CPT-I. In fact, Kolodziej and Zammit (1993) recently showed that the N-terminal peptide of CPT-I remains intact in the mature protein. Does this indicate that mitochondria outer-membrane associated proteins such as CPT-I contain a different targeting sequence? If so, where is this sequence located and is it also cleaved or does it serve a dual purpose in the mature protein? Finally, can these sequences function autonomously? Detailed deletion and mutagenesis experiments coupled with analysis of the membrane association of the resultant CPT proteins should define the location(s) and role(s) of the CPT mitochondrial targeting sequences. Similar types of experiments could also reveal the orientation of the CPT proteins with respect to the mitochondria membranes.

In addition, the recent cloning of CPT-I should now afford us the opportunity to identify the molecular basis for the different tissue isoforms of CPT-I. Unlike CPT-II which is expressed in the same form in all tissues within the species studied, CPT-I has been shown to exhibit different molecular sizes, malonyl-CoA sensitivities, and antigenicity in different tissues. In addition, patients with the hepatic form of CPT deficiency show decreased CPT-I activity in liver and fibroblast cells but normal CPTI activity in muscle. What is the basis of this tissue-specific CPT-I activity? Two obvious possibilities may be explored. One possibility is that they are products of related but different genes. Alternately, they are derived from the same gene. In this latter scenario, alternative splicing or alternate use of different promoters could generate CPT-I proteins with different size, antigenic and enzymatic characteristics. All of the above alternative mechanisms have been shown to effect tissue- and/or developmental-specific expression of distinct isoforms in other systems. Isolation and comparison of the nucleotide sequences and expression patterns of the CPT-I isoforms should be particularly useful in determining the molecular basis for their differing activities and responses to hormones and drugs as well as to different nutritional and disease states.

The availability of CPT cDNAs will also permit the isolation of their respective genes and the analysis of their transcription regulation. Foremost in this type of analysis would be to delineate the molecular mechanisms involved in the tissue-specific and developmental regulation of CPT-I gene expression. What is the basis for the differential regulation of the liver and cardiac CPT-I isoforms? Analysis of genes which are co-expressed in different tissues showed that these genes are usually regulated by shared as well as distinct regulatory elements and transcription factors. For example, recent studies in the mechanisms of cardiac muscle-specific gene regulation indicates that transcription factors distinct from those in used in skeletal muscle are required for activation and/or modulation of cardiac-specific gene expression. It is conceivable that regulation of CPT-I gene expression in cardiac muscle requires regulatory factors which are different from those employed in liver. Hepatic as well as cardiac myocyte cell cultures which are amenable to biochemical and transfection analyses are now readily available and may be used to rapidly dissect and differentiate the gene regulatory sequences and the potential transcription regulatory factors necessary for expression of the CPT gene in cardiac and hepatic tissues.

Genomic fragments purported to contain the promoter region of CPT-II have been isolated (Brady et al., 1992; G. Finocchiaro, unpublished). Initial analysis of these fragments indicates the presence of binding motifs for the general transcription factors CREB and CTF/NF1 and for C/EBF, a transcription factor previously shown to modulate expression of some hepatic enzymes involved in energy metabolism. As genes of several other β-oxidation enzymes have been cloned, it would be of interest to determine whether expression of these genes are coordinated and whether regulation of their transcription require common cis/trans regulatory sequences and factors. Understanding how expression of these mitochondrial protein genes are controlled should lead to an elucidation of how energy metabolism and mitochondria biogenesis are regulated at the molecular level.

ACKNOWLEDGMENTS

The authors would like to acknowledge Mr. Pagogh Cho for preparation of the figures, and NIH grant RO-1 HL-38863 to J.B.M.

REFERENCES

Aas, M., & Daae, L. N. W. (1971). Fatty acid activation and acyl transfer in organs from rats in different nutritional states. Biochim. Biophys. Acta 239, 208-216.

Appelkvist, E. L., & Dallner, G. (1980). Possible involvement of fatty acid binding protein in peroxisomal beta-oxidation of fatty acids. Biochim. Biophys. Acta 617, 156-160.

Awan, M. M., & Saggerson, E. D. (1993). Malonyl-CoA metabolism in cardiac myocytes and its relevance to the control of fatty acid oxidation. Biochem. J. 295, 61-66.

Bartlette, R., Bartlette, P., Bartlette, N., & Sherratt, H. S. A. (1985). Kinetics of enzymes requiring long-chain acyl-CoA ester as substrates: Effects of substrate binding to albumin. Biochem. J. 229, 559-560.

Berge, R. K., Rarsland, A., Kryvi, H., Bremer, J., & Aarsaether, N. (1989). Alkylthio acetic acids (3-thia fatty acids) - a new group of non-β-oxidizable peroxisome-inducing fatty acid analogues. Biochem. Pharm. 38, 3969-3979.

Bianchi, A., Evans, J. L., Iverson, A. J., Nordlund, A. C., Watts, T. D., & Witters, L. A. (1990). Identification of an isozymic form of acetyl-CoA carboxylase. J. Biol. Chem. 265, 1502-1509.

Bieber, L. L. (1988). Carnitine. Ann. Rev. Biochem. 57, 261-283.

Bird, M. I., & Saggerson, E. D. (1985). L carnitine decreases [^{14}C] malonyl-CoA binding to rat liver mitochondria. Biochem. Soc. Trans. 13, 156.

Boon, M. R., & Zammit, V. A. (1988). Use of a selectively permeabilized isolated rat hepatocyte preparation to study changes in the properties of overt carnitine palmitoyltransferase activity *in situ*. Biochem. J. 249, 645-652.

Brady, P. S., Parl, E. A., Liu, J.-S., Hanson, R. W., & Brady, L. J. (1992). Isolation and characterization of the promoter for the gene coding for the 68kDa carnitine palmitoyltransferase from the rat. Biochem. J. 286, 779-783.

Bremer, J. (1962a). Carnitine in intermediary metabolism. J. Biol. Chem. 237, 3628-3632.

Bremer, J. (1962b). Carnitine in intermediary metabolism. Reversible acetylation of carnitine by mitochondria. J. Biol. Chem. 237, 2228-2231.

Bremer, J. (1981). The effect of fasting on the activity of carnitine palmitoyl-transferase and its inhibition by malonyl-CoA. Biochim. Biophys. Acta 665, 628-631.

Bremer, J., & Norum, K. R. (1967a). The mechanism of substrate inhibition of palmityl coenzyme A: carnitine palmityltransferase by palmityl-CoA. J. Biol. Chem. 242, 1744-1748.

Bremer, J., & Norum, K. R. (1967b). The effects of detergents on palmityl coenzyme A: carnitine palmityltransferase. J. Biol. Chem. 242, 1749-1755.

Bremer, J., & Wojtczak, A. B. (1972). Factors controlling the rate of fatty acid β-oxidation in rat liver mitochondria. Biochim. Biophys. Acta 280, 515-530.

Brosnan, J. T., Kopec, B., & Fritz, I.B. (1973). The localization of carnitine palmitoyl-transferase on the inner membrane of bovine liver mitochondria. J. Biol Chem. 248, 4075-4082.

Christiansen, R. Z. (1977). Regulation of palmitate metabolism by carnitine and glucagon in hepatocytes isolated from fasted and carbohydrate refed rats. Biochim. Biophys. Acta 488, 249-262.

Christiansen, R., Borrebaek, B., & Bremer, J. (1976). The effect of (-) carnitine on the metabolism of palmitate in liver cells isolated from fasted and refed rats. FEBS Letters 62, 313-317.

Chung, C.-D., & Bieber, L. L. (1993). Properties of the medium chain/long chain carnitine acyltransferase purified from rat liver microsomes. J. Biol. Chem. 268, 4519-4524.

Chung, C. H., Woldegiorgis, G., Dai., Shrago, E., & Bieber, L. L. (1992). Conferral of malonyl-CoA sensitivity to purified rat heart mitochondrial carnitine palmitoyltransferase. Biochemistry 31, 9777-9783.

Clarke, P. R. H., & Bieber, L. L. (1981a). Isolation and purification of mitochondrial carnitine octanoyltransferase activities from beef heart. J. Biol. Chem. 256, 9861-9868.

Clarke, P. R. H., & Bieber, L. L. (1981b). Effect of micelles on the kinetics of purified beef heart mitochondrial carnitine palmitoyltransferase. J. Biol. Chem. 256, 9869-9873.

Cook, G.A. (1984). Differences in the sensitivity of carnitine palmitoyltransferase to inhibition by malonyl-CoA are due to differences in Ki values. J. Biol. Chem. 259, 12030-12033.

Cook, G. A., & Gamble, M. S. (1987). Regulation of carnitine palmitoyltransferase by insulin results in decreased activity and decreased apparent Ki values for malonyl-CoA. J. Biol. Chem. 262, 2050-2055.

Cook, G. A., Nielsen, R. C., Hawkins, R. A., Mehlman, M.A., Lakshmanan, M. R., & Veech, R. L. (1977). Effect of glucagon on hepatic malonyl coenzyme A concentration and on lipid synthesis. J. Biol. Chem. 252, 4421-4424.

Cook, G. A., Otto, D. A., & Cornell, N. W. (1980). Differential inhibition of keto-genesis by malonyl-CoA in mitochondria from fed and starved rats. Biochem. J. 192, 955-958.

Cook, G. A., Stephens, T. W., & Harris, R. A. (1984). Altered sensitivity of carnitine palmitoyltransferase to inhibition by malonyl-CoA in ketotic diabetic rats. Biochem. J. 219, 337-339.

Decaux, J.-F., Fere, P., Robin, D., Robin, P., & Girard, J. (1988). Decreased hepatic fatty acid oxidation at weaning in the rat is not linked to a variation in malonyl-CoA concentration. J. Biol. Chem. 263, 3284-3289.

De Clercq, P. E., Falck, J. R., Kuwajima, M., Tyminski, H., Foster, D. W., & McGarry, J. D. (1987). Characterization of the mitochondrial carnitine palmitoyltransferase enzyme system. I. Use of inhibitors. J. Biol. Chem. 262, 9812-9821.

Demaugre, F., Bonnefont, J.-P., Mitchell, G., Nguyen-Hoang, N., Pelet, A., Rimoldi, M., DiDonato, S., & Saudubray, A.-M. (1988). Hepatic and muscular presentation of carnitine palmitoyl transferase deficiency: two distinct entities. Pediatr. Res. 24, 308-311.

Derrick, J. P., & Ramsay, R. R. (1989). L-carnitine acyltransferase in intact peroxi-somes is inhibited by malonyl-CoA. Biochem. J. 262, 801-806.

DiMauro, S., & DiMauro, P. M. M. (1973). Muscle carnitine palmitoyltransferase deficiency and myoglobinuria. Science 182, 929-931.

Edwards, M. R., Bird, M. I., and Saggerson, E. D. (1985). Effects of DL-2-bromo-palmitoyl-CoA and bromoacyl-CoA in rat liver and heart mitochondria. Biochem. J. 230, 169-179.

Esser, V., Britton, C. H., Weis, B. C., Foster, D. W., & McGarry, J. D. (1993a). Cloning, sequencing and expression of a cDNA encoding rat liver carnitine palmitoyltransferase I. J. Biol. Chem. 268, 5817-5822.

Esser, V., Kuwajima, M., Britton, C. H., Krishnan, K., Foster, D. W., & McGarry, J. D. (1993b). Inhibitors of mitochondrial palmitoyltransferase I limit the action of proteases on the enzyme. J. Biol. Chem. 268, 5810-5816.

Exton, J. H., Corbin, J. G., & Park, C. R. (1969). Control of gluconeogenesis in liver. IV Differential effects of fatty acids and glucagon on ketogenesis and gluconeogenesis in the perfused rat liver. J. Biol. Chem. 244, 4095-4102.

Ferré, P., Satabin, P., Decaux, J.-F., Escriva, F., & Girard, J. (1983). Development and regulation of ketogenesis in hepatocytes isolated from newborn rats. Biochem. J. 214, 937-942.

Finocchiaro, G., Colombo, I., & DiDonato, S. (1990). Purification and partial amino acid sequences of carnitine palmitoyl-transferase from human liver. FEBS 274, 163-166.

Fiol, C. J., & Bieber, L. L. (1984). Sigmoid kinetics of purified beef heart mitochon-drial carnitine palmitoyltransferase. J. Biol. Chem. 259, 13084-13088.

Fogle, P. J., & Bieber, L. L. (1978). Evidence for carnitine acyltransferases associated with rat heart microsomes. Biochem. 9, 761-765.

Foxworthy, P., & Eacho, P. I. (1988). Inhibition of hepatic fatty acid oxidation at carnitine palmitoyltransferase I by the peroxisome proliferator 2-hydroxy-3-propyl-4-[6-(tetrazol-5-μl) hexyloxyl] acetophenone. Biochem. J. 252, 409-414.

Fritz, I. B. (1959). Action of carnitine on longchain fatty acid oxidation by the liver. Am. J. Physiol. 197, 297-304.

Fritz, I. B., & Marquis, N. R. (1965). The role of acylcarnitine esters and carnitine palmityltransferase in the transport of fatty acyl groups across the mitochondrial membrane. PNAS 54, 1226-1233.

Fritz, I. B., & McEwen, B. (1959). Effects of carnitine on fatty acid oxidation by muscle. Science 129, 334-335.

Fritz, I. B., & Yue, K. T. N. (1963). Long-chain carnitine acyltransferase and the role of acylcarnitine derivatives in the catalytic increase of fatty acid oxidation induced by carnitine. J. Lipid Res. 4, 279-288.

Gamble, M. S., & Cook, G. A. (1985). Alteration of the apparent Ki of carnitine palmitoyltransferase for malonyl-CoA by the diabetic state and reversal by insulin. J. Biol. Chem. 260, 9516-9519.

Gelb, B. D. (1993). Expression of the carnitine palmitoyltransferase II gene in the developing heart. Circulation 88, I-436.

Gerondaes, P., George, K., Alberti, M. H., & Agius, L. (1988). Interactions of carnitine palmitoyltransferase I and fibrates in cultured hepatocytes. Biochem. J. 253, 169-173.

Ghadiminejad, I., & Saggerson, E. D. (1990). Carnitine palmitoyltransferase (CPT_2) from liver mitochondrial inner membrane become inhibitable by malonyl-CoA if reconstituted with outer membrane malonyl-CoA binding protein. FEBS 269, 406-408.

Girard, J., Duée, P. H., Ferré, P., Pégorier, J. P., Escriva, F., & Decaux, J. F. (1985). Fatty acid oxidation and ketogenesis during development. Reprod. Nutr. Devel. 25, 303-319.

Girard, J., Ferré, P., Pegorier, J.-P., & Duée, P.-H. (1992). Adaptations of glucose and fatty acid metabolism during perinatal period and suckling-weaning transition. Physiol. Rev. 72, 507-562.

Grantham, B. D., & Zammit, V. A. (1986). Binding of [^{14}C] malonyl-CoA to rat liver mitochondria after blocking of the active site of carnitine palmitoyltransferase I. Displacement of low affinity binding by palmitoyl-CoA. Biochem. J. 233, 589-593.

Grantham, B. D., & Zammit, V. A. (1988). Role of carnitine palmitoyltransferase I in the regulation of hepatic ketogenesis during the onset and reversal of chronic diabetes. Biochem. J. 249, 409-414.

Gulick, T., Cresci, S., & Carter, M. E. (1993). Fatty acids regulate mitochondrial fatty acid β-oxidation gene expression through nuclear transcription factors. Circulation 88, I-234.

Guzmán, M., & Castro, J. (1991). Okadaic acid stimulates carnitine palmitoyl-transferase I activity and palmitate oxidation in isolated rat hepatocytes. FEBS 291, 105-108.

Guzmán, M., Castro, J., & Maquedano, A. (1987). Ethanol feeding to rats reversibly decreases hepatic carnitine palmitoyltransferase activity and increases enzyme sensitivity to malonyl-CoA. Biochem. Biophys. Res. Commun. 49, 443-448.

Hahn, P., Seccombe, D., & Towell, M. E. (1980). Perinatal changes in plasma carnitine levels in four species of mammal. Experientia (Basel) 36, 1341.

Harano, Y., Kashiwaji, A., Kojima, H., Suzuki, M., Hashimoto, T., & Shigeta, Y. (1985). Phosphorylation of carnitine palmitoyltransferase and activation by glucagon in isolated rat hepatocytes. FEBS Letters188, 267-272.

Hartmann, E.,Rapoport, T. A., & Lodish, H. F. (1989). Predicting the orientation of eukaryotic membrane-spanning proteins. Proc. Natl. Acad. Sci. USA 86, 5786-5790.

Herbin, C., Pegorier, J. P., Duée, P.-H., Kohl, C., & Girard, J. (1987). Regulation of fatty acid oxidation in isolated hepatocytes and liver mitochondria from newborn rabbits. Eur. J. Biochem. 165, 201-207.

Hoppel, C. L. (1991).Carnitine palmitoyltransferase In: Current Concepts in Carnitine Research (Carter, A. L., ed.), pp. 153-165. CRC Press, Boca Raton, FL.

Hoppel, C. L., & Tomec, R. J. (1972). Carnitine palmityltransferase. Location of two enzymatic activities in rat liver mitochondria. J. Biol. Chem. 247, 832-841.

Hudson, E. K., Liu, M.-H., Buja, L. M., & McMillin, J. B. (1993). Insulin increases mitochondrial carnitine palmitoyltransferase in rat neonatal cardiac myocytes in culture. Circulation 88, I-428.

Idell-Wenger, J. A., Grotyohann, L. W., & Neely, J. R. (1978). Coenzyme A and carnitine distribution in normal and ischemic hearts. J. Biol. Chem. 253, 4310-4318.

Issemann, I., & Green, S. (1990). Activation of a member of the steroid hormone receptor superfamily by peroxisomal proliferation. Nature 347, 645-650.

Kerner, J., Zaluzec, E., Gage, D., & Bieber, L.L. (1994). Characterization of the malonyl-CoA-sensititive carnitine palmitoyltransferase (CPT_o) of a rat heart mitochondrial particle. Evidence that the catalytic unit is CPT_i. J. Biol. Chem. 169,.8209-8219.

Kolodziej, M. P., Crilly, P. J., Corstorphine, C. G., & Zammit, V. A. (1992). Develop-ment and characterization of a polyclonal antibody against rat liver mitochondrial overt carnitine palmitoyltransferase. Biochem. J. 282, 415-421.

Kolodziej, M. P., & Zammit, V. A. (1990). Sensitivity of inhibition of rat liver mitochon-drial outer-membrane carnitine palmitoyltransferase by malonyl-CoA to chemical- and temperature-induced changes in membrane fluidity. Biochem. J. 272, 421-425.

Kolodziej, M. P., & Zammit, V. A. (1993). Mature carnitine palmitoyltransferase I retains the N-terminus of the nascent protein in rat liver. FEBS 327, 294-296.

Kopec, B., & Fritz, I. B. (1971). Properties of a purified carnitine palmitoyltransferase, and evidence for the existence of other carnitine acyltransferases. Can. J. Biochem. 49, 941-948.

Kopec, B., & Fritz, I. B. (1973). Comparison of properties of carnitine palmitoyl-transferase I with those of carnitine palmitoyltransferase II and preparation of antibodies to carnitine palmitoyltransferase. J. Biol. Chem. 248, 4069-4074.

Lee, A. S. (1992). Mammalian stress response: Induction of the glucose-regulated protein family. Curr. Opin. in Cell Biol 4, 267-273.

Lilly, K., Bugaisky, G. E., Umeda, P. K., & Bieber, L. L. (1990). The medium-chain carnitine acyltransferase activity associated with rat liver microsomes is malonyl-CoA sensitive. Arch. Biochem. Biophys. 280, 167-174.

Lossow, W. J., and Chaikoff, I. L. (1954). Carbohydrate sparing of fatty acid oxidation. I. The relationship of fatty acid chain length to the degree of sparing. II. The merchanism by which carbohydrate spares the oxidation of palmitic acid. Arch. Biochem. Biophys. 57, 23-40.

Markwell, M. A. K., Tolbert, N. E., & Bieber, L. L. (1976). Comparison of the carnitine acyltransferase activities from rat liver peroxisomes and microsomes. Arch. Biochem. Biophys. 176, 479-488.

Masoro, E. J., & Porter E. (1965). A comparison of fatty acid synthesis by liver and kidney. Proc. Soc. Exp. Biol. Med. 118, 1090-1095.

McGarry, J. D., & Foster, D. W. (1973) Acute reversal of experimental ketoacidosis in the rat with (+)-decanoylcarnitine. J. Clin. Invest. 52, 877-884.

McGarry, J. D., & Foster, D. W. (1974a). The metabolism of (-) octanoylcarnitine in perfused livers from fed and fasted rats. J. Biol. Chem. 249, 7984-7990.

McGarry, J. D., & Foster, D. W. (1974b). Studies with (+)-octanoylcarnitine in experimental diabetic ketoacidosis. Diabetes 23, 485-493.

McGarry, J. D., & Foster, D. W. (1980). Regulation of hepatic fatty acid oxidation and ketone body production. Ann. Rev. Biochem. 49, 395-420.

McGarry, J. D., Leatherman, G. F., & Foster, D. W. (1978b). Carnitine palmitoyl-transferase I. The site of inhibition of hepatic fatty acid oxidation by malonyl-CoA. J. Biol. Chem. 253, 4128-4136.

McGarry, J. D., Mannaerts, G. P., and Foster, D. W. (1977). A possible role for malonyl-CoA in the regulation of hepatic fatty acid oxidation and ketogenesis. J. Clin. Invest. 60, 265-270.

McGarry, J. D., Meier, J. M., & Foster, D. W. (1973). The effects of starvation and refeeding on carbohydrate and lipid metabolism *in vivo* and in the perfused rat liver. The relationship between fatty acid oxidation and esterfication in the regulation of ketogenesis. J. Biol. Chem. 248, 270-278.

McGarry, J. D., Mills, S. E., Long, C. S., & Foster, D. W. (1983). Observations on the affinity for carnitine, and malonyl-CoA sensitivity of carnitine palmitoyltransferase I in animal and human tissues. Biochem. J. 214, 21-28.

McGarry, J. D., Takabayashi, Y., & Foster, D. W. (1978a). The role of malonyl-CoA in the coordination of fatty acid synthesis and oxidation in isolated rat hepatocyte. J. Biol. Chem. 253, 8294-8300.

McGarry, J. D., Woeltje, K. F., Kuwajima, M., & Foster, D. W. (1989). Regulation of ketogenesis and the renaissance of carnitine palmitoyltransferase. Diabetes/Metab. Reviews 5, 271-284.

McGarry, J. D., Wright, P. H., & Foster, D. W. (1975). Hormonal control of keto-genesis. J. Clin. Invest. 55, 1202-1209.

McMillin, J. B., Hudson, E. K., & Van Winkle, W. B. (1992). Evidence for malonyl-CoA sensitive carnitine acyl-CoA transferase activity in sarcoplasmic reticulum of canine heart. J. Mol. Cell Cardiol. 24, 259-268.

McMillin, J., Wang, D., & Buja, L. M. (1994). Kinetic properties of carnitine palmitoyl-transferase I in cultured neonatal rat cardiac myocytes. Arch. Biochem. Biophys., in press.

Mills, S. E., Foster, D. W., & McGarry, J. D. (1983). Interaction of malonyl-CoA and related compounds with mitochondria from different rat tissues. Biochem. J. 214, 83-91.

Minoletti, F., Colombo, I., Martin, A. L., DiDonato, S., Taroni, F., Finocchiaro, G., & Pandolfo, M. (1992). Localization of the human gene for carnitine palmitoyltransferase to 1p13-p11 by nonradioactive in situ hybridization. Genomics 13, 1372-1374.

Miyazawa, S., Ozasa, H., Osumi, T., & Hashimoto, T. (1983). Purification and properties of carnitine octanoyltransferase and carnitine palmitoyltransferase from rat liver. J. Biochem. 94, 529-542.

Mobb, C. V., Fink, G., & Pfaff, D. W. (1990). HIP-70: A protein induced by estrogen in the brain and LH-RH in the pituitary. Science 247, 1477-1479.

Murthy, M. S. R., & Pande, S. V. (1993). Carnitine medium/long chain acyltransferase of microsomes seems to be the previously cloned ~54kDa protein of unknown function. Mol. Cell Biochem. 122, 133-138.

Murthy, M. S. R., Ramsay, R. R., & Pande, S. V. (1990). Acyl-CoA chain length affects the specificity of various carnitine palmitoyltransferases with respect to carnitine analogues. Biochem. J. 267, 273-276.

Murthy, M. S. R., & Pande, S. V. (1990). Characterization of a solubilized malonyl-CoA sensitive carnitine palmitoyltransferase from the mitochondrial outer membrane as a protein distinct from the malonyl-CoA insensitive carnitine palmitoyltransferase of the inner membrane. Biochem. J. 268, 599-604.

Mynatt, R. L., Lappi, M. D., & Cook, G. A. (1992). Myocardial carnitine palmitoyl-transferase of the mitochondrial outer membrane is not altered by fasting. Biochim. Biophys. Acta 1128, 105-111.

Norum, K. R. (1964). Palmityl-CoA: Carnitine-palmityltransferase purification from calf-liver mitochondria and some properties of the enzyme. Biochim. Biophys. Acta 89, 95-108.

Page, E., & McCallister, L. P. (1973). Quantitative electron microscopic description of heart muscle cells. Application to normal hypertrophied and thyroxin-stimulated hearts. Am. J. Cardiol. 31, 172-181.

Pande, S. V. (1971). On rate controlling factors of long chain fatty acid oxidation. J. Biol. Chem. 216, 5384-5390.

Pande, S. V., & Parvin, R. (1976). Characterization of carnitine acylcarnitine trans-locase system of heart mitochondria. J. Biol. Chem. 251, 6683-6691.

Parvin, R., & Pande, S. V. (1979). Enhancement of mitochondrial carnitine and carnitine acylcarnitine translocase-mediated transport of fatty acids into liver mitochondria under ketogenic conditions. J. Biol. Chem. 254, 5423-5429.

Pauly, D. F., & McMillin, J. B. (1988). Importance of acyl-CoA availability in inter-pretation of carnitine palmitoyltransferase I kinetics. J. Biol. Chem. 263, 18160-18167.

Pégovier, J. P., Duée, P.-H., Herbin, C., Laulaw, P.-Y., Blade, C., Peret, J., & Girard, J. (1988). Fatty acid metabolism in hepatocytes isolated from rats adapted to high-fat diets containing long- or medium-chain triacylglycerols. Biochem. J. 249, 801-806.

Pénicaud, L., Robin, D., Robin, P., Kaudé, J., Picon, L., Girard, J., & Ferré, P. (1991). Effect of insulin on the properties of liver carnitine palmitoyltransferase in the starved rat: Assessment by the euglycemic hyperinsulinemic clamp. Metabolism 40, 873-876.

Pfanner, N., & Neupert, W. (1990). The mitochondrial protein import apparatus. Annu. Rev. Biochem. 59, 331-353.

Prip-Buus, C., Pegorier, J.-P., Duée, P.-H., Kohl, C., & Girard, J. (1990). Evidence that the sensitivity of carnitine palmitoyltransferase I to inhibition by malonyl-CoA is an important site of regulation of hepatic fatty acid oxidation in the fetal and newborn rabbit. Biochem. J. 269, 409-415.

Ramsay, R. R. (1988). The soluble carnitine palmitoyltransferase from bovine liver. Biochem. J. 249, 239-245.

Ramsay, R. R., & Arduini, A. (1993). Perspectives in biochemistry and biophysics. The carnitine acyltransferases and their role in modulating acyl-CoA pools. Arch. Biochem. Biophys. 302, 307-314.

Ramsay, R. R., Derrick, J. P., Friend, A. S., & Tubbs, P. K. (1987). Purification and properties of the soluble carnitine palmitoyltransferase from bovine liver mitochondria. Biochem. J. 244, 271-278.

Ramsay, R. R., Mancinelli, G., & Arduini, A. (1991). Carnitine palmitoyltransferase in human erythrocyte membranes. Properties and malonyl-CoA sensitivity. Biochem. J. 275, 685-688.

Richards, E. W., Hamm, M. W., & Otto, D. A. (1991). The effect of palmitoyl-CoA binding to albumin on the apparent kinetic behavior of carnitine palmitoyltransferase I. Biochim. Biophys. Acta 1076, 23-28.

Robles-Valdes, C., McGarry, J. D., & Foster, D. W. (1976). Maternal-fetal carnitine relationships on neonatal ketosis in the rat. J. Biol. Chem. 251, 6007-6012.

Saddik, M., Gamble, J., Witter, L. A., & Lopaschuk, G. D. (1973). Acetyl-CoA carboxylase regulation of fatty acid oxidation in the heart. J. Biol. Chem. 268, 25836-25845.

Saeed, A., McMillin, J. B., Wolkowicz, P. E., & Brouillette, W. J. (1993). Carnitine acyltransferase enzymic catalysis requires a positive charge on the carnitine cofactor. Arch. Biochem. Biophys. 305, 307-312.

Saggerson, E. D., & Carpenter, C. A. (1981a). Effects of fasting, adrenalectomy and streptozotocin-diabetes on the sensitivity of hepatic carnitine acyltransferase to malonyl-CoA. FEBS Letters129, 225-228.

Saggerson, E. D., & Carpenter, C. A. (1981b). Effects of fasting and malonyl-CoA on the kinetics of carnitine palmitoyltransferase and carnitine octanyltransferase in intact rat liver mitochondria. FEBS Letters 132, 166-168.

Saggerson, E. D., & Carpenter, C. A. (1982). Regulation of hepatic carnitine palmitoyl-transferase activity during the foetal-neonatal transition. FEBS Letters 150, 177-180.

Saggerson, E. D., & Carpenter, C. A. (1983). The effect of malonyl-CoA on overt and latent carnitine acyltransferase activities in rat liver and adipoyte mitochondria. Biochem. J. 210, 591-597.

Saggerson, D., Ghadininejad, I., & Awan, M. (1992). Regulation of mitochondrial carnitine palmitoyltransferase from liver and extra hepatic tissue. Adv. Enzyme Reg. 32, 285-306.

Singh, I., Lazo, O., Dhaunsi, G. S., & Contreras, M. (1992). Transport of fatty acids into human and rat peroxisomes. J. Biol. Chem. 267, 13306-13313.

Skorin, C., Necochea, C., Johow, V., Soto, U., Grau, A. M., Bremer, J., & Leighton, F. (1992). Peroxisomal fatty acid oxidation and inhibition of the mitochondrial carnitine palmitoyltransferase I in isolated rat hepatocytes. Biochem. J. 281, 561-567.

Solberg, H. E. (1972). Different carnitine acyltransferases in calf liver. Biochim. Biophys. Acta 280, 422-433.

Stakkestad, J. A., & Bremer, J. (1983). The outer carnitine palmitoyltransferase and regulation of fatty acid metabolism in rat liver in different thyroid states. Biochim. Biophys. Acta 750, 244-252.

Taroni, F., Verderio, E., Dworzak, F., Willems, P. J., Cavadinin, P., & DiDonato, S. (1993). Identification of a common mutation in the carnitine palmitoyltransferase II gene in familial recurrent myoglobinuria patients. Nature Genetics 4, 31-320.

Taroni, F., Verderio, E., Fiorucci, S., Cavadinin, P., Finocchiaro, G., Uziel, G., Lamamtea, E., Gellera, C., & DiDonato, S. (1992). Molecular characterization of inherited carnitine palmitoyltransferase II deficiency. Proc. Natl. Acad. Sci. USA 89, 8429-8433.

Thampy, K. G. (1989). Formation of malonyl coenzyme A in rat heart. Identification and purification of an isozyme of acetyl-CoA carboxylase from rat heart. J. Biol. Chem. 264, 17631-17634.

Thomas, J., Debeer, L. J., de Schepper, P. J., & Mannaerts, G. P. (1980). Factors influencing palmitoyl-CoA oxidation by rat liver peroxisomal fractions. Substrate concentrations, organelle integrity and ATP. Biochem. J. 190, 485-494.

Tien, I., Demaugre, F., Bonnefont, J.-P., & Saudubray, A.-M. (1989). Normal muscle CTP 1 and CPT2 activities in hepatic presentation patients with CPT1 deficiency in fibroblasts. Tissue specific isoforms of CPT1. J. Neuro. Sci. 92, 229-245.

Van Tol, A., & Hiilsmann, W.C. (1969). The localization of palmitoyl-CoA: Carnitine palmitoyltransferase in rat liver. Biochim. Biophys. Acta 189, 342-353.

Williamson, J. R., Browning, E. T., Scholz, R., Kreisberg, R. A., & Fritz I. B. (1968). Inhibition of fatty acid stimulation of gluconcogenesis by (+) decanoylcarnitine in perfused rat liver. Diabetes 17, 194-208.

Wit-Peeters, E. M., Scholte, H. R., & Elenbaas, H. L. (1970). Fatty acid synthesis in heart. Biochim. Biophys. Acta 210, 360-370.

Wittels, B., & Bressler, R. (1965). Lipid metabolism in the newborn heart. J. Clin. Invest. 44, 1639-1646.

Woeltje, K. F., Esser, V., Weis, B. C., Cox, W. F., Schroeder, J. G., Liao, S.-T., Foster, D. W., & McGarry, J. D. (1990a). Inter-tissue and inter-species characteristics of the mitochondrial carnitine palmitoyltransferase enzyme system. J. Biol. Chem. 265, 10714-10719.

Woeltje, K. F., Esser, V., Weis, B. C., Sen, A., Cox, W. F., McPhane, M. J., Slaughter, C. A., Foster, D. W., & McGarry, J. D. (1990b). Cloning, sequencing and expression of a cDNA encoding rat liver mitochondrial carnitine palmitoyltransferase II. J. Biol. Chem. 265, 10720-10725.

Woeltje, K. F., Kuwajima, M., Foster, D. W., & McGarry, J. D. (1987). Character-ization of the mitochondrial carnitine palmitoyltransferase enzyme system. II. Use of detergents and antibodies. J. Biol. Chem. 262, 9822-9827.

Woldegiorgis, G., Fibich, B., Contreras, L., & Shrago, E. (1992). Restoration of malonyl-CoA sensitivity of soluble rat liver mitochondria carnitine palmitoyltransferase by reconstitution with a partially purified malonyl-CoA binding protein. Arch. Biochem. Biophys. 295, 348-351.

Wolvetang, E. J., Tager, J. M., & Wanders, R. J. A. (1990). Latency of the peroxi-somal enzyme acyl-CoA: dihydroxyacetonephosphate acyltransferase in digitonin-permeabilized fibroblasts: the effect of ATP and ATPase inhibitors. Biochem. Biophys. Res. Commun. 170, 1135-1143.

Wu, D. E., & Thornburg, K. L. (1993). Perinatal expression of carnitine palmitoyl-transferase II in newborn rat heart. Circulation 88, I-436.

Yates, D. W., & Garland, P. B. (1966). The partial latency and intramitochondrial distribution of carnitine palmitoyltransferase (e.c. 2.3.1.-) and the CoASH and carnitine permeable space of rat liver mitochondria. Biochem. Biophys. Res. Commun. 23, 460-465.

Yates, D. W., & Garland, P.B. (1970). Carnitine palmitoyltransferase activities (EC 2.3.1.-) of rat liver mitochondria. Biochem. J. 119, 547-552.

Zammit, V. A., Corstorphine, C. G., & Gray, S. R. (1984). Changes in the ability of malonyl-CoA to inhibit carnitine palmitoyltransferase I activity and to bind to rat liver mitochondria during incubation *in vitro*. Biochem. J. 222, 335-342.

Zammit, V. A., Corstorphine, G. G., & Kelliher, M. G. (1980). Evidence for distinct functional molecular sizes of carnitine palmitoyltransferase I and II in rat liver mitochondria. Biochem. J. 250, 415-420.

PROSTAGLANDIN ENDOPEROXIDE SYNTHASE ISOZYMES

William L. Smith and David L. DeWitt

Advances in Lipobiology
Volume 2, pages 227-260.

ISBN 0-7623-0205-4

ABSTRACT

PGH synthases catalyze the initial reaction common to the formation of prostanoids and are the target sites for most nonsteroidal anti-inflammatory drugs (NSAIDs). Until the recent discovery of a second PGH synthase isozyme (PGH synthase-2; PGHS-2), it was thought that NSAIDs acted by inhibiting a single enzyme, now referred to as PGH synthase-1 (PGHS-1). It is now believed that PGHS-1 produces prostaglandins that are not directly involved in inflammation, but which regulate "housekeeping" activities; this enzyme is constitutively expressed and, thus, available to respond to rapid transient increases in arachidonate brought about by hormone-induced activation of phospholipases. In contrast, PGHS-2 is the enzyme thought to produce prostaglandins involved in inflammation; this enzyme is not expressed normally, but can be induced in immunoregulatory cells, such as macrophages, and in inflamed tissues, such as fibroblasts and synoviacytes. In addition, anti-inflammatory glucocorticoids and anti-inflammatory cytokines inhibit PGHS-2 induction. It now seems likely that the anti-inflammatory effects of NSAIDs are due primarily to inhibition of PGHS-2.

Our current hypothesis is that PGHS-1 and PGHS-2 represent two prostaglandin biosynthetic systems which operate independently; that PGHS-1 and PGHS-2 derive substrate from different arachidonate pools and direct products to different extracellular (PGHS-1) and/or intracellular sites (PGHS-2). These two pathways have evolved to carry out specialized functions; the PGHS-1 pathway to produce prostaglandins that act extracellularly as local hormones to coordinate short-term cellular responses to hormonal stimulation and the PGHS-2 pathway to produce prostaglandins to coordinate prolonged physiological events such as inflammation and mitogenesis.

I. INTRODUCTION

A. Prostanoid Biosynthesis and Mechanisms of Action—An Overview

The intent of this chapter is to review recent studies on prostaglandin endoperoxide H (PGH) synthases, enzymes also familiarly referred to as "cyclooxygenases." PGH synthases catalyze the initial reaction common to the formation of prostanoids and are the target sites for most nonsteroidal anti-inflammatory drugs (NSAIDs). Interest in the prostanoid area has been rekindled by the fairly recent discovery of a second PGH synthase isozyme and indications that this second isozyme is the actual therapeutic target of NSAIDs acting in their anti-inflammatory capacities. Recent evidence also suggests that these enzymes may constitute separate prostaglandin biosynthetic pathways linked to distinct physiological processes. To place studies on PGH synthases in a broader context, we begin with an overview of prostanoid biosynthesis and mechanisms of action.

Prostanoids include the prostaglandins and thromboxanes. These compounds are oxygenated fatty acids, most commonly derived from arachidonic acid. Figure 1 shows the pathway for the biosynthesis of various prostanoids from arachidonate as it might occur in a generic cell (Smith, 1992). There are three stages to prostanoid formation: (a) release of arachidonate from phospholipid precursors, (b) oxygenation of arachidonate to prostaglandin endoperoxide H_2 (PGH_2); and (c) conversion of PGH_2 to a biologically active end-product. Prostanoid biosynthesis can be initiated by circulating hormones such as angiotensin II interacting with their cognate receptors. Hormonal stimulation usually leads to the short-term (< 15 min) mobilization of arachidonate from cellular phosphoglycerides. Arachidonate mobilization can also occur as a result of stimulation with growth factors and cytokines, although the time course of fatty acid release with these effectors is more complex and prolonged (> 2 hr) (Habenicht et al., 1985; Domin and Rozengurt, 1993). Hydrolysis of arachidonate involves the activation of one or more phospholipase A_2 enzymes (Kudo et al., 1993), which may be differentially coupled to arachidonate mobilization in response to hormonal-stimulation (acute release) and mitogen- and cytokine-stimulation (long-term release). Arachidonate is next converted by PGH synthase in two steps to PGH_2. Newly formed PGH_2 is then either reduced or isomerized by a reductase or a synthase, respectively, which catalyzes the production of an active end-product such as prostaglandin E_2 (PGE_2), $PGF_{2\alpha}$, PGI_2, or thromboxane A_2 (TXA_2) (Smith et al., 1991). All of the major pathways for PGH_2 metabolism are shown in the generic cell depicted in Figure 1; however, a given cell typically forms only one major prostanoid product. For example, platelets form almost exclusively TXA_2 from PGH_2 and arterial endothelial cells form principally PGI_2 (prostacyclin) (Arita et al., 1989; Smith, 1989, 1992). Newly synthesized prostanoids can exit the parent cell, and probably do so via carrier-mediated diffusion (Smith and Laneuville, 1994).

Prostanoids are catabolized to inactive products during their passage through the circulation by 15-hydroxy-prostaglandin dehydrogenases and Δ^{13}-prostaglandin reductases. These enzymes are found mainly in lung, kidney, and liver (reviewed in Smith and Laneuville, 1994). Because prostanoids do not survive a single pass through the circulation except under pathophysiological conditions, prostanoids are viewed as "local hormones" that must act at or near their sites of biosynthesis.

In a broad sense, prostanoids play two physiological roles. The first is to coordinate, at the local tissue level, multicellular responses to the circulating hormones which stimulate prostanoid synthesis. For example, renal collecting

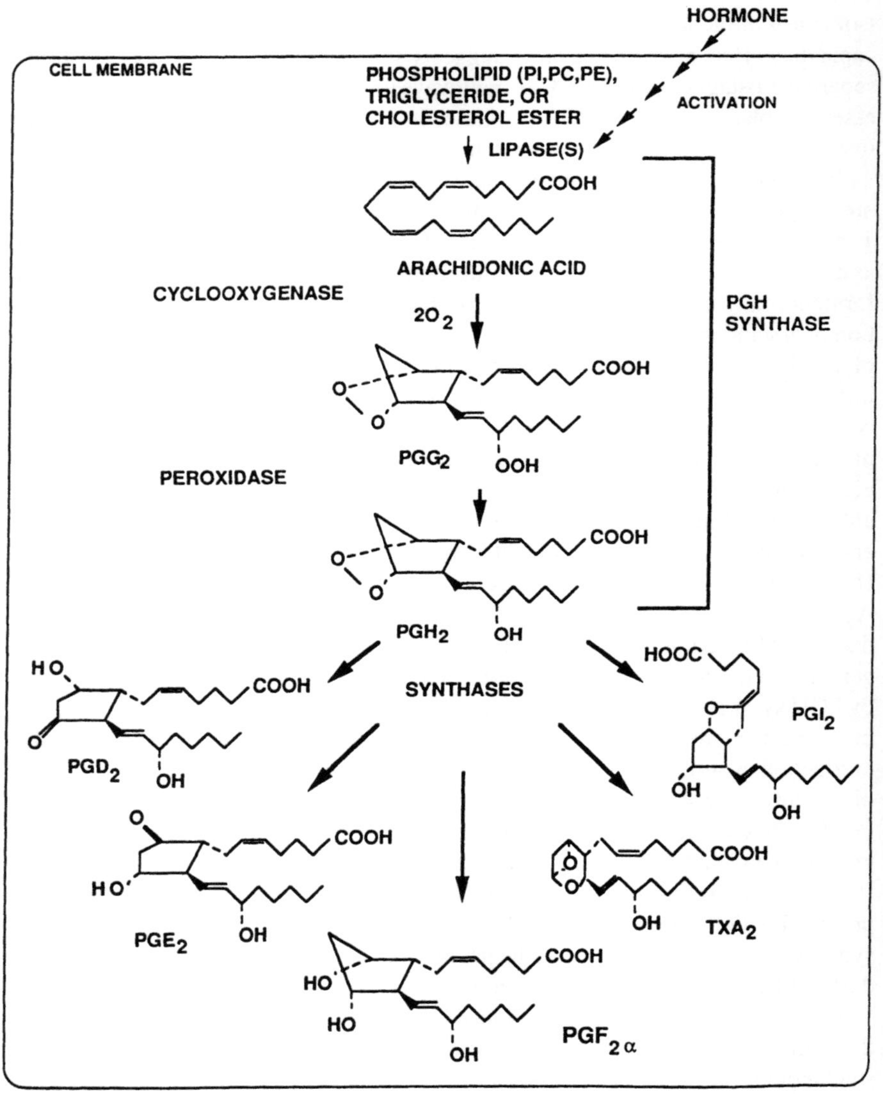

Figure 1. Prostanoid biosynthetic pathway. From Smith (1992) with permission.

tubule epithelia synthesize PGE_2 in response to circulating vasopressin. PGE_2 synthesized by collecting tubules acts on neighboring collecting tubule cells to inhibit water reabsorption, and on adjacent cells from the thick ascending limb of

Henle's loop to inhibit Na^+ reabsorption (Smith, 1989; Smith et al., 1989). A second example of a local coordinating response of prostanoids is seen in the case of platelets. TXA_2 formed by platelets in response to thrombin promotes adherence of platelets to one another and to the subendothelial layer of the arterial vasculature (Arita et al., 1989); additionally, TXA_2 acts on neighboring vascular smooth muscle cells to stimulate vasoconstriction.

The second function of prostaglandins is to coordinate longer term physiological processes. For instance, ovulation in rats in dependent on increased follicular prostaglandin synthesis (Hedin et al., 1987; Wong et al., 1989). Inflammation in man and in animals is dependent on prolonged prostaglandin synthesis by fibroblasts (Raz et al., 1989), synoviacytes (Sano et al., 1992; Crofford et al., 1994; Ristimaki et al., 1994), and macrophages (Lee et al., 1992; O'Sullivan et al., 1992a, 1992b). As we shall see, acute prostaglandin synthesis in response to circulating hormones, and extended prostaglandin synthesis that accompanies inflammation and ovulation probably result from the activities of different PGH synthase enzymes.

At the biochemical level, prostanoids function through G protein-linked receptors present both on the parent cell (autocrine action) and on adjacent, unlike cells (paracrine action) to elicit changes in intracellular cAMP or Ca^{2+} concentrations (Arita et al., 1989; Smith, 1989, 1992). It is now clear that there are one or more receptors for each of the major prostanoids shown in Figure 1. In the case of the PGE_2 receptor, there appear to be four pharmacologically distinct receptors, three of which have been cloned (Coleman et al., 1994). In addition, one of these PGE_2 receptors (the EP3 receptor) has at least four different splice variants, each of which interacts differently with different G proteins (Namba et al., 1993).

II. PGH SYNTHASE ISOZYMES

Primary Structures of PGH Synthase Isozymes

There are two PGH synthase isozymes, which are called PGH synthase-1 (PGHS-2) and PGHS-2 (PGHS-2), or commonly, cyclooxygenase (COX)-1 and -2. Both PGH synthases catalyze the same two separate reactions (Figure 1) (Miyamoto et al., 1976; Van der Ouderaa et al., 1977; Pagels et al., 1983; Fletcher et al., 1992; Meade et al., 1993) which, as discussed in detail below, occur at two separate but interactive active sites (Figure 1). The first reaction is the cyclooxygenase reaction, a *bis*-oxygenation of arachidonate producing PGG_2; the second reaction is the peroxidase reaction, a two-electron reduction of the 15-hydroperoxyl group of PGG_2 to form PGH_2. Figure 2 compares the deduced amino acid sequences of PGHS-1 and PGHS-2 from various species. PGHS-1 has been studied for almost 30 years. This enzyme was originally purified from ovine and bovine vesicular glands in the mid 1970s (Hemler et al., 1976; Miyamoto et al., 1976; Van der

Ouderaa et al., 1977). cDNAs encoding murine (DeWitt et al., 1990), human (Yokoyama and Tanabe, 1989; Funk et al., 1991), rat (Feng et al., 1993), and ovine (DeWitt and Smith, 1988; Merlie et al., 1988; Yokoyama et al., 1988) PGHS-1 have been cloned (Figure 2). PGHS-1 enzymes contain relatively long signal peptides of slightly different lengths in different species; removal of the signal peptides, however, always yields a mature PGHS-1 which contains 576 amino acids and a molecular mass of about 65,500 (Roth et al., 1980; DeWitt and Smith, 1988). The deduced amino acid sequences of PGHS-1 from sheep, human, and mouse are about 90% identical at the amino acid level. The most significant differences are found in the signal peptides and the 12 amino acids immediately preceding the common (P/S)TEL sequence at the C-terminus.

PGHS-2 was originally described in 1991 by Daniel Simmons and his colleagues as a v-*src*-inducible gene product from chicken fibroblasts (Simmons et al., 1991; Xie et al., 1991). Shortly afterwards, Harvey Herschman and his coworkers described a phorbol ester-inducible immediate early gene product they called TIS10 (which encodes PGHS-2) from murine 3T3 cells (Kujubu et al., 1991). Human PGHS-2 was subsequently cloned (Hla and Neilson, 1992). Within a species, there is about 60% amino acid identity between the deduced amino acid sequences of PGHS-1 and PGHS-2 (Figure 2). PGHS-2 differs significantly from PGHS-1 at positions prior to amino acid residue 30; mature rat PGHS-2 has the N-terminal sequence ANPCC (Sirois and Richards, 1992); this corresponds to cleavage of a 17-amino acid signal peptide from the N-terminus of the deduced sequence (Figure 2). Most notably, PGHS-2 contains an 18-amino acid insert near the C-terminus of the enzyme. This insert is absent from PGHS-1, and antibodies prepared against this peptide insert are specific for PGHS-2 (Kujubu et al., 1993; Regier et al., 1993). Excluding sequences near the N- and C-termini, the sequences of PGHS-1 and PGHS-2 are about 75% identical. All residues identified as essential for the catalytic activity of PGHS-1 are conserved in PGH synthase-2 (Figure 2); this suggests that the reactions catalyzed by PGHS-1 and PGHS-2 are fundamentally the same.

B. Physical and Chemical Properties of PGHS-1 and PGHS-2

PGHS-1 have predicted subunit molecular weights of approximately 65,500 excluding the signal peptides; however, native PGHS-1 migrate on SDS-PAGE with M_rs of about 72,000 (Hemler et al., 1976; Miyamoto et al., 1976; Van der Ouderaa et al., 1977). The difference between the calculated and experimentally determined molecular weight is due primarily to the presence of asparagine(N)-linked oligosaccharides. There are three high mannose oligosaccharides, one $Man_7(NAcGln)_2$ and two Man_9 $(NAcGln)_2$ groups, coupled to ovine PGHS-1 (Mutsaers et al., 1985). The sites of N-glycosylation in ovine PGHS-1 are Asn68, Asn144, and Asn410 (Otto et al., 1993); glycosylation at Asn410 is required for enzyme activity.

In contrast to PGHS-1, PGHS-2 typically appears as a doublet on SDS-PAGE with M_rs = 72,000 and 74,000 (Sirois and Richards, 1992; Otto et al., 1993). The

```
Chick-2    MLLPCALLAALL-----------------AAGHAANPCCSLPCQNRGECMTTGFDRYECD
Human-2    MLARALLLCAVL-----------------ALSHTANPCCSHPCQNRGVCMSVGFDQYKCD
Rat-2      MLFRAVLLCACP-----------------GLSHAANPCCSNPCQNRGECMSIGFDQTKCD
Mouse-2    MLFRAVLLCAAL-----------------GLSQAANPCCSNPCQNRGECMSTGFDQYKCD
Mouse-1    MSRRSLSLWFPLLLLLLLLPPTPSVLLADPGVPSPVNPCCYYPCQNQGVCVRFGLDNYQCD
Rat-1      MSRRSLSLQFPLLLLLLLLPPPPVLLTDAGVPSPVIPCCYYPCQNQGVCVRFGLDHYQCD
Human-1    MSR-SLLLRFLLLLLLL-PPLP-VLLADPGAPTPVNPCCYYPCQHQGICVRFGLDRYQCD
Sheep-1    MSRQSISLRFPLLLLLL-SPSP-VFSADPGAPAPVNPCCYYPCQHQGICVRFGLDRYQCD
                     10        20        30        40        50

Chick-2    CTRTGYYGENCTTPEFFTWLKLLIKPTPNTVHYILTHFKGVWNIINNSPFLRDTIMRYVL
Human-2    CTRTGFYGENCSTPEFLTRIKFLLKPTPNTVHYILTHFKGFWNVVNNIPFLRNAIMSYVL
Rat-2      CTRTGFYGENCTTPRFLTRIKLPLKPTPNTVHYILTHFKGVWNIVNNIPFLRIQSMRYVL
Mouse-2    CTRTGFYGENCTTPEFLTRLKLLLKPTPNTVHYILTHFKGVWNIVNNIPFLRSLTMKYVL
Mouse-1    CTRTGYSGPNCTIPEIWTWLRNSLRPSPSFTHFLLTHGYWLWEFVNAT-FIREVLMRLVL
Rat-1      CTRTGYSGPNCTIPEIWTWLRSSLRPSPSFTHFLLTHGYWIWEFVNAT-FIREVLMGWVL
Human-1    CTRTGYSGPNCTIPGLWTWLRNSLRPSPSFTHFLLTHGRWFWEFVNAT-FIREMLMLLVL
Sheep-1    CTRTGYSGPNCTIPEIWTWLRTTLRPSPSFIHFLLTHGRWLWDFVNAT-FIRDTLMRLVL
           60        70        80        90        100       110

Chick-2    TSRSHLIDSPPTYNSDYSYKSWEAYSNLSYYTRSLPPVGHDCPTPMGVKGKKELPDSKLI
Human-2    TSRSHLIDSPPTYNADYGYKSWEAFSNLSYYTRALPPVPDDCPTPLGVKGKKQLPDSNEI
Rat-2      TSRSHLIDSPPTYNVHYGYKSWEAFSNLSYYTRALPPVADDCPTPMGVKGNKELPDSKEV
Mouse-2    TSRSYLIDSPPTYNVHYGYKSWEAFSNLSYYTRALPPVADDCPTPMGVKGNKELPDSKEV
Mouse-1    TVRSNLIPSPPTYNSAHDYISWESFSNVSYYTRILPSVPKDCPTPMGTKGKKQLPDVQLL
Rat-1      TVRSNLIPSPPTYNTAHDYISWESFSNVSYYTRILPSVPKDCPTPMGTKGKKQLPDIHLL
Human-1    TVRSNLIPSPPTYNSAHDYISWESFSNVSYYTRILPSVPKDCPTPMGTKGKKQLPDAQLL
Sheep-1    TVRSNLIPSPPTYNIAHDYISWESFSNVSYYTRILPSVPRDCPTPMGTKGKKQLPDAEFL
            120       130       140       150       160       170

Chick-2    VEKFLLRRKFIPDPQGTNVMFTFFAQHFTHQFFKTDHKRGPGFTKAYGHGVDLNHIYGET
Human-2    VGKLLLRRKFIPDPQGSNMMFAFFAQHFTHQFFKTDHKRGPAFTNGLGHGVDLNHIYGET
Rat-2      LEKVLLRREFIPDPQGTNMMFAFFAQHFTHQFFKTDQKRGPGFTRGLGHGVDLNHVYGET
Mouse-2    LEKVLLRREFIPDPQGSNMMFAFFAQHFTHQFFKTDHKRGPGFTRGLGHGVDLNHIYGET
Mouse-1    AQRLLLRREFIPAPQGTNILFAFFAQHFTHQFFKTSGKMGPGFTKALGHGVDLGHIYGDN
Rat-1      AQRLLLRREFIPAPQGTNVLFAFFAQHFTHQFFKTSTKMGPGFTKALGHGVDLGHIYGDS
Human-1    ARRFLLRRKFIPDPQGTNLMFAFFAQHFTHQFFKTSGKMGPGFTKALGHGVDLGHIYGDN
Sheep-1    SRRELLRRKFIPDPQGTNLMFAFFAQHFTHQFFKTSGKMGPGFTKALGHGVDLGHIYGDN
            180       190       200       210       220       230

Chick-2    LERQLKLRLRKDGKLKYQMIDGEMYPPTVKDTQAEMIYPPHVPEHLQFSVGQEVFGLVPG
Human-2    LARQRKIRLFKDGKMKYQIIDGEMYPPTVKDTQAEMIYPPQVPEHLRFAVGQEVFGLVPG
Rat-2      LDRQHKLRLFQDGKLKYQVIGGEVYPPTVKDTQVDMIYPPHVPEHLRFAVGQEVFGLVPG
Mouse-2    LDRQHKLRLFKDGKLKYQVIGGEVYPPTVKDTQVEMIYPPHIPENLQFAVGQEVFGLVPG
Mouse-1    LERQYHLRLFKDGKLKYQVLDGEVYPPSVEQASVLMRYPPGVPPERQMAVGQEVFGLLPG
Rat-1      LERQYHLRLFKDGKLKYQVLDGELYPPSVEQASVKMRYPPGVPPEKQMAVAQEVFGLLPG
Human-1    LERQYQLRLFKDGKLKYQVLDGEMYPPSVEEAPVLMHYPRGIPPQSQMAVGQEVFGLLPG
Sheep-1    LERQYQLRLFKDGKLKYQMLNGEVYPPSVEEAPVLMHYPRGIPPQSQMAVGQEVFGLLPG
            240       250       260       270       280       290

Chick-2    LMMYATIWLREHNRVCDVLKQEHPEWDDEQLFQTTRLILIGETIKIVIDDYVQHLSGYHF
Human-2    LMMYATIWLREHNRVCDVLKQEHPEWGDEQLFQTSRLILIGETIKIVIDDYVQHLSGYHF
Rat-2      LMMYATIWLREHNRVCDILKQEHPEWDDERLFQTSRLILIGETIKIVIEDYVQHLRGYHF
Mouse-2    LMMYATIWLREHNRVCDILKQEHPEWGDEQLFQTSRLILIGETIKIVIDDYVQHLSGYHF
Mouse-1    LMLFSTIWLREHNRVCDLLKEEHPTWDDEQLFQTTRLILIGETIKIVIEEYVQHLSGYFL
Rat-1      LMLFSTIWLREHNRVCDLLKEEHPTWDDEQLFQTTRLILIGETIEIIIEEYVQHLSGYFL
Human-1    LMLYATLWLREHNRVCDLLKAEHPTWGDEQLFQTTRLILIGETIKIVIEEYVQQLSGYFL
Sheep-1    LMLYATIWLREHNRVCDLLKAEHPTWGDEQLFQTARLILIGETIKIVIEEYVQQLSGYFL
            300       310       320       330       340       350
```

Figure 2. Sequence comparison of various PGHS-1 and PGHS-2. Included are chicken PGHS-2 (Simmons et al., 1991), murine PGHS-2 (Kujubu et al., 1991), human PGHS-2 (Hla & Neilson, 1992), murine PGHS-1 (DeWitt et al., 1990), human PGHS-1 (Funk et al., 1991; Yokoyama & Tanabe, 1989), and ovine PGHS-1 (DeWitt & Smith, 1988; Merlie et al., 1988; Yokoyama et al., 1988). Numbering is for ovine PGHS-1 and begins with the methionine at the translation start site. Shown in bold letters are the signal peptides, asparagine residues which are N-glycosylated, catalytically essential histidine and tyrosine residues, the serine residue which is acetylated by aspirin, and the characteristic 18-amino acid insert close to the C-termini of PGHS-2.

```
Chick-2    KLKFDPELLFNQRFQYQNRIAAEFNTLYHWHPLLPDTFQIHNQEYTFQQFLYNNSIMLEH
Human-2    KLKFDPELLFNKQFQYQNRIAAEFNTLYHWHPLLPDTFQINDQKYNYQQFIYNNSILLEH
Rat-2      QLKFDPDLLFNQQFQYQNRIASEFKTLYHWHPLLPDTFNIEDQEYTFKQFLYNNSILLEH
Mouse-2    KLKFDPELLFNQQFQYQNRIASEFNTLYHWHPLLPDTFNIEDQEYSFKQFLYNNSILLEH
Mouse-1    QLKFDPELLFRAQFQYRNRIAMEFNHLYHWHPLMPNSFQVGSQEYSYEQFLFNTSMLVDY
Rat-1      QLKFDPELLFRAQFQYRNRIAMEFNHLYHWHPFMPDSFQVGSQEYSYEQFLFNTSMLVDY
Human-1    QLKFDPELLFGVQFQYRNRIATEFNHLYHWHPLMPDSFKVGSQEYSYEQFLFNTSMLVDY
Sheep-1    QLKFDPELLFGAQFQYRNRIAMEFNQLYHWHPLMPDSFRVGPQDYSYEQFLFNTSMLVDY
            360       370       380       390       400       410

Chick-2    GLSHMVKSSKRQIAGRVAGGKNVPAAVQKVAKASIDQSRQMRYQSLNEYRKRFMLKPFKS
Human-2    GITQFVESFTRQIAGRVAGGRNVPPAVQKVSQASIDQSRQMKYQSFNEYRKRFMLKPYES
Rat-2      GLAHFVESFTRQIAGRVAGGRNVPIAVQAVAKASIDQSREMKYQSLNEYRKRFSLKPYTS
Mouse-2    GLTQFVESFTRQIAGRVAGGRNVPIAVQAVAKASIDQSREMKYQSLNEYRKRFSLKPYTS
Mouse-1    GVEALVDAFSRQRAGRIGGGRNFDYHVLHVAVDVIKESREMRLQPFNEYRKRFGLKPYTS
Rat-1      GVEALVDAFSRQRAGRIGGGRNFDYHVLHVAEDVIKESREMRLQSFNEYRKRFGLKPYTS
Human-1    GVEALVDAFSRQIAGRIGGGRNMDHHILHVAVDVIRESREMRLQPFNEYRKRFGMKPYTS
Sheep-1    GVEALVDAFSRQPAGRIGGGRNIDHHILHVAVDVIKESRVLRLQPFNEYRKRFGMKPYTS
            420       430       440       450       460       470

Chick-2    FEELTGEKEMAAELEELYGDIDAMELYPGLLVEKPRPGAIFGETMVEIGAPFSLKGLMGN
Human-2    FEELTGEKEMSAELEALYGDIDAVELYPALLVEKPRPDAIFGETMVEVGAPFSLKGLMGN
Rat-2      FEELTGEKEMAAELKALYHDIDAMELYPALLVEKPRPDAIFGETMVELGAPFSLKGLMGN
Mouse-2    FEELTGEKEMAAELKALYSDIDVMELYPALLVEKPRPDAIFGETMVELGAPFSLKGLMGN
Mouse-1    FQELTGEKEMAAELEELYGDIDALEFYPGLLLEKCQPNSIFGESMIEMGAPFSLKGLLGN
Rat-1      FQEFTGEKEMAAELEELYGDIDALEFYPGLMLEKCQPNSLFGESMIEMGAPFSLKGLLGN
Human-1    FQELVGEKEMAAELEELYGDIDALEFYPGLLLEKCHPNSIFGESMIEIGAPFSLKGLLGN
Sheep-1    FQELTGEKEMAAELEELYGDIDALEFYPGLLLEKCHPNSIFGESMIEMGAPFSLKGLLGN
            480       490       500       510       520       530

Chick-2    TICSPEYWKPSTFGGKVGFEIINTASLQSLICNNVKGSPFTAFHVLNPEPTETATINVST
Human-2    VICSPAYWKPSTFGGEVGFQIINTASIQSLICNNVKGCPFTSFSVPDPELIKTVTINASS
Rat-2      PICSPQYWKPSTFGGEVGFRIINTASIQSLICNNVKGCPFASFNVQDPQATKTATINASA
Mouse-2    PICSPQYWKPSTFGGEVGFKIINTASIQSLICNNVKGCPFTSFNVQDPQPTKTATINASA
Mouse-1    PICSPEYWKPSTFGGDVGFNLVNTASLKKLVCLNTKTCPYVSFRVPDYPGDDGSVIV---
Rat-1      PICSPEYWKPSTFGGDVGFNIVNTASLKKLVCLNTKTCPYVSFRVPDYPGDDGSVRV---
Human-1    PICSPEYWKPSTFGGEVGFNIVKTATLKKLVCLNTKTCPYVSFRVPDASQDDGPAVE---
Sheep-1    PICSPEYWKASTFGGEVGFNLVKTATLKKLVCLNTKTCPYVSFHVPDPRQEDRPGVE---
            540       550       560       570       580       590

Chick-2    SNFAMEDINPTLLLKEQSAEL
Human-2    SRSGLDDINPTVLLKERSTEL
Rat-2      SHSRLDDINPTVLIKRRSTEL
Mouse-2    SHSRLDDINPTVLIKRRSTEL
Mouse-1    ---------------RRSTEL
Rat-1      ---------------RPSTEL
Human-1    ---------------RPSTEL
Sheep-1    ---------------RPPTEL
                                600
```

Figure 2. Continued

72 kDa species appears to contain three N-linked oligosaccharides, while the 74 kDa species probably contains four N-linked oligosaccharides (Otto et al., 1993). Asparagine residues homologous to those which are glycosylated in PGHS-1 are apparently glycosylated in PGHS-2. In addition, there is a fourth site of N-glycosylation in PGHS-2, located near the C terminus (i.e., at Asn580 in murine PGHS-2) (Figure 2); this latter site is glycosylated in about half of PGHS-2 molecules, but glycosylation at this position is not required for enzyme activity (Otto et al., 1993).

PGHS-1 solubilized from membranes with non-ionic detergents appears as a dimer, as judged by chemical cross-linking studies (Roth et al., 1980) and sedimentation analysis (Van der Ouderaa et al., 1977). Residues 25-70 of PGHS-1 are homologous to epidermal growth factor (EGF) (Toh, 1989). Studies of the crystal structure of ovine PGHS-1 indicate that the EGF region of the protein is one of the domains which is responsible for dimer formation (Picot et al., 1994). Attempts to generate catalytically active, monomeric forms of PGHS-1 have been unsuccessful, so it is assumed that the enzyme functions as a dimer.

C. Heme Prosthetic Group of PGH Synthase

Heme is required for both the cyclooxygenase and peroxidase activities of PGHS-1. Titration of apoenzyme with heme has indicated that there is an average of one heme per subunit (Van der Ouderaa et al., 1979; Roth et al., 1981; Kulmacz and Lands, 1984; Ruf et al., 1984). This is supported by recent X-ray crystallographic data (Picot et al., 1994). There are VIS and EPR spectral data suggesting that the heme group is liganded by imidazole groups at the fifth and sixth coordination positions of the iron (Lambeir et al., 1985; Kulmacz et al., 1987). Magnetic circular dichroism of native PGHS-1 and its cyano derivative indicates that 80% of the holoenzyme is high spin at 24o C but only 50% at low temperature (Kulmacz et al., 1987). This suggests that the heme iron equilibrates between the five and six coordinate form, with the five coordinate form predominating at room temperature. His388 is the proximal heme ligand and His207 is the distal heme ligand in ovine PGHS-1 (Shimokawa and Smith, 1991; Picot et al., 1994).

III. CYCLOOXYGENASE AND PEROXIDASE CATALYSIS

A. The Cyclooxygenase Reaction

Depicted in Figure 3 is the process thought to be involved in the synthesis of PGG_2 from arachidonate and two molecules of O_2 (Hamberg and Samuelsson, 1967; Smith and Marnett, 1991). PGH synthase is believed to interact with arachidonate having a kink in the carbon chain due to rotation about the C-9/C-10 bond (Hamberg and Samuelsson, 1967). The enzyme first abstracts the pro-S hydrogen from C-13 (Hamberg and Samuelsson, 1967). A molecule of O_2 is added at C-11 from the solvent side. Serial cyclization of the incipient 11-peroxyl radical yields an endoperoxide with aliphatic chains trans to one another. A second O_2 is then added at C-15.

There is considerable evidence that the hydrogen abstracted from the 13 *pro-S* position is abstracted as a free radical (Mason et al., 1980; Schreiber et al., 1986; Kwok et al., 1987). If so, a radical enzyme species would be required. This enzyme radical may be an enzyme tyrosyl radical. Depicted in Figure 4 is a model developed by Ruf and coworkers showing how a tyrosyl radical could be generated by the interaction of PGH synthase with an alkyl hydroperoxide (Dietz et al., 1988). According to the model, two-electron oxidation of the heme group at the peroxidase active site of PGH synthase by a hydroperoxide such as H_2O_2, 15-hydroperoxyl

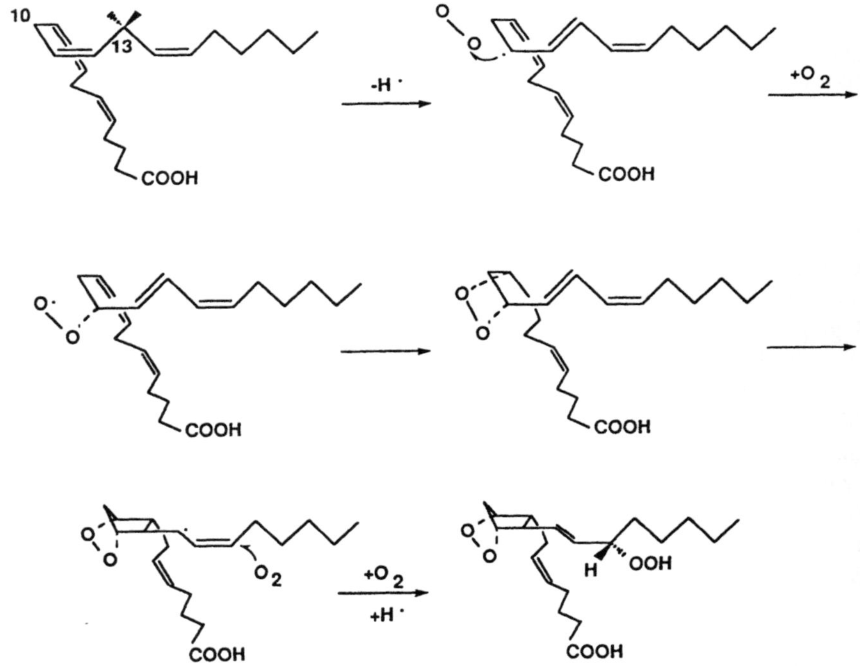

Figure 3. Proposed mechanism for the formation of PGG_2 from arachidonic acid catalyzed by PGH synthases. Adapted from Smith and Marnett (1994).

arachidonate, or PGG_2 causes formation of an Intermediate I. Intermediate I is analogous to Compound I of horseradish peroxidase (Dolphin and Felton, 1974) and has a two-electron oxidized heme in which the iron is in the 4+ state and the porphyrin group is oxidized to a radical cation. Intermediate I can abstract a hydrogen from the phenolic side-chain of a neighboring protein tyrosine residue to produce an Intermediate II with an associated protein tyrosyl radical. Intermediate II is analogous to Compound II of horseradish peroxidase and Compound ES of cytochrome *c* peroxidase in having a one-electron oxidized heme group in which the porphyrin is neutral and the iron is in its 4+ state (Dunford and Stillman, 1976; Yonetani, 1976; Nastainczyk et al., 1984). The tyrosyl radical associated with Intermediate II was proposed by Ruf and coworkers to be the species that abstracts the 13-*pro-S* hydrogen from arachidonate, initiating the cyclooxygenase reaction.

The Ruf tyrosyl radical model (Figure 4) is consistent with the following observations. Alkyl hydroperoxides are required for cyclooxygenase activity (Smith and Lands, 1972; Hemler et al., 1978a, 1978b; Hemler and Lands, 1980) and treatment of PGHS-1 with alkyl hydroperoxides causes the formation of two

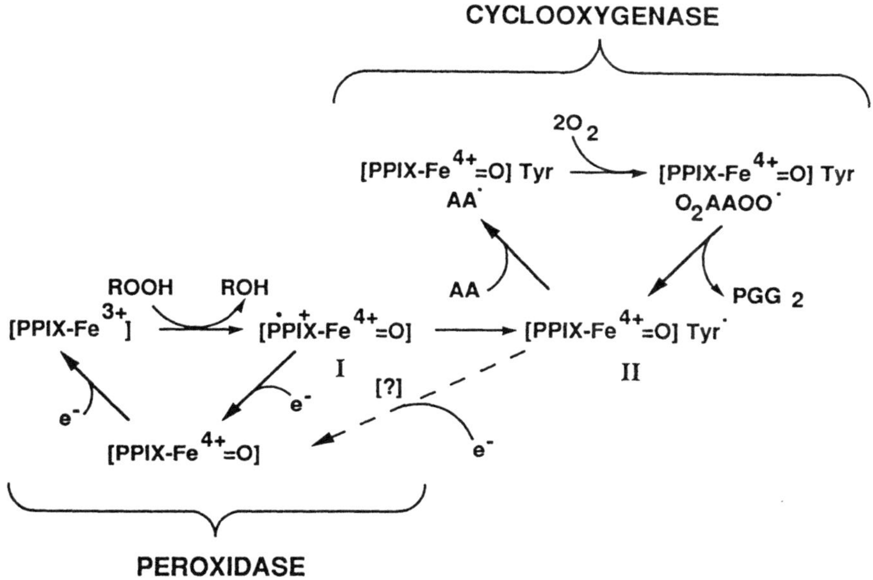

Figure 4. Model for peroxide-dependent activation of the cyclooxygenase activity of PGHS-1 by formation of an intermediate tyrosyl radical. Intermediates I and II are designated. PPIX-Fe^{3+}, heme; AA, arachidonic acid. Adapted from Dietz et al. (1988).

spectral intermediates (Intermediates I and II) with characteristics of peroxidase Compound I and Compound ES intermediates (Lambeir et al., 1985; Dietz et al., 1988; Karthein et al., 1988; Hsuanyu and Dunford, 1990). Protein tyrosyl radicals are formed when PGHS-1 is incubated with hydroperoxides (Karthein et al., 1988; Kulmacz et al., 1990), and tyrosyl radical formation and spectral intermediate II occurs concomitantly (Dietz et al., 1988; Karthein et al., 1988). A tyrosine residue—Tyr385—neighbors the heme group in the cyclooxygenase active site of PGH synthase (Van der Ouderaa et al., 1979; Shimokawa et al., 1990; Smith et al., 1990a; Picot et al., 1994).

Although the Ruf model provides an attractive description of a number of properties of PGH synthase catalysis, the question of the role of the tyrosyl radical in catalysis is still controversial. There are multiple tyrosyl radical signals formed at different times during cyclooxygenase catalysis (Karthein et al., 1988; Lassmann et al., 1991; DeGray et al., 1992; Tsai et al., 1992, 1994), and only one of these signals, a broad doublet, is temporally related to PGG_2 formation (Tsai et al., 1992). In contrast, two different types of singlet radical signals occur only when much of the activity has been lost; these singlet signals may be markers for "suicide inactivation" of PGH synthase (DeGray et al., 1992; Tsai et al., 1992). With regard to suicide inactivation of the enzyme, it is known that each molecule of PGH synthase can catalyze the synthesis of about 1300 moles of PGG_2 before the cyclooxygenase activity is lost (Smith and Lands, 1972; Kulmacz, 1987; Marshall et al., 1987). Apparently, suicide inactivation results from the decay of an unstable cyclooxygenase protein intermediate to a catalytically inactive enzyme. One possibility is that a tyrosyl radical involved with catalysis has a certain probability of changing its conformation, leading to reaction(s) with a group on the protein rather than with a molecule of arachidonate substrate. A precedent for this type of reaction occurs in the case of met-myoglobin, where oxidation of the heme group by H_2O_2 leads to generation of a tyrosyl radical which can form a covalent linkage to the heme (Catalano et al., 1989).

B. Structure of PGH Synthase Active Sites

Figure 5 is a model of the active site of ovine PGHS-1 (Smith and Marnett, 1994). An alkyl hydroperoxide is shown associated with the heme group bound to the peroxidase active site. The heme is shown liganded at the proximal position by His388 and at the distal position by His207. Oxidation of the hydroperoxide to an alcohol causes formation of a compound I-like oxidized heme (cf. Intermediate I) (Figure 4). An intramolecular electron transfer from a neighboring Tyr385 to the heme yields Intermediate II having a tyrosyl radical centered on Tyr385. The tyrosyl radical may abstract the hydrogen atom from the arachidonate bound to the cyclooxygenase active site to initiate the cyclooxygenase reaction. Ser530, the site of aspirin acetylation, is shown in close proximity to the cyclooxygenase active site.

Five general features are known about the active site of PGHS-1 (Figure 5). First, the cyclooxygenase active site is distinct from the peroxidase active site. This conclusion is based on the findings that aspirin and related NSAIDs (Van der Ouderaa et al., 1980; Mizuno et al., 1982), and docosahexaenoic acid (Marshall and Kulmacz, 1988; Meade et al., 1993; Picot et al., 1994) are competitive inhibitors of the cyclooxygenase activity of PGHS- 1, but that none of these agents appreciably affects the peroxidase reaction. Second, the cyclooxygenase active site is large enough to accommodate a structurally diverse group of NSAIDs and fatty acid substrates (Kulmacz, 1989; Smith et al., 1990b). Third, the cyclooxygenase active site becomes more accessible to NSAIDs when heme is bound to PGHS-1. For example, only holo-PGHS-1 is covalently labeled at appreciable rates with $[^3H]$aspirin (Chen and Marnett, 1989). Fourth, binding of NSAIDs induces global structural changes in PGH synthase. When indomethacin is bound to PGHS-1, the enzyme becomes less susceptible to trypsin cleavage (Kulmacz, 1989) and to covalent labeling of nonactive site tyrosines (Tyr355 and Tyr417 of ovine PGHS-1) by tetranitromethane (Shimokawa et al., 1990). And fifth, two residues, Ser530 and Tyr385, lie within or closely neighbor the cyclooxygenase active site (Picot et al., 1994). Ser530 is acetylated by aspirin, causing irreversible inactivation of cy-

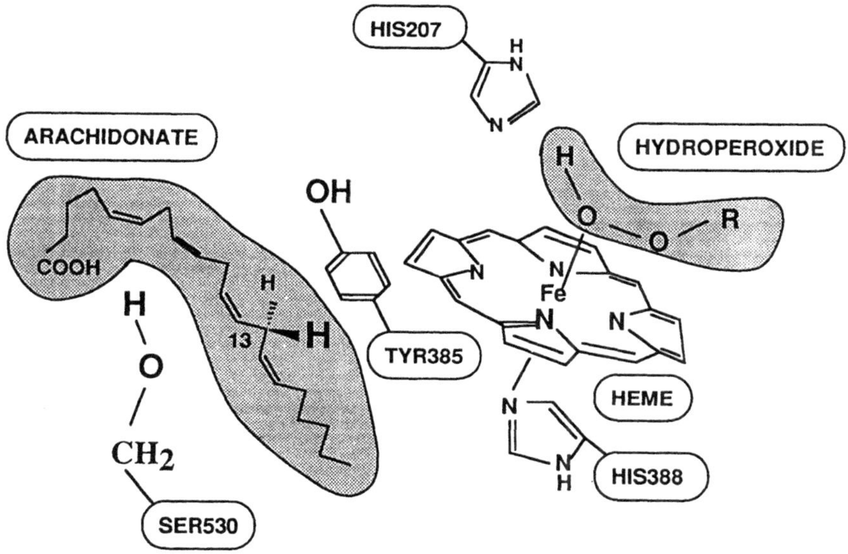

Figure 5. Model for the cyclooxygenase and peroxidase active sites of ovine PGHS-1. Residues involved in heme binding and the relative locations of the groups neighboring the arachidonate (cyclooxygenase) and hydroperoxide (peroxidase) binding sites are illustrated. From Smith and Marnett (1994) with permission.

clooxygenase activity (Roth et al., 1983). Tyr385 is required for cyclooxygenase activity and is protected from nitration by tetranitromethane by the presence of indomethacin or ibuprofen (Shimokawa et al., 1990; Smith et al., 1990a).

The recent publication of the crystal structure of ovine PGHS-1 by Garavito and coworkers (Picot et al., 1994) provides overall support for the model depicted in Figure 5. Additionally, these investigators propose that the arachidonate binding site takes the form of a hydrophobic channel in the core of what is observed as a globular protein. Arg120 is proposed to be the site at which the carboxyl group of arachidonate binds to the enzyme.

C. Cyclooxygenase Substrates and Inhibitors

The two best fatty acid substrates for the cyclooxygenase activities of PGHS-1 are arachidonic (20:4 [n-6]) and dihomo-γ-linolenic acids (20:3 [n-6]). The K_m values for both substrates are about 5 μM (Lands et al., 1973; Flower, 1974; Meade, 1992). PGHS-1 catalyzes the oxygenation of 5,8,11,14,17-eicosapentaenoic acid (EPA), the fish oil fatty acid at about 5% the rate of arachidonate (Meade, 1992).

A large number of fatty acids having different combinations of double-bond positions and chain lengths have been tested as substrates and inhibitors for the cyclooxygenase activity of PGHS-1. In general, n-3 and n-9 polyunsaturated fatty acids containing 18-22 carbons are poor substrates but efficient cyclooxygenase inhibitors (Lands et al., 1973). Fatty acid inhibitors of PGHS-1 include eicosatetraynoic acid, known as ETYA (Vanderhoek and Lands, 1973), docosahexaenoic acid (22:6 [n-3]) (Lands et al., 1973; Marshall and Kulmacz, 1988; Meade et al., 1993); and 5c-, 8c-, 12t-, 15c-eicosatetraenoic acid (Flower, 1974).

The substrate specificities of PGHS-2 differ somewhat from those of PGHS-1. Arachidonic and 8,11,14-eicosatrienoic acids are the best substrates for PGHS-2 (Meade, 1992; Breuer et al., 1994). Moreover, the K_m values for the two isozymes for arachidonate are very similar (Meade et al., 1993; Laneuville et al., 1994). Although a large number of substrates have not been tested, it appears that PGHS-2 oxygenates a larger variety of fatty acids more efficiently than PGHS-1 (Breuer et al., 1994). Docosahexaenoic acid is an equally effective inhibitor of both PGHS-1 and PGHS-2 (Meade et al., 1993; Laneuville et al., 1994).

D. Inhibition of Cyclooxygenase Activity by Nonsteroidal Anti-inflammatory Drugs

Most nonsteroidal anti-inflammatory drugs (NSAIDs) are potent inhibitors of the cyclooxygenase activities of PGHS-1 and PGHS-2 (Table 1) (Rome and Lands, 1975; Carty et al., 1980; Meade et al., 1993; Laneuville et al., 1994). Until recently, it was thought that these agents acted by inhibiting PGHS-1. The fairly recent discovery of PGHS-2 has led to a reinvestigation of this issue. Most of the current data support the concept discussed above that PGHS-2 is the therapeutic target of NSAIDs acting in their anti-inflammatory capacities (O'Banion et al., 1991; Lee et al., 1992; O'Sullivan, et al., 1992a, 1992b; Futaki et al., 1993, 1994). The anti-thrombogenic activity of aspirin (Hennekens et al., 1989; Patrono et al., 1990), however, appears to be due solely to inhibition of the cyclooxygenase activity of platelet PGHS-1 because PGHS-1 appears to be the only isozyme in platelets (Funk et al., 1991).

Aspirin and other NSAIDs block the cyclooxygenase activities of PGH synthases without affecting peroxidase activity. This is consistent with the concept that the cyclooxygenase and peroxidase active sites are physically distinct. Many NSAIDs are conventional competitive cyclooxygenase inhibitors which compete reversibly with arachidonate for binding to the enzyme; included in this class of compounds are piroxicam (Carty et al., 1980), flufenamate (Rome and Lands, 1975), sulindac sulfide (Meade et al., 1993), and ibuprofen (Rome and Lands, 1975; Meade et al., 1993). Other NSAIDs, including indomethacin, flurbiprofen, meclofenamate, and diclofenac, exhibit more complex kinetics and are known as time-dependent competitive inhibitors (Rome and Lands, 1975; Kulmacz and Lands, 1985; Walenga et al., 1986; Laneuville et al., 1994). These agents bind

Table 1. Inhibition of Human PGH Synthases by Nonsteroidal Anti-inflammatory Drugs.

NSAID	*IC^{50} for PGHS-1 (μM)*	*IC^{50} for PGHS-2 (μM)*
Indomethacin	13.5	> 1000.0
Sulindac Sulfide	1.3	50.7
Piroxicam	17.7	> 500.0
Diclofenac	2.7	20.5*
Flurbiprofen	0.5	3.2
Meclofenamate	1.5	9.7
Phenylbutazone	16.0	> 100.0
Naproxen	4.8	28.4*
Ibuprofen	4.0	12.5
Ketorolac Tromethamine	31.5	60.5
DHA (22:6)	25.6	41.0
6-MNA	64.0	93.5*
Etodolac	74.4	60.0*
Salicylic acid	> 1000.0	> 1000.0

Notes: Values are for instantaneous inhibition of cyclooxygenase activity of human PGHS-1 and human PGHS-2 expressed in transfected cos-1 cells. From Lanueville et al. (1994) with permission.

rapidly and reversibly in a first phase but, if retained for a sufficient time in the active site, cause a conformational change in the protein associated with tighter (but noncovalent) binding. Once bound in this tighter form, these time-dependent NSAIDs only slowly dissociate from the cyclooxygenase active site (Rome and Lands, 1975; Kulmacz and Lands, 1985; Walenga et al., 1986; Laneuville et al., 1994).

Table 1 compares the ability of a variety of common NSAIDs to cause "instantaneous" inhibition of PGHS-1 and PGHS-2 *in vitro* (Laneuville et al., 1994). This assay system provides a measure of the affinity of various NSAIDs for each of the isozymes and circumvents the phenomenon of time-dependent inhibition. When time-dependent inhibition is factored in, compounds such as indomethacin and flurbiprofen become highly effective inhibitors of PGHS-2. These studies have indicated (a) that PGHS-1 and PGHS-2 are pharmacologically distinct and (b) that the cyclooxygenase active site of PGHS-2 is somewhat larger than that of PGHS-1. A major side-effect associated with the use of NSAIDs is gastrointestinal bleeding and ulcer formation. One fairly selective inhibitor of PGHS-2, called NS398, has recently been described (Futaki et al., 1993, 1994). This agent is an effective anti-inflammatory but is stomach-sparing. Thus, it appears that it will be possible to develop a new generation of NSAIDs without the gastrointestinal effects associated with common NSAIDs.

Aspirin is an unusual NSAID because, upon binding to the cyclooxygenase active site, aspirin causes acetylation of an "active site" serine residue (Roth et al., 1983) (Figure 6). In ovine PGHS-1, this is Ser530 (DeWitt and Smith, 1988; Merlie et al., 1988; Yokoyama et al., 1988; Shimokawa and Smith, 1992); an homologous

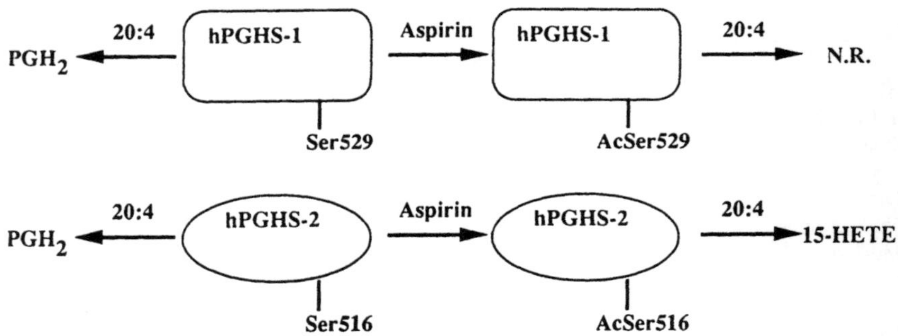

Figure 6. Schematic representation of the acetylation of human (h) PGHS-1 and PGHS-2 by aspirin and the effects of acetylation on product formation by each of the enzymes. From Smith and Marnett (1994) with permission.

serine in PGHS-2 (Ser516) of human PGHS-2 is also acetylated (Lecomte et al., 1994). Acetylation of PGHS-1 by aspirin causes complete inhibition of cyclooxygenase activity but does not affect peroxidase activity (Figure 6) (Van der Ouderaa et al., 1980; Mizuno et al., 1982; Roth et al., 1983). Acetylation of PGHS-2 by aspirin converts this isozyme to a form which catalyzes the formation of 15*R*-hydroxyeicosatetraenoic acid (15-HETE) instead of PGG_2 (Figure 6) (Holtzman et al., 1992; Meade et al., 1993; Lecomte et al., 1994). The so-called "active site" serine residue of PGHS-1 is not required for enzyme activity because replacement of this group with an alanine yields an enzyme which is active and has a K_m for arachidonate which is very close to that of the native enzyme (DeWitt et al., 1990). Aspirin acetylation of PGHS-1 places a bulky group at Ser530 which prevents arachidonate from binding (DeWitt et al., 1990; Shimokawa and Smith, 1992). Aspirin acetylation of PGHS-2 appears to prevent access of incoming O_2 to C-11, so that a reaction at C-15 occurs instead (Lecomte et al., 1994).

IV. REGULATION OF EXPRESSION OF THE GENES FOR PGHS-1 AND PGHS-2

A. Differential Regulation of the Expression of PGH Synthase Genes—An Overview

The two PGH synthase enzymes appear to have distinct physiological roles, resulting in part from their dissimilar expression. PGHS-1 is constitutive in most tissues, and some specialized cells express quite high levels of this isozyme. In contrast, PGHS-2 is not present in resting, unstimulated tissues, although it can be rapidly induced with the appropriate stimulus. Hormonal activation of tissues that express PGHS-1 results in acute short-term prostaglandin production that helps to coordinate cellular responses to these hormones. Tissues and cells which constitutively contain high levels of PGHS- 1 invariably perform unique prostaglandin-mediated functions. Induction of PGHS-2 expression, on the other hand, is usually correlated with specialized physiological events requiring continuous prostaglandin synthesis over several hours or days, processes such as inflammation, mitogenesis, or ovulation.

B. Regulation of PGHS-1 Expression

Both PGHS-1 mRNA and protein are present in most tissues (Simmons et al., 1991; Feng et al., 1993; Kargman et al., 1994), although some cells, notably seminal vesicles (DeWitt et al., 1981), vascular endothelia (DeWitt et al., 1983), platelets (Funk et al., 1991), renal collecting tubule epithelia (Huslig et al., 1979), and monocytes (Lee et al., 1992; O'Sullivan, et al., 1992a, 1992b), express higher than average amounts of the isozyme. Generally, PGHS-1 is elevated most in highly differentiated cells, and prostaglandin synthesis by PGHS-1 plays a role in the specialized functions of these cells. For example, prostacyclin and thromboxane A_2 produced by vascular endothelia and platelets, respectively, regulate vascular homeostasis; PGE_2 produced by collecting tubules regulates renal water reabsorption; and monocytes/macrophages produce prostaglandins that are important regulators of B and T cell maturation and function (Phipps et al., 1991).

Cells utilize PGHS-1 when acute, on-demand synthesis of prostaglandins is required. For example, intermittent but immediate PGE_2 production is required to regulate water resorption in the renal collecting duct in response to changes in circulating vasopressin (Smith, 1989). In the vasculature, instantaneous thromboxane A_2 production by platelets is used to promote timely clot formation and to limit blood loss at sites of injury. Endothelial cells must also respond quickly to limit platelet aggregation in adjacent intact vessels by producing prostacyclin. Both platelets and endothelial cells express high levels of PGHS-1 (Smith et al., 1983).

That PGHS-1 is constitutively expressed in most issues, and is elevated only in differentiated cells which employ prostaglandins for specialized physiological functions, suggests that PGHS-1 expression is regulated primarily developmentally; and indeed, no examples of acute regulation have been reported for PGHS-1. While circumstantial evidence supports the idea of the developmental regulation of PGHS-1, direct evidence for such regulation has been obtained only in two model systems, pro-monocytic cell lines and in the developing sheep. TPA-induced differentiation of pro-monocytic THP-1 cells to a macrophage phenotype (Smith et al., 1993) leads to a dramatic increase in PGHS-1 protein expression; to a somewhat lesser extent, PGHS-1 also increases in TPA-differentiated U937 cells (Hoff et al., 1993). Between parturition and four weeks after birth, there is about a 6-fold increase in newborn sheep pulmonary arterial endothelial cell PGHS-1 levels

(Brannon et al., 1994). Developmental regulation in these model systems is characterized by a permanent change in PGHS-1 expression that occurs simultaneously with growth or differentiation; while acute regulation, such as occurs with PGHS-2, is usually more rapid (0.5-4 hr), and temporary.

Although PGHS-1 protein levels are maintained constant in most tissues and cell lines (DeWitt and Meade, 1993), 3- to 4-fold increases in PGHS-1 mRNA levels are frequently observed in response to various stimuli, including EGF (Hamasaki et al., 1993), IL-1 (Maier et al., 1990), TPA (Hamasaki et al., 1993; Hoff et al., 1993; Smith et al., 1993), serum (DeWitt et al., 1991; DeWitt and Meade, 1993), and PDGF (Lin et al., 1989). While PGHS-1 mRNA levels change, PGHS-1 protein levels do not; in none of the examples cited above has a corresponding increase in PGHS-1 protein been shown to accompany the increase in PGHS-1 mRNA. It cannot be ruled out, however, that increased PGHS-1 mRNA expression may lead to increased synthesis and turnover of PGHS-1 protein.

C. Regulation of PGHS-2 Expression

In contrast to PGHS-1, PGHS-2 protein is not expressed in tissues under normal physiological conditions. PGHS-2 mRNA, on the other hand, does accumulate in some tissues, although message levels vary considerably. By northern blot analysis, significant PGHS-2 mRNA levels have been detected in mouse prostate and brain (Simmons et al., 1991). More sensitive analysis using reverse transcriptase-polymerase chain reaction amplification has also detected relatively high levels of PGHS-2 mRNA in human prostate and in human lung (O'Neill and Ford-Hutchinson, 1993). In addition, intermediate to low levels of PGHS-2 mRNA were detected by RT-PCR in human uterus, small intestine, mammary gland, and stomach, thymus, liver, kidney, testis pancreas, and brain. Cell-specific expression of PGHS-2 mRNA in these tissues was not examined, so it was not determined whether these tissues express uniformly low levels of PGHS-2, or if individual cells, such as tissue macrophages, express relatively high levels of PGHS-2 mRNA. Regardless of the cell type or level of PGHS-2 mRNA expression, PGHS-2 protein does not appear to be expressed to any significant degree, except in response to physiological stimuli. Western blotting experiments with PGHS-2-specific antisera could not detect any PGHS-2 protein in any unstimulated rat tissue (Kargman et al., 1994). However, PGHS- 2 could be detected following carrageenan injection in rat paw (Kargman et al., 1994) and in articular tissue during staphylococcal cell wall or adjuvant-induced arthritis in rat (Sano et al., 1992). PGHS-2 is also expressed in joints of individuals with rheumatoid arthritis (Sano et al., 1992; Crofford et al., 1994; Hulkower et al., 1994), in rat follicles preceding ovulation (Sirois and Richards, 1992; Sirois et al., 1992), and in neurons following synaptic activity (Yamagata et al., 1993). A similar pattern of expression is observed *in vitro* in experiments examining PGHS-2 regulation using cultured cells. PGHS-2 expression can be both rapidly (1-3 hr) and dramatically (20- to 80-fold) induced in a number of lines. Growth factors, phorbol esters, and IL-1β induce PGHS-2 in fibroblasts, synoviacytes, and endothelial cells (Kujubu et al., 1991; Xie et al., 1991; Fletcher et al., 1992; Hla and Neilson, 1992; O'Banion et al., 1992; Evett et al., 1993); and LPS, IL-1, and TNFα stimulate PGHS-2 expression *ex vivo* in monocytes and macrophages (Lee et al., 1992; O'Sullivan, et al., 1992a, 1992b; Crofford et al., 1994; Riese et al., 1994). While only a limited number of tissues and cell types have been examined, it is likely that PGHS-2 can be induced in almost any cell or tissue with the appropriate stimuli. Importantly, PGHS-2 expression, but not PGHS-1 expression, can be completely inhibited by anti-inflammatory glucocorticoids such as dexamethasone (Kujubu and Herschman, 1992; Lee et al., 1992; O'Banion et al., 1992; Sano et al., 1992; DeWitt and Meade, 1993; Kujubu et al., 1993; Yamagata et al., 1993; Hulkower et al., 1994).

D. PGH Synthase Isozyme Gene Structure and Regulation

PGHS-1 and PGHS-2 are encoded by separate genes; the PGHS-1 gene is on chromosome 2 in mouse (Wen et al., 1993) and chromosome 9 in humans (Funk et al., 1991), while PGHS-2 is located on chromosome 1 in both mouse and humans (Jones et al., 1993; Kosaka et al., 1994). Although these genes are separate, the primary sequences and exon structures of PGHS-1 and PGHS-2 are highly conserved, suggesting that they were derived from a single ancestral gene (Figure 7). The single structural difference between the two PGH synthase genes is that the first and second exons of PGHS-1, which contain the transcription and translational start site and signal peptide, are fused to form a single exon in PGHS-2. Thus, the PGHS-2 gene contains 10 exons, while the PGHS-1 gene has 11. This condensation results in one significant difference between the primary structures of the PGHS-1

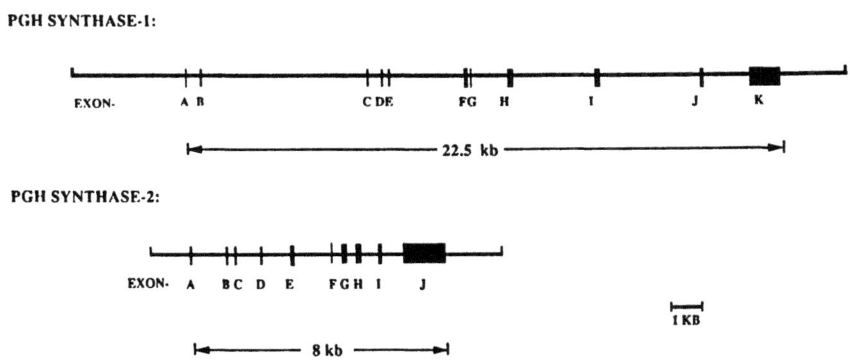

Figure 7. Comparison of the structures of the genes encoding murine PGHS-1 and PGHS-2. Exons are denoted with letters. PGHS-1 and PGHS-2 gene structures are from Kraemer et al. (1992) and Fletcher et al. (1992), respectively.

and PGHS-2 proteins; the signal peptide of the PGHS-2 is 6-9 amino acids shorter than that of PGHS- 1. While the exon splice sites and sizes are conserved between the two genes, the intron sizes are not. The PGHS-2 gene is about 8 kb in length (Fletcher et al., 1992), while the PGHS-1 gene is about 22 kb (Kraemer et al., 1992). Primary response genes are generally ≤ 8 kb (Herschman, 1991), which is consistent with the characterization of PGHS-2 as an immediate early gene (Herschman, 1991).

The organization of the PGHS-1 and PGHS-2 promoters are indicative of their differential modes of regulation. The TATA-less PGHS-1 promoter (Kraemer et al., 1992; Wang et al., 1993) is typical of many constitutively expressed "housekeeping genes." Reporter plasmids constructed with the 5′-upstream region of the PGHS-1 gene have failed to demonstrate transcriptional control from this promoter (Kraemer, 1994), supporting the concept that expression of PGHS-1 is situationally dependent on cell type and not on external stimuli. The PGHS-2 promoter, on the other hand, contains a TATA box, and experiments with reporter plasmids containing the PGHS-2 promoter and upstream 5′-flanking sequence have demonstrated that PGHS-2 transcription is highly regulatable (Fletcher et al., 1992).

Transcriptional activation of the PGHS-2 gene appears to be one important mechanism for increasing PGHS-2 expression (DeWitt and Meade, 1993; Evett et al., 1993). Transcription of PGHS-2 is unique, in that it can be controlled by multiple signaling pathways, including the cAMP pathway (LH, dibutyryl cAMP) (Wong et al., 1989; Kujubu et al., 1991; Sirois and Richards, 1993; Sirois et al., 1993), the protein kinase C pathway (phorbol esters) (Goerig et al., 1987; Kujubu et al., 1991, 1993; Hla and Neilson, 1992; Pilbeam et al., 1993; Crofford et al., 1994), by viral transformation (*src*) (Simmons et al., 1989; Han et al., 1990; Evett et al., 1993), and by other pleiotropic pathways such as those activated by growth factors (PDGF, EGF) (Habenicht et al., 1981; Kujubu et al., 1991; Hamasaki et al., 1993), bacterial endotoxin (LPS) (Hla and Neilson, 1992; Lee et al., 1992; O'Sullivan, et al., 1992a, 1992b; Feng et al., 1993; Jones et al., 1993), and inflammatory cytokines (IL-1) (Feng et al., 1993; Jones et al., 1993; Crofford et al., 1994; Hulkower et al., 1994). While the primary structures of the human (Hla, 1993), mouse (Fletcher et al., 1992; Herschman et al., 1993; Wen et al., 1993), and rat (Sirois and Richards, 1993; Sirois et al., 1993) PGHS- 2 genes and 5′-flanking regions have been determined, the complex analysis of cis-elements responsible for regulation of this gene are, as yet, in their early stages (Fletcher et al., 1992).

Fletcher et al. (1992) have demonstrated, using luciferase gene reporter plasmids, that transcriptional control elements necessary for activation of the mouse PGHS-2 gene by phorbol esters and serum are located within the first 371 nucleotides upstream of the mouse PGHS-2 transcription start site. Sirois et al. (1993) have further demonstrated that an NF-IL6/C/EBP regulatory element in the rat promoter centered at position -131 is responsible, at least in part, for increased PGHS-2 gene transcription in rat follicular cells following exposure to cAMP or the physiologically relevant gonadotrophins LH and FSH (Sirois and Richards,

1993). The C/EBPβ (also called NF-IL6) transcription factor was also shown to be induced *in vivo* by ovulatory doses of human chorionic gonadotrophin, and gel shift assays demonstrated that this protein interacted with PGHS-2 NF-IL6/C/EBP elements. Point mutations in the NF-IL6/C/EBP DNA sequence reduced C/EBPβ binding to the promoter, and the mutated promoter was less responsive to activation than the native promoter, further confirming the role of the C/EBPβ in regulating PGHS- 2 gene transcription.

Only about 50% of the forskolin-induced transcription from the PGHS-2 promoter in rat follicular cells was attributable to the NF-IL6/C/EBP enhancer, suggesting that other DNA elements also cooperate in the cAMP-response pathway. Additional cis-acting DNA elements undoubtedly also participate in the multiple other effector pathways that regulate PGHS-2 expression. Those cis-acting elements have yet to be characterized, but comparison of the 5′-flanking region of the mouse, rat, and human PGHS-2 genes reveals some striking homologies and provides some likely candidate regulatory sequences (Figure 8). Putative regulatory elements for NFκB, Sp1, Ets, and NF-IL6/C/EBP are all present in the upstream

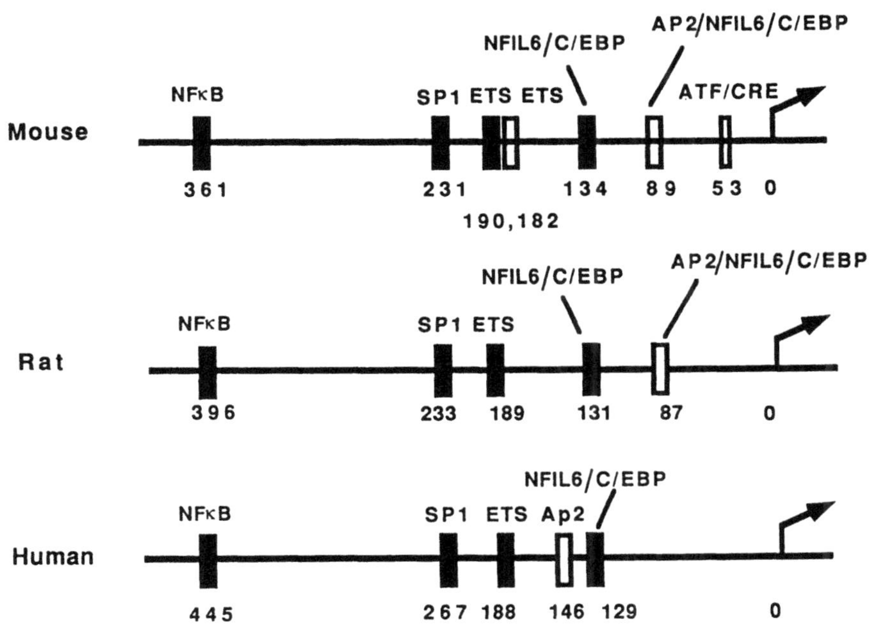

Figure 8. Diagram of the 5′-untranslated region of the mouse (Fletcher et al., 1992), rat (Sirois & Richards, 1993) and human PGHS-2 genes (Kosaka et al., 1994). Common putative transcriptional regulatory elements are represented by black boxes. Regulatory elements that are not shared are represented by open boxes.

regulatory regions of each of these genes (Fletcher et al., 1992; Hla, 1993; Sirois et al., 1993), and at approximately the same location relative to the transcriptional start sites. NFκB and NF-IL6 cis-elements are frequently found in tandem in the 5′-flanking regions of certain inflammatory response genes, including those for IL-8, IL-6, TNFα, MIP-1α, and G-CSF (Akira and Kishimoto, 1992; Grove and Plumb, 1993). NFκB and NF-IL6/C/EBP elements alone, or in tandem, confer inducibility to these genes for the same effectors; IL-1, LPS, TNFα, mitogens, and phorbol esters, that stimulate PGHS-2 expression (Hensel et al., 1989; Isshiki et al., 1990; Zhang et al., 1990; Muller et al., 1993). The Ets cis-regulatory element is frequently found in phorbol ester and serum-regulated genes and is a key element for the regulation of inflammatory proteins such as stromelysin (Macleod et al., 1992). Sp1 is a basal transcription factor that often works synergistically with other transcriptional regulators (Kadonaga et al., 1986; Pugh and Tijan, 1990). It is likely that each of these cis-regulatory elements play some role in the complex transcriptional regulation of the PGHS-2.

While PGHS-2 expression is regulated acutely by transcriptional activation, post-transcriptional regulation also occurs. PGHS-2 mRNA is unstable compared to PGHS-1 mRNA (DeWitt and Meade, 1993), a feature predicted from the presence of multiple RNA instability sequences (AUUUA) in its 3′-untranslated region. PGHS-2 mRNA is translated as soon as it is synthesized; therefore, the short mRNA half-life limits PGHS-2 production post-transcriptionally (DeWitt and Meade, 1993). Dexamethasone can reduce PGHS-2 by at least two mechanisms. Glucocorticoids inhibit PGHS-2 expression by suppressing PGHS-2 gene transcription (DeWitt and Meade, 1993), but they also appear to reduce PGHS-2 mRNA stability and may block translation of the PGHS-2 mRNA (Lee et al., 1992; DeWitt and Meade, 1993; Evett et al., 1993). PGHS-2 protein is also much less stable in fibroblasts than is PGHS-1, a post-translational regulatory mechanism that limits PGHS-2 levels in these cells (DeWitt and Meade, 1993). What accounts for the different protein stabilities of PGHS-1 and PGHS-2 is not known, but increased protein turnover of PGHS-2 may be mediated via the carboxy-terminal protein sequences that are unique to PGHS-2.

E. Why Are There Two PGH Synthase Isozymes?

Our current hypothesis is that PGHS-1 and PGHS-2 represent two prostaglandin biosynthetic systems which operate independently. Recent evidence suggests that arachidonate used for prostaglandin synthesis by PGHS-1 and by PGHS-2 is derived from different cellular (or extracellular) sources, and that prostaglandins produced by PGHS-1 and PGHS-2 are channeled through separate effector pathways, coincident with their respective functions. Thus, differential regulation, subcellular localization, and substrate utilization allow these two enzymes to produce prostaglandins which play unique physiological roles.

PGHS-1 is thought to regulate "housekeeping" activities because it is constitutively expressed and, thus, available to respond to rapid transient increases in arachidonate brought about by hormone-induced activation of phospholipases. In contrast, PGHS-2 is expressed, at the earliest, only several hours following exposure to mitogenic or inflammatory stimuli.

PGHS-2 is thought to produce prostaglandin involved in inflammation because it is induced in immunoregulatory cells and in inflamed tissues, and because anti-inflammatory glucocorticoids and anti-inflammatory cytokines inhibit it's induction.

PGHS-2 also appears to be tied to mitogenesis in some manner. PGHS-2 is an immediate early gene in cycling fibroblasts and, like *c-fos* and *c-jun*, PGHS-2 is expressed only transiently during G_0 and early G_1 (DeWitt and Meade, 1993; Evett et al., 1993; Kujubu et al., 1993). Long-term PGHS-2 expression occurs only in terminally differentiated cells such as macrophages or synoviacytes, suggesting some incompatibility with PGHS-2 expression and cell division (Lee et al., 1992; Crofford et al., 1994,). The exact role of PGHS-2 in mitogenesis is unknown, but PGH synthase products have been tied to proliferation in a number of cell types. PGE_2 or $PGF_{2\alpha}$ are mitogens (Jimenez-de-Asua et al., 1981, 1983; Otto et al., 1982), and oxygenated linoleic derivatives have also been implicated in growth factor-mediated mitogenesis (Glasgow and Eling, 1994). PGHS-2, unlike PGHS-1, can efficiently oxygenate fatty acids such as linoleic (Breuer et al., 1994). Preliminary experiments in this laboratory (Morita et al., 1994) indicate that PGHS-2 directs prostaglandin products into the nucleus (see below), where they may affect cellular processes differently than products of PGHS-1, which are strictly cytoplasmic and which likely exit and signal extracellularly.

Several observations suggest that PGHS-2 and PGHS-1 function independently. First, induction of PGHS-2 does not significantly enhance cellular PGH synthase activity in most cells, so it does not seem logical that regulation of this enzyme is used as a simple amplification mechanism. While increased PGHS-2 expression does usually lead to significantly elevated prostaglandin production, PGHS-2 raises total cellular PGH synthase activity relatively little. In LPS-stimulated macrophages (Lee et al., 1992) and in IL-1-, serum-, or PDGF- stimulated fibroblasts (Habenicht et al., 1985; Raz et al., 1988; Lin et al., 1989; Evett et al., 1993), total cellular PGH synthase activity increases, at most, about 2-fold. It also appears that from an activity standpoint, PGHS-2 induction is not always even necessary. Unstimulated 3T3 fibroblasts have sufficient PGHS-1 to account several times over for total prostaglandin synthesis which occurs following PDGF-stimulated induction of PGHS-2 (Lin et al., 1989). Of course, some tissues, such as ovarian follicles, contain little constitutive PGHS-1, and induction of the PGHS-2 does result in a gain of function (Hedin et al., 1987), but most cells and tissues in which up-regulation of PGHS-2 appears to have important physiological consequences have significant endogenous activity from PGHS-1 that is only modestly elevated by induction of PGHS-2.

A second observation that suggests distinct roles for PGHS-1 and PGHS-2 are their subcellular locations. We initially believed that PGHS-2 co-localized to the same nuclear and endoplasmic reticular membranes as PGHS-1 (Regier et al., 1993). However, more precise experiments in this laboratory using confocal microscopy have since determined that PGHS-2 is preferentially targeted to the nuclear membrane, and that PGHS-1 is primarily located in the endoplasmic reticulum (ER). We have observed that serum-stimulated synthesis of PGHS-2 is initially directed to the nuclear membrane, and only at later times does PGHS-2 appear in the ER (Morita et al., 1994). Since these membranes are contiguous, it appears as if PGHS-2 is targeted to the nuclear membrane but spills out to the ER as the enzyme concentrations increase. We have further substantiated that the nuclear location of PGHS-2 is different from PGHS-1 by using a new technique to localize prostaglandin production in live cells. The indicator probe, 2,7-dichlorofluoresceine diacetate (DCFDA), is readily taken up by cells and fluoresces when oxidized by PGH synthase during oxygenation of arachidonic acid. Using fluorescent confocal microscopy, we have observed that prostaglandin synthesis by PGHS-2 occurs in the nuclear space and also in the cytoplasm, while prostaglandin synthesis by PGHS-1 occurs solely in the cytoplasm (Morita et al., 1994).

Targeting of PGHS-1 and PGHS-2 to different membrane systems may account for the recent observation of Reddy and Herschman (1994) that indicates the two enzymes utilize separate pools of arachidonate. These researchers demonstrated that antisense RNA to PGHS-2 mRNA inhibited PGHS-2 protein and prostaglandin synthesis in mitogen- or phorbol ester-stimulated fibroblasts and in lipopolysaccharide-stimulated macrophages. Inhibition of prostaglandin synthesis occurred even though antisense RNA did not diminish mitogen- or lipopolysaccharide-stimulated arachidonate release from these cells, and even though the cells contained sufficient active PGHS-1. Reddy and Herschman reasoned from these data that a separate pool of arachidonate must exist that is available only to PGHS-2 (presumably, another pool exists for PGHS-1).

An alternative interpretation of the Reddy and Herschman experiments is that specific proteins, possibly phospholipases or other fatty acid binding proteins, physically interact with one or the other of the PGH synthases to provide arachidonate substrate. A precedent for fatty acid channelling occurs in the leukotriene biosynthetic pathway. The 5-lipoxygenase (5-LO) is dependent on the 5-lipoxygenase activating protein (FLAP) for leukotriene synthesis (Abramovitz et al., 1993; Mancini et al., 1993). FLAP binds arachidonate, apparently making it available to 5-LO. By coincidence, 5-LO has also been localized to the nuclear membrane. At least three phospholipase A_2 enzymes have been characterized which appear to be involved in arachidonate metabolism (Kudo et al., 1993), but little is known about their association with PGH synthases.

In summary, the physiological purpose of PGHS-2 is likely more complex than to simply enhance prostaglandin synthesis by PGHS-1. These two isozymes appear to be separated into distinct prostaglandin biosynthetic systems that allows them to subserve different physiological processes. As demonstrated by Reddy and

Herschman, these pathways derive substrate from different arachidonate pools. As shown in this laboratory, these pathways direct products to different extracellular (PGHS-1) and/or intracellular (PGHS-2) sites (Figure 9). These two pathways apparently have evolved to carry out specialized functions, the PGHS-1 pathway to produce prostaglandins that act extracellularly as local hormones to coordinate short-term cellular responses to hormonal stimulation and the PGHS-2 pathway to produce prostaglandins to coordinate prolonged physiological events such as inflammation and ovulation. The PGHS-2 pathway produces prostaglandins that are directed to the nucleus that may regulate mitogenesis, or other nuclear events.

F. Future Work

In terms of regulation of prostanoid formation, there are two areas in which we know very little about the PGHS-1 and PGHS-2 isozymes. The first involves

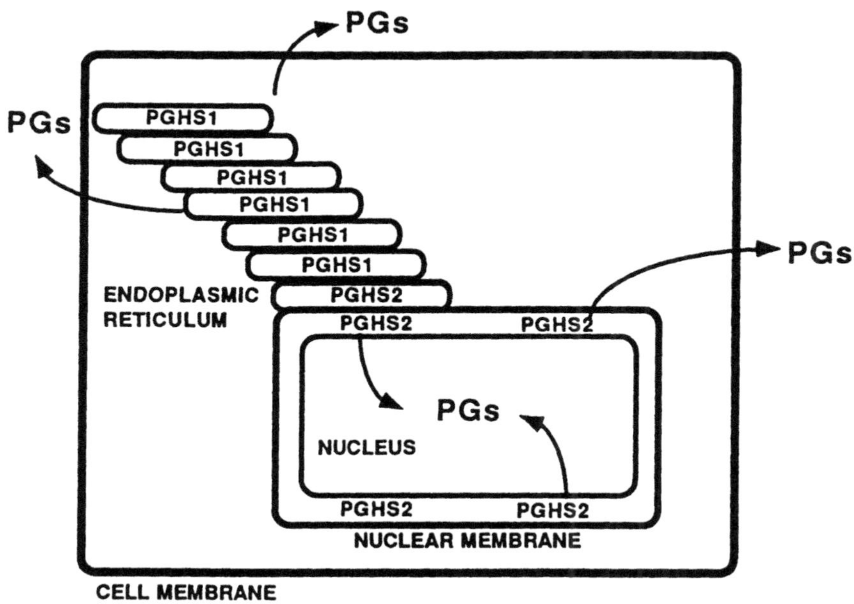

Figure 9. Diagram of the proposed scheme for prostaglandin synthesis by PGHS-1 and PGHS-2. PGHS-1 has been localized to the endoplasmic reticulum and produces prostaglandins that are released into the cytoplasm from where they likely exit to signal extracellularly. PGHS-2 has been localized to the nuclear membrane and produces prostaglandins that are released both into the nucleus and into the cytoplasm. While prostaglandins formed by PGHS-2 likely also signal through extracellular receptors, they may additionally signal in the nucleus, by a second, as yet undetermined mechanism.

regulation of expression of the two isozymes: (a) How is PGHS-1 regulated developmentally, such that selected cells express high levels of PGHS-1? (b) How is PGHS-2 regulated acutely? and (c) What are the important transcriptional regulatory elements for PGHS-2 (and PGHS-1) expression? The second area which now requires further investigation is to differentiate between the functions of PGs produced via the PGHS-1 and PGHS-2 systems. The development of isozyme-specific inhibitors and mouse knockout models for PGHS- 1 and PGHS-2 in the next few years should help answer these important questions. What function do PGHS-1 and PGHS-2 play in inflammation and cell growth? Aspirin use has been associated with a reduced risk of the development of some cancers (Thun et al., 1991). Are PGHS-1 and/or PGHS-2 involved in the progression of cancer? The most exciting hypothesis is that these two isozymes belong to separate prostaglandin-synthesizing pathways, and our biggest challenge will be to understand how these pathways are separated within the cell.

ACKNOWLEDGMENTS

Studies in the authors' laboratories described in this chapter were supported in part by NIH grants DK42509 (WLS), DK22042 (WLS), and GM40713 (DLD).

REFERENCES

Abramovitz, M., Wong, E., Cox, M. E., Richardson, C. D., Li, C., & Vickers, P. J. (1993). 5-lipoxygenase-activating protein stimulates the utilization of arachidonic acid by 5-lipoxygenase. Eur. J. Biochem. 215, 105-111.

Akira, S., & Kishimoto, T. (1992). IL-6 and NF-IL6 in acute-phase response and viral infection. Immunol. Rev. 127, 25-50.

Arita, H., Nakano, T., & Hanasaki, K. (1989). Thromboxane A_2: its generation and role in platelet activation. Prog. Lipid Res. 28, 273-301.

Brannon, T. S., North, A. J., Wells, L. B., & Shaul, P. W. (1994). Prostacyclin synthesis in ovine pulmonary artery is developmentally regulated by changes in cyclooxygenase-1 gene expression. J. Clin. Invest. 93, 2230-2235.

Breuer, D., Laneuville, O., DeWitt, D., & Smith, W. L. (1994). Fatty acid substrate specificities of prostaglandin H synthases-1 and -2. Biochem. Biophys. Res. Commun., in press.

Carty, T. J., Stevens, J. S., Lombardino, J. G., Parry, M. J., & Randall, M. J. (1980). Piroxicam, a structurally novel anti-inflammatory compound. Mode of prostaglandin synthesis inhibition. Prostaglandins 19, 671-682.

Catalano, C. E., Choe, Y. S., & Ortiz-de-Montellano, P. R. (1989). Reactions of the protein radical in peroxide-treated myoglobin. Formation of a heme-protein cross-link. J. Biol. Chem. 264, 10534-10541.

Chen, Y. N., & Marnett, L. J. (1989). Heme prosthetic group required for acetylation of prostaglandin H synthase by aspirin. FASEB J. 3, 2294-2297.

Coleman, R. A., Smith, W. L., & Narumiya, S. (1994). Classification of prostanoid receptors: Properties, distribution and structure of the receptors and their subtypes. Pharm. Rev. 46, 205-229.

Crofford, L. J., Wilder, R. L., Ristimaki, A. P., Sano, H., Remmers, E. F., Epps, H. R., & Hla, T. (1994b). Cyclooxygenase-1 and -2 expression in rheumatoid synovial tissues. Effects of interleukin-1β, phorbol ester, and corticosteroids. J. Clin. Invest. 93, 1095-1101.

DeGray, J. A., Lassmann, G., Curtis, J. F., Kennedy, T. A., Marnett, L. J., Eling, T. E., & Mason, R. P. (1992). Spectral analysis of the protein-derived tyrosyl radicals from prostaglandin H synthase. J. Biol. Chem. 267, 23583-23588.

DeWitt, D. L., & Meade, E. A. (1993). Serum and glucocorticoid regulation of gene transcription and expression of the prostaglandin H synthase-1 and prostaglandin H synthase-2 isozymes. Arch. Biochem. Biophys. 306, 94-102.

DeWitt, D. L., & Smith, W. L. (1988). Primary structure of prostaglandin G/H synthase from sheep vesicular gland determined from the complementary DNA sequence. Proc. Natl. Acad. Sci. USA 85, 1212-1416.

DeWitt, D. L., Rollins, T. E., Day, J. S., Gauger, J. A., & Smith, W. L. (1981). Orientation of the active site and antigenic determinants of prostaglandin endoperoxide synthase in the endoplasmic reticulum. J. Biol. Chem. 256, 10375-10382.

DeWitt, D. L., Day, J. S., Sonnenburg, W. K., & Smith, W. L. (1983). Concentrations of prostaglandin endoperoxide synthase and prostaglandin I2 synthase in the endothelium and smooth muscle of bovine aorta. J. Clin. Invest. 72, 1882-1888.

DeWitt, D. L., El-Harith, E. A., Kraemer, S. A., Andrews, M. J., Yao, E. F., Armstrong, R. L., & Smith, W. L. (1990). The aspirin and heme-binding sites of ovine and murine prostaglandin endoperoxide synthases. J. Biol. Chem. 265, 5192-5198.

DeWitt, D. L., Kraemer, S. A., & Meade, E. A. (1991). Serum induction and superinduction of PGH synthase mRNA levels in 3T3 fibroblasts. In Prostaglandins and Related Compounds II, Vol. 21, B. (Samuelsson et al., eds.), pp 75-90. Raven Press, Ltd.NY.

Dietz, R., Nastainczyk, W., & Ruf, H. H. (1988). Higher oxidation states of prostaglandin H synthase. Rapid electronic spectroscopy detected two spectral intermediates during the peroxidase reaction with prostaglandin G2. Eur. J. Biochem. 171, 321-328.

Dolphin, D., & Felton, R. H. (1974). The biochemical significance of porphyrin π cation radicals. Acc. Chem. Res. 7, 26-32.

Domin, J., & Rozengurt, E. (1993). Platelet-derived growth factor stimulates a biphasic mobilization of arachidonic acid in Swiss 3T3 cells. J. Biol. Chem. 268, 8927-8934.

Dunford, H. B., & Stillman, J. S. (1976). On the function and mechanism of action of peroxidases. Coord. Chem. Rev. 19, 187-193.

Evett, G. E., Xie, W., Chipman, J. G., Robertson, D. L., & Simmons, D. L. (1993). Prostaglandin G/H isoenzyme 2 expression in fibroblasts: regulation by dexamethasone, mitogens, and oncogenes. Arch. Biochem. Biophys. 306, 169-177.

Feng, L., Sun, W., Xia, Y., Tang, W. W., Chanmugam, P., Soyoola, E., Wilson, C. B., & Hwang, D. (1993). Cloning two isoforms of rat cyclooxygenase: Differential regulation of their expression. Arch. Biochem. Biophys. 307, 361-368.

Fletcher, B. S., Kujubu, D. A., Perrin, D. M., & Herschman, H. R. (1992). Structure of the mitogen-inducible TIS10 gene and demonstration that the TIS10-encoded protein is a functional prostaglandin G/H synthase. J. Biol. Chem. 267, 4338-4344.

Flower, R. J. (1974). Drugs which inhibit prostaglandin biosynthesis. Pharmacol. Rev. 26, 33-67.

Funk, C. D., Funk, L. B., Kennedy, M. E., Pong, A. S., & Fitzgerald, G. A. (1991). Human platelet/erythroleukemia cell prostaglandin G/H synthase: cDNA cloning, expression, and gene chromosomal assignment. FASEB J. 5, 2304-2312.

Futaki, N., Yoshikawa, K., Hamasaka, Y., Arai, I., Higuchi, S., Iizuka, H., & Otomo, S. (1993). NS-398, a novel non-steroidal anti-inflammatory drug with potent analgesic and antipyretic effects, which causes minimal stomach lesions. Gen. Pharmacol. 24, 105-110.

Futaki, N., Takahishi, S., Yokoyama, M., Arai, I., Higuchi, S., & Otomo, S. (1994). NS398, a new anti-inflammatory agent, selectively inhibits a novel prostaglandin G/H synthase/cyclooxygenase (COX-2) activity *in vitro*. Prostaglandins 47, 55-59.

Glasgow, W. C., & Eling, T. E. (1994). Structure-activity relationship for potentiation of EGF-dependent mitogenesis by oxygenated metabolites of linoleic acid. Arch. Biochem. Biophys. 311, 286-292.

Goerig, M., Habenicht, A. J. R., Heitz, R., Zeh, W., Katus, H., Kommerell, B., Ziegler, R., & Glomset, J. A. (1987). sn-1,2-Diacyglycerols and phorbol diesters stimulate thromboxane synthesis by *de novo* synthesis of prostaglandin H synthase in human promyelocytic leukemia cells. J. Clin. Invest. 79, 903-907.

Grove, M., & Plumb, M. (1993). C/EBP, NF-kappa B, and c-Ets family members and transcriptional regulation of the cell-specific and inducible macrophage inflammatory protein 1α immediate-early gene. Mol. Cell. Biol. 13, 5276-5289.

Habenicht, A. J. R., Glomset, J. A., King, W. C., Nist, C., Mitchell, C. D., & Ross, R. (1981). Early changes in phosphatidylinositol and arachidonic acid metabolism in quiescent swiss 3T3 cells stimulated to divide by platelet-derived growth factor. J. Biol. Chem. 256, 12329-12335.

Habenicht, A. J. R., Goerig, M., Grulich, J., Rothe, D., Gronwald, R., Loth, U., Schettler, G., Kommerell, B., & Ross, R. (1985). Human platelet-derived growth factor stimulates prostaglandin synthesis by activation and by rapid *de novo* synthesis of cyclooxygenase. J. Clin. Invest. 75, 1381-1387.

Hamasaki, Y., Kitzler, J., Hardman, R., Nettesheim, P., & Eling, T. E. (1993). Phorbol ester and epidermal growth factor enhance the expression of two inducible prostaglandin H synthase genes in rat tracheal epithelial cells. Arch. Biochem. Biophys. 304, 226-234.

Hamberg, M., & Samuelsson, B. (1967). On the mechanism of the biosynthesis of prostaglandins E_1 and F_α. J. Biol. Chem. 242, 5336-5343.

Han, J., Sadowski, H., Young, D. A., & Macara, I. G. (1990). Persistent induction of cyclooxygenase in p60v-*src*-transformed 3T3 fibroblasts. Proc. Natl. Acad. Sci. USA 87, 3373-3377.

Hedin, L., Gaddy-Kurten, D., Kurten, R., DeWitt, D. L., Smith, W. L., & Richards, J. S. (1987). Prostaglandin endoperoxide synthetase in rat ovarian follicles: content, cellular distribution and evidence for hormonal induction preceding ovulation. Endocrinology 121, 722-731.

Hemler, M. E., & Lands, W. E. M. (1980). Evidence for a peroxide-initiated free radical mechanism of prostaglandin biosynthesis. J. Biol. Chem. 255, 6253-6261.

Hemler, M., Lands, W. E. M., & Smith, W. L. (1976). Purification of the cyclooxygenase that forms prostaglandins. Demonstration of two forms of iron in the holoenzyme. J. Biol. Chem. 251, 5575-5581.

Hemler, M. E., Crawford, C. G., & Lands, W. E. M. (1978a). Lipoxygenation activity of purified prostaglandin-forming cyclooxygenase. Biochemistry 17, 1772-1779.

Hemler, M. E., Graff, G., & Lands, W. E. M. (1978b). Accelerative autoactivation of prostaglandin biosynthesis by PGG_2. Biochem. Biophys. Res. Commun. 85, 1325-1331.

Hennekens, C. H., Buring, J. E., Sandercock, P., Collins, R., & Peto, R. (1989). Aspirin and other antiplatelet agents in the secondary and primary prevention of cardiovascular disease. Circulation. 80, 749-756.

Hensel, G., Meichle, A., Pfizenmaier, K., & Kronke, M. (1989). PMA-responsive 5′ flanking sequences of the human TNF gene. Lymphokine. Res. 8, 347-351.

Herschman, H. R. (1991). Primary response genes induced by growth factors and tumor promoters. In: Annual Review of Biochemistry, Vol. 60 (Richardson, C. C., Abelson, J. N., Meister, A., & Walsh, C. T. W. eds.), pp. 281-319. Annual Reviews Inc.,Palo Alto, CA.

Herschman, H. R., Fletcher, B. S., & Kujubu, D. A. (1993). TIS10, a mitogen-inducible glucocorticoid-inhibited gene that encodes a second prostaglandin synthase/cyclooxygenase enzyme. J. Lipid Mediators 6, 89-99.

Hla, T. personal communication.

Hla, T., & Neilson, K. (1992). Human cyclooxygenase-2 cDNA. Proc. Natl. Acad. Sci. USA 89, 7384-7388.

Hoff, T., DeWitt, D., Kaever, V., Resch, K., & Goppelt-Struebe, M. (1993). Differentiation-associated expression of prostaglandin G/H synthase in monocytic cells. FEBS Lett. 320, 38-42.

Holtzman, M. J., Turk, J., & Shornick, L. P. (1992). Identification of a pharmacologically distinct prostaglandin H synthase in cultured epithelial cells. J. Biol. Chem. 267, 21438-21445.

Hsuanyu, Y., & Dunford, H. B. (1990). Reactions of prostaglandin H synthase in the presence of the stabilizing agents diethyldithiocarbamate and glycerol. Biochem. Cell. Biol. 68, 965-972.

Hulkower, K. I., Wertheimer, S. J., Levin, W., Coffey, J. W., Anderson, C. M., Chen, T., DeWitt, D., Crowl, M., Hope, W. C., & Morgan, D. W. (1994). Interleukin-1β induces cytosolic phospholipase A_2 and prostaglandin H synthase in rheumatoid synovial fibroblasts: Evidence for their roles in the production of prostaglandin E_2. Arthritis Rheum. 37, 653-661.

Huslig, R. L., Fogwell, R. L., & Smith, W. L. (1979). The prostaglandin forming cyclooxygenase of ovine uterus: relationship to luteal function. Biol. Reprod. 21, 589-600.

IP0,5Isshiki, H., Akira, S., Tanabe, O., Nakajima, T., Shimamoto, T., Hirano, T., & Kishimoto, T. (1990). Constitutive and interleukin-1 (IL-1)-inducible factors interact with the IL-1-responsive element in the IL-6 gene. Mol. Cell. Biol. 10, 2757-2764.

Jimenez-de-Asua, L., Richmond, K. M., & Otto, A. M. (1981). Two growth factors and two hormones regulate initiation of DNA synthesis in cultured mouse cells through different pathways of events. Proc. Natl. Acad. Sci. USA 78, 1004-1008.

Jimenez-de-Asua, L., Otto, A. M., Lindgren, J. A., & Hammarstrom, S. (1983). The stimulation of the initiation of DNA synthesis and cell division in Swiss mouse 3T3 cells by prostaglandin $F_{2\alpha}$ requires specific functional groups in the molecule. J. Biol. Chem. 258, 8774-8780.

Jones, D. A., Carlton, D. P., McIntyre, T. M., Zimmerman, G. A., & Prescott, S. M. (1993). Molecular cloning of human prostaglandin endoperoxide synthase type II and demonstration of expression in response to cytokines. J. Biol. Chem. 268, 9049-9054.

Kadonaga, J. T., Jones, K. A., & Tijan, R. (1986). Promoter-specific activation of RNA polymerase II transcription by Sp1. TIBS 11, 20-23.

Kargman, S., Chan, S., Evans, J., Vickers, P., & O'Neill, G. (1994). Tissue distribution of prostaglandin G/H synthase-1 and -2 (PGHS-1 and PGHS-2) using specific anti-peptide antibodies. J. Cell. Biochem. Supplement 18B, 319 (Abstract O109).

Karthein, R., Dietz, R., Nastainczyk, W., & Ruf, H. H. (1988). Higher oxidation states of prostaglandin H synthase. EPR study of a transient tyrosyl radical in the enzyme during the peroxidase reaction. Eur. J. Biochem. 171, 313-320.

Kosaka, T., Miyata, A., Ihara, H., Hara, S., Takeda, O., Takahashi, E., & Tanabe, T. (1994). Characterization of the human gene (PTGS2) coding for prostaglandin endoperoxide synthase-2. Eur. J. Biochem., in press.

Kraemer, S. A. (1994). The prostaglandin synthase-1 gene: Regulation of the prostaglandin synthase-2 gene by dioxin [Ph.D. Dissertation]. Michigan State University, East Lansing, MI.

Kraemer, S. A., Meade, S. A., & DeWitt, D. L. (1992). Prostaglandin endoperoxide synthase gene structure: Identification of the transcriptional start site and 5′-flanking regulatory sequences. Arch. Biochem. Biophys. 293, 391-400.

Kudo, I., Murakami, M., Hara, S., & Inoue, K. (1993). Mammalian non-pancreatic phospholipases A_2. Biochim. Biophys. Acta 1170, 217-231.

Kujubu, D. A., & Herschman, H. R. (1992). Dexamethasone inhibits mitogen induction of the TIS10 prostaglandin synthase/cyclooxygenase gene. J. Biol. Chem. 267, 7991-7994.

Kujubu, D. A., Fletcher, B. S., Varnum, B. C., Lim, R. W., & Herschman, H. R. (1991). TIS10, a phorbol ester tumor promoter inducible mRNA from Swiss 3T3 cells, encodes a novel prostaglandin synthase/cyclooxygenase homologue. J. Biol. Chem. 266, 12866-12872.

Kujubu, D. A., Reddy, S. T., Fletcher, B. S., & Herschman, H. R. (1993). Expression of the protein product of the prostaglandin synthase-2/TIS10 gene in mitogen-stimulated Swiss 3T3 cells. J. Biol. Chem. 268, 5425-5430.

Kulmacz, R. J. (1987). Prostaglandin G_2 levels during reaction of prostaglandin H synthase with arachidonic acid. Prostaglandins 34, 225-240.

Kulmacz, R. J. (1989). Topography of prostaglandin H synthase. Antiinflammatory agents and the protease-sensitive arginine 253 region. J. Biol. Chem. 264, 14136-14144.

Kulmacz, R. J., & Lands, W. E. (1984). Prostaglandin H synthase. stoichiometry of heme cofactor. J. Biol. Chem. 259, 6358-6363.

Kulmacz, R. J., & Lands, W. E. M. (1985). Stoichiometry and kinetics of the interaction of prostaglandin H synthase with anti-inflammatory agents. J. Biol. Chem. 260, 12572-12578.

Kulmacz, R. J., Tsai, A. L., & Palmer, G. (1987). Heme spin states and peroxide-induced radical species in prostaglandin H synthase. J. Biol. Chem. 262, 10524-10531.

Kulmacz, R. J., Ren, Y., Tsai, A. L., & Palmer, G. (1990). Prostaglandin H synthase: spectroscopic studies of the interaction with hydroperoxides and with indomethacin. Biochemistry 29, 8760-8771.

Kwok, P.-Y., Muellner, F. W., & Fried, J. (1987). Enzymatic conversion of 10,10-difluoroarachidonic acid with PGH synthase and soybean lipoxygenase. J. Am. Chem. Soc. 109, 3692-3698.

Lambeir, A. M., Markey, C. M., Dunford, H. B., & Marnett, L. J. (1985). Spectral properties of the higher oxidation states of prostaglandin H synthase. J. Biol. Chem. 260, 14894-14896.

Lands, W. E. M., LeTellier, P. R., Rome, L. H., & Vanderhoek, J. Y. (1973). Inhibition of prostaglandin biosynthesis. Adv. Biosci. 9, 15-28.

Laneuville, O., Breuer, D. K., DeWitt, D. L., Hla, T., Funk, C. D., & Smith, W. L. (1994). Differential inhibition of human prostaglandin endoperoxide H synthases-1 and -2 by nonsteroidal anti-inflammatory drugs. J. Pharm. Exp. Therap. in press.

Lassmann, G., Odenwaller, R., Curtis, J. F., DeGray, J. A., Mason, R. P., Marnett, L. J., & Eling, T. E. (1991). Electron spin resonance investigation of tyrosyl radicals of prostaglandin H synthase. Relation to enzyme catalysis. J. Biol. Chem. 266, 20045-20055.

Lecomte, M., Laneuville, O., Chuan, J., DeWitt, D. L., & Smith, W. L. (1994). Acetylation of human prostaglandin endoperoxide synthase-2 (cyclooxygenase-2) by aspirin. J. Biol. Chem. 269, 13207-13215.

Lee, S. H., Soyoola, E., Chanmugam, P., Hart, S., Sun, W., Zhong, H., Liou, S., Simmons, D., & Hwang, D. (1992). Selective expression of mitogen-inducible cyclooxygenase in macrophages stimulated with lipopolysaccharide. J. Biol. Chem. 267, 25934-25938.

Lin, A. H., Bienkowski, M. J., & Gorman, R. R. (1989). Regulation of prostaglandin H synthase mRNA levels and prostaglandin biosynthesis by platelet-derived growth factor. J. Biol. Chem. 264, 17379-17383.

Macleod, K., Leprince, D., & Stehelin, D. (1992). The Ets gene family. TIBS 17, 251-256.

Maier, J. A. M., Hla, T., & Maciag, T. (1990). Cyclooxygenase is an immediate-early gene induced by interleukin-1 in human endothelial cells. J. Biol. Chem. 265, 10805-10808.

Mancini, J. A., Abramovitz, M., Cox, M. E., Wong, E., Charleson, S., Perrier, H., Wang, Z., Prasit, P., & Vickers, P. J. (1993). 5-lipoxygenase-activating protein is an arachidonate binding protein. FEBS Lett. 318, 277-281.

Marshall, P. J., & Kulmacz, R. J. (1988). Prostaglandin H synthase: distinct binding sites for cyclooxygenase and peroxidase substrates. Arch. Biochem. Biophys. 266, 162-170.

Marshall, P. J., Kulmacz, R. J., & Lands, W. E. M. (1987). Constraints on prostaglandin biosynthesis in tissues. J. Biol. Chem. 262, 3510-3517.

Mason, R. P., Kalyanaraman, B., Tainer, B. E., & Eling, T. E. (1980). A carbon-centered free radical intermediate in the prostaglandin synthetase oxidation of arachidonic acid. Spin trapping and oxygen uptake studies. J. Biol. Chem. 255, 5019-5022.

Meade, E. A. (1992). Kinetic and pharmacological characterization of murine PGH synthase-1 and PGH synthase-2 [Ph.D. Dissertation]. Michigan State University, East Lansing, MI.

Meade, E. A., Smith, W. L., & DeWitt, D. L. (1993). Differential inhibition of prostaglandin endoperoxide synthase (cyclooxygenase) isozymes by aspirin and other non-steroidal anti-inflammatory drugs. J. Biol. Chem. 268, 6610-6614.

Merlie, J. P., Fagan, D., Mudd, J., & Needleman, P. (1988). Isolation and characterization of the complementary DNA for sheep seminal vesicle prostaglandin endoperoxide synthase (cyclooxygenase). J. Biol. Chem. 263, 3550-3553.

Miyamoto, T., Ogino, N., Yamamoto, S., & Hayaishi, O. (1976). Purification of prostaglandin endoperoxide synthetase from bovine vesicular gland microsomes. J. Biol. Chem. 251, 2629-2636.

Mizuno, K., Yamamoto, S., & Lands, W. E. M. (1982). Effects of non-steroidal anti-inflammatory drugs on fatty acid cyclooxygenase and prostaglandin hydroperoxidase activities. Prostaglandins 23, 743-757.

Morita, I., DeWitt, D. L., Schindler, M. L., & Smith, W. (in preparation). Differential localization of prostaglandin endoperoxide H synthase-1 and -2 proteins and activities by using fluorescence confocal image analysis in murine NIH-3T3 cells.

Muller, J. M., Ziegler-Heitbrock, H. W., & Baeuerle, P. A. (1993). Nuclear factor kappa B, a mediator of lipopolysaccharide effects. Immunobiology. 187, 233-256.

Mutsaers, J. H., van-Halbeek, H., Kamerling, J. P., & Vliegenthart, J. F. (1985). Determination of the structure of the carbohydrate chains of prostaglandin endoperoxide synthase from sheep. Eur. J. Biochem. 147, 569-574.

Namba, T., Sugimoto, Y., Negishi, M., Irie, A., Ushikubi, F., Kakizuka, A., Ito, S., Ickikawa, A., & Narumiya, S. (1993). Alternative splicing of C-terminal tail of prostaglandin E receptor subtype EP3 determines G-protein specificity. Nature 365, 166-170.

Nastainczyk, W., Schuhn, D., & Ullrich, V. (1984). Spectral intermediates of prostaglandin hydroperoxidase. Eur. J. Biochem. 144, 381-385.

O'Banion, M. K., Sadowski, H. B., Winn, V., & Young, D. A. (1991). A serum- and glucocorticoid-regulated 4 kb mRNA encodes a cyclooxygenase-related protein. J. Biol. Chem. 266, 23261-23267.

O'Banion, M. K., Winn, V. D., & Young, D. A. (1992). cDNA Cloning and functional activity of a glucocorticoid-regulated inflammatory cyclooxygenase. Proc. Natl. Acad. Sci. USA 89, 4888-4892.

O'Neill, G. P., & Ford-Hutchinson, A. W. (1993). Expression of mRNA for cyclooxygenase-1 and cyclooxygenase-2 in human tissues. FEBS Lett. 330, 156-160.

O'Sullivan, M. G., Chilton, F. H., Huggins Jr., E. M., & McCall, C. E. (1992). Lipopolysaccharide priming of alveolar macrophages for enhanced synthesis of prostanoids involves induction of a novel prostaglandin H synthase. J. Biol. Chem. 267, 14547-14550.

O'Sullivan, M. G., Huggins Jr., E. M., Meade, E. A., DeWitt, D. L., & McCall, C. E. (1992). Lipopolysaccharide induces prostaglandin H synthase-2 in alveolar macrophages. Biochem. Biophys. Res. Commun. 187, 1123-1127.

Otto, A. M., Nilsen-Hamilton, M., Boss, B. D., Ulrich, M. O., & Jimenez-De-Asua, L. (1982). Prostaglandins E_1 and E_2 interact with prostaglandin $F_{2\alpha}$ to regulate initiation of DNA replication and cell division in swiss 3T3 cells. Proc. Natl. Acad. Sci. USA. 79, 4992-4996.

Otto, J. C., DeWitt, D. L., & Smith, W. L. (1993). N-glycosylation of prostaglandin endoperoxide synthases-1 and -2 and their orientations in the endoplasmic reticulum. J. Biol. Chem. 268, 18234-18242.

Pagels, W. R., Sachs, R. J., Marnett, L. J., Dewitt, D. L., Day, J. S., & Smith, W. L. (1983). Immunochemical evidence for the involvement of prostaglandin H synthase in hydroperoxide-dependent oxidations by ram seminal vesicle microsomes. J. Biol. Chem. 258, 6517-6523.

Patrono, C., Ciabattoni, G., & Davi, G. (1990). Thromboxane biosynthesis in cardiovascular diseases. Stroke 21(12 Suppl), IV130-IV133.

Phipps, R. P., Stein, S. H., & Roper, R. L. (1991). A new view of prostaglandin E regulation of the immune response. Immunol. Today 12, 349-352.

Picot, D., Loll, P. J., & Garavito, M. (1994). The X-ray crystal structure of the membrane protein prostaglandin H_2 synthase-1. Nature 367, 243-249.

Pilbeam, C. C., Kawaguchi, H., Hakeda, Y., Vosnesensky, O., Alander, C. B., & Raisz, L. G. (1993). Differential regulation of inducible and constitutive prostaglandin endoperoxide synthase in osteoblastic MC3T3-E1 cells. J. Biol. Chem. 268, 25643-25649.

Pugh, B. F., & Tijan, R. (1990). Mechanism of transcriptional activation by Sp1: Evidence for coactivators. Cell 61, 1187-1197.

Raz, A., Wyche, A., Siegel, N., & Needleman, P. (1988). Regulation of fibroblast cyclooxygenase synthesis by interleukin-1. J. Biol. Chem. 263, 3022-3028.

Raz, A., Wyche, A., Fagan, D., & Needleman, P. (1989). The cell biology of Fibroblast Cyclooxygenase. In: Advances in Experimental Medicine and Biology: Renal Eicosanoids, Vol. 259 (Dunn, M. J., Patrono, C., & Cinotti, G. A., eds.), pp. 1-21. Plenum Press, NY.

Reddy, S. T., & Herschman, H. R. (1994). Ligand-induced prostaglandin synthesis requires expression of the TIS10/PGS-2 prostaglandin synthase gene in murine fibroblasts and macrophages. J. Biol. Chem. 269, 15473-15480.

Regier, M. K., DeWitt, D. L., Schindler, M. S., & Smith, W. L. (1993). Subcellular localization of prostaglandin endoperoxide synthase-2 in murine 3T3 cells. Arch. Biochem. Biophys. 301, 439-444.

Riese, J., Hoff, T., Nordhoff, A., DeWitt, D. L., Resch, K., & Kaever, V. (1994). Transient expression of prostaglandin endoperoxide synthase-2 during mouse macrophage activation. J. Leukocyte Biol., in press.

Ristimaki, A., Garfinkel, S., Wessendorf, J., Maciag, T., & Hla, T. (1994). Induction of cyclooxygenase-2 by interleukin-$_{1\alpha}$: Evidence for post-transcriptional regulation. J. Biol. Chem. 269, 11769-11777.

Rome, L. H., & Lands, W. E. M. (1975). Structural requirements for time-dependent inhibition of prostaglandin biosynthesis by anti-inflammatory drugs. Proc. Natl. Acad. Sci. USA 72, 4863-4865.

Roth, G. J., Siok, C. J., & Ozols, J. (1980). Structural characteristics of prostaglandin synthetase from sheep vesicular gland. J. Biol. Chem. 255, 1301-1304.

Roth, G. J., Machuga, E. T., & Strittmatter, P. (1981). The heme-binding properties of prostaglandin synthetase from sheep vesicular gland. J. Biol. Chem. 256, 10018-10022.

Roth, G. J., Machuga, E. T., & Ozols, J. (1983). Isolation and covalent structure of the aspirin-modified, active-site region of prostaglandin synthetase. Biochemistry 22, 4672-4675.

Ruf, H. H., Schuhn, D., & Nastainczyk, W. (1984). EPR titration of ovine prostaglandin H synthase with hemin. FEBS Lett. 165, 293-296.

Sano, H., Hla, T., Maier, J. A. M., Crofford, L. J., Case, J. P., Maciag, T., & Wilder, R. L. (1992). *In vivo* cyclooxygenase expression in synovial tissue of patients with rheumatoid arthritis and osteoarthritis and rats with adjuvant and streptococcal cell wall arthritis. J. Clin. Invest. 89, 97-108.

Schreiber, J., Eling, T. E., & Mason, R. P. (1986). The oxidation of arachidonic acid by the cyclooxygenase activity of purified prostaglandin H synthase: spin trapping of a carbon-centered free radical intermediate. Arch. Biochem. Biophys. 249, 126-136.

Shimokawa, T., & Smith, W. L. (1991). Essential histidines of prostaglandin endoperoxide synthase. J. Biol. Chem. 266, 6168-6173.

Shimokawa, T., & Smith, W. L. (1992). Prostaglandin endoperoxide synthase: The aspirin acetylation region. J. Biol. Chem. 267, 12387-12392.

Shimokawa, T., Kulmacz, R. J., DeWitt, D. L., & Smith, W. L. (1990). Tyrosine 385 of prostaglandin endoperoxide synthase is required for cyclooxygenase catalysis. J. Biol. Chem. 265, 20073-20076.

Simmons, D. L., Levy, D. B., Yannoni, Y., & Erikson, R. L. (1989). Identification of a phorbol ester-repressible v-*src*-inducible gene. Proc. Natl. Acad. Sci. USA 86, 1178-1182.

Simmons, D. L., Xie, W., Chipman, J. G., & Evett, G. E. (1991). Multiple cyclooxygenases: cloning of a mitogen-inducible form. In: Prostaglandins, Leukotrienes, Lipoxins, and PAF (Bailey, J. M., ed.), pp. 67-78. Plenum, NY.

Sirois, J., & Richards, J. S. (1992). Purification and characterization of a novel, distinct isoform of prostaglandin endoperoxide synthase induced by human chorionic gonadotropin in granulosa cells of rat preovulatory follicles. J. Biol. Chem. 267, 6382-6388.

Sirois, J., & Richards, J. S. (1993). Transcriptional regulation of the rat prostaglandin endoperoxide synthase 2 gene in granulosa cells. J. Biol. Chem. 268, 21931-21938.

Sirois, J., Simmons, D. L., & Richards, J. (1992). Hormonal regulation of messenger ribonucleic acid encoding a novel isoform of prostaglandin endoperoxide H synthase in rat preovulatory follicles. J. Biol. Chem. 267, 11586-11592.

Sirois, J., Levy, L. O., Simmons, D. L., & Richards, J. S. (1993). Characterization and hormonal regulation of the promoter of the rat prostaglandin endoperoxide synthase 2 gene in granulosa cells. Identification of functional and protein-binding regions. J. Biol. Chem. 268, 12199-12206.

Smith, C. J., Morrow, J. D., Roberts, L. J., & Marnett, L. J. (1993). Differentiation of monocytoid THP-1 cells with phorbol ester induces expression of prostaglandin endoperoxide synthase-1 (Cox-1). Biochem. Biophys. Res. Commun. 192, 787-793.

Smith, W. L. (1989). The eicosanoids and their biochemical mechanisms of action. Biochem. J. 259, 315-324.

Smith, W. L. (1992). Prostanoid biosynthesis and mechanisms of action. Am. J. Physiol. 263(2 Pt 2), F181-F191.

Smith, W. L., & Lands, W. E. M. (1972). Oxygenation of polyunsaturated fatty acids during prostaglandin biosynthesis by sheep vesicular glands. Biochemistry 11, 3276-3282.

Smith, W. L., & Laneuville, O. (1994). Cyclooxygenase and lipoxygenase pathways of arachidonic acid metabolism. In: Eicosanoids and Their Inhibitors in Cancer Immunology and Immunotherapy (Anderson, K. M., & Harris, J., eds.), pp. xx-xx. CRC Press, Boca Raton, FL.

Smith, W. L., & Marnett, L. J. (1991). Prostaglandin endoperoxide synthase: structure and catalysis. Biochim. Biophys. Acta 1083, 1-17.

Smith, W. L., & Marnett, L. J. (1994). Prostaglandin endoperoxide synthases. In: Metal Ions in Biological Systems, Vol. 30 (Sigel, H., & Sigel, A., eds.), pp. 163-199. Marcel Dekker), NY.

Smith, W. L., DeWitt, D. L., & Allen, M. L. (1983). Bimodal distribution of the prostaglandin I_2 synthase antigen in smooth muscle cells. J. Biol. Chem. 258, 5922-5926.

Smith, W. L., Sonnenburg, W. K., Allen, M. L., Watanabe, T., Zhu, J., & El-Harith, E. A. (1989). The biosynthesis and actions of prostaglandins in the renal collecting tubule and thick ascending limb. In: Renal Eicosanoids (Dunn, M. J., Patrono, C., & Cinotti, G. A., eds.), pp. 131-147.Plenum Press, NY.

Smith, W. L., DeWitt, D. L., Kraemer, S. A., Andrews, M. J., Hla, T., Maciag, T., & Shimokawa, T. (1990a). Structure-Function Relationships in sheep, mouse, and human prostaglandin endoperoxide G/H synthases. In: Advances in Prostaglandin, Thromboxane, and Leukotriene Research, Vol. 20 (Samuelsson, B., Dahlen, S. E., Fritsch, J., & Hedqvist, P., eds.), pp. 14-21. Raven Press, Ltd., NY.

Smith, W. L., DeWitt, D. L., Shimokawa, T., Kraemer, S. A., & Meade, E. A. (1990b). Molecular basis for the inhibition of prostanoid biosynthesis by nonsteroidal anti-inflammatory agents. Stroke 21, IV24-IV28.

Smith, W. L., Marnett, L. J., & DeWitt, D. L. (1991). Prostaglandin and thromboxane biosynthesis. Pharmacol. Ther. 49, 153-179.

Thun, M. J., Namboodiri, M. M., & Heath, C. W. (1991). Aspirin use and reduced risk of fatal colon cancer. N. Eng. J. Med. 325, 1593-1596.

Toh, H. (1989). Prostaglandin endoperoxide synthase contains an EGF-like domain. FEBS Lett. 258, 317-319.

Tsai, A. L., Palmer, G., & Kulmacz, R. J. (1992). Prostaglandin H synthase: kinetics of tyrosyl radical formation and of cyclooxygenase catalysis. J. Biol. Chem. 267, 17753-17759.

Tsai, A.-L., Hsi, L. C., Kumacz, R. J., & Smith, W. L. (1994). Characterization of the tyrosyl radicals in ovine prostaglandin H synthase by isotope replacement and site-directed mutagenesis. J. Biol. Chem. 269, 5085-5091.

Vanderhoek, J. Y., & Lands, W. E. M. (1973). Acetylenic inhibitors of sheep vesicular gland oxygenase. Biochim. Biophys. Acta 296, 374-381.

Van der Ouderaa, F. J., Buytenhek, M., Nugteren, D. H., & Van-Dorp, D. A. (1977). Purification and characterization of prostaglandin endoperoxide synthetase from sheep vesicular glands. Biochim. Biophys. Acta 487, 315-331.

Van der Ouderaa, F. J., Buytenhek, M., & Slikkerveer, F. J. (1979). On the hemoprotein character of prostaglandin endoperoxide synthetase. Biochim. Biophys. Acta 572, 29-42.

Van der Ouderaa, F. J., Buytenhek, M., Nugteren, D. H., & Van-Dorp, D. A. (1980). Acetylation of prostaglandin endoperoxide synthetase with acetylsalicylic acid. Eur. J. Biochem. 109, 1-8.

Walenga, R. W., Wall, S. F., Setty, B. N., & Stuart, M. J. (1986). Time-dependent inhibition of platelet cyclooxygenase by indomethacin is slowly reversible. Prostaglandins 31, 625-637.

Wang, L., Hajubeigi, A., Xu, X., Loose-Mitchell, D., & Wu, K. K. (1993). Characterization of the promoter of human prostaglandin H synthase-1 gene. Biochem. Biophys. Res. Commun. 190, 406-411.

Wen, P. Z., Warden, C., Fletcher, B. S., Kujubu, D. A., Herschman, H. R., & Lusis, A. J. (1993). Chromosomal organization of the inducible and constitutive prostaglandin synthase/cyclooxygenase genes in mouse. Genomics. 15, 458-460.

Wong, W. Y. L., DeWitt, D. L., Smith, W. L., & Richards, J. S. (1989). Rapid induction of prostaglandin endoperoxide synthase in rat preovulatory follicles by luteinizing hormone and cAMP is blocked by inhibitors of transcription and translation. Mol. Endocrinol. 3, 1714-1723.

Xie, W., Chipman, J. G., Robertson, D. L., Erikson, R. L., & Simmons, D. L. (1991). Expression of a mitogen-responsive gene encoding prostaglandin synthase is regulated by mRNA splicing. Proc. Natl. Acad. Sci. USA 88, 2692-2696.

Yamagata, K., Andreasson, K. I., Kaufmann, W. E., Barnes, C. A., & Worley, P. F. (1993). Expression of a mitogen-inducible cyclooxygenase in brain neurons: regulation by synaptic activity and glucocorticoids. Neuron. 11, 371-386.

Yokoyama, C., & Tanabe, T. (1989). Cloning of the human gene encoding prostaglandin endoperoxide synthase and primary structure of the enzyme. Biochem. Biophys. Res. Commun. 165, 888-894.

Yokoyama, C., Takai, T., & Tanabe, T. (1988). Primary structure of sheep prostaglandin endoperoxide synthase deduced from cDNA sequence. FEBS Lett. 231, 347-351.

Yonetani, T. (1976). Cytochrome c peroxidase. In: The Enzymes, Vol. 13 (Boyer, P., ed.), pp. 345-362. Academic Press, NY.

Zhang, Y. H., Lin, J. X., & Vilcek, J. (1990). Interleukin-6 induction by tumor necrosis factor and interleukin-1 in human fibroblasts involves activation of a nuclear factor binding to a kappa B-like sequence. Mol. Cell. Biol. 10, 3818-3823.

PLASMALOGENS: THEIR METABOLISM AND CENTRAL ROLE IN THE PRODUCTION OF LIPID MEDIATORS

Fred Snyder, Ten-ching Lee, and Merle L. Blank

Advances in Lipobiology
Volume 2, pages 261-286.

ISBN 0-7623-0205-4

ABSTRACT

Considerable interest in plasmalogens has recently been rekindled because of their important linkage to the production of potent bioactive lipid mediators such as platelet-activating factor (PAF) and the various oxygenated metabolites of arachidonic acid. Moreover, much progress has been made towards attaining a better understanding of the metabolism and functional role of plasmalogens in mammalian cells, including the establishment of the direct metabolic relationship between the ethanolamine- and choline-containing plasmalogens. This chapter reviews the early studies that established the enzymatic reactions leading to the synthesis of the ethanolamine plasmalogens by a Δ1-alkyl desaturase and discusses more recent results that have shown the O-alk-1-enyl linkage in choline plasmalogens can be derived from the ethanolamine plasmalogens by enzymatic reactions involving the remodeling of substituents at both the *sn*-2 and/or *sn*-3 positions. Other topics covered include a critical discussion of criteria and methods required for establishing proof of the plasmalogenic structure isolated from biological materials and an updated view of how plasmalogens participate in the trafficking of arachidonate among membrane phospholipids. A number of studies suggest the ethanolamine plasmalogens serve as a repository for arachidonate and other polyunsaturates in that they appear to be the final destination in the movement of polyenoic acids through the choline and ethanolamine-containing subclasses of glycerophospholipids. Ethanolamine-containing lysoplasmalogens have been shown to be involved in the biosynthesis of PAF by serving as acyl acceptors for the arachidonoyl moiety from alkylarachidonoyl-glycerophosphocholine (a membrane precursor of PAF) in a reaction catalyzed by a CoA-independent transacylase. The lyso-PAF intermediate formed by the transacylase can then be acetylated to form PAF by an acetyl-CoA acetyltransferase. The coupled phospholipase A_2/transacylase pathway also is closely associated with the production of eicosanoid mediators since arachidonic acid and other polyenoic fatty acids are released from the ethanolamine plasmalogens in the initial hydrolytic step that forms the lyso plasmalogen acceptor.

I. INTRODUCTION

Although the specific pathophysiological functions of plasmalogens still remain somewhat aloof, it is apparent that these membrane components are closely associated with the production of lipid mediators such as platelet-activating factor (PAF) and eicosanoid metabolites. In fact, it appears that the ethanolamine plasmalogens occupy an obligatory intermediary position in the metabolic processing and release of arachidonic acid and lyso-PAF, both of which are the immediate precursors of extremely potent bioactive lipids. Furthermore, the ethanolamine-

and choline-containing plasmalogens associated with membranes also serve as precursors of the plasmalogen analogs of PAF.

The biosynthesis and catabolism of plasmalogens and their significance as a component of membrane structure in mammalian cells have previously been reviewed in this series. Certain other aspects related to the dynamics of plasmalogens in cellular membranes (Paltauf chapter) and phospholipid turnover (Schmid chapter) appear in this volume. However, during the past several years, it has become increasingly obvious that the ethanolamine plasmalogens play an important central role as the final destination site in the trafficking of arachidonate among membrane phospholipids (see reviews by MacDonald and Sprecher, 1991; Snyder et al., 1992) and in the biosynthesis of PAF (Snyder, 1994a); both topics wil be addressed in this chapter.

Additional areas that will be discussed concern the criteria required for proof of the chemical structure of native plasmalogens and a brief look at the current concepts regarding the enzymatic synthesis of both the ethanolamine and choline plasmalogens. The need to revisit reliable quantitative analytical approaches for identifying plasmalogenic structures in glycerolipids become extremely important when one considers the pitfalls that can occur, especially in view of their instability and the general necessity of analyzing extremely small quantities of mass or radiolabeled samples (e.g., the plasmalogen analog of PAF) containing the O-alk-1-enyl moiety. New information about the enzymatic pathways for the biosynthesis of choline plasmalogens have also emerged from recent studies that clearly indicate the ethanolamine plasmalogens are their precursor.

II. REQUIREMENTS FOR PROOF OF CHEMICAL STRUCTURE

In biochemical studies involving plasmalogens, it is first necessary to establish presumptive evidence for the presence of the vinyl ether structure and then to provide sufficient chemical evidence of structural proof. Ultimately, quantitative determination of both the amount of plasmalogens present and composition of the vinyl ether chains are also usually required.

Plasmalogen analyses are complicated by the fact that under most chromatographic conditions [open column chromatography, TLC, and HPLC], phospholipids containing vinyl ether groups are not easily separable from phospholipids that contain the same phosphobase group but possess acyl and/or alkyl moieties instead of the plasmalogenic configuration. In fact, researchers often ignore the possible presence of plasmalogen structures in the choline- and ethanolamine glycerophospholipid fractions, which authors sometimes incorrectly refer to as "phosphatidylcholine" and "phosphatidylethanolamine" (i.e., the diacyl forms). However, this practice is not as prevalent in current scientific publications as in the past. Proof of the plasmalogen structure, like other molecular structures, becomes less equivocal when the identification is based on the use of a variety of different chemical/chromatographic procedures. Examples of what we believe to be convincing analytical schemes for identifying radiolabeled ethanolamine plasmalogens are outlined in papers by Snyder et al. (1971) for those with *sn*-2 acyl groups and by Tessner and Wykle (1987) for those containing an *sn*-2 acetyl moiety. This section summarizes these and other methods that have been successfully used to analyze plasmalogens. Although the procedures described have been used primarily for analysis of plasmenylethanolamine, most can also be applied to the generally less abundant plasmenylcholines as well as other glycerolipids containing O-alk-1-enyl moieties.

A. Treatment of Plasmalogens with Acids and/or Bases

Mild alkaline hydrolysis of diradylglycerophosphatides (Dawson et al., 1972; Horrocks and Ansell, 1967) removes acyl groups but leaves the alkyl and alk-1-enyl ether moieties intact. Products from the base catalyzed hydrolysis are soaps (or fatty acids if acid neutralization is used), glycerophosphobases, and alk-1-enyl(and alkyl)lysoglycerophosphatides. Quantitative deacylation of phospholipids can also be accomplished with an alcohol-water solution of monomethylamine (Clarke and Dawson, 1981) and has the advantage of being able to recover the products by simple evaporation of the solvent. Products of alkaline hydrolysis are usually separated by TLC and the phosphorus content of the lysophospholipid bands is used to determine the amount of total ether-containing (alk-1-enyl plus alkyl groups) phosphatides present.

In contrast to the stability of the O-alkyl ether linkage, the vinyl ether bond of plasmalogens is quite susceptible to hydrolysis with mineral acids. In fact, the first detection of plasmalogens in tissues in 1924 was due to a reaction involving the release of aldehydes by acidic treatment of plasmalogens in situ (see review by Debuch and Seng, 1972). Lability of the vinyl ether group to cleavage by mineral acids is the most widely used single initial test for the presence of plasmalogens and forms the basis for several quantitative methods. The acid catalyzed hydrolysis of plasmalogens converts the vinyl ether side chain into fatty aldehydes when conducted in an aqueous environment (Frosolono and Rapport, 1969) or into dimethylacetals if hydrolysis is done in methanol (Morrison and Smith, 1964). A two-dimensional TLC system, or modifications thereof, with hydrolysis of phospholipid plasmalogens by exposure to HCl between the two chromatographic developments was described years ago (Horrocks, 1968) and this technique has been extensively utilized for qualitative and quantitative analysis of plasmalogens in mixtures of phospholipids. A similar approach utilizing another two-dimensional TLC system uses a mercuric chloride spray reagent to cleave the vinyl ether bond of phospholipid plasmalogens (Owens, 1966). The vinyl ether bond in neutral lipids can also be quantitatively hydrolyzed with HCl during a two-dimensional separation-reaction-separation TLC sequence (Schmid and Mangold, 1966). A combination of acid-base treatment, in conjunction with multisolvent development of the products by TLC, has also been described for the quantitatation of plasmalogensinphospholipidmixtures(Dembitsky,1988).

B. Treatment of Plasmalogens with $LiAlH_4$ or $NaAlH_2(OCH_2CH_2OCH_3)_2$ [Vitride]

Both $LiAlH_4$ (Horrocks and Cornwell, 1962) and Vitride (Snyder et al., 1971) reduction have been used to remove the phosphobase and acyl moieties from glycerolipids; the acyl groups are converted to free fatty alcohols in this reaction. Alkyl and alk-1-enyl ether groups remain attached to the glycerol molecule when treated with either reagent; the products being alkylglycerols and alk-1-enylglycerols. In our hands (Snyder et al., 1971), the Vitride reagent provides a somewhat better quantitative yield of glyceryl ethers (both alkyl- and alk-1-enylglycerols) from the intact ether-containing phospholipids than obtained with $LiAlH_4$. However, $LiAlH_4$ is adequate for many applications and, unlike Vitride, it has the advantge of being readily available from numerous commercial sources.

Quantitation of the alk-1-enyl- and alkyl-glycerols has been accomplished by photodensitometry of charred TLC plates that contain the separated components after reduction of the lipid sample with $LiAlH_4$ (Wood and Snyder, 1968). A quantitative spectrophotometric method based on a fuchsin dye reaction with aldehydes produced by acid hydrolysis of the alk-1-enylglycerols from the Vitride reduction has also been described (Blank et al., 1975). Composition of the alk-1-enylglycerol side chain has been determined by GLC of either the aldehydes (Farquhar, 1962; Anderson et al., 1969) or the dimethylacetals (Farquhar, 1962; Morrison and Smith, 1964; Snyder et al., 1971). GLC and mass spectral analyses of bismethyl ether derivatives of alk-1-enylglycerols, produced from the products of $LiAlH_4$ reduction of plasmalogens, have also been used to locate and identify the vinyl ether chains (Knorr and Spiteller, 1990). We have determined the composition of alk-1-enylglycerol side chains by GLC of both aldehydes (Anderson et al., 1969) and dimethylacetals (Snyder et al., 1971). Hydrogenation of alk-1-enylglycerols to form alkylglycerols (Snyder and Blank, 1969) permits a variety of other methods of derivatization (Hanahan, 1972).

C. Treatment of Plasmalogens with Phospholipases A_2 and C

Hydrolysis of plasmalogens by phospholipase A_2 using a method described by Okuyama and Nojima (1965) yields the acid labile 1-alk-1′-enyl-2-lyso-GPE (or GPC) and free fatty acids that formerly occupied the sn-2 position. These products can be analyzed by using techniques previously described for products formed by base-catalyzed hydrolysis of plasmalogens. The advantage of using phospholipase A_2 instead of simple saponification to remove the acyl moeity is that phospholipase A_2 is stereospecific for hydrolysis of *sn*-2 acyl groups of phospholipids and can therefore verify the stereoconfiguration of the plasmalogens.

Phospholipase C from *Bacillus cereus* is capable of hydrolyzing the phosphobase group from phospholipids yielding diradylglycerols (Ansell and Spanner, 1965; Mavis et al., 1972) that can be derivatized into compounds that are easily separated into the diacyl, alkylacyl, and alk-1-enylacyl subclasses by TLC and HPLC. Different derivatives of the diradylglycerols used for these analyses include: acetates (Nakagawa and Horrocks, 1983), dinitrobenzoates (Takamura et al., 1986), anthroyl groups (Takamura and Kito, 1991), and benzoates (Blank et al., 1984, 1990). Except for the acetates, the other derivatives of diradylglycerols can be quantitated by their UV absorbance which is extremely useful during separations by HPLC.

Underivatized alk-1-enylacylglycerols, produced by phospholipase C hydrolysis of glycerophospholipids, have been separated from the other two subclasses (alkylacylglycerols and diacylglycerols) by column chromatography (Curstedt, 1977) or by TLC (Myher and Kuksis, 1984). The distribution of molecular species in the isolated alk-1-enylacylglycerols are then determined by GLC analysis of their trimethylsilyl ether derivatives (Curstedt, 1977; Myher and Kuksis, 1984) or their t-butyldimethylsilyl ether derivatives (Myher and Kuksis, 1984). In one instance (Curstedt, 1977), additional information about the alk-1-enyl chains was obtained by coupling the GLC to a mass spectrometer.

D. Other Techniques Applicable to Analysis of Plasmalogens

A recent review by Blank and Snyder (1994) covers several other methods for plasmalogen analyses, in addition to providing a more detailed description of many of those systems mentioned previously in this section. Renkonen (1968) described derivatization of the diradyl-GPE fraction with diazomethane (phosphate group) and fluorodinitrobenzene (amino group) so that the three subclasses could be separated by multiple solvent developments on TLC plates. This procedure is very time-consuming and the loss of some plasmalogen species probably occurred during the treatment of the lipid sample with diazomethane. Nevertheless, the dinitrobenzene derivative has been extremely valuable for verifying the presence of the ethanolamine group in plasmalogens (Snyder et al., 1971; Tessner and Wykle, 1987).

As mentioned earlier, intact, underivatized plasmalogen phospholipids are normally not well-separated from the other two subclasses (alkylacyl and diacyl) by TLC or normal-phase HPLC. However, a two-step HPLC separation scheme has been devised for resolving intact ethanolamine and choline plasmalogens from their diacyl counterparts (Dugan et al., 1986). Lysoplasmalogens (alk-1-enyllyso-GPE/GPC) can also be separated to some extent from the corresponding monoacyl and monoalkyl subclasses by reversed-phase HPLC (Creer and Gross, 1985).

Acid lability of the vinyl ether linkage in plasmalogens results in the production of aldehydes and 1-lyso-2-acyl-GPE/GPC during normal-phase HPLC of phospholipids using elution solvents containing phosphoric (Chen and Kou, 1982; Hoving et al., 1988) or sulfuric (Yandrasitz et al., 1981) acids. This lability to acid has been used for cleaving the vinyl ether linkage either as a separate step in a two-step HPLC procedure (Murphy et al., 1993) or during the actual HPLC run (Bjerrum et al., 1989) as a method for quantitation of plasmalogens.

Mass spectrometry should provide an excellent tool for plasmalogen analysis since it has been utilized in the identification of several classes of phospholipids (Benfenati and Reginato, 1985; Murphy, 1993). However, the practical use of mass spectral data for measuring intact plasmalogens are still somewhat limited and will require further development before becoming a useful routine tool.

High-resolution ^{31}P nuclear magnetic resonance spectroscopy (Meneses and Glonek, 1988) has been utilized to identify the presence and measure the amount of choline plasmalogens in the surfactant fraction from beef and fetal lamb lungs (Rana et al., 1993). This recently developed methodology could prove valuable especially for the determination of plasmalogens in mixtures of phospholipids.

III. CURRENT STATUS OF BIOSYNTHETIC PATHWAYS

A. Ethanolamine Plasmalogens

Only a brief synopsis is provided about the early investigations that led to the characterization of the Δ 1 alkyl desaturase system responsible for the synthesis of the O-alk-1-enyl linkage in the ethanolamine plasmalogens. The reason is that essentially nothing new has been reported about this desaturation step since its discovery in the early 1970s and the subject has already been extensively reviewed (Snyder et al., 1985). The first cell-free system for the biosynthesis of plasmalogens was described in studies that used [^{14}C]hexadecanol as the precursor and Ehrlich ascites cells or mouse preputial gland tumors as the enzyme source (Wykle et al., 1970; Snyder et al., 1971). Subsequent experiments with the ascites cell system (Blank et al., 1971) demonstrated that 1-[9,10-^{3}H]hexadecyl-2-acyl-*sn*-[U^{14}C]glycero-3-P could be incorporated into the ethanolamine plasmalogens with no change in the ^{3}H/^{14}C ratio. Ultimately, it became clear that 1-alkyl-2-acyl-GPE was the substrate for the Δ 1 alkyl desaturase (Figure 1), a mixed function oxidase, that forms the O-alk-1-enyl linkage at the *sn*-1 position in the ethanolamine plasmalogens; the desaturase requires NADH or NADPH and is cyanide sensitive (Wykle et al., 1972; Paltauf and Holasek, 1973; Wykle and Schremmer Lockmiller, 1975). Cytochrome b_5 was also shown to be a component of this novel desaturase system (Paltauf et al., 1974).

Although 1-alkyl-2-lyso-GPE is a better precursor of the ethanolamine plasmalogens than 1-alkyl-2-acyl-GPE, kinetic data firmly established that the desaturation at the Δ 1 position occurs only after the lysophospholipid is first acylated (Wykle et al., 1972; Paltauf and Holasek, 1973). This acylation step appears to be catalyzed by a CoA-independent transacylase since no CoA or acyl-CoAs are required.

In all respects, except for the unique position for the desaturation of the hydrocarbon chain and the requirement for an intact phospholipid being the substrate, the properties of the Δ 1 alkyl desaturase are similar to those of the more

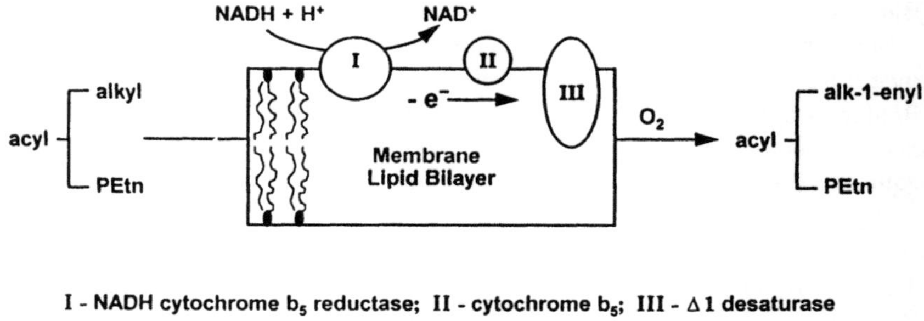

Figure 1. Reaction scheme depicting the biosynthesis of ethanolamine plasmalogens from 1-alkyl-2-acyl-*sn*-glycero-3-phosphoethanolamine by a Δ1 alkyl desaturase. Except for the substrate and specificity of the desaturase, the other components and possible mechanism of electron transfer are thought to be the same as for the Δ9 acyl-CoA desaturase.

extensively studied Δ 9 stearoyl-CoA desaturase system (Oshino et al., 1966; Holloway and Katz, 1972; Oshino and Omura, 1973; Strittmatter et al., 1974). However, unlike the Δ 9 desaturase, the Δ 1 alkyl desaturase complex has not yet been purified into its individual component parts.

B. Choline Plasmalogens

While it has been known for almost twenty years that the alk-1-enyl bond in plasmenylethanolamine originates from plasmanylethanolamine by the action of a microsomal cytochrome b_5-dependent desaturation, the origin of the alk-1-enyl linkage of plasmenylcholine has only recently begun to be understood. The earliest indication which suggests that the Δ 1 alkyl desaturase may be base-specific for ethanolamine came from the observation that 1-O-alkyl-2-acyl-*sn*-glycero-3-phospho-(N-dimethyl) ethanolamine was not a substrate for the Δ 1 alkyl desaturase of the mucosa of hamster small intestine (Paltauf and Holasek, 1973). Similarly, Snyder et al. (1985) showed that no synthesis of choline plasmalogens from alkyllyso-GPC or alkylacyl-GPC occurred in a cell-free system where 50% of the corresponding ethanolamine precursor was converted to ethanolamine plasmalogens. However, the first conclusive evidence to establish that plasmenylcholine is not directly derived from plasmanylcholine or lysoplasmanylcholine in intact cells is based on results obtained from experiments where [1′,2′-^{3}H]hexadecyllyso-[N-methyl-^{14}C]-GPC (^{3}H/^{14}C=2.7) was incubated with neonatal rat myocytes for various times up to 24 hr (Lee et al., 1991). Under these conditions, the ^{3}H/^{14}C ratios remained relatively constant in alkyllyso-GPC and alkylacyl-GPC throughout the incubation times, but decreased in the alk-1-enylacyl-GPC to less than the expected ratio of 1.35, assuming half of the radiolabel is lost from the [1′,2′-^{3}H]alkylacyl-GPC due to desaturation and that there is no isotope effect of the substrate on the desaturase. These results further indicate the incorporation of the [N-methyl-^{14}C]choline moiety of alkyllyso-GPC into alk-1-enylacyl-GPC is considerably faster than that of the [^{3}H]alkyl portion of alkyllyso-GPC. Strum et al. (1992) also showed that [^{3}H]alkyllyso-GPC is not a direct precursor to [^{3}H]alk-1-enylacyl-GPC; they found that [^{3}H]alkyllyso-GPC is converted into [^{3}H]alkyl-linked and [^{3}H]alk-1-enyl-linked ethanolamine-containing phosphoglycerides, but not into the [^{3}H]alk-1-enylacyl-GPC throughout 24 hr incubation with MDCK cells.

The notion that the vinyl ether linkage in the plasmenylcholines originates from the plasmenylethanolamines is based on the results obtained by several laboratories (Lee et al., 1991; Strum and Daniel, 1993; Ford and Gross, 1994). When equal concentrations of [^{3}H]hexadecyllyso-GPC or [^{3}H]hexadecyllyso-GPE were incubated under identical conditions with neonatal rat myocytes for 4, 12, and 24 hr, the rate of plasmenylcholine formation is faster from [^{3}H]hexadecyllyso-GPE than from [^{3}H]hexadecyllyso-GPC, whereas the generation of alk-1-enylacyl-GPC lags behind the production of alk-1-enylacyl-GPE (Lee et al., 1991). Also, both [^{3}H]alkyllyso-GPE and [^{3}H]alk-1-enyllyso-GPE are converted to [^{3}H]alk-1-enylacyl-GPC during 2 hr incubations with MDCK cells (Strum and Daniel, 1993). Furthermore, a precursor-product relationship was demonstrated between plasmenylethanolamines and plasmenylcholines with pulse-chase experiments in rabbit hearts using [1-^{3}H]hexadecanol as the precursor (Ford and Gross, 1994).

Possible metabolic pathways that could lead to the biosynthesis of plasmenylcholines from plasmenylethanolamines are outlined in Figures. 2 and 3. Even though enzymatic methylation of plasmenylethanolamines has been reported in rabbit myocardial membranes (Mogelson and Sobel, 1981) and human platelet lysates (Mozzi et al., 1987), the activity rates were relatively low (3.9 pmol/mg/h) in rabbit myocardial membranes and diacyl-GPE was shown to be the preferred substrate for the methylation rather than alk-1-enylacyl-GPE. In addition, when guinea pig hearts were perfused with radiolabeled ethanolamine for up to 120 min, plasmenylcholine contained only 0.3% of the radioactivity incorporated into plasmenylethanolamine (Wientzek et al., 1987). Furthermore, when [^{14}C]choline or [^{14}C]methionine was incubated with neonatal rat myocytes, the amount of methionine incorporated into alk-1-enylacyl-GPC was only 7-17% that of choline found in the alk-1-enylacyl-GPC (Lee et al., 1991). Also, when MDCK cells were incubated with [^{3}H]alk-1-enyllyso-[^{32}P]GPE for up to 24 hr, only 0.4% of the total cellular [^{32}P] radioactivity was recovered in alk-1-enylacyl-GPC (Strum and Daniel, 1993). Collectively, these results suggest that the reactions catalyzed by N-methyltransferase and the base-exchange enzyme do not contribute significantly to the synthesis of alk-1-enylacyl-GPC from alk-1-enylacyl-GPE.

There are three possible metabolic pathways that could lead to the production of alk-1-enylacyl-Gro from alk-1-enylacyl-GPE which in turn is then converted to alk-1-enylacyl-GPC. These three metabolic pathways involve a plasmenylethano-

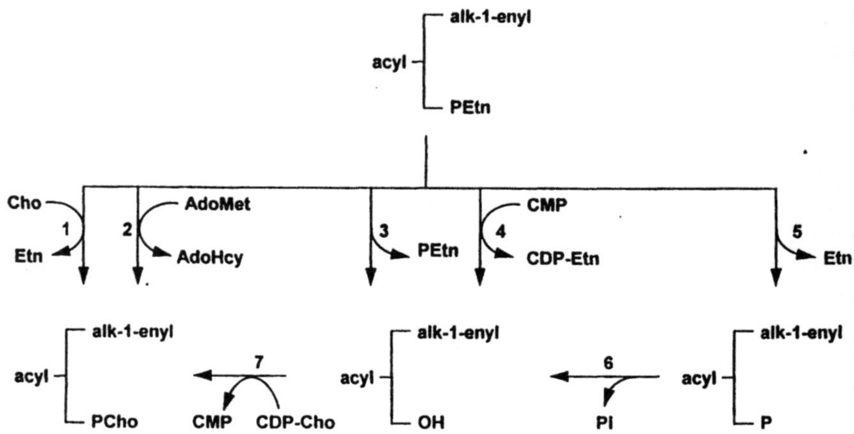

Figure 2. Possible metabolic pathways for the biosynthesis of plasmenylcholines from plasmenylethanolamine which involve direct polar-head group modifications. The enzymes involved include: (1) base-exchange enzyme, (2) N-methyltransferase, (3) phospholipase C, (4) reverse reaction of ethanolaminephosphotransferase, (5) phospholipase D, (6) phosphohydrolase, and (7) cholinephosphotransferase. Abbreviations used are: AdoMet, S-adenosyl-L-methionine; AdoHcy, S-adenosyl-2-homocysteine.

lamine-specific phospholipase C (PLC) (Figure 2, reaction 3), a reverse reaction of ethanolaminephosphotransferase (Figure 2, reaction 4), and a plasmenylethanolamine-specific phospholipase D (PLD)/phosphohydrolase (Figure 2, reactions 5 and 6).

Huang et al. (1992) have identified a PLD activity in the 400 × g pellet of MDCK cells which can hydrolyze ethanolamine-containing phosphoglycerides at 1/10 of the rate over that of choline phosphoglycerides, but the ability of PLD to cleave plasmenylethanolamine in the cell-free system was not assessed in this study. However, the hydrolysis of alk-1-enylacyl-GPE with the liberation of ethanolamine (an indication of PLD activity) was shown to be stimulated by 12-O-tetradecanoylphorbol-13-acetate in NIH 3T3 cells (Kiss and Anderson, 1989) and MDCK cells (Daniel et al., 1993). Additionally, Strum and Daniel (1993) used a phosphono analog of alkyllyso-GPE ([3H]alkyllyso-*sn*-glycero-3-phosphonoethanolamine) incubated with MDCK cells to investigate the contribution of PLD/phosphohydrolase to the synthesis of choline plasmalogen. This phosphono analog was actively metabolized by the MDCK cells and converted to phosphodiester phospholipids in the decreasing order of alkylacyl-GPE, alk-1-enylacyl-GPE, alkylacyl-GPC, diacyl-GPC, and alk-1-enylacyl-GPC. These results suggested PLD/phosphohydrolase is not a required step for the conversion of alk-1-enylacyl-GPE to alk-1-enylacyl-GPC since the phosphono analog is not hydrolyzed by PLD. Nevertheless, without quantitative evaluation of the data, the

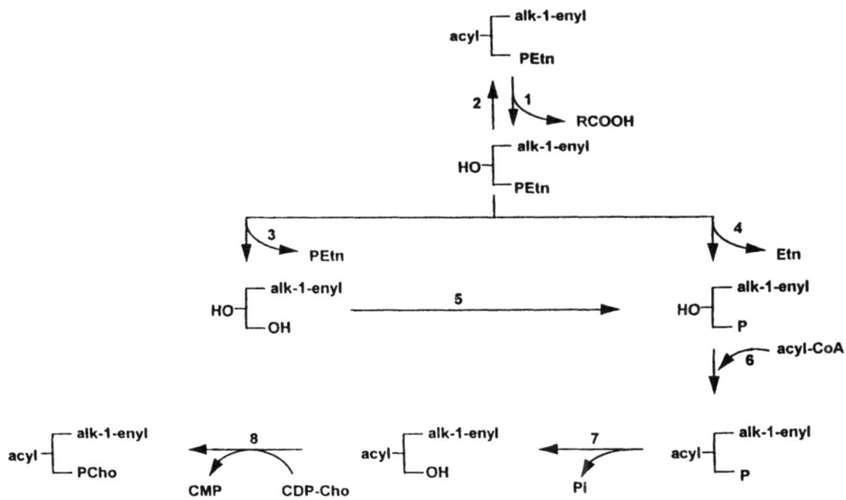

***Figure 3**.* Possible metabolic pathways for the biosynthesis of plasmenylcholines from plasmenylethanolamine which involve the *sn*-2 and polar-head group modifications by the following enzymes: (1) phospholipase A_2, (2) CoA-independent transacylase, (3) lysophospholipase C, (4) lysophospholipase D, (5) phosphotransferase, (6) acyl-CoA acyltransferase, (7) phosphohydrolase, and/or (8) cholinephosphotransferase.

possible participation of PLD/phosphohydrolase in the biosynthesis of plasmenylcholine from plasmenylethanolamine can not be totally ruled out.

The limited data available to suggest the possible involvement of the reverse step catalyzed by ethanolaminephosphotransferase (Figure 2, reaction 4) in the biosynthesis of choline plasmalogen from ethanolamine plasmalogen were obtained from an experiment where microsomes of MDCK cells prelabeled with $[^3H]$alkylacyl-GPE and $[^3H]$alk-1-enylacyl-GPE were incubated with CMP; under these conditions a 10-fold increase in $[^3H]$diradyl-Gro was observed (Strum and Daniel, 1993). However, since the K_m for CMP and the K_i for CDP-ethanolamine in this reaction have not been established, it is difficult to estimate whether the reverse reaction of ethanolaminephosphotransferase would proceed at a sufficiently high enough rate to participate in the biosynthesis of plasmenylcholine under physiological concentrations of CMP and CDP-ethanolamine.

Wolf and Gross (1985) have identified a neutral PLC in the canine myocardium after several steps of partial purification from the cytosolic fraction that can use both phosphatidylcholine and plasmenylcholine as substrates to generate diradyl-Gro and phosphocholine. It is implied the same PLC can also use plasmenylethanolamine as a substrate (Figure 2, reaction 3), but unfortunately, no data were provided to support this conclusion. When a similar method was used to assay PLC activity in the cytosolic and microsomal fractions of MDCK cells, hydrolysis of hexadecyloleoyl-GPC was not detectable (Strum et al., 1992); however, the possible presence of a potent inhibitor for PLC in MDCK cells as in the case for PLC in the cytosol of the myocardium (Wolf and Gross, 1985) was not discussed.

In order to show the validity for reactions 3-5 (Figure 2) participating in the biosynthesis of plasmenylcholine, it is first necessary to demonstrate the existence of endogenous levels of alk-1-enylacyl-Gro and an alk-1-enylacyl-Gro:CDP-choline cholinephosphotransferase activity (Figure 2, reaction 7). The pool size of alk-1-enylacyl-Gro has been reported to be 17 nmol/g in guinea pig hearts (Wientzek et al., 1987) and 0.46 μg/g (≃ 0.7 pmol/g) in rabbit hearts, with only the hexadec-1′-enylacyl-Gro molecular species being detected (Ford and Gross, 1988). Alk-1-enylacyl-Gro:CDP-choline cholinephosphotransferase has been shown to be present in the particulate fractions of rat liver (Kiyasu and Kennedy, 1960), ox heart (Poulos et al., 1968), and guinea pig heart (Wientzek et al., 1987), rabbit platelet membranes (Morikawa et al., 1987), rabbit myocardial microsomes (Ford and Gross, 1988), and a microsomal fraction from MDCK cells (Strum and Daniel, 1993). Furthermore, myocardial cholinephosphotransferase (Ford and Gross, 1988) along with ethanolaminephosphotransferase (Ford et al., 1992) exhibit marked substrate selectivity for the utilization of endogenous alk-1-enylacyl-Gro in comparison to diacyl-Gro. However, when using rabbit platelet membranes, endogenous substrates, and CDP-[^{3}H]ethanolamine or CDP-[^{14}C]choline as co-substrates, Morikawa et al. (1987) found the newly synthesized phospholipids were mainly diacyl and alk-1-enylacyl species and rarely the alkylacyl type. The relative ratios of the synthesized alk-1-enylacyl to diacyl type of phospholipids depend on the concentrations of CDP-ethanolamine and CDP-choline. When 1, 10, and 30 μM CDP-[^{3}H]ethanolamine were used, the labeled ethanolamine phosphoglycerides contained 53, 37, and 27% of the alk-1-enylacyl type, respectively. In contrast, when 1, 10, and 30 μM CDP-[^{14}C]choline were used, the labeled choline phosphoglycerides contained 10, 17, and 24% alk-1-enylacyl type, respectively. In addition, the synthesis of alk-1-enylacyl-GPC was severely inhibited by unlabeled CDP-ethanolamine. Therefore, besides the availability of alk-1-enylacyl-Gro controlled by the reactions 3-5 (Figure 2), the relative endogenous levels of CDP-ethanolamine and CDP-choline also undoubtedly have an important role in regulating the amounts of alk-1-enylacyl-GPC synthesized. Therefore, based on current available data, it is difficult at present to assess the quantitative contribution of each metabolic pathway listed in Figure 2 for the conversion of ethanolamine plasmalogens to choline plasmalogens. Nevertheless, as will be discussed below, 30% of the choline plasmalogens in HL-60 cells is estimated to be formed from ethanolamine plasmalogens through the reactions involving direct polar head group modification as summarized in Figure 2 (Blank et al., 1993).

The likelihood that choline plasmalogens can be synthesized from ethanolamine plasmalogens through both *sn*-2 and polar head group modifications as depicted in

Figure 3 and by only the polar head group transformations described in Figure 2 is based on results obtained by Blank et al. (1993) since dissimilarities were found in the distribution of molecular species between [^{3}H]alk-1-enylacyl-GPC and [^{3}H]alk-1-enylacyl-GPE when HL-60 cells were incubated with [^{3}H]alk-1-enyllyso-GPE. In contrast, the distribution of molecular species of [^{3}H]alk-1-enylacyl-GPC matched that of theoretically produced [^{3}H]alk-1-enylacyl-GPC if one assumed 30% of the [^{3}H]alk-1-enylacyl-GPC is derived from only the polar head group modifications of [^{3}H]alk-1-enylacyl-GPE (via a combination of the reactions) listed in Figure 2 and 70% of [^{3}H]alk-1-enylacyl-GPC is formed from [^{3}H]alk-1-enylacyl-GPE through a series of reactions involving either alk-1-enylglycerol or alk-1-enylglycerophosphate as an intermediate (Figure 3).

Among all the enzymes participating in the metabolic conversion of plasmenylethanolamine to plasmenylcholine summarized in Figure 3, the existence and properties of the transacylase (reaction 2), lysophospholipase D (reaction 4), phosphotransferase (reaction 5), acyl-CoA acyltransferase (reaction 6), and phosphohydrolase (reaction 7) have been reviewed (Snyder et al., 1985; Snyder 1994b; MacDonald and Sprecher, 1991; Snyder et al., 1992). A diversified group of phospholipase A_2 activities can liberate fatty acids from phospholipids (see recent review, Dennis, 1994), however, the phospholipase A_2 responsible for the hydrolysis of alk-1-enylacyl-GPE (reaction 1/Figure 3) has not been firmly established. A calcium-independent phospholipase A_2 purified from canine myocardial cytosol displays selectivity toward arachidonic acid-containing plasmenylcholine (subclass rank order: plasmenylcholine > plasmanylcholine > phosphatidylcholine) (Hazen et al., 1990), and a 85 kDa cytosolic phospholipase A_2 purified from rabbit platelets can cleave plasmenylethanolamine, but does not discriminate between arachidonic acid-containing alk-1-enylacyl-GPE and diacyl-GPE (Shikano et al., 1994). Also, a calcium-independent phospholipase A_2 from rabbit lung microsomes can hydrolyze alk-1-enylarachidonoyl-GPE (Angle et al., 1988). The requirement of such specificity for a phospholipase A_2 is particularly important since both Chilton and Connell (1988) and Tessner et al. (1990) have indirectly shown the majority of endogenous arachidonic acid is lost from alk-1-enylarachidonoyl-GPE during neutrophil stimulation by ionophore A23187.

Recently, a Mg^{2+}-dependent lysophospholipase C (reaction 3/Figure 3) was detected in the microsomes of MDCK cells (Strum and Daniel, 1993). In parallel, the conversion of alk-1-enylacyl-GPE to alk-1-enylacyl-GPC in [^{3}H]alk-1-enyllyso-GPE prelabeled MDCK cells was stimulated with a concomitant increase in alk-1-enylglycerol following with 12-O-tetradecanoylphorbol-13-acetate (Strum and Daniel, 1993). Nevertheless, it is still unknown whether the metabolic pathways illustrated in Figs. 2 and 3 converge at the point where alk-1-enylacyl-GroP is generated or whether a different set or compartmental localization of phosphohydrolases and alk-1-enylacyl-Gro:CDP-choline cholinephosphotransferases exist.

IV. INVOLVEMENT OF PLASMALOGENS IN THE TRAFFICKING OF ARACHIDONATE AMONG PHOSPHOLIPIDS

Two recent reviews (MacDonald and Sprecher, 1991; Snyder et al., 1992) have reported the extensive literature dealing with the intracellular trafficking of arachidonic acid. Therefore, this section will concentrate primarily on a pathway that may help explain some of the observations related to the role of plasmalogens in the intracellular movement of arachidonate.

Intracellular levels of free arachidonic acid are normally extremely low; however, significant amounts of arachidonate are found esterified to various glycerolipids. Phospholipids with O-alkyl or O-alk-1′-enyl ether groups at the *sn*-1 position of glycerol contain, when compared to their diacyl counterparts, enriched levels of arachidonate esterified at the *sn*-2 position of glycerol (MacDonald and Sprecher, 1991; Snyder et al., 1992). In most mammalian tissues, plasmenylethanolamine represents the largest pool of plasmalogens (Horrocks, 1972; Snyder et al., 1989); however, several tissues, particularly heart, also contain abundant amounts of plasmenylcholine. The much larger quantities of ethanolamine plasmalogens in most tissues compared to choline plasmalogens, but not necessarily their greater importance from a biological standpoint, undoubtedly explains the reason why plasmenylethanolamines have generally received more attention by researchers in the lipid field. However, because of their relatively high concentrations and their large content of arachidonate, the ethanolamine plasmalogens do represent one of the most significant sources of arachidonic acid in many tissues. It also appears the ethanolamine plasmalogens, as well as the choline plasmalogens, play an important role in the trafficking of arachidonic acid (Figure 4).

Arachidonic acid is initially incorporated into cellular glycerolipids via an acyl-CoA acyltransferase during *de novo* synthesis of glycerolipids from *sn*-glycero-3-P and dihydroxyacetone-P. Arachidonate can then be shuttled into various phospholipids by three different enzymatic mechanisms involving lysophospholipids: (1) an acyl-CoA acyltransferase, originally described for fatty acids other than arachidonic acid (Lands, 1960), (2) a CoA-dependent transacylase initially described by Irvine and Dawson (1979); and (3) a CoA-independent transacylase first reported by Kramer and Deykin, 1983. The literature describing all three types of acylation reactions has recently been reviewed in considerable detail (MacDonald and Sprecher, 1991; Snyder et al., 1992) and it appears that, after the initial incorporation into phospholipids, arachidonate and certain other polyunsaturated acyl groups are moved among phospholipids primarily by the CoA-independent transacylase.

More than twenty years ago it was suggested that plasmenylethanolamine may serve as a repository for storage of arachidonic acid because it was the last phospholipid class to be depleted of arachidonate in testes of rats placed on fat-free diets (Blank et al., 1973). Ethanolamine plasmalogens have also been suggested to

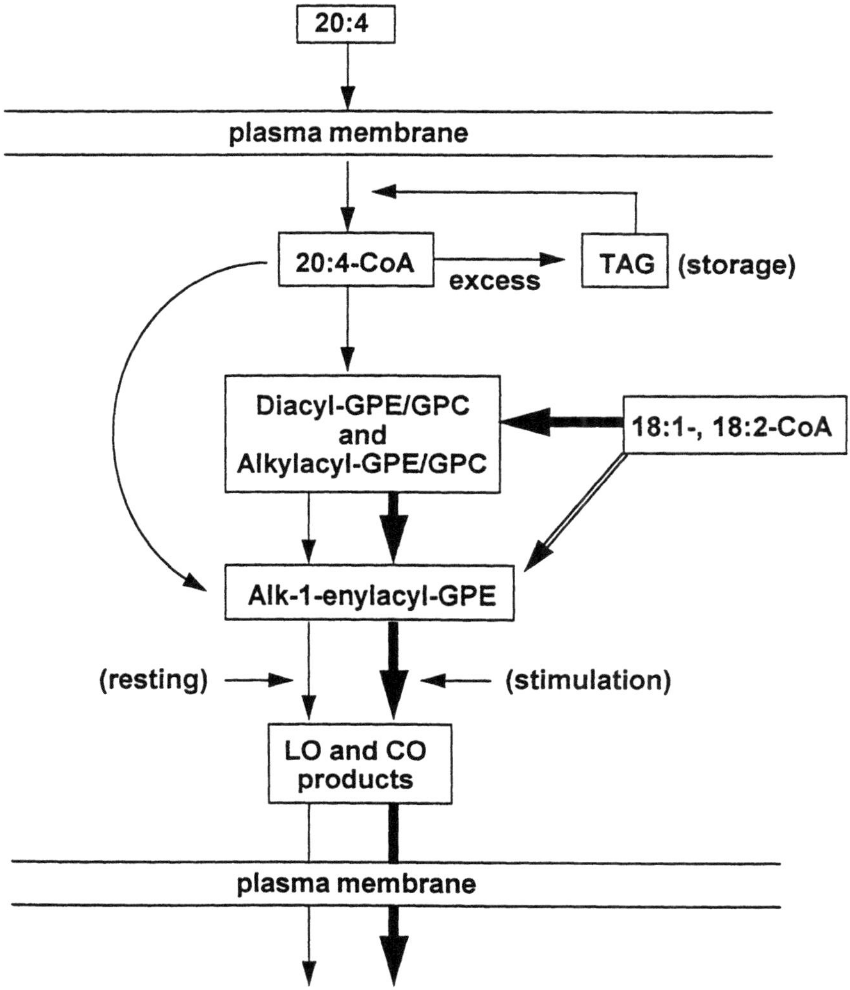

Figure 4. Proposed scheme for the intracellular trafficking of arachidonic acid among membrane phospholipids. Heavy arrows represent increased movement of arachidonic acid (20:4) among phospholipids induced by stimuli that increase phospholipase A_2 type and/or CoA-independent transacylase activities. Lighter arrows indicate the normal physiological turnover of intracellular arachidonate. LO and CO designate lipoxygenase and cyclooxygenase, respectively.

function as the source of arachidonic acid for production of bioactive eicosanoids (see reviews MacDonald and Sprecher, 1991; Snyder et al., 1992; Garg and Haerdi, 1993). The stimulated release of arachidonic acid and concomitant formation of alk-1-enyllyso-GPE can also act as a mechanism to initiate PAF biosynthesis (see next section).

Much of the data and function of alk-1-enylarachidonoyl-GPE would fit the arachidonic acid trafficking scheme proposed in Figure 4. In this simplified scheme arachidonic acid flows through ethanolamine plasmalogens as the last stop on its way to becoming a product of the cyclooxygenase and/or lipoxygenase enzyme systems. Arachidonate removed from plasmenylethanolamine is constantly and rapidly replaced by arachidonic acid from other diacyl (and sometimes alkylacyl) phospholipids thus making it difficult to pinpoint plasmalogens as the terminal source for arachidonic acid used in cellular functions. In fact, if replacement of arachidonic acid in plasmenylethanolamine by a CoA-independent transacylase is at least as rapid as the removal of arachidonate by a phospholipase A_2 activity, analysis of lipids after cellular stimulation would indicate that the donor phospholipid in the transacylase reaction, rather than plasmenylethanolamine, is the target molecule of the phospholipase A_2 activity.

The 18:1- and 18:2-CoA derivatives (Figure 4) represent acyl-CoA acyltransferase activities that would increase when arachidonate replacement is not sufficient to compensate for losses of 20:4 from plasmenylethanolamine. This would be another explanation for accumulation of 18:1 and 18:2 in alkylacyl-GPC during cellular stimulation observed by Chilton et al. (1984). Plasmenylethanolamine would also be the last phospholipid class to be depleted of arachidonate if the supply of arachidonic acid to the cell (or its precursor linoleic acid) is cut off as previously done by placing rats on a fat-free diet (Blank et al., 1973). It will be interesting to see if, and how well, this model scheme will hold up in future experiments.

V. ROLE OF PLASMALOGENS IN THE BIOSYNTHESIS OF PAF AND ITS ANALOGS

The biosynthesis of PAF and related analogs (e.g., acyl and ethanolamine plasmalogen analogs of PAF) appears to be closely linked to a CoA-independent transacylase that utilizes ethanolamine-containing lysoplasmalogens or other ethanolamine/choline lysoglycerophosphatides as acyl acceptor molecules (Sugiura et al., 1990; Uemura et al., 1991; Venable et al., 1991; Nieto et al., 1991; Blank et al., 1994). In this reaction, the arachidonoyl moiety at the *sn*-2 position of alkylacyl-GPC is transferred to a lysoplasmalogen which leads to the generation of lyso-PAF, the immediate precursor of PAF (Figure 5). The lyso-PAF or its analogs can then be acetylated by acetyl-CoA:lyso-PAF acetyltransferase to form PAF or one of the commonly encountered analogs of PAF; PAF transacetylase is also thought to play a major role in the biosynthesis of the various PAF analogs (Lee et al., 1992). Direct

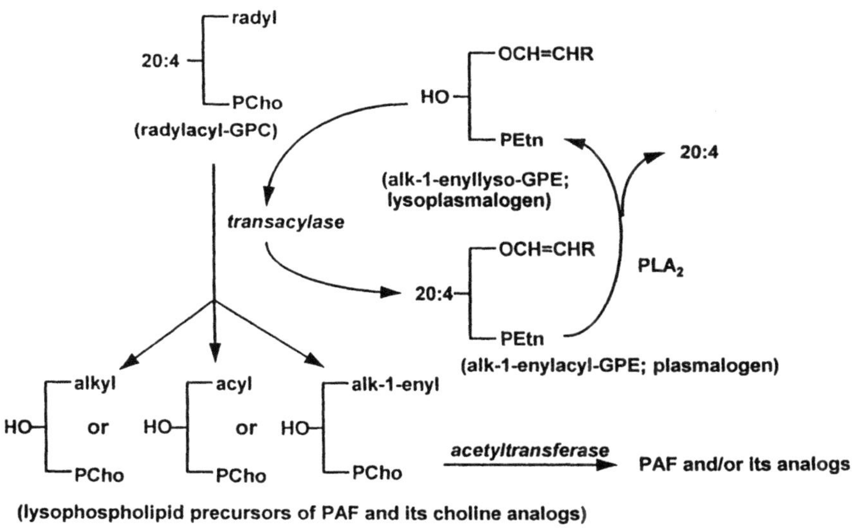

Figure 5. Lysoplasmalogen as the initiator for the biosynthesis of PAF and its choline-containing analogs via a transacylation reaction that produces lyso-PAF as an intermediate. The ethanolamine-containing analogs could be produced via a similar set of reactions but with 1-radyl-2-arachidonoyl-GPE as the membrane precursor.

coupling of the reaction catalyzed by the CoA-independent transacylase to the production of PAF has been demonstrated in HL-60 cells differentiated into a neutrophil-type form (Uemura et al., 1991) and in human neutrophils (Nieto et al., 1991). Acyl and plasmalogen analogs of PAF can also be formed by generating lysophospholipids via the transacylase from the corresponding diacyl-GPC/GPE and choline- or ethanolamine-containing plasmalogens (Figure 5), respectively, since these lyso-phospholipid intermediates have been shown to be produced via transacylation reactions (MacDonald and Sprecher, 1991; Snyder et al., 1992). On the other hand, the lysophospholipid acyl acceptors could also be produced by the direct hydrolysis of the *sn*-2 acyl moieties of the radylacyl-GPC/GPE precursors through the action of a phospholipase A_2.

The transacylase responsible for forming lyso-PAF possesses both hydrolytic phospholipase A_2 and acyl transfer type activities. However, it is not known whether these two activities reside in a single protein or whether they exist as two separate proteins in a very tightly integrated complex. Two recent reviews (Snyder, 1994a, b) contain more detailed information about how PAF and its analogs can be biosynthesized via the reactions catalyzed by the CoA-independent transacylase or directly by a phospholipase A_2.

Studies done with human neutrophils (Tessner et al., 1990) have implicated the ethanolamine plasmalogens as the principal source of the lysophospholipid acceptor that triggers the transcylation reaction involved in PAF biosynthesis. In these cells, roughly two-thirds of the ethanolamine-containing glycerophosphatides are plasmalogens (Gottfried, 1967; Walsh et al., 1983; Mueller et al., 1984) and 70-80% of this subclass contains arachidonate at the *sn*-2 position (Mueller et al., 1984; Chilton and Connell, 1988). Moreover, the arachidonoyl species of the ethanolamine plasmalogens in human neutrophils are selectively hydrolyzed following agonist stimulation (Chilton and Connell, 1988; Tessner et al., 1990). Thus, the fact that significant quantities of lysoplasmalogens accumulate under these conditions, make them a logical candidate as the probable lysophospholipid acceptor molecule in the reaction involving the transfer of the arachidonoyl moiety from alkylacyl-GPC to the lysoplasmalogen in the production of lyso-PAF as the substrate for acetyltransferase-catalyzed step in PAF biosynthesis.

The biggest puzzle about the role of lysoplasmalogens in PAF biosynthesis concerns the existence of an arachidonoyl-specific phospholipase A_2 required to produce the lysoplasmalogen or other lysophospholipid that serves as the acyl acceptor in the formation of lyso-PAF by the transacylase. The need to identify such a selective phospholipase A_2 is particularly important in view of the fact that the availability of lyso-PAF appears to be rate-limiting in the synthesis of PAF in the remodeling pathway (Venable et al., 1991; Sugiura et al., 1990; Ninio et al., 1991; Joly et al., 1990; Leyravaud et al., 1989; Sugiura et al., 1991; Blank et al., 1994). A number of phospholipase A_2s with a high degree of specificity for arachidonoyl glycerophospholipids containing choline or ethanolamine polar head groups have been described in the literature (Alonzo et al., 1986; Angle et al., 1988; Channon and Leslie, 1990; Clark et al., 1990; deCarvalho et al., 1993; Diez and Mong, 1990; Gronich et al., 1990; Hazen et al., 1990; Kim et al., 1991; Kramer et al., 1991; Leslie, 1991; Leslie et al., 1988; Wijkander and Sundler, 1991; Wolf and Gross, 1985; Hazen et al., 1991; Ford et al., 1991; Hazen et al., 1993; Wong et al., 1992; Shikano et al., 1994; also see Section III B of this chapter), but none of these have been directly linked to PAF biosynthesis. Furthermore, both microsomal (Wolf and Gross, 1985) and cytosolic (Wolf and Gross, 1985; Hazen et al., 1993) forms of a calcium-independent phospholipase A_2 from myocardial tissue have been shown to be highly selective for the choline plasmalogen species containing arachidonate, but their preference for ethanolamine plasmalogens has not yet been established. Nevertheless, such findings are encouraging since they support the contention that a highly selective phospholipase A_2, with a substrate preference for arachidonoyl ethanolamine plasmalogens, will ultimately be linked to the transacylation of lyso-PAF and the subsequent production of PAF. The generation of lysophospholipid acyl acceptors by a phospholipase A_2 for use as substrates by the transacylase linked to PAF production takes on added importance when one considers that for each molecule of lyso-PAF formed in the transacylation process, another molecule of arachidonate would be released for possible channeling into potent eicosanoid mediators.

VI. ASSESSMENT OF FUTURE STRATEGIES

Purification of the Δ1 alkyl desaturase responsible for the biosynthesis of ethanolamine plasmalogens is a difficult but high priority task needed to be done in order to fully understand the mechanism of this unique reaction. Such an accomplishment would also make it possible to develop antibody probes to further extend our knowledge of the plasmalogens that are prominent components of many membrane systems. Equally important is the need to establish the quantitative significance of the potential pathways for the biosynthesis of the choline plasmalogens, since the present concensus of recent studies is that they are derived from the ethanolamine plasmalogen pool rather than by a desaturase-catalyzed reaction involving alkylacylglycerophosphocholine as a substrate.

At this time the regulatory controls and the precise function of plasmalogens are still obscure, although it is clear they can play a significant role in the production of lipid mediators and are critical membrane components of myocardial, vascular, and neural tissues. In the future, more detailed sophisticated studies than those currently in the literature are required to better understand the combined actions of the phospholipase A_2 and the CoA-independent transacylase that release arachidonic acid during the formation of lyso-PAF via the transacylation reaction linked to the formation of both PAF and eicosanoid metabolites.

We hope this brief review will provide an up-to-date status report of background information and a critical assessment of emerging new concepts to further help promote a better understanding of the role of plasmalogens in mammalian cells. The greatest obstacle for future progress in the plasmalogen field will be the task of obtaining the various plasmalogen-related enzymes in sufficient purity to investigate their properties and to develop the necessary probes for pursuing imporant regulatory studies. It is our belief that the tools of molecular biology are already available to conduct such experiments and it is our hope that sufficient scientific interest can be generated in this small field of research to make these possibilities come true.

ACKNOWLEDGMENTS

This work was supported by the Office of Energy Research, U.S. Department of Energy (Contract No. DE-AC05-760R00033), The American Cancer Society (Grant BE-26Y), and The National Heart, Lung, and Blood Institute (Grant HL27109-13A1).

APPENDIX

Abbreviations

Cho; choline
Etn; ethanolamine

GLC; gas-liquid chromatography
GPC; *sn*-glycero-3-phosphocholine
GPE; *sn*-glycero-3-phosphoethanolamine
Gro; *sn*-glycerol
HPLC; high-performance liquid chromatography
MDCK; Madin Darby canine kidney cells
PAF; platelet-activating factor
phosphatidyl; diacylglycerophospho radical
plasmanyl; alkylacylglycerophospho radical
plasmenyl; alk-1-enylacylglycerophospho radical
PLD; phospholipase D
TAG; triacylglycerols
TLC; thin-layer chromatography

REFERENCES

Alonso, F., Henson, P. M., & Leslie, C. C. (1986). A cytosolic phospholipase in human neutrophils that hydrolyzes arachidonoyl-containing phosphatidylcholine. Biochim. Biophys. Acta 878, 273-280.

Anderson, R. E., Garrett, R. D., Blank, M. L., & Snyder, F. (1969). The quantitative production of aldehydes from O-alk-1-enyl glycerols. Lipids 4, 327-330.

Angle, M. J., Paltauf, F., & Johnston, J. M. (1988). Selective hydrolysis of ether-containing glycerophospholipids by phospholipase A_2 in rabbit lung. Biochim. Biophys. Acta 962, 234-240.

Ansell, G. B., & Spanner, S. (1965). The action of phospholipase C on ethanolamine plasmalogen (2-acyl-1-alkenylglycerylphosphoryl-ethanolamine). Biochem. J. 97, 375-379.

Benfenati, E., & Reginato, R. (1985). A comparison of three methods of soft ionization mass spectrometry of crude phospholipid extracts. Biomed. Mass Spectrometry 12, 643-651.

Bjerrum, O. W., Nielsen, H., & Borregaard, N. (1989). Quantitative analysis of phospholipids and demonstration of plasmalogens in human neutrophil subcellular fractions by high-performance liquid chromatography. Scand. J. Clin. Lab. Invest. 49, 613-622.

Blank, M. L., Cress, E. A., Fitzgerald, V., & Snyder, F. (1990). Thin-layer and high-performance liquid chromatographic separation of glycerolipid subclasses as benzoates. J. Chromatogr. 508, 382-385.

Blank, M. L., Cress, E. A., Piantadosi, C., & Snyder, F. (1975). A method for the quantitative determination of glycerolipids containing O-alkyl and O-alk-1-enyl moieties. Biochim. Biopys. Acta 380, 208-218.

Blank, M. L., Fitzgerald, V., Lee, T-c., & Snyder, F. (1993). Evidence for biosynthesis of plasmenylcholine from plasmenylethanolamine in HL-60 cells. Biochim. Biophys. Acta 1166, 309-312.

Blank, M. L., Robinson, M., Fitzgerald, V., & Snyder, F. (1984). Novel quantitative method for determination of molecular species of phospholipids and diglycerides. J. Chromatogr. 298, 473-482.

Blank, M.L,. Smith, Z.L., Fitzgerald, V., & Snyder, F. (1994). The CoA-independent transacylase in PAF biosynthesis: Tissue distribution and molecular species selectivity. Biochim. Biophys. Acta. 1254, 295-301.

Blank, M. L., & Snyder, F. (1994).Chromatographic analysis of ether-linked glycerolipids, including platelet-activating factor and related cell mediators. In: Lipid Chromatographic Analysis (Shibamoto, T., ed.), Chap. 9, pp. 291-316. Marcel Dekker, NY.

Blank, M. L., Wykle, R. L., & Snyder, F. (1972). The biosynthesis of ethanolamine plasmalogens by a postmitochondrial fraction from rat brain. Biochem. Biophys. Res. Commun. 47, 1203-1208.

Blank, M. L., Wykle, R. L., & Snyder, F. (1973). The retention of arachidonic acid in ethanolamine plasmalogens of rat testes during essential fatty acid deficiency. Biochim. Biophys. Acta 316, 28- 34.

Channon, J. Y., & Leslie, C. C. (1990). A calcium-dependent mechanism for associating a soluble arachidonoyl-hydrolyzing phospholipase A_2 with membrane in the macrophage cell line RAW 264.7. J. Biol. Chem. 265, 5409-5413.

Chen, S. S-H., & Kou, A. Y. (1982). Improved procedure for the separation of phospholipids by high-performance liquid chromatography. J. Chromatogr. 227, 25-31.

Chilton, F. H., & Connell, T. R. (1988). 1-Ether-linked phosphoglycerides: Major endogenous sources of arachidonate in the human neutrophil. J. Biol. Chem. 263, 5260-5265.

Chilton, F. H., Ellis, J. M., Olson, S.C., & Wykle, R. L. (1984). 1-O-Alkyl-2-arachidonoyl-*sn*-glycero-3-phosphocholine a common source of platelet-activating factor and arachidonate in human polymorphonuclear leukocytes. J. Biol. Chem. 259, 12014-12019.

Clark, J. D., Milona, N., & Knopf, J. L. (1990). Purification of a 110-kilodalton cytosolic phospholipase A_2 from the human monocytic cell line U937. Proc. Natl. Acad. Sci. 87, 7708-7712.

Clarke, N. G., & Dawson, R. M. C. (1981). Alkaline *O*→*N*-transacylation. Biochem. J. 195, 301-306.

Creer, M. H., & Gross, R. W. (1985). Reversed-phase high-performance liquid chromatographic separation of molecular species of alkyl ether, vinyl ether, and monoacyl lysophospholipids. J. Chromatogr. 338, 61-69.

Curstedt, T. (1977). Analysis of molecular species of ether analogues of phosphatidylcholines from biological samples. Biochim. Biophys. Acta 489, 79-88.

Daniel, L. W., Huang, C., Strum, J. C., Smitherman, P. K., Greene, D., & Wykle, R. L. (1993). Phospholipase D hydrolysis of choline phosphoglycerides is selective for the alkyl-linked subclass of Madin-Darby canine kidney cells. J. Biol. Chem. 268, 21519-21526.

Dawson, R. M. C., Hemington, N., & Davenport, J. B. (1972). Improvements in the method of determining individual phospholipids in a complex mixture by successive chemical hydrolyses. Biochem. J. 84, 497-501.

Debuch, H., & Seng, P. (1972). The history of ether-linked lipids through 1960. In: Ether Lipids, Chemistry and Biology (Snyder, F., ed.), Chap. 1, pp 1-24. Academic Press, NY.

de Carvalho, M. S., McCormack, F. X, & Leslie, C. C. (1993). The 85-kDa, arachidonic acid-specific phospholipase A_2 is expressed as an activated phosphoprotein in Sf9 cells. Arch. Biochem. Biophys. 306, 534-540.

Dembitsky, V. M. (1988). Quantification of plasmalogen, alkylacyl and diacylglycerophospholipids by micro-thin-layer chromatography. J. Chromatogr. 436, 467-473.

Dennis, E. A. (1994). Diversity of group types, regulation, and function of phospholipase A_2. J. Biol. Chem. 269, 13057-13060.

Diez, E., & Mong, S. (1990). Purification of a phospholipase A_2 from human monocytic leukemic U937 cells. J. Biol. Chem. 265, 14654-14661.

Dugan, L. L., Demediuk, P., Pendley, C. E., II, & Horrocks, L. A. (1986). Separation of phospholipids by high-performance liquid chromatography: All major classes including ethanolamine and choline plasmalogens, and most minor classes, including lysophosphatidyl ethanolamine. J. Chromatogr. 378, 317-327.

Farquhar, J. W. (1962). Identification and gas-liquid chromatographic behavior of plasmalogen aldehydes and their acetal, alcohol, and acetylated alcohol derivatives. J. Lipid Res. 3, 21-30.

Ford, D. A., & Gross, R. W. (1988). Identification of endogenous 1-*O*-alk-1′-enyl-2-acyl-*sn*-glycerol in myocardium and its effective utilization by choline phosphotransferase. J. Biol. Chem. 263, 2644-2650.

Ford, D. A., & Gross, R. W. (1994). The discordant rates of *sn*-1 aliphatic chain and polar head group incorporation into plasmalogen molecular species demonstrate the fundamental importance of polar head group remodeling in plasmalogen metabolism in rabbit myocardium. Biochemistry 33, 1216-1222.

Ford, D. A., Hazen, S. L., Saffitz, J. E., & Gross, R. W. (1991). The rapid and reversible activation of a calcium-dependent plasmalogen-selective phospholipase A_2 during myocardial ischemia. J. Clin. Invest. 88, 331-335.

Ford, D. A., Rosenbloom, K. B., & Gross, R. W. (1992). The primary determinant of rabbit myocardial ethanolamine phosphotransferase substrate selectivity is the covalent nature of the sn-1 aliphatic group of diradyl glycerol acceptors. J. Biol. Chem. 267, 11222-11228.

Frosolono, M. F., & Rapport, M. M. (1969). Reactivity of plasmalogens: kinetics of acid-catalyzed hydrolysis. J. Lipid Res. 10, 504-506.

Garg, M. L., and Haerdi, J. C. (1993). The biosynthesis and functions of plasmalogens. J. Clin. Biochem. Nutrition 14, 71-82.

Gottfried, E. L. (1967). Lipids of human leukocytes: relation to cell type. J. Lipid Res. 8, 321-327.

Gronich, J. H., Bonventre, J. V., & Nemenoff, R. A. (1990). Purification of a high-molecular-mass form of phospholipase A_2 from rat kidney activated at physiological calcium concentration. Biochem. J. 271, 37-43.

Hanahan, D. J. (1972). Ether-linked lipids: Chemistry and methods of measurement. In: Ether Lipids, Chemistry and Biology (Snyder, F., ed.), Chap. 2, pp 25-50. Academic Press, NY.

Hazen, S. L., Ford, D. A., & Gross, R. W. (1991). Activation of a membrane-associated phospholipase A_2 during rabbit myocardial ischemia which is highly selective for plasmalogen substrate. J. Biol. Chem. 266, 5629-5633.

Hazen, S. L., Hall, C. R., Ford, D. A., & Gross, R. W. (1993). Isolation of a human myocardial cytosolic phospholipase A_2 isoform. Fast atom bombardment mass spectroscopic and reverse-phase high pressure liquid chromatography identification of choline and ethanolamine glycerophospholipid substrates. J. Clin. Invest. 91, 2513-2522.

Hazen, S. L., Stuppy, R. J., & Gross, R. W. (1990). Purification and characterization of canine myocardial cytosolic phospholipase A_2. J. Biol. Chem. 265, 10622-10630.

Holloway, P. W., & Katz, J. T. (1972). A requirement for cytochorme b_5 in microsomal stearyl coenzyme A desaturation. Biochemistry 11, 3689-3695.

Horrocks, L. A. (1968). The alk-1-enyl group content of mammalian myelin phosphoglycerides by quantitative two-dimensional thin-layer chromatography. J. Lipid Res. 9, 469-472.

Horrocks, L. A. (1972). Content, composition, and metabolism of mammalian and avian lipids that contain ether groups. In: Ether Lipids Chemistry and Biology (Snyder, F., ed.), Chap. 9, pp. 177-272. Academic Press, NY.

Horrocks, L. A., & Ansell, G. B. (1967). Studies on the phospholipids of rat brain which contain glyceryl ethers. Biochim. Biophys. Acta 137, 90-97.

Horrocks, L. A., & Cornwell, D. G. (1962). The simultaneous determination of glycerol and fatty acids in glycerides by gas-liquid chromatography. J. Lipid Res. 3, 165-169.

Hoving, E. B., Prins, J., Rutgers, H. M., & Muskiet, F. A. J. (1988). Behavior of plasmalogens during high-performance liquid chromatography on a silica column with a mobile phase containing phosphoric acid. J. Chromatogr. 434, 411-416.

Huang, C., Wykle, R. L., Daniel, L. W., & Cabot, M. C. (1992). Identification of phosphatidylcholine-selective and phosphatidylinositol-selective phospholipase D in Madin-Darby canine kidney cells. J. Biol. Chem. 267, 16859-16865.

Irvine, R.F., & Dawson, R. M. C. (1979). Transfer of arachidonic acid between phospholipids in rat liver microsomes. Biochem. Biophys. Res. Commun. 91, 1399-1405.

Joly, F., Vigrain, I., Bossant, M. J., Bessou, G., Benveniste, J., & Ninio, E. (1990). Biosynthesis of PAF-acether. Activators of protein kinase C stimulate cultured mast cell acetyltransferase without stimulating PAF-acether synthesis. Biochem. J. 271, 501-507.

Kim, D. K., Kudo, I., & Inoue, K. (1991). Purification and characterization of rabbit platelet cytosolic phospholipase A_2. Biochim. Biophys. Acta 1083, 80-88.

Kiss, Z., & Anderson, W. B. (1989). Phorbol ester stimulates the hydrolysis of phosphatidylethanolamine in leukemic HL-60, NIH 3T3, and baby hamster kidney cells. J. Biol. Chem. 264, 1483-1487.

Kiyasu, J. Y., & Kennedy, E. P. (1960). The enzymatic synthesis of plasmalogens. J. Biol. Chem. 235, 2590-2594.

Knörr, W., & Spiteller, G. (1990). Simple method for the analysis of glycerol enol ethers derived from plasmalogens in complex lipid mixtures and subsequent determination of the aldehydic components by gas chromatography-mass spectrometry. J. Chromatogr. 526, 303-318.

Kramer, R. M., & Deykin, D. (1983). Arachidonoyl transacylase in human platelets. J. Biol. Chem. 258, 13806-13811.

Kramer, R. M., Roberts, E. F., Manetta, J., & Putnam, J. E. (1991). The Ca^{2+}-sensitive cytosolic phospholipase A_2 is a 100-kDa protein in human monoblast U937 cells. J. Biol. Chem. 266, 5268-5272.

Lands, W. E. M. (1960). Metabolism of glycerolipids. II. The enzymatic acylation of lysolecithin. J. Biol. Chem. 235, 2233-2237.

Lee, T-c., Qian, C., & Snyder, F. (1991). Biosynthesis of choline plasmalogens in neonatal rat myocytes. Arch. Biochem. Biophys. 286, 498-503.

Lee, T-c., Uemura, Y., & Snyder, F. (1992). A novel CoA-independent transacetylase produces the ethanolamine plasmalogen and acyl analogs of platelet-activating factor (PAF) with PAF as the acetate donor in HL-60 cells. J. Biol. Chem. 267, 19992-20001.

Leslie, C. C. (1991). Kinetic properties of a high molecular mass arachidonoyl-hydrolyzing phospholipase A_2 that exhibits lysophospholipase activity. J. Biol. Chem. 266, 11366-11371.

Leslie, C. C., Voelker, D. R., Channon, J. Y., Wall, M. M., & Zelarney, P. T. (1988). Properties and purification of an arachidonoyl-hydrolyzing phospholipase A_2 from a marcophage cell line, RAW 264.7. Biochim. Biophys. Acta 963, 476-492.

Leyravaud, S., Bossant, M. J., Joly, F., Bessou, G., Benveniste, J., & Ninio, E. (1989). Biosynthesis of PAF-acether. X. Phorbol myristate acetate-induced PAF-acether biosynthesis and acetyltransferase activation in human neutrophils. J. Immunol. 143, 245-249.

MacDonald, J. L. S., & Sprecher, H. (1991). Phospholipid fatty acid remodeling in mammalian cells. Biochim. Biophys. Acta 1084, 105-121.

Mavis, R. D., Bell, R. M., & Vagelos, P. R. (1972). Effect of phospholipase C hydrolysis of membrane phospholipids on membranous enzymes. J. Biol. Chem. 247, 2835-2841.

Meneses, P., & Glonek, T. (1988). High resolution ^{31}P NMR of extracted phospholipids. J. Lipid Res. 29, 679-689.

Mogelson, S., & Sobel, B. E. (1981). Ethanolamine plasmalogen methylation by rabbit myocardial membranes. Biochim. Biophys. Acta 666, 205-211.

Morikawa, S., Taniguchi, S., Fujii, K., Mori, H., Kumada, K., Fujiwara, M., & Fujiwara, M. (1987). Preferential synthesis of diacyl and alkenylacyl ethanolamine and choline glycerophospholipids in rabbit platelet membranes. J. Biol. Chem. 262, 1213-1217.

Morrison, W. R., & Smith, L. M. (1964). Preparation of fatty acid methyl esters and dimethylacetals from lipids with boron fluoride-methanol. J. Lipid Res. 5, 600-608.

Mozzi, R., Gresele, P., Siepi, D., Goracci, G., Nenci, G. G., & Porcellati, G. (1987). Choline plasmalogen biosynthesis by transmethylation in human platelets. Thromb. Res. 45, 687-693.

Mueller, H. W., O'Flaherty, J. T., Greene, D. G., Samuel, M. P., & Wykle, R. L. (1984). 1-O-alkyl-linked glycerophospholipids of human neutrophils: Distribution of arachidonate and other acyl residues in the ether-linked and diacyl species. J. Lipid Res. 25, 383-388.

Murphy, R. C. (1993). In: Mass Spectrometry of Lipids (Murphy, R. C., ed.), Chap. 7, pp. 213-252, Plenum Press, New York, NY.

Murphy, E. J., Stephens, R., Jurkowitz-Alexander, M., & Horrocks, L. A. (1993). Acidic hydrolysis of plasmalogens followed by high-performance liquid chromatography. Lipids 28, 565-568.

Myher, J. J., & Kuksis, A. (1984). Resolution of alkenylacylglycerol moieties of natural glycerophospholipids by gas-liquid chromatography on polar capillary columns. Can. J. Biochem. Cell Biol. 62, 352-362.

Nakagawa, Y., & Horrocks, L. A. (1983). Separation of alkenylacyl, alkylacyl, and diacyl analogues and their molecular species by high performance liquid chromatography. J. Lipid Res. 24, 1268-1275.

Nieto, M.L., Venable, M.E., Bauldry, S.A., Greene, D.G., Kennedy, M., Bass, D.A., & Wykle, R.L. (1991). Evidence that hydrolysis of ethanolamine plasmalogens triggers synthesis of platelet-activating factor via a transacylation reaction. J. Biol. Chem. 266, 18699-18706.

Ninio, E., Breton, M., Bidault, J., & Colard, O. (1991). Biosynthesis of paf-acether. XVII. Regulation by the CoA-independent transacylase in human neutrophils. FEBS Lett. 289, 138-140.

Okuyama, H., & Nojima, S. (1965). Studies on hydrolysis of cardiolipin by snake venom phospholipase A. J. Biochem. 57, 529-538.

Oshino, N., Imai, Y., & Sato, R. (1966). Electron-transfer mechanism associated with fatty acid desaturation catalyzed by liver microsomes. Biochim. Biophys. Acta 149, 369-377.

Oshini, N., & Omura, T. (1973). Immunochemical evidence for the participation of cytochrome b_5 in microsomal stearyl-CoA desaturation reaction. Arch. Biochem. Biophys. 157, 395-404.

Owens, K. (1966). A two-dimensional thin-layer chromatographic procedure for the estimation of plasmalogens. Biochem. J. 100, 354-361.

Paultauf, F., & Holasek, A. (1973). Enzymatic synthesis of plasmalogens. Characterization of the 1-*O*-alkyl-2-acyl-*sn*-glycero-3-phosphorylethanolamine desaturase from mucosa of hamster small intestine. J. Biol. Chem. 248, 1609-1651.

Paultauf, F., Prough, R. A., Masters, B. S. S., & Johnston, J. M. (1974). Evidence for the participation of cytochrome b_5 in plasmalogen biosynthesis. J. Biol. Chem. 249, 2661-2662.

Poulos, A., Hughes, B. P., & Cumings, J. N. (1968). The biosynthesis of choline plasmalogen by ox heart. Biochim. Biophys. Acta 152, 629-632.

Rana, F. R., Harwood, J. S., Mautone, A. J., & Dluhy, R. A. (1993). Identification of phosphocholine plasmalogen as a lipid component in mammalian pulmonary surfactant using high-resolution ^{31}P NMR spectroscopy. Biochem. 32, 27-31.

Renkonen, O. (1968). Chromatographic separation of plasmalogenic, alkyl-acyl, and diacyl forms of ethanolamine glycerophosphatides. J. Lipid Res. 9, 34-39.

Schmid, H. H. O., & Mangold, H. K. (1966). Alkoxylipids. II. "Neutral plasmalogens" in the liver oil of the ratfish (*Hydrolagus colliei*). Biochim. Biophys. Acta 125, 182-184.

Shikano, M., Masuzawa, Y., Yazawa, K., Takayama, K., Kudo, I., & Inoue, K. (1994). Complete discrimination of docosahexaenoate from arachidonate by 85 kDa cytosolic phospholipase A_2 during the hydrolysis of diacyl- and alkenylacylglycerophosphoethanolamine. Biochim. Biophys. Acta 1212, 211-216.

Snyder, F. (1994a). Platelet-activating factor and its analogs: Metabolic pathways and related intracellular processes. Biochim. Biophys. Acta. 1254, 231-249.

Snyder, F. (1994b). Platelet-activating factor: the enzymes and their regulatory controls. Biochem. J. 305, 689-705.

Snyder, F., & Blank, M. L. (1969). Relationships of chain lengths and double bond locations in O-alkyl, O-alk-1-enyl, acyl, and fatty alcohol moieties in preputial glands of mice. Arch. Biochem. Biophys. 130, 101-110.

Snyder, F., Blank, M. L., & Wykle, R. L. (1971). The enzymic synthesis of ethanolamine plasmalogens. J. Biol. Chem. 246, 3639-3645.

Snyder, F., Lee, T-c. & Blank, M.L. (1989). In: Phosphatidylcholine Metabolism (Vance, D. E., ed.), Chap. 9, pp. 143-164, CRC Press, Boca Raton, FL.

Snyder, F., Lee, T-c. & Blank, M.L. (1992). The role of transacylases in the metabolism of arachidonate and platelet activating factor. Prog. Lipid Res. 31, 65-86.

Snyder, F., Lee- T-c., & Wykle, R. L. (1985). Ether-linked glycerolipids and their bioactive species: Enzymes and metabolic regulation. In: The Enzymes of Biological Membranes, (Martonosi, A. N. ed), Vol. 2, pp. 1-58. Plenum Publishing Co., NY.

Strittmatter, P., Spatz, L., Corcoran, D., Rogers, M. J., Setlow, B., & Redline R. (1974). Purification and properties of rat liver microsomal stearyl coenzyme A desaturase. Proc. Nat. Acad. Sci. 71, 4565-4569.

Strum, J. C., & Daniel, L. W. (1993). Identification of a lysophospholipase C that may be responsible for the biosynthesis of choline plasmalogens by Madin-Darby canine kidney cells. J. Biol. Chem. 268, 25500-25508.

Strum, J. C., Emilsson, A., Wykle, R. L., & Daniel, L. W. (1992). Conversion of 1-*O*-alkyl-2-acyl-*sn*-glycero-3-phosphocholine to 1-*O*-alk-1′-enyl-2-acyl-*sn*-glycero-3-phosphoethanola min e. J. Biol. Chem. 267, 1576-1583

Sugiura, T., Fukuda, T., Cheng, N-n., & Waku, K. (1991). Transient activation of 1-O-alkyl-*sn*-glycero-3-phosphocholine: Acetyl-CoA acetyltransferase during the incubation of macrophages. Lipids 26, 861-865.

Sugiura, T., Fukuda, T., Masuzawa, Y., & Waku, K. (1990). Ether lysophospholipid-induced production of platelet-activating factor in human polymorphonuclear leukocytes. Biochim. Biophys. Acta 1047, 223-232.

Takamura, H., & Kito, M. (1991). A highly sensitive method for quantitative analysis of phospholipid molecular species by high-performance liquid chromatography. J. Biochem. 109, 436-439.

Takamura, H., Narita, H., Urade, R., & Kito, M. (1986). Quantitative analysis of polyenoic phospholipid molecular species by high performance liquid chromatography. Lipids 21, 356-361.

Tessner, T. G., Greene, D. G., & Wykle, R. L., (1990). Selective deacylation of arachidonate-containing ethanolamine-linked phosphoglycerides in stimulated human neutrophils. J. Biol. Chem. 265, 21032-21038.

Tessner, T. G., & Wykle, R. L., (1987). Stimulated neutrophils produce an ethanolamine plasmalogen analog of platelet-activating factor. J. Biol. Chem. 262, 12660-12664.

Uemura, Y., Lee, T-c., & Snyder, F. (1991). A CoA-independent transacylase is linked to the formation of platelet activating factor (PAF) by generating the lyso-PAF intermediate in the remodeling pathway. J. Biol. Chem. 266, 8268-8272.

Venable, M.E., Nieto, M.L., Schmitt, J.D., & Wykle, R.L. (1991). Conversion of 1-O-[^{3}H]alkyl-2-arachidonoyl-*sn*-glycero-3-phosphorylcholine to lyso platelet-activating factor by the CoA-independent transacylase in membrane fractions of human neutrophils. J. Biol. Chem. 266, 18691-18698.

Walsh, C. E., DeChatelet, L. R., Chilton, F. H., Wykle, R. L., & Waite, M. (1983). Mechanism of arachidonic acid release in human polymorphonuclear leukocytes. Biochim. Biophys. Acta 750, 32-40.

Wientzek, M., Man, R. Y. K., & Choy, P. C. (1987). Choline glycerophospholipid biosynthesis in the guinea pig heart. Biochem. Cell Biol. 65, 860-868.

Wijkander, J., & Sundler, R. (1991). A 100-kDa arachidonate-mobilizing phospholipase A_2 in mouse spleen and the macrophage cell line J774. Purification, substrate interaction and phosphorylation by protein kinase C. Eur. J. Biochem. 202, 873-880.

Wolf, R. A., & Gross, R. W. (1985). Identification of neutral active phospholipase C which hydrolyzes choline glycerophospholipids and plasmalogen selective phospholipase A_2 in canine myocardium. J. Biol. Chem. 260, 7295-7303.

Wong, B., Tang, W., & Ziboh, V. A. (1992). Identification of a membrane-associated 1-O-alkyl-2-arachidonoyl-*sn*-glycero-3-phosphocholine hydrolyzing phospholipase A_2 in guinea pig 1 epidermis. FEBS 305, 213-216.

Wood, R., & Snyder, F. (1968). Quantitative determination of alk-1-enyl-and alkyl-glyceryl ethers in neutral lipids and phospholipids. Lipids 3, 129-135.

Wykle, R. L., Blank, M. L., Malone, B., & Snyder, F. (1972). Evidence for a mixed function oxidase in the biosynthesis of ethanolamine plasm alogens from 1-alkyl-2-acyl-sn-glycero-3-phosphorylethanolamine. J. Biol. Chem. 247, 5442-5447.

Wykle, R. L., Blank, M. L., & Snyder, F. (1970). The biosynthesis of plasmalogens in a cell-free system. FEBS Lett. 12, 57-60.

Wykle, R. L., & Lockmiller, J. M. (1975). The biosynthesis of plasmalogens by rat brain: Involvement of the microsomal eletron transport system. Biochim. Biophys. Acta 380, 291-298.

Yandrasitz, J. R., Berry, G., & Segal, S. (1981). High-performance liquid chromatography of phospholipids with UV detection: optimization of separations on silica. J. Chromatogr. 225, 319-328.

THE CDP-ETHANOLAMINE PATHWAY IN MAMMALIAN CELLS

P. Sebastiaan Vermeulen, Math J.H. Geelen, Lilian B.M. Tijburg, and Lambert M.G. van Golde

Advances in Lipobiology
Volume 2, pages 287-322.

ISBN 0-7623-0205-4

ABSTRACT

The CDP-ethanolamine pathway is one of the major routes for the biosynthesis of phosphatidylethanolamine in mammalian cells in vivo. In this Chapter we will discuss current knowledge on the enzymes that catalyze the various steps in this pathway. The properties and subcellular organization of these enzymes will be contrasted with those of the corresponding enzymes in the CDP-choline pathway leading to phosphatidylcholine. Although the regulation of phosphatidylethanolamine biosynthesis can take place at multiple sites current evidence suggests that, under most conditions, the supply of CDP-ethanolamine and diacylglycerol are the principal factors controlling the rate of this process. We suggest that reversible binding of phosphoethanolamine cytidylytransferase to the endoplasmic reticulum could contribute to regulating the amount of CDP-ethanolamine in the cell.

I. INTRODUCTION

Phosphatidylethanolamine (PE) is a phospholipid class that seems to be universally present in biological membranes (Strickland, 1973). It is perhaps not always realized that PE occurs more widely in nature than its choline counterpart, phosphatidylcholine (PC). Prokaryotes, and particularly Gram-negative bacteria, are generally very rich in PE , but - with a few exceptions - hardly ever contain PC. Although PC is undoubtedly the principal phospholipid class in eukaryotic cells (Pelech and Vance, 1984; Kent, 1990), PE is usually the second most abundant phospholipid in most animals and plants. In some specialized mammalian membrane systems, such as myelin for example, the content of PE is even more prominent than that of PC (White, 1973). In many mammalian tissues PE exists in two major subclasses: 1,2-diacyl-*sn*-glycero-3-phosphoethanolamines and 1-alk-1′-enyl-2-acyl-*sn*-glycero-3-phosphoethanolamines, also called ethanolamine plasmalogens or plasmenylethanolamines. A third subclass, the 1-alkyl-2-acyl form of PE or plasmanylethanolamine, is found in only small quantities in most tissues (Snyder, 1985).

It is well known that the neutral amphipathic phospholipids PE and PC both serve a fundamental role as key building stones of cellular and subcellular membranes. In this respect the two phospholipids are certainly not interchangeable. This is already apparent from their very asymmetric distribution between the outer and inner leaflet of most plasma membranes: whereas PE is predominantly localized in the inner leaflet, PC appears to favor the outer monolayer (Zachowski, 1993). Another example is the different partitioning of PE and PC between the inner and outer membrane of mitochondria. Whereas the inner membrane is enriched in cardiolipin and PE, PC is concentrated in the outer membrane (Hovius et al., 1990).

In addition to a structural role as constituent of membranes or circulating lipoproteins, PC has recently taken on a new and exciting role as a substrate for the production of second messengers (Exton, 1990; Billah and Anthes, 1991), and evidence is forthcoming for a potentially analogous role of PE (Kiss and Anderson, 1989; Hii et al., 1991; Kiss, 1992). Furthermore, plasmenylethanolamines represent the largest endogenous storage depot of arachidonic acid mass in most mammalian cells (Daniel et al., 1981; Chilton and Connell, 1988; Ford and Gross, 1989). During cellular activation arachidonic acid can be mobilized from this storage depot by the action of phospholipase(s), and subsequently converted into prostaglandins and leukotrienes (Chilton and Connell, 1988). These, as well as other observations discussed by Ford et al. (1992), strongly suggest that plasmenylethanolamine metabolism plays a crucial role in important signal transduction processes in mammalian cells.

The observations are intriguing that PE is also involved in the formation of cell surface glycoproteins that are associated with the plasma membrane through a glycosylphosphatidylinositol membrane anchor. Strong evidence was provided that intracellular PE is the donor of the ethanolamine residue to the phosphoethanolamine bridge that links the glycosylphosphatidylinositol anchor to the protein (Menon and Stevens, 1992).

These important findings have undoubtedly intensified interest in studies on cellular metabolism of PC and PE and its regulation. So far, most investigations have been focused on PC. Judging from the vast amount of literature that has accumulated in the past two decades on PC metabolism and its regulation (Vance, 1989a, b; Kent, 1990), and the far more limited information that is available on PE metabolism (Tijburg et al., 1989a), it is fair to say that in this respect the latter is clearly the "Cinderella" of these two phospholipids.

In this contribution we will attempt to review current knowledge of the biogenesis of PE and its regulation in mammalian cells. The following aspects will be highlighted. First we will discuss the three major pathways that could be involved in the biosynthesis of PE, and try to evaluate current knowledge with respect to their relative importance to overall PE synthesis in mammalian cells. Subsequently we will focus in more detail on the enzymes catalyzing the individual steps of one of these pathways, the CDP-ethanolamine route, and outline differences with the comparable route leading to PC, the CDP-choline pathway. Although we are still pretty ignorant about the regulation of PE synthesis in mammalian cells when compared to that of PC, we will then discuss possible rate-regulatory steps in the CDP-ethanolamine route and the cellular organization of the enzymes involved in this pathway, including its possible implications for regulation.

II. PATHWAYS OF PHOSPHATIDYLETHANOLAMINE BIOSYNTHESIS

Three pathways are available in mammalian cells to generate PE. This phospholipid can be synthesized by base-exchange of free ethanolamine with preexisting phospholipids (Borkenhagen et al., 1961; Bjerve, 1973, 1985; Kanfer, 1980, 1989;

Kennedy, 1986), by decarboxylation of phosphatidylserine (PS) (Borkenhagen et al., 1961; Kennedy, 1986) or by the incorporation of ethanolamine via the CDP-ethanolamine pathway (Kennedy and Weiss, 1956; Kennedy, 1986).

A. Formation by Base-Exchange

Determination of the quantitative role of base-exchange in the biosynthesis of PE in rat hepatocytes demonstrated that in the presence of physiological ethanolamine concentrations, 8-9% of the ethanolamine incorporated into PE could be attributed to the direct base-exchange reaction. This contribution rose to 30-40% when the ethanolamine concentration was raised twenty-fold (Sundler et al., 1974a). It now seems generally accepted that base-exchange represents a minor pathway for the synthesis of most phospholipids except PS (Kanfer, 1980; Kanfer, 1989). It is interesting, however, that a recent paper demonstrated the existence of base-exchange activity in plasma membranes of rat liver (Siddiqui and Exton, 1992). The phospholipid specificity of the base-exchange activity in plasma membranes was different from that reported earlier for microsomes (Bjerve, 1973). Interestingly, the plasma membrane base-exchange activity was stimulated by the same agonists that catalyze G-protein-mediated PC hydrolysis. The authors speculated that this base-exchange mechanism could play an important role in replenishing PC after the often substantial and prolonged breakdown of PC that occurs upon stimulation of cells with hormones, neurotransmittors, and growth factors. In contrast to resynthesis via the CDP-choline pathway, replenishment by base-exchange would be fast and not require immediate expenditure of energy. Although the study of Siddiqui and Exton focused on PC, it can be speculated that plasma membrane base-exchange activity may play a role in compensating for agonist-induced PE hydrolysis as well.

B. Formation by Decarboxylation of Phosphatidylserine

The formation of PE by the decarboxylation pathway obviously first requires the synthesis of PS. The formation of the latter phospholipid proceeds by calcium-dependent base exchange with either choline or ethanolamine (Figure 1, reactions 8 and 9). Genetic studies with mutant strains of Chinese hamster ovary (CHO) cells defective in PS synthesis provided evidence that the serine base-exchange reaction is catalyzed by at least two different phosphatidylserine synthases. The first (phosphatidylserine synthase I) utilizes PC as a phosphatidyl donor (Figure 1, reaction 9), whereas the second (phosphatidylserine synthase II) can only utilize PE as a phosphatidyl donor (reaction 8) (Kuge et al., 1985; Kuge et al., 1986a, b; Voelker and Frazier, 1986; Kuge et al., 1991). These studies also demonstrated that PC functions as the main precursor of PS. Phosphatidylserine synthase I can also catalyze the energy-independent incorporation of choline and ethanolamine into PC and PE, respectively, while phosphatidylserine synthase II cannot catalyze the

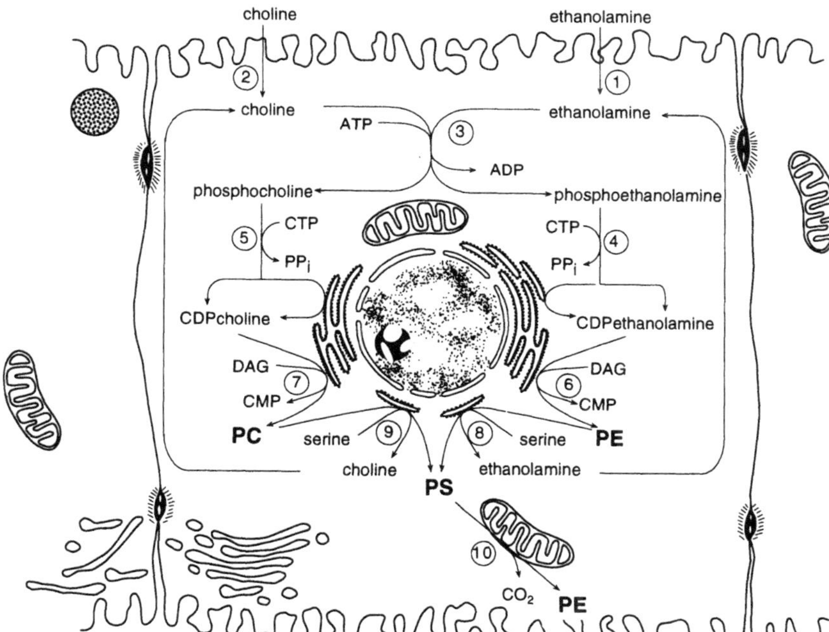

Figure 1. Biosynthesis of phosphatidylethanolamine (PE) and phosphatidylcholine (PC). The CDP-ethanolamine pathway comprises the reactions 3, 4 and 6; the CDP-choline pathway the reactions 3, 5, and 7. Phosphatidylserine (PS) is synthesized by base-exchange from either PE (reaction 8) or PC (reaction 9); PS can be converted to PE by decarboxylation (reaction 10). Cellular uptake of ethanolamine and choline is represented as 1 and 2. Enzymes: 3: choline/ethanolamine kinase; 4: CTP: phosphoethanolamine cytidylyltransferase (ET); 5: CTP:phosphocholine cytidylyltransferase (CT); 6: ethanolaminephosphotransferase (EPT); 7: cholinephosphotransferase (CPT); 8: phosphatidylserine synthase II; 9: phosphatidylserine synthase I; 10: phosphatidylserine decarboxylase.

choline base-exchange reaction (Kuge et al., 1986a). An enzyme with very similar properties as phosphatidylserine synthase II of CHO cells has been purified from rat-brain microsomes (Suzuki and Kanfer, 1985). Both enzymes use PE but not PC as substrate and do not catalyze the base-exchange reaction of phospholipids with choline.

The use of PS as a primary precursor of PE presents the cell with a formidable lipid transport requirement (Bishop and Bell, 1988). Whereas the formation of PS proceeds at the endoplasmic reticulum (Dennis and Kennedy, 1972; Bjerve, 1973,

Jelsema and Morré, 1978; Kanfer 1980; Suzuki and Kanfer, 1985), its decarboxylation (Figure 1, reaction 10) takes place in the mitochondria (Dennis and Kennedy, 1972), more specifically in the mitochondrial inner membrane (van Golde et al., 1974; Hovius et al., 1992). The PS formed from exogenous serine is translocated to the mitochondria by a two-step process (Voelker, 1985, 1989, 1991). The first step requires ATP and may involve vesicle budding or mechanical movement of the endoplasmic reticulum. The ATP-independent second step may require formation of collision complexes between the endoplasmic reticulum vesicles and the mitochondrion. Interestingly, cytosolic factors such as lipid transfer proteins appear not to be required for the flow of PS to the mitochondria. Subsequent studies by Voelker (1993) provided evidence that the ATP-dependent synthesis and translocation of PS to the mitochondria could be reconstituted in a cell-free system. Mixing experiments using heterologous donor and acceptor cell extracts provided evidence for selective and restricted transport of nascent PS to the mitochondria of autologous cells. These findings were consistent with an earlier work by Vance (1990) who reported the isolation of a special fraction of endoplasmic reticulum membranes that was associated with mitochondria. This fraction was enriched in the enzymes involved in the synthesis of PS, PE, and PC from serine. It is interesting to speculate that this particular membrane may provide a bridge that would facilitate the transfer of newly synthesized phospholipids to the mitochondria. Quite recently Cui et al. (1993) reported the cloning and expression of a novel phosphatidylethanolamine *N*-methyltransferase (PEMT-2). Interestingly, this enzyme appeared to be a specific biochemical and cytological marker for the specific mitochondria-associated fraction of the endoplasmic reticulum isolated by Vance (1990).

Studies by Hovius et al. (1992) suggested that PS flows from the outer membrane to the inner membrane through contact sites between inner and outer membrane, to become decarboxylated, and that the formed PE flows back to the outer membrane without mixing with inner membrane PE. It has been shown that the PE formed via decarboxylation of PS is preferentially exported from the mitochondria to the endoplasmic reticulum (Vance, 1991), possibly via interacting specialized microdomains of both membranes (Ardail et al., 1993). It is attractive to speculate that the special PEMT-2 containing membrane (Cui et al., 1993) is also involved in transferring PE out of mitochondria so that PS-derived PE can be methylated to PC.

In the older literature, decarboxylation of PS was not considered to contribute significantly to the biosynthesis of PE in mammalian cells (van den Bosch, 1974; Bell and Coleman, 1980). However, more recently Voelker (1984) reported that PS decarboxylation could account for approximately all PE synthesized in baby hamster kidney (BHK) cells when the supply of ethanolamine in the culture medium was restricted. A similar conclusion was drawn by Miller and Kent (1986). The latter authors isolated a CHO mutant cell line, which was reduced more than two-fold in the rate of the flux through the CDP-ethanolamine pathway, as measured by the incorporation of labeled ethanolamine. However, the levels of PE and the incorporation of ^{32}Pi into PE and that of labeled serine into PS and PE were the same in mutant and parent cells. This led the authors to the conclusion that the CDP-ethanolamine pathway did not contribute significantly to PE synthesis in CHO cells cultured in the absence of ethanolamine. Further evidence came from studies with two other CHO cell lines, both defective in PC-dependent synthesis of PS (Kuge et al., 1986a, b; Voelker and Frazier, 1986). These mutant strains showed lowered levels of both PS and PE. Taken collectively, these findings indeed strongly suggested that PE synthesis via PS and PS decarboxylase contributes significantly to membrane biogenesis in mammalian cells.

C. Synthesis via the CDP-Ethanolamine Pathway

In interpreting the significance of the findings described in the previous section it should be kept in mind that the cell lines were usually cultured in the absence of exogenous ethanolamine. The general perception is that the ethanolamine required for the CDP-ethanolamine pathway is derived intracellularly from the decarboxylation of PS, followed by base-exchange with serine and release of ethanolamine (Bremer et al., 1960). Although normal epithelial cells and some of their transformed cells will require exogenous ethanolamine to proliferate normally in culture (Kano-Sueoka and Errick, 1981; Kano-Sueoka and King, 1987), it is well known that many other cell lines indeed do not require the presence of ethanolamine in the culture medium. On the other hand, ethanolamine concentrations ranging from 8-50 μM have been measured in the circulation of different species (Sundler and Åkesson, 1975b; Baba et al., 1984; Kruse et al. 1985; Milakolfsky et al. 1985; Houweling et al., 1992), implying that cells in vivo are exposed to a constant supply of ethanolamine. Furthermore, specific transport systems for the uptake of extracellular ethanolamine have been reported for several tissues (Zelinski and Choy, 1982a; Pu and Anderson, 1984; Yorek et al., 1985, 1986; Lipton et al., 1988) (see also following section) and extracellular ethanolamine can readily be channeled into PE via the CDP-ethanolamine pathway (Zelinsky and Choy 1982a, Pu and Anderson 1984, Tijburg et al., 1987a; Lipton et al., 1990). As mentioned earlier the concentration of ethanolamine present in the media of cultured cells is often much less than that encountered *in vivo*. It is possible that this causes the cells to rely on decarboxylation of PS for PE synthesis rather than on synthesis of PE via CDP-ethanolamine since a free ethanolamine source may be limited (Lipton et al., 1990). This may indeed be an important factor as studies with freshly isolated hepatocytes indicated that in the presence of a physiological concentration of extracellular ethanolamine the biosynthesis of PE via the CDP-ethanolamine pathway was strongly enhanced relative to the synthesis of PE by decarboxylation of PS (Tijburg et al., 1989b). These findings were corroborated and extended by studies *in vivo* with labeled serine and ethanolamine (Arthur and Page, 1991). These authors demonstrated that the CDP-ethanolamine pathway is an important route in the synthesis of PE in rat heart, kidney, and liver. In a recent study with keratinocytes in culture Arthur and Lu (1993) demonstrated that the synthesis of PE via decarboxylation appeared to decrease when ethanolamine was made available to the cells and the CDP-ethanolamine pathway was functioning.

The presence of extracellular ethanolamine is probably not the only factor that determines the relative importance of the CDP-ethanolamine pathway versus the decarboxylation route. Another factor may certainly be the cell type or tissue under investigation. Xu et al. (1991) recently investigated the utilization of serine as a precursor of PS and PE synthesis in a glioma cell line. The authors provided strong evidence that in this cell line, in contrast to BHK and CHO cells (Voelker, 1984; Miller and Kent, 1986; Kuge et al., 1986a,b), exogenous serine maximally was only 2-5% as effective as ethanolamine as an exogenous precursor of PE. This was thought to be due to the fact that serine was rapidly converted to other metabolites thus limiting its use as a direct phospholipid precursor.

It seems fair to say that the studies described above (Tijburg et al., 1989b; Lipton et al., 1990; Xu et al., 1991; Arthur and Page, 1991) make it very likely that the CDP-ethanolamine pathway does play an important role in the synthesis of PE in several mammalian tissues in vivo. However, it remains difficult to assess the actual contribution of the CDP-ethanolamine pathway and that of the route via PS to overall PE synthesis *in vivo*. A perhaps more fundamental and more interesting question is why cells have developed the use of two important pathways for PE synthesis, despite the fact that at least one of these routes presents them with such significant lipid transport problem. Studies with liver have shown that PE destined for secretion as component of lipoproteins is synthesized preferentially via the PS decarboxylation route (Vance and Vance, 1986; Vance, 1988). In this regard it is interesting to mention that the studies of Arthur and Page (1991) showed that PE synthesis via decarboxylation of PS was much more active in liver than in tissues that do not secrete lipoproteins, such as heart and kidney.

It has been suggested that formation of plasmenylethanolamine would be the physiological function of the CDP-ethanolamine pathway (Miller and Kent, 1986). This suggestion was supported by earlier studies of Polokoff et al. (1981), who observed that in a CHO cell mutant with defective ethanolaminephosphotransferase the incorporation of ^{32}Pi into plasmenylethanolamine was much more reduced than that in the diacyl subclass of PE. Yorek et al. (1985) studied the incorporation of labeled ethanolamine and serine into both subclasses of PE in retinoblastoma cells. Whereas the cells incorporated labeled ethanolamine into diacyl PE and plasmenylethanolamine at equivalent rates, only a small amount of label from serine was recovered in plasmenylethanolamine. These findings suggested that in retinoblastoma cells the PS decarboxylation pathway hardly contributed to the formation of plasmenylethanolamine. Similar findings were reported by Arthur and Page (1991) who followed the *in vivo* incorporation of labeled ethanolamine and serine into the diacyl subclass of PE and its plasmalogen analog of rat heart, kidney and liver. In all three tissues the CDP-ethanolamine pathway was utilized for the synthesis of both subclasses of PE, whereas the decarboxylation pathway only produced the diacyl subclass. On the other hand, studies with glioma cells in culture demonstrated that both ethanolamine and serine could be used as precursor of the head group of plasmenylethanolamine (Xu et al., 1991).

While these studies indeed suggest that the CDP-ethanolamine pathway is a major pathway for the generation of plasmenylethanolamine, it is important to emphasize that they also show that the importance of the CDP-ethanolamine pathway is not limited to the production of plasmenylethanolamines as suggested by Miller and Kent (1986). In several mammalian cells *in vivo* this pathway undoubtedly contributes to the biosynthesis of the diacyl subclass of PE as well.

III. INDIVIDUAL STEPS OF THE CDP-ETHANOLAMINE PATHWAY

A. Sources of Ethanolamine

Before discussing the individual enzymes involved in the conversion of ethanolamine into PE via the CDP-ethanolamine route, we will first summarize the possible sources of ethanolamine that are available for PE synthesis by mammalian cells. It is generally accepted that the ethanolamine required for PE synthesis can be generated intracellularly by decarboxylation of PS, followed by base-exchange with serine and release of free ethanolamine (Bremer et al., 1960; Kennedy, 1986). It is obvious that ethanolamine released by intracellular degradation of PE, for example by phospholipase D (Kiss, 1991; Hii et al., 1991), or by base-exchange with choline (Kanfer, 1980, 1989) can also be reutilized for PE synthesis.

However, as discussed in the preceding section, circulating ethanolamine is another potentially important source of ethanolamine for PE synthesis by mammalian cells. This extracellular ethanolamine can originate from intracellular metabolism in tissues but also from dietary sources. Experiments with rats demonstrated that diets containing 2% PE enhanced circulating ethanolamine concentrations from 36 to 52 μM (Imaizumi et al., 1989). Earlier work of that group had shown that the base moiety of dietary PE was transported mainly through the portal circulation rather than as chylomicron-phospholipid through the mesenteric lymph system (Ikeda et al., 1987). In addition, serine from multiple dietary sources can contribute indirectly, i.e., via the PS decarboxylation cycle, to the circulating pool of ethanolamine.

The fact that specific uptake systems of ethanolamine have been described for a number of cell types and tissues further suggests that circulating ethanolamine may play an important role as substrate for PE synthesis. High- and low-affinity uptake systems for ethanolamine were reported by Pu and Anderson (1984) for photoreceptor cells of the retina. The authors showed that ethanolamine accumulated by the high-affinity uptake process was not used for neurotransmission by the photoreceptor cells, but was used for the formation of PE. The presence of a two-component transport system for ethanolamine was also reported for several other cell types such as, for example, human Y79 retinoblastoma cells (Yorek et al., 1985), a cultured cell line of retinal origin that retains many neural characteristics, and cultured endothelial cells (Lipton et al., 1988, 1990). On the other hand, only one uptake system for ethanolamine was identified in the hamster heart (Zelinsky and Choy, 1982a). In view of the physiological concentration of serum ethanolamine (8-50 μM) and the apparent Km values that have been reported for the high affinity uptake system of ethanolamine, 2 μM in photoreceptor cells (Pu and Anderson, 1984), 3-4.7 μM in cultured endothelial cells (Lipton et al., 1988, 1990) and 40.6 μM in human retinoblastoma cells (Yorek et al., 1985) it seems highly likely that the uptake of ethanolamine by mammalian cells occurs through a saturable, high-affinity carrier-mediated mechanism (Ishidate, 1989). A rather high Km value (170 μM) has been reported for the uptake of ethanolamine by the hamster heart (Zelinski and Choy, 1982a), but it should be kept in mind that the concentration of serum ethanolamine in hamsters is also exceptionally high (0.9 mM [Zelinski and Choy, 1982a] versus 8-50 μM in other species).

The uptake of choline by noncholinergic cells also occurs largely through a specific carrier-mediated transport mechanism (Ishidate, 1989). It has been repeatedly suggested that the uptake of ethanolamine and choline occurs through specific and distinct transport systems (Zelinski et al., 1980; Sundler and Åkesson, 1975a; Yorek et al., 1985 and 1986) (Figure 1, reactions 1 and 2). On the other hand, there is also evidence from studies with cultured endothelial (Lipton et al., 1988) and glioma cells (Xu et al., 1993) that the transport of ethanolamine and choline may occur through the same or very similar systems. Most of these conclusions were largely derived from studies on the effects of physiological and supraphysiological concentrations of ethanolamine on choline uptake and vice versa. It is possible that these discrepant findings may reflect differences in the uptake systems between species and cell types (Lipton et al., 1988). However, it is probable that isolation and characterization of the choline and ethanolamine transport systems from several mammalian cell types will be required to establish with more certainty whether or not ethanolamine and choline can share the same transporter.

B. Formation of Phosphoethanolamine

The phosphorylation of ethanolamine by ATP is catalyzed by ethanolamine kinase (EC 2.7.1.82) (Figure 1, reaction 3). It has been a point of controversy for many years whether in mammalian tissues the phosphorylation of choline and ethanolamine is catalyzed by one single enzyme (a choline/ethanolamine kinase) or whether choline kinase (EC 2.7.1.32) and ethanolamine kinase are separate enzymes. The two activities are clearly separable in most lower eukaryotes and plants (Kent, 1990 and references cited therein). Early studies with partially purified enzyme preparations suggested that also rat liver contained distinct choline and ethanolamine kinase activities (Weinhold and Rethy, 1974; Brophy et al.,

1977). On the other hand, kinetic studies on choline and ethanolamine kinase activities of lactating bovine mammary gland suggested that the two activities may have distinct active sites on the same protein (Infante and Kinsella, 1976).

Subsequent studies with pure choline/ethanolamine kinases (Table 1) provided strong evidence that choline and ethanolamine kinase are indeed identical enzymes in mammalian tissues (Figure 1, reaction 3). The first evidence came from studies by Ulane et al. (1978, 1982) who purified choline kinase 1000-fold from adult African green monkey lung. The final preparation was stated to be pure on basis of SDS-PAGE analysis. The ratio of choline to ethanolamine kinase activity remained constant throughout the entire purification procedure. The authors concluded that primate lung tissue contained only one enzyme for the phosphorylation of ethanolamine and choline. Much stronger evidence was provided by later studies with enzymes purified to homogeneity from rat kidney (Ishidate et al., 1984, 1985a; Ishidate, 1989; Ishidate and Nakazawa, 1992), and rat liver (Porter and Kent, 1990, 1992). These studies demonstrated copurification of choline and ethanolamine activities to a single homogeneous protein and a constant ratio of activities throughout the purification. Furthermore, antibodies raised against the purified enzymes were able to precipitate choline and ethanolamine kinase activities with completely overlapping titration curves from crude cytosols of a variety of rat tissues. Also the enzyme purified to homogeneity from rat brain (Uchida and Yamashita, 1990, 1992a) utilized choline, ethanolamine as well as several analogous amino alcohols. Draus et al. (1990) recently reported the purification to homogeneity of an ethanolamine kinase from human liver. Although the highly purified enzyme also displayed activity towards choline, the authors did not find complete copurification of ethanolamine and choline kinase activities. They suggested the possibility of several isoforms that may differ with respect to affinity towards ethanolamine and choline, respectively. Taken collectively, the studies described above provide overwhelming evidence that at least in a number of different tissues choline and ethanolamine activities reside on the same protein.

Table 1. Choline/Ethanolamine Kinases Purified to Homogeneity from Mammalian Tissues

	Purification	*Molecular Size*		*Km (mM)*	
Source	*Factor (fold)*	*SDS-PAGE*	*Gel filtration*	*Choline*	*Ethanolamine*
Monkey lung[a]	1000		80,000	0.03	1.2
Rat kidney[b]	1500	42,000	75,000-80, 000	0.1	0.6
Rat brain[c]	15000	44,000	90,000	0.015	0.787
Rat liver[d]	26000	47,000	160,000	0.013	1.2
Human liver[e]	1051	42,000	87,000	n.d.	0.25

Notes: [a] Ulane et al., 1978, 1982
[b] Ishidate et al., 1984, 1985a
[c] Uchida & Yamashita, 1990
[d] Porter & Kent, 1990
[e] Draus et al., 1990

However, the possibility that in some tissues or species kinases may exist that are specific for either ethanolamine or choline cannot yet be entirely excluded.

The molecular masses estimated by gel filtration ranged between 80,000 and 90,000 for the choline/ethanolamine kinases purified from monkey lung, human liver, rat kidney, and rat brain, strongly suggesting a dimeric structure for the enzyme as the molecular masses on SDS-PAGE gels appeared to be between 42,000 and 44,000. Interestingly, the rat liver enzyme purified by Porter and Kent (1990) appears to have a tetrameric structure.

Although considerable differences have been found for the Km values towards choline and ethanolamine for the different purified enzymes (Table 1), the data clearly show that the Km values for choline (ranging between 13 and 100 μM) are much smaller than those for ethanolamine (0.25-1.2 mM). The reported Km values for ATP vary considerably from tissue to tissue and appear to be highly dependent on the free Mg^{2+} concentration (Ishidate and Nakazawa, 1992).

Porter and Kent (1990) reported that choline and ethanolamine were mutually competitive inhibitors. Their respective Km values (0.013 and 1.2 mM) were similar to the Ki values (0.014 and 2.2 mM). This suggested that choline and ethanolamine kinase activities did not only reside on the same protein but also occurred at the same site. A different conclusion was reached by Ishidate et al. (1985a) who found for the enzyme purified from rat kidney that ethanolamine was a weak competitive inhibitor (Ki 4.8 mM) of choline kinase activity whereas choline was a very strong competitive inhibitor (Ki 7.5 μM) of ethanolamine kinase activity. This would suggest that the choline and ethanolamine kinase do not share a common active site on the same enzyme protein, which is in agreement with the earlier kinetic studies of Infante and Kinsella (1976) with supernatant of bovine mammary gland.

The enzyme choline/ethanolamine kinase exists as at least two isoelectric forms in all tissues examined (Tadokoro et al., 1985; Ishidate et al., 1984 and 1985b; Porter and Kent, 1990; Uchida and Yamashita, 1990), and a third form is inducible with polycyclic aromatic hydrocarbons and hepatoxic agents in liver (Tadokoro et al., 1985; Ishidate, 1989; Ishidate and Nakazawa, 1992). This induced form differed from the constitutive forms in molecular weight, isoelectric point, and immunochemical properties, suggesting that it may be encoded by a different gene. The inducibility of choline/ethanolamine kinase may play a role in the long-term regulation of PE and PC synthesis.

Uchida and Yamashita (1992b) recently reported the cloning of a cDNA encoding one of the choline kinase isoforms present in mammalian tissues. This clone was successfully expressed in *Escherichia coli*. The deduced amino acid sequence significantly resembled that published earlier for yeast choline kinase. This cDNA will undoubtedly be of great value for studies on choline/ethanolamine kinase isoenzymes and their regulation in mammalian tissues. Interestingly, the cloned choline kinase was most highly expressed in testis, suggesting a role of this enzyme in spermatogenesis. A human choline kinase cDNA has also been cloned (Hosaka et al., 1992). The deduced sequence of human choline/ethanolamine kinase resembled that of the rat liver enzyme over the entire sequence.

Studies on the reaction mechanism indicated that the forward reaction catalyzed by choline/ethanolamine kinase follows a sequentially ordered mechanism with ATP-Mg^{2+} (or the base) binding to the enzyme first, followed by choline or ethanolamine (ATP-Mg^{2+}), and then activation of the ternary complex by free Mg^{2+}. The release of phosphocholine or phosphoethanolamine occurs prior to that of ADP-Mg^{2+}. The overall rate of the reaction is probably limited by the release of ADP-Mg^{2+} from the complex (Ishidate, 1989; Ishidate and Nakazawa, 1992).

C. Formation of CDP-Ethanolamine

The second step in the CDP-ethanolamine pathway (Figure 1, reaction 4) involves the formation of CDP-ethanolamine and pyrophosphate from CTP and phosphoethanolamine, a reaction that is catalyzed by CTP:phosphoethanolamine cytidylyltransferase (ethanolamine-phosphate cytidylyltransferase, ET). Already in 1975 ET had been purified 1100-fold from a post-microsomal fraction of rat liver (Sundler, 1975). More recently a new procedure was developed to purify the rat-liver enzyme to homogeneity (Tijburg et al., 1992; Vermeulen et al., 1993). This new procedure, comprising hydrophobic interaction chromatography on Octyl Sepharose CL-4B and pseudo-affinity chromatography on Matrex gels Red A and Blue A as key steps, consistently yielded pure ET with a final specific activity (6.5 μmol/min per mg protein) that was a factor of five higher than that reported by Sundler (1975). Polyacrylamide gel electrophoresis under denaturing and reducing conditions showed the presence of a single protein with a molecular mass of 49.6 kDa. Gel filtration experiments revealed a molecular weight of the native enzyme of 100, 000-120, 000 (Sundler, 1975; Tijburg et al., 1992), indicating that the enzyme occurs as a dimer. ET has one sharp pH optimum at pH 7.8 and another one, with lower maximal activity around pH 6-6.5 (Sundler, 1975; Vermeulen et al., 1994). The apparent Km value for CTP is 53 μM. Interestingly, two apparent Km values of 70 and 185 μM were found for phosphoethanolamine (Vermeulen et al., 1994). This could imply that ET contains a second binding site for phosphoethanolamine. Whether this may be related to the dimeric structure of ET and the existence of two pH optima remains to be investigated.

In contrast to choline and ethanolamine kinase, ET and CTP:phosphocholine cytidylyltransferase (choline-phosphate cytidylyltransferase, CT), the enzyme that catalyzes the corresponding step in the formation of PC via the CDP-choline pathway (Figure 1, reaction 5), appear to be separate activities residing on different proteins (Vermeulen et al., 1993). This notion is supported by the following observations: (i) antibodies raised against ET do not cross-react with CT and vice versa, (ii) CT activity is absent in purified ET (and vice versa), (iii) whereas the activities of cytosolic (Choy and Vance, 1978; Feldman et al., 1978; Vermeulen et al., 1993) and purified CT (Cornell, 1991a,b; Vermeulen et al., 1993) are highly dependent on the presence of exogenous lipids, the activity of cytosolic and purified ET was not markedly affected by lipids (Vermeulen et al., 1993), and (iv) the subcellular localization of ET differs considerably from that of CT, which will be discussed in detail in a following section.

Early studies showed that CT and ET do not only catalyze the formation of CDP-choline and CDP-ethanolamine, respectively, but also that of the corresponding deoxyribonucleotides dCDP-choline and dCDP-ethanolamine (Kennedy et al., 1959). Later studies with a mutant cell line derived from CHO cells provided convincing evidence that the formation of CDP-choline and dCDP-choline is catalyzed by a single enzyme (Esko et al., 1981). Likewise it is highly likely, but not yet proven, that one single enzyme (ET) is responsible for the formation of both CDP-ethanolamine and its deoxy-analog. The biological function of dCDP-choline and dCDP-ethanolamine remains an enigma (Kennedy, 1986). The presence of relatively high levels of these deoxyribonucleotides in rapidly metabolizing cells, may simply reflect the high levels of dCTP in the cytoplasma of such cells. On the other hand, selective utilization of dCDP-ethanolamine and dCDP-choline e.g., for the formation of distinct pools or distinct molecular species of PE and PC, respectively, cannot be excluded.

A recent study by Vermeulen et al. (1994) focused on the specificity of ET for phosphorylated bases with a varying degree of *N*-methylation. Increasing the number of *N*-methyl groups on phosphoethanolamine led to a strong increase in the apparent Km value and to an even more pronounced decrease of the Vmax of ET. Phosphomono-methylethanolamine, phosphodimethylethanolamine and phosphocholine appeared to be weak competitive inhibitors of the reaction catalyzed by ET when phosphoethanolamine was used as a substrate, with Ki values of 7.0, 6.8, and 52.9 mM, respectively. The results of that study show that ET is highly specific for phosphoethanolamine. The kinetic parameters of pure ET responded in completely opposite directions to those of CT (Jamil and Vance, 1991) upon changing the number of *N*-methyl groups, further underscoring the view that ET and CT, unlike ethanolamine and choline kinase, are two separate proteins (see Table 2 for a summary of a number of parameters and properties of these two enzymes).

D. Conversion of CDP-Ethanolamine into Phosphatidylethanolamine

The last step in the CDP-ethanolamine pathway (Figure 1, reaction 6) involves the transfer of phosphoethanolamine from CDP-ethanolamine to diacylglycerol resulting in the formation of PE and CMP. The enzyme catalyzing this reaction, ethanolaminephosphotransferase (EC 2.7.8.1; EPT), is distinct from cholinephosphotransferase (EC 2.7.8.2; CPT), which catalyzes the transfer of phosphocholine to diacylglycerol in the CDP-choline pathway. (Figure 1, reaction 7). There is a considerable amount of indirect evidence suggesting that CPT and EPT activities indeed reside on separate proteins, such as different degrees of stimulation of CPT

Table 2. Comparison of some Characteristics of Phosphoethanolamine-and Phosphocholine Cytidylyltransferase of Rat Liver

	ET	*CT*
Monomeric molecular mass (kDa)	49-50[a,b]	42; 44.5[c,d,e]
Native structure	dimers[a,f]	dimers[e,g]
Purified to homogeneity	yes[b]	yes[c,d,e,h,i]
Purification factor	1430[b]	8333[c]
Specific activity (μmol/min/mg)	6.5[b]	47.5[c]
Antibodies available	yes[b]	yes
pH-optima	7.8 and 6-6.5[a,j]	7.0[c,k]
Michaelis constants (mM)		
CTP	0.050-0.053[a,j]	0.22[c]
phosphocholine	6.2[j]	0.17[l]
phosphoethanolamine	0.072[j]	69.4[l]
Specificity constants (K_{cat}/Km)		
phosphoethanolamine	492[j]	5.7[l]
phosphomonomethylethanolamine	48	84
phosphodimethylethanolamine	1.4	430
phosphocholine	0.0024	3222
Maximal activity requires lipids	no[b]	yes[b,c,m,n]
Subcellular localization	cytosol and E.R.[o]	cytosol and E.R.[p] nuclei[q]
Cloned and expressed	no	yes[r,s,t]

Notes: [a] Sundler, 1975; [b] Vermeulen et al., 1993; [c] Weinhold & Feldman, 1992; [d] Sanghera & Vance, 1989; [e] Cornell, 1989b; [f] Tijburg et al., 1992; [g] Weinhold et al., 1989; [h] Watkins & Kent, 1990; [i] Weinhold et al., 1986; [j] Vermeulen et al., 1994; [k] Feldman & Weinhold, 1987; [l] Jamil & Vance, 1991; [m] Choy & Vance, 1978; [n] Cornell, 1991a,b; [o] Van Hellemond et al., 1994; [p] Vance, 1989; [q] Wang et al., 1993; [r] Kalmar et al., 1990; [s] Johnson et al., 1992; [t] Luche et al. 1993.

and EPT by exogenous diacylglycerols, different selectivities of the two enzymes towards various species of diacylglycerols, a different selectivity for Mg^{2+} and Mn^{2+} in some tissues, and the observations that CPT is more sensitive to delipidation than EPT (for review see Cornell, 1989a). More conclusive evidence has come from studies with mutants defective in either CPT or EPT. Polokoff et al. (1981) isolated mutant CHO cells in which the EPT activity was six- to tenfold lower than in the parent cell, whereas the CPT activity was unchanged.

Further evidence that CPT and EPT are encoded by different genes was provided by the isolation of yeast mutants defective in either CPT (Hjelmstad and Bell, 1987) or EPT (Hjelmstad and Bell, 1988) activities. Genetic and biochemical analysis of both classes of mutants strongly suggested that the *CPT1* and *EPT1* loci represent the structural genes for CPT and EPT, respectively. The wild-type genes were cloned by genetic complementation, and enzymological analysis demonstrated that the *CPT1* gene product is predominantly a CPT activity. The *EPT1* gene product was less specific and exhibited both CPT and EPT activity. Sequencing of the *CPT1* gene revealed an open reading frame which encodes 407 amino acids, with a short intron near the 5′-end (Hjelmstad and Bell, 1990). The CPT protein predicted by the open reading frame is 50% hydrophobic, which is consistent with the well known fact that this enzyme is an integral membrane protein. Analysis of the amino acid sequence predicted seven transmembrane α-helical segments. In a subsequent paper Hjelmstadt and Bell (1991a) reported the complete nucleotide sequence of the *EPT1* gene. The sequenced region encoded a 391-amino acid protein product which exhibited 54% amino acid sequence similarity to the *CPT1* gene product. Predictive structural analysis of the yeast *EPT1* gene product revealed close similarities to the *CPT1* gene product with respect to membrane topography, features of secondary structure, and transmembrane asymmetry. Hjelmstad and Bell (1991b) demonstrated that the *CPT1* and *EPT1* gene products are the only choline- and ethanolaminephosphotransferases present in yeast. They constructed a *cpt1 ept1* double null mutant that was devoid of both activities. The activities of the *CPT1* and *EPT1* gene products could then be studied independently in membranes prepared from strains bearing null mutations in the structural gene encoding the cognate gene product. These studies convincingly showed that the two enzymes exhibited intrinsic differences in their substrate specificities towards CDP-choline and CDP-ethanolamine, dioleoylglycerol dependencies, activation by Mg^{2+}, and CMP inhibition profiles. It is still uncertain as to whether these fascinating findings can be extended to CPT and EPT of mammalian cells, although circumstantial evidence certainly suggests that the mammalian and yeast CPT and EPT may be analogous (Hjelmstad and Bell, 1991b).

The purification of EPT and CPT from mammalian tissues has been severely hampered by the fact that these enzymes are integral membrane proteins that are resistant to extraction with high ionic strength medium, EDTA, and with low concentrations of detergent. Kanoh and Ohno (1976) were the first to report a successful solubilization of CPT and EPT from rat liver microsomes using sodium deoxycholate as detergent. This resulted in a 4-5 fold purification of both CPT and EPT. Whereas EPT could be further purified (8.5-fold) by sucrose gradient density centrifugation, the majority of CPT activity was lost during this subsequent step. Several other investigators have reported solubilization of CPT from different membrane sources by use of various detergents (Cornell and MacLennan, 1985; O and Choy, 1990; Ishidate et al., 1993), but further efforts to purify the enzyme have not been very successful due to the susceptibility of this enzyme to detergents. More progress has been made with the solubilization and purification of EPT as this enzyme is relatively stable in the presence of some nonionic detergents. Roberti et al. (1989) reported the solubilization of EPT from rat brain microsomes with a buffer containing 1% Triton X-100 and 0.01% of the antioxidant 2,6-di-*tert*-butyl-4-methylphenol. Purification of the solubilized preparation by means of DEAE Bio-gel A and Hydroxyapatite chromatography yielded an EPT preparation with a specific activity of about 1200-times higher than that of the crude solubilized enzyme. SDS polyacrylamide gel electrophoresis showed that the final preparation still contained five protein bands. Interestingly, the apparent Km of the detergent-solubilized enzyme for CDP-ethanolamine is very low (0.8 μ) in comparison with that of the membrane-bound enzyme (86 μM). The activity of the detergent-solubilized EPT and the affinity of the enzyme for CDP-ethanolamine appeared to be strongly modulated by phospholipids. These observations led Roberti et al. (1989) to the suggestion that PE synthesis is not only controlled by the availabilty of CDP-ethanolamine but also by the lipid composition of the membrane in which the enzyme is embedded.

Cornell (1989a) proposed that in addition to "classic brute force" the use of photo-affinity labels may offer perspective for the purification of CPT and EPT. Following up on this idea Ishidate et al. (1992) studied photo-affinity labeling of CPT from rat liver microsomes directly by radioactive CDP-choline analogs. This approach could facilitate identification of the enzyme on SDS polyacrylamide gels. The authors concluded from their experiments that CPT probably exists in rat liver microsomes as a 55-kDa peptide, which is somewhat higher than the molecular mass reported for the yeast enzyme (Hjelmstad and Bell, 1990). This approach, which seems very promising, has not yet been followed for the identification of EPT.

It is well known that the molecular species compositions of PC and PE in animal tissues show considerable differences, despite the fact that these phospholipids share 1,2-diacyl-*sn*-glycerols as immediate precursor. Many investigators have studied the question as to whether these differences can be explained by different specificities of CPT and EPT towards various molecular species of diacylglycerols. Such studies have included the use of a large variety of exogenous diacylglycerols as well as endogenous membrane-bound diacylglycerols as substrates for EPT and CPT. It would be beyond the scope of this review to discuss these studies in detail, but the major conclusions can be briefly summarized as follows: Whereas CPT does not seem to discriminate between diacylglycerol classes differing with respect to the degree of unsaturation, EPT shows a marked selectivity towards hexaenoic diacylglycerols (see e.g., Kanoh and Ohno, 1975 and references cited therein). Further studies by Morimoto and Kanoh (1978) with solubilized rat liver CPT and EPT and by Holub (1978) with rat liver microsomes demonstrated that CPT and EPT possess a marked acyl chain length dependency with regard to the saturated fatty acids located at the C-1 position of diacylglycerol. Whereas 1-palmitoyl-2-unsaturated species of diacylglycerol were favored by CPT, EPT showed selectivity towards 1-stearoyl-2-unsaturated diacylglycerols over their 1-palmitoyl homologs. Although these findings demonstrate that the selectivities of both CPT and EPT can contribute to maintaining the characteristic molecular compositions of PC and PE, it is obvious that deacylation-reacylation and transacylation mechanisms play an important role in the fine tuning of the composition of these phospholipids (MacDonald and Sprecher, 1991; Schmid et al., 1991). It is interesting that at least in hepatocytes PE appears to be much less subject to remodeling than PC (Samborski et al., 1990).

In many mammalian tissues the plasmalogen content of PE is higher than that of PC (White, 1973; Snyder, 1985). Morikawa et al. (1987) showed that CPT and

EPT of rabbit platelet membranes showed opposite substrate specificities towards alkenylacyl-and diacylglycerol when incubated in the presence of CDP-choline and CDP-ethanolamine. Especially in the presence of CDP-choline, EPT showed a high preference to synthesis of plasmenylethanolamine whereas CPT preferentially synthesized the diacyl form of PC. The authors concluded that this selectivity of EPT and CPT certainly contributes to the uneven alkenylacyl phospholipid distribution in platelets. Ford et al. (1992) studied the substrate selectivity of EPT in rabbit myocardial microsomes. Although these membranes contained a more than 20-fold molar excess of endogenous diacylglycerol over 1-*O*-alk-1′-enyl-2-acylglycerol, incubation with CDP-ethanolamine resulted in the highly selective synthesis of plasmenylethanolamine containing arachidonate at the 2-position. As the endogenous alkenylacylglycerols were similarly enriched in arachidonate, the authors concluded that the predominant determinant of the substrate specificity of rabbit myocard EPT is the nature of the covalent linkage at the *sn*-1 position of the diradylglycerol acceptor.

There is convincing evidence that the CPT catalyzed reaction can operate in the forward as well as in the reversed direction in vivo (Bjornstad and Bremer, 1966; Åkesson and Sundler, 1977). These studies have been extended by experiments in vitro showing that incubation of microsomes from liver (Kanoh and Ohno, 1973 a,b), lung (Sarzala and van Golde, 1976) and brain (Goracci et al., 1981) with CMP promoted the back reaction resulting in the enlargement of the endogenous diacylglycerol pool size. The reaction catalyzed by EPT can also proceed in the reversed direction, as was shown by studies in vitro using microsomes from rat liver (Kanoh and Ohno, 1973b) and rat brain (Goracci et al., 1986). However, both groups reported that the rate of the back reaction was considerably slower with EPT than CPT. In a later study Roberti et al. (1992) studied the reversibility of the reactions catalyzed by EPT and CPT using a solubilized enzyme preparation. The authors showed that the EPT activity catalyzing the forward reaction and that catalyzing the reverse reaction co-eluted during anion exchange chromatography of the solubilized microsomes on DEAE Bio-Gel A, strongly suggesting that both activities indeed reside on one protein.

An important question regards a possible physiological role of the back reactions catalyzed by EPT and CPT. It has been suggested by Goracci et al. (1981) that the reverse reactions of PC and PE synthesis may be responsible for the increase of the cerebral concentrations of CDP-choline and CDP-ethanolamine occurring during brain ischemia. Studies of Sundler and Åkesson (1974b) with isolated hepatocytes provided evidence that diacylglycerols generated by reversal of the CPT reaction contributed about 26 and 13% of the total diacylglycerols that were utilized for PE and PC synthesis, respectively. Recently, Strum et al. (1992) studied the conversion of 1-*O*-alkyl-2-acyl-*sn*-glycero-3-phosphocholine into 1-*O*-alk-1′-enyl-2-acyl-*sn*-glycero-3-phosphoethanolamine in a kidney cell line. Their results strongly suggested that the reverse reaction of CPT may play a crucial role in this conversion by catalyzing the generation of 1-alkyl-2-acylglycerol that can be further metabolized to 1-alkyl-2-acyl-*sn*-glycero-3-phosphoethanolamine and plasmenylethanolamine.

IV. SUBCELLULAR ORGANIZATION OF THE CDP-ETHANOLAMINE PATHWAY

There has been much interest in the subcellular localization of the enzymes involved in the CDP-choline and CDP-ethanolamine pathway. In the majority of these studies the various subcellular fractions were isolated either by classical differential centrifugation or via a variety of density-gradient centrifugation methods. In such approaches characterization of the isolated subcellular fractions has to rely on measurements of appropriate established marker enzymes and, if applicable, on electron microscopical analysis. It should be realized that *bona fide* marker enzymes are not always available for every subcellular fraction. A typical example of such fraction is *cis* Golgi (Vance and Vance, 1988). In addition, there is sometimes uncertainty about the *in situ* localization of the so-called cytosolic enzymes: Are they really cytosolic or are they released from membrane structures during the mechanical fractionation of the tissue or during cell permeabilization employing cholesterol sequestring agents such as digitonin? As will be exemplified below, classical subcellular distribution studies should, if possible, be complemented with immuno-electron microscopy studies. Fortunately, antibodies against the enzymes involved in the CDP-choline and CDP-ethanolamine pathway are becoming available, except for CPT and EPT.

It is generally assumed that choline/ethanolamine kinase is a cytosolic enzyme in mammalian tissues, although there are a few reports showing that in rat brain this enzyme may be partly membrane-associated (for review see Ishidate, 1989). To the best of our knowledge the subcellular localization of choline/ethanolamine kinase has not yet been investigated using immuno-electron microscopy. This may be relevant in relation to the metabolic compartmentation of the enzymes involved in the CDP-choline and CDP-ethanolamine pathway that was recently proposed by George and his associates (George et al., 1989, 1991a,b).

It was generally accepted that ET, the second enzyme of the CDP-ethanolamine pathway, was a soluble enzyme that did not associate with cellular organelles. This conclusion was based on differential centrifugation studies as well as enzyme-release measurements from digitonin-permeabilized hepatocytes (Vermeulen et al., 1993 and references cited therein). However, Vermeulen et al. noted that purified ET had a high content of hydrophobic amino acids, which suggested that *in situ* the enzyme might be associated with some kind of cellular structure. Subsequent immuno-electron microscopy studies using an affinity-purified antibody against ET convincingly showed that the enzyme was not randomly distributed in hepatocytes (Van Hellemond et al., 1994). The majority of ET-label was found in areas that contained cisternae of the endoplasmic reticulum. Within these areas the

ET-label showed a bimodal distribution between the cisternae of the endoplasmic reticulum and the cytosol. Other cellular organelles, including nuclei, mitochondria, plasma membranes, and Golgi, were only marginally labeled.

Until recently it was virtually unanimously accepted that CT is an enzyme that is found in both the cytosolic and endoplasmic reticulum compartment of all animal cells (Vance, 1989b). Although evidence has been provided that CT also occurs in Golgi prepared from rat liver (Vance and Vance, 1988), the great majority of membrane-bound CT was -like ET- believed to be present at the endoplasmic reticulum. However, recent subfractionation and immunofluorescence studies of Kent and her colleagues suggested that in phosphatidylcholine-deficient CHO cells membrane-bound CT was associated with the nuclear membrane (Morand and Kent, 1989; Watkins and Kent, 1992). Further immunofluorescence microscopy studies of Wang et al. (1993) localized also soluble CT to the nucleus of a variety of cell lines, suggesting that CT is predominantly an intranuclear enzyme in these cells. These intriguing findings, which would indicate that the subcellular distribution of CT is entirely different from that of ET, should obviously be complemented with immuno-electron microscopy data.

The terminal enzymes of the CDP-choline and CDP-ethanolamine pathway, CPT and EPT, are localized predominantly in the endoplasmic reticulum (Van Golde et al., 1971; Vance and Vance, 1988). There has been some uncertainty about the question as to whether CPT activity was present in Golgi complex as well. Whereas van Golde et al. (1971) had reported that CPT was present exclusively at the rough and smooth endoplasmic reticulum, Jelsema and Morré (1978) found that the activity of CPT measured in their Golgi preparation could not be accounted for by microsomal contamination. In a very careful study Vance and Vance (1988) compared several different methods to isolate Golgi fractions. Using the method of Fleischer and Karina (1974) that had also been employed by van Golde et al. (1971), they completely confirmed the findings of the latter authors that CPT was absent in the Golgi fraction. However, with the method that had been developed by Croze and Morré (1984), Vance and Vance did find significant CPT (and EPT) activity in the Golgi fraction that could certainly not be explained by contaminating microsomes. As both methods yielded Golgi fractions that were similarly enriched in the *trans* Golgi marker enzyme, Vance and Vance (1988) suggested that the Golgi fraction isolated by the method of Fleischer and Karina would lack *cis* Golgi elements. Unfortunately, a good marker for *cis* Golgi was not available to test this assumption. Although it is currently believed that the Golgi has some capacity to synthesize phospholipids, it was stated by Vance and Vance that unambiguous proof of the presence of CPT and EPT (and other phospholipid biosynthetic enzymes) in Golgi should be obtained by immuno-electron microscopy. In a later paper (Vance, 1990) convincing evidence was provided that highly purified mitochondria contained extremely low CPT and EPT activities. The claims in the older literature for the presence of these enzymes in mitochondrial preparations (for references see Vance, 1990) are probably due to the fact that these preparations contained the mitochondria-associated endoplasmic reticulum fraction.

Several groups (Vance et al., 1977; Coleman and Bell, 1978; Ballas and Bell, 1980) performed studies on the topography of CPT and EPT within the transverse plane of rat liver microsomes and were able to show that the active sites of these enzymes are located on the external surface of the microsomal vesicles. This would support the view that the biosynthesis of PC and PE occurs asymmetrically on the cytoplasmic surface of the endoplasmic reticulum. A later study by Freysz et al. (1982) provided evidence that the situation may be different in chicken brain microsomes. In these microsomes CPT was localized on the outer face of the microsomal vesicles, whereas EPT could be a transmembrane protein or be situated on the inner face of the membrane. The authors suggested that through this asymmetric distribution the two phosphotransferases may participate in forming and maintaining the asymmetric distribution of PC and PE between the inner and outer leaflet of this membrane.

V. REGULATION OF THE CDP-ETHANOLAMINE PATHWAY

A. Potential Regulatory Sites

The regulation of the rate of synthesis of the product of a metabolic path is often considered to be determined by the activity of one so-called rate-limiting enzyme in that pathway. However, as pointed out by Kacser and Burns (1973) it is more useful to determine quantitatively the contribution to overall regulation which is exerted at each individual step of the biosynthetic pathway. It is obvious that this pattern of regulation may differ between various cell types and tissues. An enzyme that is capable of exerting significant control over the pathway is often called a regulatory enzyme; the kinetic properties of such enzymes are controlled by factors other than the concentrations of its substrates (Rolleston, 1972). In other words, the reaction catalyzed by such regulatory enzyme should not be at equilibrium under cellular conditions. Again, enzymes other than regulatory ones may also contribute to the overall control of the pathway, even if they catalyze reactions that are at equilibrium in the cell.

Studies on the regulation of pathways of lipid synthesis are often limited to determination of the incorporation of radioactively labeled substrates or to measurement of the specific and total activities of the enzymes catalyzing the individual steps of the pathway. The latter are performed *in vitro* under Vmax conditions, which do not necessarily reflect the situation *in vivo*. An additional complicating factor is that the substrates of the enzymes under investigation are often not water soluble. Determination of the rate of incorporation of labeled precursors is not free of risks either. It presumes homogeneous labeling of biological specimens and the pool sizes of the intermediates are supposed to remain constant during the experimental period. Hence, it is imperative that such studies include determination of the pool sizes of the intermediates of the pathway.

The overall regulation of PC synthesis via the CDP-choline pathway has been extensively studied (for excellent reviews see Vance, 1989a,b; Ishidate, 1989; Cornell, 1989a; Kent, 1990; Kent et al., 1991). Much less information is available on the control of PE synthesis. Although it has long been tacitly assumed that the CDP-choline and CDP-ethanolamine pathways would be similarly regulated, there is circumstantial evidence that these two pathways are under independent control (for review see Tijburg et al., 1989b). For example, early studies by Groener et al. (1979) indicated that hepatic PC and PE synthesis in rats responded differently to fasting and fasting followed by refeeding of the animals. Later experiments by Tijburg et al. (1988) demonstrated that the hepatic levels of PC and choline-containing precursors of PC were more susceptible to alterations in the dietary state of the rats than the levels of PE and ethanolamine-containing precursors. Observations that the incorporation of labeled choline into PC and that of labeled ethanolamine into PE responded in opposite directions when hepatocytes were exposed to either norepinephrine (Haagsman et al., 1984) or vasopressin (Tijburg et al., 1987b) also pointed to independent control of the CDP-choline and CDP-ethanolamine pathways.

Several studies have been focused on the question as to which steps in the CDP-ethanolamine pathway could play an important regulatory role. Abundant evidence is available that increasing the level of extracellular ethanolamine in the medium from very low levels to concentrations exceeding the physiological level in the circulation by several fold, generally increases the rate of PE synthesis via the CDP-ethanolamine pathway (Sundler and Åkesson, 1975b; Tijburg et al., 1987a; Ronen et al., 1991; McMaster and Choy, 1992a, Houweling et al., 1992). Apart from the fact that it is not known whether ethanolamine transport is at equilibrium under cellular conditions, there is no convincing experimental evidence available that ethanolamine transport is a rate-limiting step for PE synthesis. It is much more likely that the availability of extracellular ethanolamine, rather than its transport, may limit PE synthesis under these conditions.

Kinetic data led Infante (1977) to the conclusion that both the reaction catalyzed by ethanolamine kinase (choline/ethanolamine kinase) and that catalyzed by ET could be regulatory steps in the CDP-ethanolamine pathway in rat liver. Sundler and Åkesson (1975b) demonstrated that exposure of hepatocytes to increasing concentrations of ethanolamine enhanced the incorporation of labeled glycerol into PE. This effect was accompanied by a considerable increase of the pool size of phosphoethanolamine, whereas the pool size of CDP-ethanolamine remained constant. These observations strongly suggested that ET should indeed be considered as a possible regulatory site in PE biosynthesis. Further evidence for such role of ET was provided by studies of Tijburg et al. (1987a) who showed that exposure of hepatocytes to phorbol ester led to an increased rate of incorporation of labeled ethanolamine into PE with a concomitant increase in the activity of ET and a reduction in the pool size of phosphoethanolamine. Interestingly, the phorbol treatment also caused an enhanced conversion of CDP-ethanolamine into PE by EPT. The latter effect could be explained by the increased amount of cellular diacylglycerol that was generated by the treatment of the cells with phorbol ester. Further evidence for a potentially important role of diacylglycerol supply to the regulation of PE synthesis was provided by a subsequent study of Tijburg et al. (1989c) on the effect of glucagon on PE synthesis in hepatocytes. The authors showed that exposure of hepatocytes to glucagon or cyclic AMP analogs decreased the rate of incorporation of labeled ethanolamine into PE without affecting the activity of ET or the pool size of phosphoethanolamine. The inhibitory effect of glucagon on PE synthesis could, however, be attributed to a diminished supply of diacylglycerol resulting in a decreased conversion of CDP-ethanolamine into PE. A similar conclusion was recently reached for the inhibitory effect of glucagon on hepatic PC synthesis (Jamil et al., 1992). These authors showed that the cyclic AMP-mediated inhibition of PC biosynthesis by glucagon was not due to an effect on the phosphorylation of CT but to a diminished supply of diacylglycerol as substrate for CPT. Recent studies of Zimmerman et al. (1994) with maturing type II pneumocytes also showed that CT is not regulated by cAMP-dependent phosphorylation. The studies discussed above allow the conclusion that, apart from ethanolamine availability to the cell, the rate of PE synthesis via the CDP-ethanolamine pathway is largely determined by, perhaps coordinate, control of the activity of ET on the one hand, and the supply of diacylglycerol on the other.

At least in liver there is very little experimental evidence suggesting important control of PE synthesis at the site of ethanolamine kinase. Experiments of Sundler and Åkesson (1975b) and Houweling et al. (1992) with hepatocytes showed that at low ethanolamine concentrations in the medium the rate of PE synthesis is controlled by the supply of ethanolamine, whereas at higher ethanolamine concentrations (> 30-50 μM) phosphoethanolamine started to accumulate indicating that ET had become rate-limiting. As is evident from a recent paper by McMaster and Choy (1992a) the situation may be different in heart tissue. These authors perfused isolated hamster hearts with medium containing varying concentrations of labeled ethanolamine. At low concentrations in the perfusate (< 0.1 μM), the majority of the radioactivity was associated with the phosphoethanolamine fraction, indicating that ET was rate-limiting under these conditions. At higher ethanolamine concentrations in the perfusate (0.4-1000 μM), the majority of the label was in the ethanolamine fraction, suggesting that the conversion of ethanolamine to phosphoethanolamine had become rate-limiting. Interestingly, the intracellular pool size of ethanolamine did not change, suggesting that the newly imported ethanolamine, which was preferentially utilized for PE synthesis, did not equilibrate with the endogenous ethanolamine pool. In another study McMaster and Choy (1992b) suggested that circulating serine may play an important role in modulating PE synthesis via the CDP-ethanolamine pathway in the hamster heart. The authors provided evidence that serine not only attenuated ethanolamine uptake but also reduced the formation of PE from labeled ethanolamine. They suggested that the increased intracellular pool of serine inhibited ethanolamine kinase activity. Interestingly, serine did not modulate choline kinase activity suggesting that serine might be a specific modulator of PE synthesis. The authors showed that serine did not affect the contribution of the decarboxylase pathway for PE formation. In an earlier study Zelinsky and Choy (1982b) had shown that perfusion of isolated hamster heart with (rather high) choline concentrations did not affect the uptake of labeled ethanolamine, but also led to reduced PE synthesis by inhibition of ethanolamine kinase.

Houweling and his colleagues (1991, 1992, 1993) studied the biosynthesis of PC and PE in the regenerating rat liver after partial (70%) hepatectomy. Between 4 and 96 h after partial hepatectomy, the mass of PE increased from 30% to 80% of sham-operation values. In line with the increase in PE mass, the rate of synthesis of PE *in vivo* from labeled ethanolamine was stimulated 1.6- and 1.3-fold at 22 and 48 h after surgery, respectively. Surprisingly, the activity of ET was not altered after partial hepatectomy. In addition, neither ethanolamine kinase nor EPT showed any changes in activity. However, the hepatic levels of ethanolamine and particularly that of phosphoethanolamine drastically increased after partial hepatectomy. A similar strong increase in the pool size of phosphoethanolamine in regenerating liver was also established via a completely different approach (^{31}P-NMR analysis) by Murphy et al. (1992). Interestingly, Houweling et al. (1992) noticed that the concentration of ethanolamine in the serum of the hepatectomized rats had increased from 29 μM to 50 μM during the first day after surgery. They suggested that increased substrate pressure owing to the elevated serum levels of ethanolamine led to enhanced PE synthesis and accumulation of ethanolamine and phosphoethanolamine in the liver remnant. The origin of the enhanced level of circulating ethanolamine was not established. Interestingly, strongly increased levels of phosphoethanolamine have not only been reported in regenerating liver, but also in various tumor tissues (Dixon and Tian, 1993 and references cited therein) and in developing tissues (Gyulai et al., 1984).

The changes in PE metabolism in the regenerating liver (Houweling et al., 1992) are quite different from those occurring in PC metabolism after partial hepatectomy (Houweling et al., 1991, 1993). The restoration of PC mass after partial hepatectomy was not only accompanied by an increased rate of synthesis of PC from labeled choline, but also by a significant increase in the activity of both soluble and microsomal CT and decrease of the pool size of phosphocholine. Interestingly, the stimulating effect on cytosolic CT was mainly due to an increase in the number of enzyme molecules as was shown by immunotitration of the amount of cytosolic CT protein. In a later paper (Houweling et al., 1993) evidence was provided that in regenerating liver PC biosynthesis and CT are regulated at a pretranslational level.

The last enzyme in the CDP-ethanolamine pathway in mammals is thought to be at equilibrium *in vivo* and as such is in itself not a regulatory enzyme. However, as already discussed, there is considerable evidence that the availability of diacylglycerol can be limiting PE (and PC) synthesis under several conditions. In addition, Roberti et al. (1989) showed that lipids can modulate the apparent affinity of EPT for CDP-ethanolamine. They actually suggested that the affinity of membrane-bound EPT is back-regulated by the PE content of the membrane. Ishidate (1993) recently reported the isolation from rat liver microsomes of a heat-labile, nondialyzable factor (presumably a protein) that acted as an inhibitor on CPT but not on EPT. The authors suggested that the inhibition occurred at or near the binding site between the enzyme and diacylglycerol. They postulated that this inhibitory protein may be involved in regulating the availability of diacylglycerol to CPT (but not to EPT) in microsomal membranes. Studies by Mantel et al. (1993) suggested a regulatory role for CDP-choline concentration in the selection of diacylglycerol species by CPT resulting in the de novo synthesis of different molecular species of PC. It would be of interest to investigate whether CDP-ethanolamine concentration would exert a similar regulatory role in PE synthesis. Recently O and Choy (1993) reported that fasting of hamsters for 24 and 48 h resulted in significant reduction in PC biosynthesis, corroborating earlier studies with rats (Groener et al., 1979, Tijburg et al., 1988). Interestingly, O and Choy demonstrated that this reduction of PC synthesis resulted entirely from attenuation of CPT activity by an endogenous inhibitor that was identified as argininosuccinate. The effect of this metabolite on EPT was not investigated in that study.

Summarizing it can be stated that regulation of PE synthesis can take place at multiple sites in the CDP-ethanolamine pathway, although under most conditions the supply of CDP-ethanolamine and diacylglycerol appear to be principal factors governing the rate of this process.

B. Is Translocation of Ethanolamine-Phosphate Cytidylyltransferase Involved in Controlling Phosphatidylethanolamine Synthesis?

The major mechanism identified for the control of PC biosynthesis involves the regulation of the subcellular distribution of CT. The enzyme appears to exist in the soluble fraction as an inactive reservoir and can be translocated rapidly and reversibly to cellular membranes where CT becomes activated by interactions with lipids (see Vance (1989b) and Kent (1990) for review). Most investigators believe that this translocation takes place between the cytosol and the endoplasmic reticulum (e.g., Vance, 1989b; Tercé et al., 1991). As mentioned earlier, Kent and her colleagues recently proposed that soluble CT is located in the nucleus and is translocated to the nuclear membrane (Watkins and Kent, 1992; Wang et al., 1993).

A number of stimuli have been demonstrated to promote translocation of CT to membranes whereas other stimuli cause release of the enzyme from membranes to the cytosol (Vance, 1989b; Kent, 1990; Jamil et al., 1993 and references cited therein). The over-riding common factor in regulation CT binding to membranes may be the ratio of bilayer to non-bilayer lipids in that membrane (Jamil et al.,

1993). In all cases there appeared to be an inverse relationship between the binding of CT to membranes and the ratio of bilayer- to non-bilayer-forming lipids. When the ratio decreases there is enhanced binding of CT to membranes and the enzyme is released when the ratio increases.

Until recently it was generally assumed that the activity of ET, the putative key-regulatory step in the CDP-ethanolamine pathway, is not regulated by a translocation mechanism as it was commonly accepted that ET was a fully soluble enzyme that did not occur in any membrane fraction. However, as mentioned in a preceding section, recent immuno-electron microscopy studies by van Hellemond et al. (1994) suggested that *in situ* ET partitions between endoplasmic reticulum membranes and cytosol. It is certainly feasible that binding of ET to the endoplasmic reticulum may be an important event in regulation of PE synthesis. Possibly there is an equilibrium between the amount of membrane-bound and free ET that is influenced by the amount of binding factor that is exposed at the cytosolic phase of the endoplasmic reticulum. Interestingly, a factor capable of specifically binding CT was recently identified by Feldman and Weinhold (1993). A comparable factor binding ET would fit our hypothesis.

There is increasing interest in the topodynamic regulation of functionally related enzymes (Kaprelyants, 1988). In such a model, enzymes and substrates are arranged in highly organized systems through adsorption to membranes or cytoskeletal elements. Such high degree of functional organization of related enzymes can serve substrate channeling, a process that can markedly increase the overall flux of a reaction chain (Kholodenko and Westerhoff, 1993). Rapid control through ambiquitous distribution of an element of the compartment can regulate the extent of functional organization and thereby the overall flux. In a series of intriguing papers (George et al., 1989, 1991a,b) evidence was presented that the enzymes of the CDP-choline pathway are highly compartmentalized in cultured glioma cells. Only choline that was taken up by the high affinity uptake system was converted into PC, whereas choline, phosphocholine, and CDP-choline that were administered to the cells by electropermeabilization, did not enter into PC. It would be of great interest if this compartmentation could be demonstrated for other cell types and by other techniques of introducing phosphocholine, CDP-choline, and their ethanolamine analogs into the cells. With respect to ET, the subcellular distribution seems compatible with the channeling model. The observation that ET is particularly present in endoplasmic reticulum-rich regions implies that the enzyme is in the proximity of EPT, which is embedded in the endoplasmic reticulum membrane. The early observation by Sundler (1973) that phosphoethanolamine formed from exogenous ethanolamine did not freely equilibrate with the endogenous liver phosphoethanolamine pool also suggests that the product of the ethanolamine kinase reaction is specifically channeled towards ET. In such a compartmentalized system reversible binding of ET to the membrane could play a keyrole in the topodynamic regulation of the entire CDP-ethanolamine pathway.

ACKNOWLEDGMENTS

The research in the authors laboratory is supported by the Netherlands Foundation for Chemical Research (S.O.N.) with financial aid from the Netherlands Organization of Scientific Research (N.W.O.).

REFERENCES

Åkesson, B., & Sundler, R. (1977). Factors controlling the biosynthesis of individual phosphoglycerides in liver. Biochem. Soc. Trans. 5, 43-48.

Ardail, D., Gasnier, F., Lermé, F., Simonot, C., Louisot, P., & Gateau-Roesch, O. (1993). Involvement of mitochondrial contact sites in the subcellular compartmentalization of phosholipid biosynthetic enzymes. J. Biol. Chem. 268, 25985-25992.

Arthur, G., & Page, L. (1991). Synthesis of phosphatidylethanolamine and ethanolamine plasmalogen by the CDP-ethanolamine and decarboxylase pathways in rat heart, kidney and liver. Biochem. J. 273, 121-125.

Arthur, G., & Lu, X. (1993). The ethanolamine requirement of keratinocytes for growth is not due to defective synthesis of ethanolamine phosphoacylglycerols by the decarboxylation pathway. Biochem. J. 293, 125-130.

Baba, S., Watanabe, Y., Gajyo, F., & Arakawa, M. (1984). High-performance liquid chromatographic determination of serum aliphatic amine in chronic renal failure. Clin. Chim. Acta 136, 49-56.

Ballas, L.M., & Bell, R.M. (1980). Topography of phosphatidylcholine, phosphatidylethanolamine and triacylglycerol biosynthetic enzymes in rat liver microsomes. Biochim. Biophys. Acta 602, 578-590.

Bell, R.M., & Coleman, R.A. (1980). Enzymes of glycerolipid synthesis in eukaryotes. Annu. Rev. Biochem. 49, 459-487.

Billah, M.M., & Anthes, J.C. (1991). The regulation and cellular functions of phosphatidylcholine hydrolysis. Biochem. J. 269, 281-291.

Bishop, W.P., & Bell, R.M. (1988). Assembly of phospholipids into cellular membranes. Biosynthesis, transmembrane movement and intracellular translocation. Ann. Rev. Cell Biol. 4, 579-610.

Bjerve, K.S. (1973). The phospholipid substrates in the Ca^{2+}-stimulated incorporation of nitrogen bases into microsomal phospholipids. Biochim. Biophys. Acta 306, 396-402.

Bjerve, K.S. (1985). The biosynthesis of phosphatidylserine and phosphatidylethanolamine from L-[3-^{14}C]serine in isolated rat hepatocytes. Biochim. Biophys. Acta 833, 396-405.

Björnstad, P., & Bremer, J. (1966). *In vivo* studies on pathways for the biosynthesis of lecithin in the rat. J. Lipid Res. 7, 38-45.

Borkenhagen, L.F., Kennedy, E.P., & Fielding, L. (1961). Enzymatic formation and decarboxylation of phosphatidylserine. J. Biol. Chem. 236, 28-29.

Bremer, J., Figard, P.H., & Greenberg, D.M. (1960). The biosynthesis of choline and its relation to phospholipid metabolism. Biochim. Biophys. Acta 43, 477-488.

Brophy, P.J., Choy, P.C., Toone, J.R., & Vance, D.E. (1977). Choline kinase and ethanolamine kinase are separate, soluble enzymes in rat liver. Eur. J. Biochem. 78, 491-495.

Chilton, F.H., & Connell, T.R. (1988). 1-Ether-linked phosphoglycerides. Major endogenous sources of arachidonate in the human neutrophil. J. Biol. Chem. 263, 5260-5265.

Choy, P.C., & Vance, D.E. (1978). Lipid requirements for activation of CTP:phosphocholine cytidylyltransferase from rat liver. J. Biol. Chem. 253, 5163-5167.

Coleman, R., & Bell, R.M. (1978). Evidence that biosynthesis of phosphatidylethanolamine, phosphatidylcholine, and triacylglycerol occurs on the cytoplasmic side of microsomal versicles. J. Cell Biol. 76, 245-253.

Cornell, R. (1989a). Cholinephosphotransferase. In: Phosphatidylcholine Metabolism (Vance, D.E., ed.), pp. 47-64. CRC Press Inc., Boca Raton, FL.

Cornell, R. (1989b). Chemical cross-linking reveals a dimeric structure for CTP:phosphocholine cytidylyltransferase. J. Biol. Chem. 264, 9077-9082.

Cornell, R.B. (1991a). Regulation of CTP:phosphocholine cytidylyltransferase by lipids. 1. Negative surface charge dependence for activation. Biochemistry 30, 5873-5880.

Cornell, R.B. (1991b). Regulation of CTP:phosphocholine cytidylyltransferase by lipids. 2. Surface curvature, acyl chain length, and lipid-phase dependence for activation. Biochemistry 30, 5881-5888.

Cornell, R., & MacLennan, D.H. (1985). Solubilization and reconstitution of cholinephosphotransferase from sarcoplasmic reticulum: Stabilization of solubilized enzyme by diacylglycerol and glycerol. Biochim. Biophys. Acta 821, 97-105.

Croze, E.M., & Morré, D.J. (1984). Isolation of plasma membrane, Golgi apparatus and endoplasmic reticulum fractions from single homogenates of mouse liver. J. Cell. Physiol. 119, 46-57.

Cui, Z., Vance, J.E., Chen, M.H., Voelker, D.R., & Vance, D.E. (1993). Cloning and expression of a novel phosphatidylethanolamine *N*-methyltransferase. A specific biochemical and cytological marker for a unique membrane fraction in rat liver. J. Biol. Chem. 268, 16655-16663.

Daniel, L.W. King, L., & Waite, M. (1981). Source of arachidonic acid for prostaglandin synthesis in Madin-Darby canine kidney cells. J. Biol. Chem. 24, 12830-12835.

Dennis, E.A., & Kennedy, E.P. (1972). Intracellular sites of lipid synthesis and the biogenesis of mitochodria. J. Lipid Res. 13, 263-267.

Dixon, R.M., & Tian, M. (1993). Phospholipid synthesis in the lymphomatous mouse liver studied by ^{31}P nuclear magnetic resonance spectroscopy *in vitro* and by administration of ^{14}C-radiolabelled compounds in vivo. Biochim. Biophys. Acta 1181, 111-121.

Draus, E., Niefind, J., Victor, K., & Havsteen, B. (1990). Isolation and characterization of the human liver ethanolamine kinase. Biochim. Biophys. Acta 1045, 195-204.

Esko, J.D., Wermuth, M.M., & Raetz, C.R.M. (1981). Thermolabile CDP-choline synthetase in an animal cell mutant defective in lecithin formation. J. Biol. Chem. 256, 7388-7393.

Exton, J.H. (1990). Signalling through phosphatidylcholine breakdown. J. Biol. Chem. 265, 1-4.

Feldman, D.A., Kovac, C.R., Dranginis, P.L., & Weinhold, P.A. (1978). The role of phosphatidylglycerol in the activation of CTP:phosphocholine cytidylyltransferase from rat lung. J. Biol. Chem. 253, 4980-4986.

Feldman, D.A., & Weinhold, P.A. (1987). CTP:phosphorylcholine cytidylyltransferase from rat liver. Isolation and characterization of the catalytic subunit. J. Biol. Chem. 262, 9075-9081.

Feldman, D.A., & Weinhold, P.A. (1993). Identification of a protein complex between choline-phosphate cytidylyltransferase and a 112-kDa protein in rat liver. J. Biol. Chem. 268, 3127-3135.

Fleischer, S., & Kervina, M. (1974). Subcellular fractionation of rat liver. Methods Enzymol. 31, 6-41.

Ford, D.A., & Gross, R.W. (1989). Plasmenylethanolamine is the major storage depot for arachidonic acid in rabbit vascular smooth muscle and is rapidly hydrolyzed after angiotesin II stimulation. Proc. Natl. Acad. Sci. USA 86, 3479-3483.

Ford, D.A., Rosenbloom, K.B., & Gross, R.W. (1992). The primary determinant of rabbit myocardial ethanolamine phosphotransferase substrate selectivity is the covalent nature of the *sn*-1 aliphatic group of diradyl glycerol acceptors. J. Biol. Chem. 267, 11222-11228.

Freysz, L., Harth, S., & Dreyfus, H. (1982). Topographic distribution of enzymes synthesizing phosphatidylcholine and phosphatidylethanolamine in chicken brain microsomes. J. Neurochem. 38, 582-587.

George, T.P., Morash, S.C., Cook, H.W., Byers, D.M., Palmer, F.B.St.C., & Spence, M.W. (1989). Phosphatidylcholine biosynthesis in cultured glioma cells: Evidence for channeling of intermediates. Biochim. Biophys. Acta 1004, 283-291.

George, T.P., Cook, H.W., Byers, D.M., Palmer, F.B.St.C., & Spence, M.W. (1991a). Channeling of intermediates in the CDP-choline pathway of phosphatidylcholine biosynthesis in cultured glioma cells is dependent on intracellular Ca^{++}. J. Biol. Chem. 266, 12419-12423.

George, T.P., Cook, H.W., Byers, D.M., Palmer, F.B.St.C., & Spence, M.W. (1991b). Inhibition of phosphatidylcholine and phosphatidylethanolamine biosynthesis by cytochalasin B in cultured glioma cells: potential regulation of biosynthesis by Ca^{2+}-dependent mechanisms. Biochim. Biophys. Acta 1084, 185-193.

Goracci, G., Francescangeli, E., Horrocks, L.A., & Porcellati, G. (1981). The reverse reaction of cholinephosphotransferase in rat brain microsomes: a new pathway for degradation of phosphatidylcholine. Biochim. Biophys. Acta 664, 373-379.

Goracci, G., Francescangeli, E., Horrocks, L., & Porcellati, G. (1986). A comparison of the reversibility of phosphoethanolamine transferase and phosphocholine transferase in rat brain microsomes. Biochim. Biophys. Acta 876, 387-391.

Groener, J.E.M., Klein, W., & van Golde, L.M.G. (1979). The effect of fasting and refeeding on the composition and synthesis of triacylglycerols, phosphatidylcholines, and phosphatidylethanolamines in rat liver. Arch. Biochem. Biophys. 198, 287-295.

Gyulai, L., Bolinger, L., Leigh, J.S.,Jr., Barlow, C., & Chance, B. (1984). Phosphorylethanolamine—the major constituent of the phosphomonoester peak observed by ^{31}P-NMR in developing dog brain. FEBS Lett. 178, 137-142.

Haagsman, H.P., van den Heuvel, J.M., van Golde, L.M.G., & Geelen, M.J.H. (1984). Synthesis of phosphatidylcholines in rat hepatocytes. Possible regulation by norepinephrine via an α-adrenergic mechanism. J. Biol. Chem. 259, 11273-11278.

Hii, C.S.T., Edwards, Y.S., & Murray, A.W. (1991). Phorbol ester-stimulated hydrolysis of phosphatidylcholine and phosphatidylethanolamine by phospholipase D in HeLa cells: Evidence that the basal turnover of phosphoglycerides does not involve phospholipase D. J. Biol. Chem. 266, 20238-20243.

Hjelmstad, R.H., & Bell, R.M. (1987). Mutants of *Saccharomyces cerevisiae* defective in *sn*-1,2-diacylglycerol cholinephosphotransferase. Isolation, characterization, and cloning of the *CPT1* gene. J. Biol. Chem. 262, 3909-3917.

Hjelmstad, R.H., & Bell, R.M. (1988). The *sn*-1,2-diacylglycerol ethanolaminephosphotransferase activity of *Saccharomyces cerevisiae*. Isolation of mutants and cloning of the *EPT1* gene. J. Biol. Chem. 263, 19748-19757.

Hjelmstad, R.H., & Bell, R.M. (1990). The sn-1,2-diacylglycerol cholinephosphotransferase of *Saccharomyces cerevisiae*. Nucleotide sequence, transcriptional mapping, and gene product analysis of the *CPT1* gene. J. Biol. Chem. 265, 1755-1764.

Hjelmstad, R.H., & Bell, R.M. (1991a). *sn*-1,2-Diacylglycerol choline- and ethanolaminephosphotransferases in *Saccharomyces cerevisiae*. Nucleotide sequence of the *EPT1* gene and comparison of the *CPT1* and *EPT1* gene products. J. Biol. Chem. 266, 5094-5103.

Hjelmstad, R.H., & Bell, R.M. (1991b). *sn*-1,2-Diacylglycerol choline- and ethanolaminephosphotransferases in *Saccharomyces cerevisiae*. Mixed micellar analysis of the *CPT1* and *EPT1* gene products. J. Biol. Chem. 266, 4357-4365.

Holub, B.J. (1978). Differential utilization of 1-palmitoyl and 1-stearoyl homologues of various unsaturated 1,2-diacyl-*sn*-glycerols for phosphatidylcholine and phosphatidylethanolamine synthesis in rat liver microsomes. J. Biol. Chem. 253, 691-696.

Hosaka, K., Tanaka, S., Nikawa, J., & Yamashita, S. (1992). Cloning of a human choline kinase cDNA by complementation of the yeast cki mutation. FEBS Lett. 304, 229-232.

Houweling, M., Tijburg, L.B.M., Jamil, H., Vance, D.E., Nyathi, C.B., Vaartjes, W.J., & van Golde, L.M.G. (1991). Phosphatidylcholine metabolism in rat liver after partial hepatectomy. Evidence for increased activity and amount of CTP:phosphocholine cytidylyltransferase. Biochem. J. 278, 347-351.

Houweling, M., Tijburg, L.B.M., Vaartjes, W.J., & Van Golde, L.M.G. (1992). Phosphatidylethanolamine metabolism in rat liver after partial hepatectomy. Control of biosynthesis of phosphatidylethanolamine by the availability of ethanolamine. Biochem. J. 283, 55-61.

Houweling, M., Tijburg, L.B.M., Vaartjes, W.J., Batenburg, J.J., Kalmar, G.B., Cornell, R.B., & van Golde, L.M.G. (1993), Evidence that CTP:choline-phosphate cytidylyltransferase is regulated at a pretanslational level in rat liver after partial hepatectomy. Eur. J. Biochem. 214, 927-933.

Hovius, R., Lambrechts, H., Nicolay, K., & De Kruijff, B. (1990). Improved methods to isolate and subfractionate rat liver mitochondria. Lipid composition of the inner and outer membrane. Biochim. Biophys. Acta 1021, 217-226.

Hovius, R., Faber, B., Brigot, B., Nicolay, K., & De Kruijff, B. (1992). On the mechanism of the mitochrondrial decarboxylation of phosphatidylserine. J. Biol. Chem. 267, 16790-16795.

Ikeda, I., Imaizumi, K., & Sugano, M. (1987). Absorption and transport of base moieties of phosphatidylcholine and phosphatidylethanolamine in rats. Biochim. Biophys. Acta 921, 245-253.

Imaizumi, K., Sakono, M., Mawatari, K., Murata, M., & Sugano, M. (1989). Effect of phosphatidylethanolamine and its constituent base on the metabolism of linoleic acid in rat liver. Biochim. Biophys. Acta 1005, 253-259.

Infante, J.P., & Kinsella, J.E. (1976). Phospholipid synthesis in mammary tissue. Choline and ethanolamine kinases: Kinetic evidence for two discrete active sites. Lipids 11, 727-735.

Infante, J.P. (1977). Rate-limiting steps in the cytidine pathway for the synthesis of phosphatidylcholine and phosphatidylethanolamine. Biochem. J. 167, 847-849.

Ishidate, K. (1989). Choline transport and choline kinase. In: Phosphatidylcholine Metabolism (Vance, D.E., ed.), pp. 9-32. CRC Press Inc., Boca Raton, FL.

Ishidate, K., Nakagomi, K., & Nakazawa, Y. (1984). Complete purification of choline kinase from rat kidney and preparation of rabbit antibody against rat kidney choline kinase. J. Biol. Chem. 259, 14706-14710.

Ishidate, K., Furusawa, K., & Nakazawa, Y. (1985a). Complete co-purification of choline kinase and ethanolamine kinase from rat kidney and immunological evidence for both activities residing on the same enzyme protein(s) in rat tissue. Biochim. Biophys. Acta 836, 119-124.

Ishidate, K., Iida, K., Tadokoro, K., & Nakazawa, Y. (1985b). Evidence for the existence of multiple forms of choline (ethanolamine) kinase in rat tissues. Biochim. Biophys. Acta 833, 1-8.

Ishidate, K., & Nakazawa, Y. (1992). Choline/ethanolamine kinase from rat kidney. Meth. Enzymol. 209, 121-134.

Ishidate, K., Matsuo, R., & Nakazawa, Y. (1992). CDPcholine:1,2-diacylglycerol cholinephosphotransferase from rat liver microsomes. II Photoaffinity labeling by radioactive CDP-choline analogs. Biochim. Biophys. Acta 1124, 36-44.

Ishidate, K., Matsuo, R., & Nakazawa, Y. (1993). CDPcholine:1,2-diacylglycerol cholinephosphotransferase from rat liver microsomes. 1. Solubilization and characterization of the partially purified enzyme and the possible existence of an endogenous inhibitor. Lipids 28, 89-96.

Jamil, H., & Vance, D.E. (1991). Substrate specificity of CTP:phosphocholine cytidylyltransferase. Biochim. Biophys. Acta 1086, 335-339.

Jamil, H., Utal. A.K., & Vance, D.E. (1992). Evidence that cyclic AMP-induced inhibition of phosphatidylcholine biosynthesis is caused by a decrease in cellular diacylglycerol levels in cultured rat hepatocytes. J. Biol. Chem. 267, 1752-1760.

Jamil, H., Hatch, G.M., & Vance, D.E (1993). Evidence that binding of CTP:phosphocholine cytidylyltransferase to membranes in rat hepatocytes is modulated by the ratio of bilayer- to non-bilayer-forming lipids. Biochem. J. 291, 419-427.

Jelsema, C.J., & Morré, D.J. (1978). Distribution of phospholipid biosynthetic enzymes among cell components of rat liver. J. Biol. Chem. 253, 7960-7971.

Johnson, J.E., Kalmar, G.B., Sohal, P.S., Walkey, C.J., Yamashita, S., & Cornell, R.B. (1992). Comparison of the lipid regulation of yeast and rat CTP:phosphocholine cytidylyltransferase expressed in COS cells. Biochem. J. 285, 815-820.

Kacser, H., & Burns, J.A. (1973). The control of flux. Symp. Soc. Exp. Biol. 27, 65-104.

Kalmar, G.B., Kay, R.J., Lachance, A., Aebersold, R., & Cornell, R.B. (1990). Cloning and expression of rat liver CTP:phosphocholine cytidylyltransferase: An amphipatic protein that controls phosphatidylcholine synthesis. Proc. Natl. Acad. Sci. USA 876, 6029-6033.

Kanfer, J.N. (1980). The base exchange enzymes and phospholipase D of mammalian tissues. Can. J. Biochem. 58, 1370-1380.

Kanfer, J.N. (1989). Phospholipase D and the base exchange enzyme. In: Phosphatidylcholine Metabolism (Vance, D.E., ed.). pp. 65-86. CRC Press Inc., Boca Raton, FL

Kano-Sueoka, T., & Errick, J.E. (1981). Effects of phosphoethanolamine on growth of mammary carcinoma cells in culture. Exp. Cell Res. 136, 137-145.

Kano-Sueoka, T., & King, D.M. (1987). Phosphatidylethanolamine biosynthesis in rat mammary carcinoma cells that require and do not require ethanolamine for proliferation. J. Biol. Chem. 262, 6074-6081.

Kanoh, H., & Ohno, K. (1973a). Utilization of endogenous phospholipids by the back-reaction of CDPcholine (-ethanolamine):1,2-diglyceride choline (ethanolamine)-phosphotransferase in rat liver microsomes. Biochim. Biophys. Acta 306, 203-217.

Kanoh, H., & Ohno, K. (1973b). Studies on 1,2-diglycerides formed from endogenous lecithins by the back-reaction of rat liver microsomal CDPcholine:1,2-diacylglycerol cholinephosphotransferase. Biochim. Biophys. Acta 326, 17-25.

Kanoh, H., & Ohno, K. (1975). Substrate-selectivity of rat liver microsomal 1,2-diacylglycerol:CDPcholine (ethanolamine) choline (ethanolamine) phosphotransferase in utilizing endogenous substrates. Biochim. Biophys. Acta 380, 199-207.

Kanoh, H., & Ohno, K. (1976). Solubilization and purification of rat liver microsomal 1,2-diacylglycerol: CDPcholine cholinephosphotransferase and 1,2-diacylglycerol: CDP-ethanolamine ethanolaminephosphotransferase. Eur. J. Biochem. 66. 201-210.

Kaprelyants, A.S. (1988). Dynamic spatial distribution of proteins in the cell. Trends Biochem. Sci. 13, 43-46.

Kennedy, E.P., & Weiss, S.B. (1956). The function of cytidine coenzyme in the biosynthesis of phospholipids. J. Biol. Chem. 22. 193-214.

Kennedy, E.P., Borkenhagen, L.F., & Smith, S.W. (1959). Possible metabolic functions of deoxycytidine diphosphate choline and deoxycytidine diphosphate ethanolamine. J. Biol. Chem. 234. 1998-2000.

Kennedy, E.P. (1986). The biosynthesis of phospholipids. In: Lipids and Membranes. Past, Present and Future (Op den Kamp, J.A.F., Roelofsen, B., & Wirtz, K.W.A., eds.), pp. 171-206. Elsevier, Amsterdam.

Kent, C. (1990). Regulation of phosphatidylcholine biosynthesis. Prog. Lipid Res. 29, 87-105.

Kent, C., Carman, G.M., Spence, M.W., & Dowhan, W. (1991). Regulation of eukaryotic phospholipid metabolism. FASEB J. 5, 2258-2266.

Kholodenko, B.N., & Westerhoff, H.V. (1993). Metabolic channeling and control of the flux. FEBS Lett. 320, 71-74.

Kiss, Z, & Anderson, W.B. (1989). Phorbol ester stimulates the hydrolysis of phosphatidylethanolamine in leukemic HL-60, NIH 3T3, and baby hamster kidney cells. J. Biol. Chem. 264. 1483-1487.

Kiss, Z. (1991). Determination of phospholipase D-mediated hydrolysis of phosphatidylethanolamine. Lipids 26, 321-323.

Kiss, Z. (1992). Differential effects of platelet-derived growth factor, serum and bombesin on phospholipase D-mediated hydrolysis of phosphatidylethanolamine in NIH 3T3 fibroblasts. Biochem. J. 285, 229-233.

Kruse, T., Reiber, H., & Neuhoff, V. (1985). Amino acid transport across the human blood-CSF barrier. J. Neurol. Sci. 70, 129-138.

Kuge, O., Nishijima, M., & Akamatsu, Y. (1985). Isolation of a somatic-cell mutant defective in phosphatidylserine biosynthesis. Proc. Natl. Acad. Sci. USA 82, 1926-1930.

Kuge, O., Nishijima, M., & Akamatsu, Y. (1986a). Phosphatidylserine biosynthesis in cultured Chinese hamster ovary cells. II. Isolation and characterization of phosphatidylserine auxotrophs. J. Biol. Chem. 261, 5790-5794.

Kuge, O., Nishijima, M., & Adamatsu, Y. (1986b). Phosphatidylserine biosynthesis in cultured Chinese hamster ovary cells. III. Genetic evidence for utilization of phosphatidylcholine and phosphatidylethanolamine as precursors. J. Biol. Chem. 261, 5795-5798.

Kuge, O., Nishijima, M., & Akamatsu, Y. (1991). A Chinese hamster cDNA encoding a protein essential for phosphatidylserine synthase I activity. J. Biol. Chem. 266, 24184-24189.

Lipton, B.A., Yorek, M.A., & Ginsberg, B.H. (1988). Ethanolamine and choline transport in cultured bovine aortic endothelial cells. J. Cell Physiol. 137, 571-576.

Lipton, B.A., Davidson, E.P., Ginsberg, B.H., & Yorek, M.A. (1990). Ethanolamine metabolism in cultured bovine aortic endothelial cells. J. Biol. Chem. 265, 7195-7201.

Luche, M.M., Rock, C.O., & Jackowski, S. (1993). Expression of rat CTP:phosphocholine cytidylyltransferase in insect cells using a Baculovirus vector. Arch. Biochem. Biophys. 301, 114-118.

MacDonald, J.I.S., & Sprecher, H. (1991). Phospholipid fatty acid remodeling in mammalian cells. Biochim. Biophys. Acta 1084, 105-121.

Mantel, C.R., Schulz, A.R., Miyazawa, K., & Broxmeyer, H.E. (1993). Kinetic selectivity of cholinephosphotransferase in mouse liver: the Km for CDP-choline depends on diacylglycerol structure. Biochem. J. 289, 815-820.

McMaster, C.R., & Choy, P.C. (1992a). Newly imported ethanolamine is preferentially utilized for phosphatidylethanolamine biosynthesis in the hamster heart. Biochim. Biophys. Acta 1124, 13-16.

McMaster, C.R., & Choy, P.C. (1992b). Serine regulates phosphatidylethanolamine biosynthesis in the hamster heart. J. Biol. Chem. 267, 14586-14591.

Menon, A.K., & Stevens, V.L. (1992). Phosphatidylethanolamine is the donor of the ethanolamine residue linking a glycosylphosphatidylinositol anchor to protein. J. Biol. Chem. 267, 15277-15280.

Milakolfsky, L., Hare, T.A., Miller, J.M., & Vogel, W.H. (1985). Rat plasma levels of amino acids and related compounds during stress. Life Sci. 36, 753-761.

Miller, M.A., & Kent, C. (1986). Characterization of the pathways for phosphatidylethanolamine biosynthesis in Chinese hamsters ovary mutant and parental cell lines. J. Biol. Chem. 261, 9753-9761.

Morand, J.N., & Kent, C. (1989). Localization of the membrane-associated CTP:phosphocholine cytidylyltransferase in Chinese hamster ovary cells with an altered membrane composition. J. Biol. Chem. 264, 13785-13792.

Morikawa, S., Taniguchi, S., Fujii, K., Mori, H., Kumada, K., Fujiwara, M., & Fujiwara M. (1987). Preferential synthesis of diacyl and alkenylacyl ethanolamine and choline glycerophospholipids in rabbit platelet membranes. J. Biol. Chem. 262, 1213-1217.

Morimoto, K., & Kanoh, H. (1978). Acyl chain length dependency of diacylglycerol cholinephosphotransferase and diacylglycerol ethanolaminephosphotransferase. Effect of different saturated fatty acids at the C-1 or C-2 position of diacylglycerol on solubilized rat liver microsomal enzymes. J. Biol. Chem. 253, 5056-5060.

Murphy, E.J., Brindle, K.M., Rorison, C.J., Dixon, R.M., Rajagopalan, B., & Radda, G.K. (1992). Changes in phosphatidylethanolamine metabolism in regenerating rat liver as measured by ^{31}P-NMR. Biochim. Biophys. Acta 1135, 27-34.

O, K-M., & Choy, P.C. (1990). Solubilization and partial purification of cholinephosphotransferase in hamster tissues. Lipids 25, 122-124.

O, K-M., & Choy, P.C. (1993). Effects of fasting on phosphatidylcholine biosynthesis in hamster liver: regulation of cholinephosphotransferase activity by endogenous argininosuccinate. Biochem. J. 289, 727-733.

Pelech, S.L., & Vance, D.E. (1984). Regulation of phosphatidylcholine biosynthesis. Biochim. Biophys. Acta 799, 217-251.

Polokoff, M.A., Wing, D.C., & Raetz, C.H.R. (1981). Isolation of somatic cell mutants defective in the biosynthesis of phosphatidylethanolamine. J. Biol. Chem. 256, 7687-7690.

Porter, T.J., & Kent, C. (1990). Purification and characterization of choline/ethanolamine kinase from rat liver. J. Biol. Chem. 265, 414-422.

Porter, T.J., & Kent, C. (1992). Choline/ethanolamine kinase from rat liver. Meth. Enzymol. 209, 134-146.

Pu, G.A-W., & Anderson, R.E. (1984). Ethanolamine accumulation by photoreceptor cells of the rabbit retina. J. Neurochem. 42, 182-191.

Roberti, R., Vecchini, A., Freysz, L., Masoom, M., & Binaglia, L. (1989). An improved procedure for the purification of ethanolaminephosphotransferase. Reconstitution of the purified enzyme with lipids. Biochim. Biophys. Acta 1004, 80-88.

Roberti, R., Mancini, A., Freysz, L., & Binaglia, L. (1992). Reversibility of the reactions catalyzed by cholinephosphotransferase and ethanolaminephosphotransferase solubilized from rat-brain microsomes. Biochim. Biophys. Acta 1165, 183-188.

Rolleston, F.S. (1972). Metabolic regulation. In: Current Topics in Cellular Regulation, Vol. 5 (Horecker, B.L., & Stadman, E.R., eds.), pp. 47-75. Academic Press, NY.

Ronen, S.M., Rushkin, E., & Degani, H. (1991). Lipid metabolism in T47D human breast cancer cells: ^{31}P and ^{13}C-NMR studies of choline and ethanolamine uptake. Biochim. Biophys. Acta 1095, 5-16.

Samborski, R.W., Ridgway, N.D., & Vance, D.E. (1990). Evidence that only newly made phosphatidylethanolamine is methylated to phosphatidylcholine and that phosphatidylethanolamine is not significantly deacylated-reacylated in rat hepatocytes. J. Biol. Chem. 265, 18322-18329.

Sanghera, J.S., & Vance, D.E. (1989). CTP:phosphocholine cytidylyltransferase is a substrate for cAMP-dependent protein kinase *in vitro*. J. Biol. Chem. 264, 1215-1223.

Sarzala, M.G., & van Golde, L.M.G. (1976). Selective utilization of endogenous unsaturated phosphatidylcholines and diacylglycerols by choline-phosphotransferase of mouse lung microsomes. Biochim. Biophys. Acta 441, 423-432.

Schmid, P.C., Johnson, S.B., & Schmid, H.H.O. (1991). Remodeling of rat hepatocyte phospholipids by selective acyl turnover. J. Biol. Chem. 266, 13690-13697.

Siddiqui, R.A., & Exton, J.H. (1992). Phospholipid base exchange activity in rat liver plasma membranes. Evidence for regulation by G-protein and P_{2y}-purinergic receptor. J. Biol. Chem. 267, 5755-5761.

Snyder, F. (1985). Metabolism, regulation and function of ether-linked glycerolipids. In: Biochemistry of Lipids and Membranes (Vance, D.E., & Vance, J.E., eds.), pp. 273-295. Benjamin Cummings Publishing Company, Menlo Park, CA.

Srere, P.A. (1987). Complexes of sequential metabolic enzymes. Ann. Rev. Biochem. 56, 89-124.

Strickland, K.P. (1973). The chemistry of phospholipids In: Form and function of Phospholipids (Ansell, G.B. Hawthorne, J.N., & Dawson, R.M.C., eds.), pp. 9-42. Elsevier Scientific Publishing Company, Amsterdam.

Strum, J.C., Emilsson, A., Wykle, R.L., & Daniel, L.W. (1992). Conversion of 1-*O*-alkyl-2-acyl-*sn*-glycero-3-phosphocholine to 1-*O*-alk-1'-enyl-2-acyl-*sn*-glycero-3-phosphoethanolamine. A novel pathway for the metabolism of ether-linked phosphoglycerides. J. Biol. Chem. 267, 1576-1583.

Sundler, R. (1973). Biosynthesis of rat liver phosphatidylethanolamines from intraportally injected ethanolamine. Biochim. Biophys. Acta 306, 218-226.

Sundler, R., Åkesson, B., & Nilsson, Å. (1974a). Quantitative role of base exchange in phosphatidylethanolamine synthesis in isolated rat hepatocytes. FEBS Lett. 43, 303-307.

Sundler, R., Åkesson, B., & Nilsson, Å. (1974b). Sources of diacylglycerols for phospholipid synthesis in rat liver. Biochim. Biophys. Acta 337, 248-254.

Sundler, R. (1975). Ethanolaminephosphate cytidylyltransferase. Purification and characterization of the enzyme from rat liver. J. Biol. Chem. 250, 8585-8590.

Sundler, R., & Åkesson, B. (1975a). Biosynthesis of phosphatidylethanolamines and phosphatidylcholines from ethanolamine and choline in rat liver. Biochem. J. 146, 309-315.

Sundler. R., & Åkesson, B. (1975b). Regulation of phospholipid biosynthesis in isolated rat hepatocytes. Effect of different substrates. J. Biol. Chem. 250, 3359-3367.

Suzuki, T.T., & Kanfer, J.N. (1985). Purification and properties of an ethanolamine-serine base exchange enzyme of rat brain membranes. J. Biol. Chem. 260, 1394-1399.

Tadokoro, K., Ishidate, K., & Nakazawa, Y. (1985). Evidence for the existence of isoenzymes of choline kinase and their selective induction in 3-methylcholanthrene- or carbon tetrachloride-treated rat liver. Biochim. Biophys. Acta 835, 501-513.

Tercé, F., Record, M., Tronchère, H., Ribbes, G., & Chap, H. (1991). Cytidylyltransferase translocation onto endoplasmic reticulum and increased de novo synthesis without phosphatidylcholine accumulation in Krebs-II ascite cells. Biochim. Biophys. Acta 1084, 69-77.

Tijburg, L.B.M., Houweling, M., Geelen, M.J.H., & Van Golde, L.M.G. (1987a). Stimulation of phosphatidylethanolamine synthesis in isolated rat hepatocytes by phorbol 12-myristate 13-acetate. Biochim. Biophys. Acta 922, 184-190.

Tijburg, L.B.M., Schuurmans, E.A.J.M., Geelen, M.J.H., & van Golde, L.M.G. (1987b). Effects of vasopressin on the synthesis of phosphatidylethanolamines and phosphatidylcholines by isolated rat hepatocytes. Biochim. Biophys. Acta 919, 49-57.

Tijburg, L.B.M., Houweling, M., Geelen, M.J.H., & van Golde, L.M.G. (1988). Effects of dietary conditions on the pool sizes of precursors of phosphatidylcholine and phosphatidylethanolamine synthesis in rat liver. Biochim. Biophys. Acta 959, 1-8.

Tijburg, L.B.M., Geelen, M.J.H., & Van Golde, L.M.G. (1989a). Regulation of the biosynthesis of triacylglycerol, phosphatidylcholine and phosphatidylethanolamine in the liver. Biochim. Biophys. Acta 1004, 1-19.

Tijburg, L.B.M., Geelen, M.J.H., & Van Golde, L.M.G. (1989b). Biosynthesis of phosphatidylethanolamine via the CDP-ethanolamine route is an important pathway in isolated rat hepatocytes. Biochem. Res. Commun. 160, 1275-1280.

Tijburg, L.B.M., Houweling, M., Geelen, M.J.H., & van Golde, L.M.G. (1989c). Inhibition of phosphatidylethanolamine synthesis by glucagon in isolated rat hepatocytes. Biochem. J. 257, 645-650.

Tijburg, L.B.M., Vermeulen, P.S., & Van Golde, L.M.G. (1992). Ethanolamine-phosphate cytidylyltransferase. Meth. Enzymol. 229, 258-263.

Uchida, T., & Yamashita, S. (1990). Purification and properties of choline kinase from rat brain. Biochim. Biophys. Acta 1043, 281-288.

Uchida, T., & Yamashita, S. (1992a). Choline/ethanolamine kinase from rat brain. Meth. Enzymol. 209, 147-153.

Uchida, T., & Yamashita, S. (1992b). Molecular cloning, characterization, and expression in Escherichia coli of a cDNA encoding mammalian choline kinase. J. Biol. Chem. 267, 10156-10162.

Ulane, R.E., Stephenson, L.L., & Farrell, P.M. (1978). Evidence for the existence of a single enzyme catalyzing the phosphorylation of choline and ethanolamine in primate lung. Biochim. Biophys. Acta 531, 295-300.

Ulane, R.E. (1982). The copcholine pathway: choline kinase. In: Lung Development: Biological and Clinical Perspectives (Farrell, P.M., ed.), Vol. 1, pp. 295-316. Academic Press, NY.

Vance, D.E., Choy, P.C., Bloke Farren, S., Lein, P.H., & Schneider, W.J. (1977). Asymmetry of phospholipid biosynthesis in rat liver microsomes. Nature 270, 268-269.

Vance, D.E. (1989a). Regulatory and functional aspects of phosphatidylcholine metabolism. In: Phosphatidylcholine Metabolism (Vance, D.E., ed.), pp. 225-239. CRC Press Inc., Boca Raton, FL.

Vance, D.E. (1989b). CTP: cholinephosphate cytidylyltransferase. In: Phosphatidylcholine Metabolism (Vance, D.E., ed.), pp. 33-45. CRC Press Inc., Boca Raton, FL.

Vance, J.E., & Vance, D.E. (1986). Specific pools of phospholipids are used for lipoprotein secretion by cultured rat hepatocytes. J. Biol. Chem. 261, 4486-4491.

Vance, J.E. (1988). Compartmentalization of phospholipids for lipoprotein assembly on the basis of molecular species and biosynthetic origin. Biochim. Biophys. Acta 963, 70-81.

Vance, J.E, & Vance, D.E. (1988). Does rat liver Golgi have the capacity to synthesize phospholipids for lipoprotein secretion? J. Biol. Chem. 263, 5898-5909.

Vance, J.E. (1990). Phospholipid synthesis in a membrane fraction associated with mitochondria. J. Biol. Chem. 265, 7248-7256.

Vance, J.E. (1991). Newly made phosphatidylserine and phosphatidylethanolmaine are preferentially translocated between rat liver and endoplasmic reticulum. J. Biol. Chem. 266, 89-97.

Van den Bosch, H. (1974). Phosphoglyceride metabolism. Ann. Rev. Biochem. 43, 243-277.

Van Golde, L.M.G., Fleischer, B., & Fleischer, S. (1971). Some studies on the metabolism of phospholipids in Golgi complex from bovine and rat liver in comparison to other subcellular fractions. Biochim. Biophys. Acta 249, 318-330.

Van Golde, L.M.G., Raben, J., Batenburg, J.J., Fleischer, B., Zambrano, F., & Fleischer, S. (1974). Biosynthesis of lipids in Golgi complex and other subcellular fractions from rat liver. Biochim. Biophys. Acta 360, 179-192.

Van Hellemond, J.J., Slot, J.W., Geelen, M.J.H., van Golde, L.M.G., & Vermeulen, P.S. (1994). Ultrastructural localization of CTP:phosphoethanolamine cytidylyltransferase in rat liver, J. Biol. Chem. 269, 15415-15418.

Vermeulen, P.S., Tijburg, L.B.M., Geelen, M.J.H., & Van Golde, L.M.G. (1993). Immunological characterization, lipid dependence, and subcellular localization of CTP:phosphoethanolamine cytidylyltransferase purified from rat liver. Comparison with CTP:phoshocholine cytidylyltransferase. J. Biol. Chem. 268, 7458-7464.

Vermeulen, P.S., Geelen, M.J.H., & Van Golde, L.M.G. (1994). Substrate specificity of CTP: phosphoethanolamine cytidylyltransferase purified from rat liver. Biochim. Biophys. Acta, 1211, 343-349.

Voelker, D.R. (1984). Phosphatidylserine functions as the major precursor of phosphatidylethanolamine in cultured BHK-21 cells. Proc. Natl. Acad. Sci. USA 81, 2669-2673.

Voelker, D.R. (1985). Disruption of phosphatidylserine translocation to the mitochondria in baby hamster kindey cells. J. Biol. Chem. 260, 14671-14676.

Voelker, D.R., & Frazier, J.L. (1986). Isolation and characterization of a Chinese hamster ovary cell line requiring ethanolamine or phosphatidylserine for growth and exhibiting defective phosphatidylserine synthase activity. J. Biol. Chem. 261, 1002-1008.

Voelker, D.R. (1989). Reconstitution of phosphatidylserine import into rat liver mitochondria. J. Biol. Chem. 264, 8019-8025.

Voelker, D.R. (1991). Organelle biogenesis and intracellular lipid transport in eukaryotes. Microbiol. Revs. 55, 543-560.

Voelker, D.R. (1993). The ATP-dependent translocation of phosphatidylserine to the mitochondria is a process that is restricted tot the autologous organelle. J. Biol. Chem. 268, 7069-7074.

Wang, Y., Sweitzer, T.D., Weinhold, P.A., & Kent, C. (1993). Nuclear localization of soluble CTP:phosphocholine cytidylyltransferase. J. Biol. Chem. 268, 5899-5904.

Watkins, J.D., & Kent, C. (1990). Phosphorylation of CTP:phosphocholine cytidylyltransferase in vivo. Lack of effect of phorbol ester treatment in HeLa cells. J. Biol. Chem. 265, 2190-2197.

Watkins, J.D., & Kent, C. (1992). Immunolocalization of membrane-associated CTP:phosphocholine cytidylyltransferase in phosphatidylcholine-deficient Chinese Hamster Ovary cells. J. Biol. Chem. 267, 5886-5892.

Weinhold, P.A., & Rethy, V.B. (1974). Separation, purification, and characterization of ethanolamine kinase and choline kinase from rat liver. Biochemistry 13, 5135-5141.

Weinhold, P.A., Rounsifer, M.E., & Feldman, D.A. (1986). The purification and characterization of CTP:phosphorylcholine cytidylyltransferase from rat liver. J. Biol. Chem. 261, 5104-5110.

Weinhold, P.A., Rounsifer, M.E., Charles, L., & Feldman, D.A. (1989). Characterization of cytosolic forms of CTP:choline-phosphate cytidylyltransferase in lung, isolated alveolar type II cells, A549 cell and Hep G2 cells. Biochim. Biophys. Acta 1006, 299-310.

Weinhold, P.A., & Feldman, D.A. (1992). Choline-phosphate cytidylyltransferase. Meth. Enzymol. 209, 248-258.

White, D.A. (1973). The phospholipid composition of mammalian tissues. In: Form and Function of Phospholipids (Ansell, G.B. Hawthorne, J.N., & Dawson, R.M.C., eds.), pp. 441-482. Elsevier Scientific Publishing Company, Amsterdam.

Xu, Z., Byers, D.M. Palmer, F.B.St.C., Spence, M.W., & Cook, H.W. (1991). Serine utilization as a precursor of phosphatidylserine and alkenyl-(plasmenyl)-, alkyl-, and acylethanolamine phosphoglycerides in cultured glioma cells. J. Biol. Chem. 266, 2143-2150.

Xu, Z., Byers, D.M., Palmer, F.B.St.C., Spence, M.W., & Cook, H.W. (1993). Limited metabolic interaction of serine with ethanolamine and choline in the turnover of phosphatidylserine, phosphatidylethanolamine and plasmalogens in cultured glioma cells. Biochim. Biophys. Acta 1168, 167-174.

Yorek, M.A., Rosario, R.T., Dudley, D.T., & Spector, A.A. (1985). The utilization of ethanolamine and serine for ethanolamine phosphoglycerine synthesis by human Y79 retinoblastoma cells. J. Biol. Chem. 260, 2930-2936.

Yorek, M.A., Dunlap, J.A., Spector, A.A., & Ginsberg, B.H. (1986). Effect of ethanolamine on choline uptake and incorporation into phosphatidylcholine in human Y79 retinoblastoma cells. J. Lipid Res. 27, 1205-1213.

Zachowski, A. (1993). Phospholipids in animal eukaryotic membranes: Transverse asymmetry and movement. Biochem. J. 294, 1-14.

Zelinski, T.A., Savard, J.D., Man, R.Y.K., & Choy, P.C. (1980). Phosphatidylcholine biosynthesis in isolated hamster heart. J. Biol. Chem. 255, 11423-11428.

Zelinski, T.A., & Choy, P.C. (1982a). Phosphatidylethanolamine biosynthesis in isolated hamster heart. Can. J. Biochem. 60, 817-823.

Zelinski, T.A., & Choy, P.C. (1982b). Choline regulates phosphatidylethanolamine biosynthesis in isolated hamster heart. J. Biol. Chem. 257, 13201-13204.

Zimmerman, L.J., Lee, W-S., Smith, B.T., & Post, M. (1994). Cyclic AMP-dependent protein kinase does not regulate CTP:phosphocholine cytidylyltransferase activity in maturing type II cells. Biochim. Biophys. Acta 1211, 44-50.

OF PHOSPHOLIPIDS AND PHOSPHOLIPASES

Moseley Waite

ABSTRACT

This chapter is designed to bring together two research areas, phospholipase enzymology and lipid biophysics. A true understanding of phospholipases, their function, and their physiologic and pharmacologic regulation requires both approaches. From

Advances in Lipobiology
Volume 2, pages 323-350.

ISBN 0-7623-0205-4

the technical point of view, it is important for the investigator to match the goal of the phospholipase study with the proper technical approach for substrate preparation. The best technical approach can be chosen with a basic understanding of the physical as well as the chemical properties of the phospholipid. Here a number of studies on the properties of phospholipids will be cited including some new work from our laboratory that is designed to further explain the basis for methods of phospholipase measurement. More detailed information on these subjects can be seen in works by Small (1986), Marsh (1990), Birdi (1989), Davies and Rideal (1963), Lindblom and Rilfors (1989), de Kruijff, et al. (1985), Chachaty (1987), and Tanford (1980).

I. CHARACTERISTICS OF PHOSPHOLIPASES

Phospholipases comprise a class of ubiquitous enzymes that share two fundamental characteristics. First, they are esterases, although two types of esterase activity exist. One type has acyl hydrolase activity (PLA_1, PLA_2, PLB, and lysoPL) while the second type has phosphodiesterase activity (PLC and PLD; Figure 1). Second, they share common substrates, phospholipid and water. While they all hydrolyze phospholipid, some will also degrade neutral glycerides and some use agents other than water to break the ester bond such as alcohols. As far as is known now though, all substrates used by phospholipases to break ester bonds have a free-hydroxyl group to replace water. In the case of the acyl hydrolases, a transacylation reaction can occur in which an acyl group is transferred from one phospholipid to the hydroxyl of an acceptor hydroxyl group of a lysoPL or other lipid. These transacylase reactions are catalyzed by some but certainly not all of the acyl hydrolases. One of the phosphodiesterases, PLD, has transferase activity. This class of enzymes act via a transphosphatidylation reaction in which it is thought that a phosphatidyl-enzyme intermediate is formed. This phosphatidyl group, covalently linked to the enzyme, subsequently can be transferred to a wide range of hydroxyl containing compounds such as those found on phospholipids (ethanolamine, serine, choline, glycerol) as well as ethanol and other primary alcohols.

The functions of the phospholipases are widely divergent as are the substrates for the phospholipases (Waite, 1987). Some but probably not all of the functions that phospholipases carry out are:

1. Digestion of extracellular phospholipid (nutrient sources)
2. Formation of bioactive molecules or the precursors of bioactive molecules
3. Membrane remodeling
4. Toxicity

1. The target for the first listed function, digestion, has either emulsified or membranous lipid as substrate. Examples of the former are bile salt micelles in the

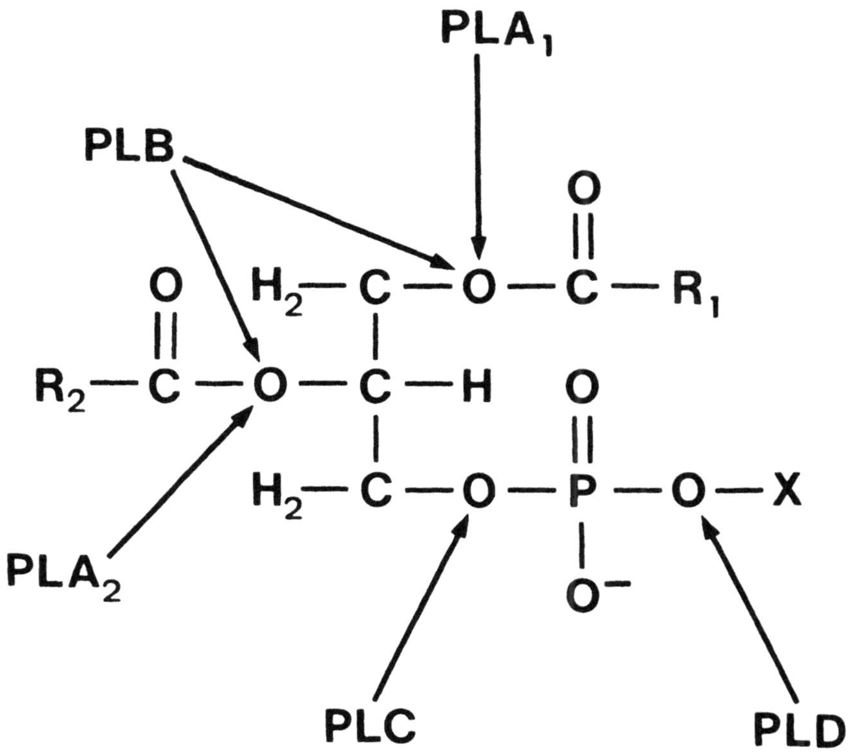

Figure 1.

intestine and lipoproteins in the circulatory system. In these cases the degradation products are taken into the cell for energy sources, for membrane renewal, or repackaging for further transport. Membranous phospholipid can be degraded by phospholipase's secreted from bacteria, often PLC's. It appears as if the action of bacterial PLC on extracellular phospholipid can provide a source of phosphate required for bacterial growth.

2. The role of phospholipases in signal transduction pathways is clearly established and involves PLA_2, PLC and PLD (Liscovitch, 1992). These enzymes are present in cells and are activated by cell signaling events that can involve multiple regulatory components (kinases, phosphatases, G proteins, etc.). (See Figure 2). The action of these phospholipases on cellular membranes is limited and carefully regulated. As a consequence, relatively little membrane lipid is degraded and mechanisms within the cell rapidly restore depleted membrane lipid to maintain

Phospholipases in Signal Transduction

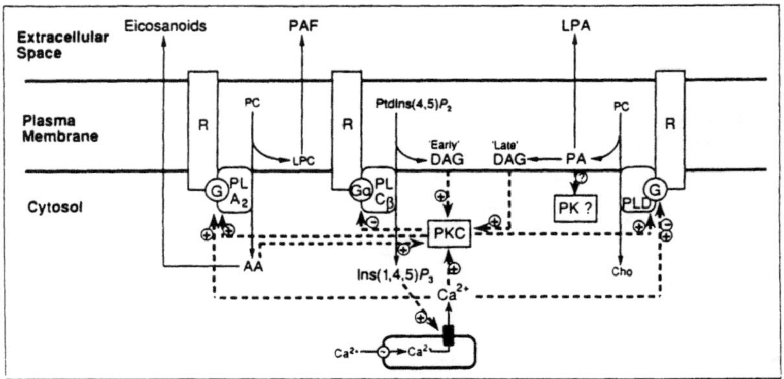

Figure 2. A schematic and simplified representation of coordinate, receptor-mediated activation of multiple phospholipases in signal transduction. Note that all the signal activated phospholipases depicted are not necessarily activated by a given agonist in one particular cell. Likewise, many of the interactions shown have not been established beyond doubt. Here only G protein-mediated receptor-phospholipase coupling is depicted. AcCoA, fatty acyl-coenzyme A; CDPCho, CDP-choline; CDP-DAG, CDP-diacylglycerol; Cho, choline; FFA, free fatty acid; Gro, glycerol; Ins, D-*myo*-inositol; LPA, lysophosphatidic acid; LPC, Lysophosphatidylcholine; l-MAG, l-monoacylglycerol; 2-MAG, 2-monoacylglycerol; PtdIns4*P*, phosphatidylinositol-4-phosphated; PK, protein kinase; AA, arachidonic acid; R, receptor. (From Liscovitch, 1992.)

cellular homeostasis. The complexities of subcellular organization of the events (i.e., organelle membrane targets involved and translocation events) remain elusive.

3. Biological membranes contain a host of phospholipid molecules that differ in the polar head group as well as their acyl composition. It can be estimated that hundreds of different molecular species of phospholipid exist in a mammalian cell. Given the specificity of enzymes involved in the incorporation of the acyl components during *de novo* phospholipid synthesis, it is clear that mechanisms must exist for remodeling the acyl content of phospholipids. Specifically, phospholipids formed *de novo* are comprised mainly of mono-and diunsaturated fatty acids at position-2 of the glycerol backbone. Examination of the composition of membranous phospholipid shows a high content of polyunsaturated acyl chains that are thought to be incorporated by a deacylation-reacylation pathway (Lands, 1960).

Studies in intact cells using ^{18}O incorporation into phospholipids provide strong evidence that the deacylation-reacylation cycle replacement of acyl chains occurs at both position 1 and 2 of the glycerol backbone, a finding that implicates both PLA_1, and PLA_2 in membrane phospholipid remodeling (Schmid, 1991). Transacylases are also thought to be involved in these remodeling events also, especially during cell stimulation (Neito, et al., 1990). The relative contribution of the deacylation-reacylation versus. transacylation pathways in remodeling is unknown.

4. The fourth function of phospholipases as toxins will not be a focus of this chapter since these events appear to involve phospholipase-receptor interaction as well as bulk lipid hydrolysis. This area of research has taken on a new dimension, however, since it has been reported that phospholipases that are secreted from the cell can bind to cell surface receptors, an event that is not cytotoxic but may be a part of the signal transduction mechanism (Arita, 1991). It is possible that PLA_2 or PLA_2-like toxins from venoms mimic cellular PLA_2 binding to receptors but cause a pathologic cellular response.

Despite the different molecular mechanisms, functions and origins, all share the basic characteristic of acting at a lipid-water interface. This implies that all phospholipases have the capacity to interact with bulk lipid. Although not investigated in all cases, the action of phospholipase's is regulated by the nature of the interface that includes the surface potential and charge and lipid packing. These, plus other factors, are of major importance to the understanding of phospholipase action. It is well recognized, though, that many but not all phospholipases are activated by interaction with the lipid interface. This was most vividly demonstrated by de Haas and his coworkers (Van Eijk, 1983) when they compared the activity of a number of PLA2 on substrates above and below the cmc (Figure 3).

A number of causes for interfacial activation have been advanced for this observed activation:

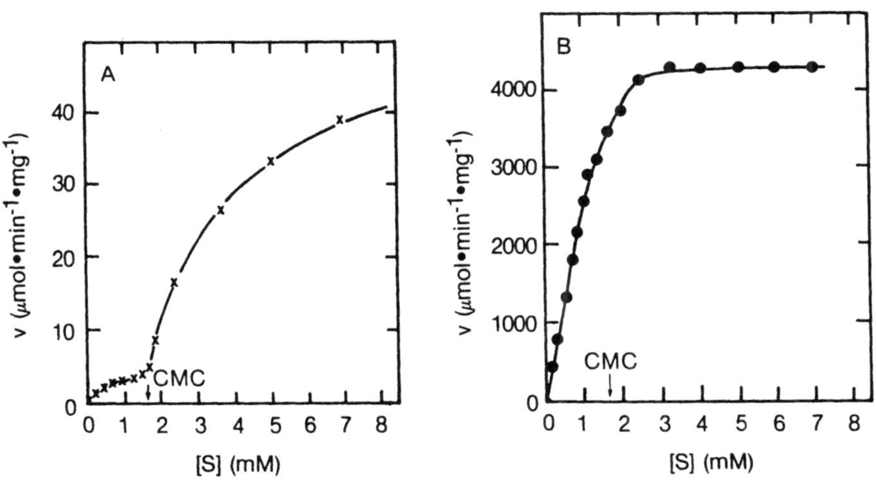

Figure 3. Hydrolysis of diheptanoyl glycerophosphatidylcholine catalyzed by (A) porcine pancreatic or (B) *N. melanoleuca* phospholipase A_2. (From van Eijk et al., 1983.)

1. Induction of a conformational change in the enzyme molecule
2. Increased effective concentration of the substrate
3. Enhanced diffusion of product from the enzyme

The extent to which each of these factors may regulate a given phospholipase is yet to be established. Since many of the phospholipases and lipases studied thus far have a "lipid-binding" site it is probable that conformational changes do occur to some extent when lipolytic enzymes bind to the lipid interface.

II. ASSAY OF PHOSPHOLIPASE

Many different assays are used to assay phospholipases and these have been discussed by several authors in adequate detail (summarized in "The Phospholipases" Waite, 1987). Improvements and refinements have been introduced since 1987 but the principles involved remain much the same. The following is a summary of techniques commonly used with a comment on the pros and cons for each.

Methods	*Pro's and Con's*
Titration, H^+ release	Good continuous assay but lacks sensitivity, specificity
Colorimetric, release of thiol from thioester bond	Sensitive but substrate difficult to synthesize and properties are not identical to native substrate
Radiolabel, release of radiolabeled product	Sensitive and employs natural substrates but methods are usually laborious and less well suited to kinetic studies
Fluorimetric, release of fluorophore that changes fluorescence of lipid	Rapid and sensitive but substrate is not identical to native substrate

There is an exception to the use of a fluorescently labeled phospholipid, an assay system in which the fatty acid released from the substrate displaces a fluorescent fatty acid bound to protein (Kinkaid and Wilton, 1991).

III. TYPES OF SUBSTRATE FORMS (AGGREGATED STATES) USED IN PHOSPHOLIPASE ASSAYS

In general, the natural substrates for phospholipases are either membranous or in detergent emulsions. The goal in understanding the physiologic activity of a phospholipase is to establish how it degrades lipid in the natural substrate form. A number of investigators have used natural substrates effective to study basic mechanisms of phospholipase action. A commonly used biological membrane system is radiolabeled *E. coli* membranes that have been autoclaved to inactivate endogenous enzymes (Elsbach et al., 1985). Likewise, bile salt-phospholipid emulsions have been used extensively to study pancreatic phospholipases since these emulsions mimic the natural substrate rather faithfully (de Haas et al., 1963). It is difficult, on the other hand, to study the detailed mechanisms of the phospholipases with these systems since they each have complex and uncontrolled characteristics. Realizing that any assay system used will be limited to the characteristics of the aggregated state of phospholipid chosen, it is important to appreciate the characteristics of the system as fully as possible. The information of phospholipases in a physically well defined system facilitates understanding of their action in more complex but natural substrate systems. Figure 4 cartoons most of the structures to be described here.

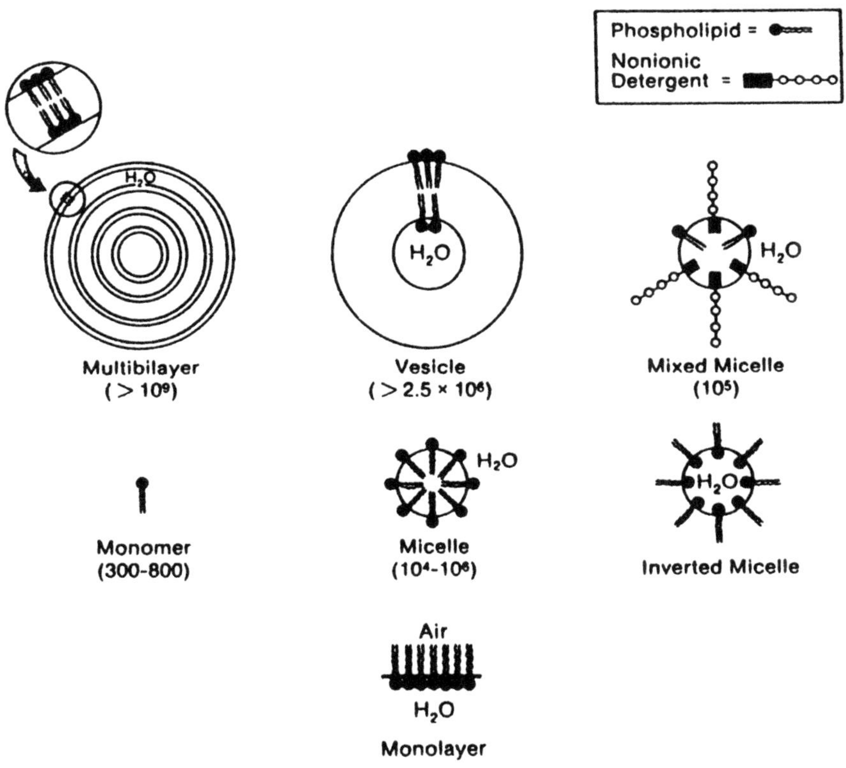

Figure 4. Schematic representation of possible aggregation states of phospholipids and apparent molecular weights. (Adapted from Hazlett et al., 1990.)

A. Monomer-Aggregated Substrate Systems

As stated earlier, it is a characteristic of many phospholipases to have a higher turnover number when the concentration of phospholipid substrate exceeds the cmc, when compared with monomeric substrate. Often these studies are carried out with short to medium chain length acyl groups on phosphatidylcholine so that the final aggregated state of the substrate is a micelle. The selection of phospholipid substrates that form micelles rather than bilayer vesicles is based on the relatively high cmc of micelle forming substrate, often in the mMolar range. The high solubility (high cmc) of the substrates allows a more accurate phospholipase assay on the monomeric substrate. The generality of phospholipase activation by the formation of a lipid-water interface is not universal, however, for example, the PLA_2 from *Naja melanolenca* venom is fully active on monomeric substrate and does not attack micellar substrate effectively (Figure 3). Likewise, chemical modification of venom phospholipases (α amino to keto group conversion) reduces their ability to attach phospholipid micelles (Verheij, 1981a).

The consideration of the lipid aggregation as a simple system in which monomers are in equilibrium with a single and stable form of micelle (or any aggregated form, for that matter) may be too simplistic. A number of studies on the process of monomer to micelle conversion suggest that intermediate states may exist and that the degree to which an intermediate state exists is dependent on the amphipath studied. The process of aggregate formation of a lipid, L, can be considered by two general formula. (Modified from Kresheck, et al.)

$$L+L \leftrightarrow L_2+L \leftrightarrow L_3+L \leftrightarrow L_4 \cdots \leftrightarrow L_{n-1}+L \leftrightarrow L_n \quad (1)$$

Where L_n is the largest sized stable aggregate and single monomers add to a growing micelle

$$nL \leftrightarrow L_n \quad (2)$$

Where no intermediate stages exist

Undoubtedly there are a number of variations on these two formulations. An example of one such variation was proposed by Allgyer and Wells (1978) to explain anomalous phospholipase activity at concentrations of substrate just below the cmc. In modeling of their substrate, dihexanoylphosphatidylcholine, they proposed that two micellar states existed. The "phase transition", as they termed this conversion of one micellar form to the other, was heterogeneous and dependent upon the length of the acyl chains. As it was discussed by Allgyer and Wells, the true nature of these two forms is still ill defined. The "heterogeneity" of the two micelle forms postulated might well fit with formulation 1 with the modification that relatively stable intermediate forms may exist.

$$L+L \leftrightarrow L_2+d \cdots hL \leftrightarrow L_{2+d \cdots h} \leftrightarrow L_n \qquad (3)$$

As described here, an indefinite number of monomers (2+d•••h) would aggregate to a heterogenous but stable intermediate size micelle ($L_{2+d \cdots h}$) that would, as the total phospholipid concentration increased, form the final stable sized micelle Ln. At higher chain length, bilayer rather than micellar structures are formed, as is found when acyl chains of nine carbons are present on phosphatidylcholine. No distinction is made in either formulation 1 or 3 as to whether the intermediate sized micelles would aggregate to form the final sized particle L_n. It would seem most reasonable, however, that the micelles form by addition of monomers since these conversions are concentration dependent.

Studies using NMR techniques have been useful in probing the mechanism of the process of micelle formation. While many of the studies carried out thus far do not employ phospholipids, the results reported using other amphipaths give important insight into the events that occur during the micelle formation. The mechanism of micellization is not uniform and depends on the amphipath studied and the conditions of the solvent (temperature, ionic strength, ph, etc.). In an NMR study by Persson et al. (1979), the formation of small aggregates of [^{13}C]octanoate with 4-5 monomers first occurred as the octanoate concentration was increased. The aggregation number increased to 10-11 monomers as the octanoate concentration was further increased. Reduction in the acyl chain by two methylene groups (hexanoate) lead to the formation of a small aggregate only (4-5 monomers). Likewise, the addition of two methylene groups to hexadecylphosphocholine increased the aggregation number from about 155 to 200 monomers per micelle for octadecylphosphocholine. Proton NMR studies of the Ó methylene protons have shown that dihexanoylphosphatidylcholine undergoes a micelle rearrangement between 1 and 4 times the concentration of its cmc. In this case the phospholipid molecules went from a fast exchange between monomers and micelles to a slow exchange system.

While no uniform single model can be drawn, it is clear that for many phospholipid substrates one must consider more that simple phospholipase action on monomers vs. single micelle population. As described earlier, Wells (1974a) found that the activity of phospholipases A2 was parabolic rather than hyperbolic when studied as a function of substrate concentration in the substrate range between monomers and maximal sized micelles. This is most easily seen in Figure 5 when the data are plotted as the change in substrate concentration divided by the observed velocity ([S]/V).

If the kinetics of hydrolysis were only hyperbolic on both monomers and micelles (with an activated rate on micelles), there would be no intermediate rates observed between about 10 and 40 mm dihexanoylphosphatidylcholine. This study fits very well with the proton NMR study of dihexanoylphosphatidylcholine micelle formation. One can conclude from these works that the PLA_2 from venom or pancreas act optimally on the more tightly packed (slower monomer micelle

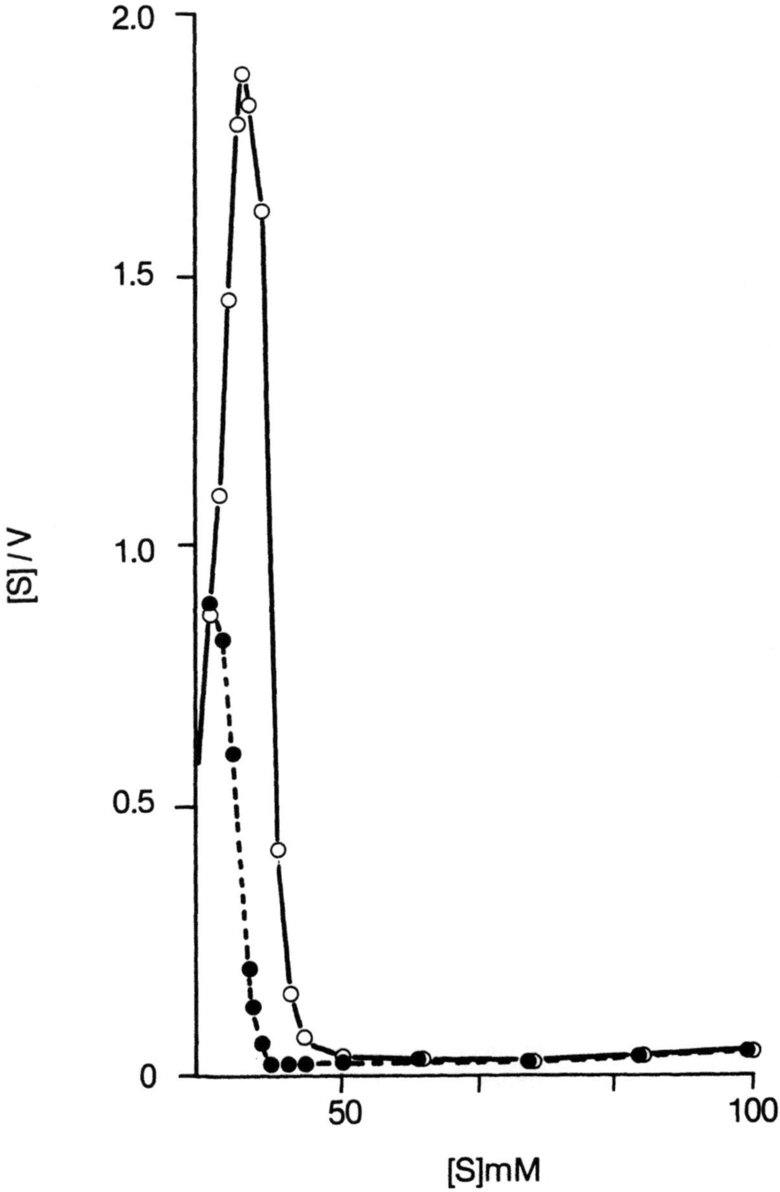

Figure 5. Effect of the concentration of dihexanoyllecithin on the rate of hydrolysis by phospholipase A_2. Reaction was carried out at 45°C and pH 8.0; o, reaction in the presence of 1 mM Ca^{2+}; _ reaction in the presence of 1 mM Ca^{2+} and 2 M KCl. (From Wells, 1974.)

exchange rate) than the loosely packed aggregate formed at 10-40 mM substrate. The poorest substrate, however, is the monomer.

Comparison of the precursor zymogen form of the pancreatic PLA_2 with its processed, mature form provides evidence that the loosely formed aggregate can lead to interfacial activation of the enzyme (Volwerk et al., 1979). The zymogen PLA_2 that does not undergo interfacial activation exhibits hyperbolic kinetics on the monomer only, whereas the processed phospholipase shows parabolic activity in the concentration range just below the cmc of substrate (Figure 6).

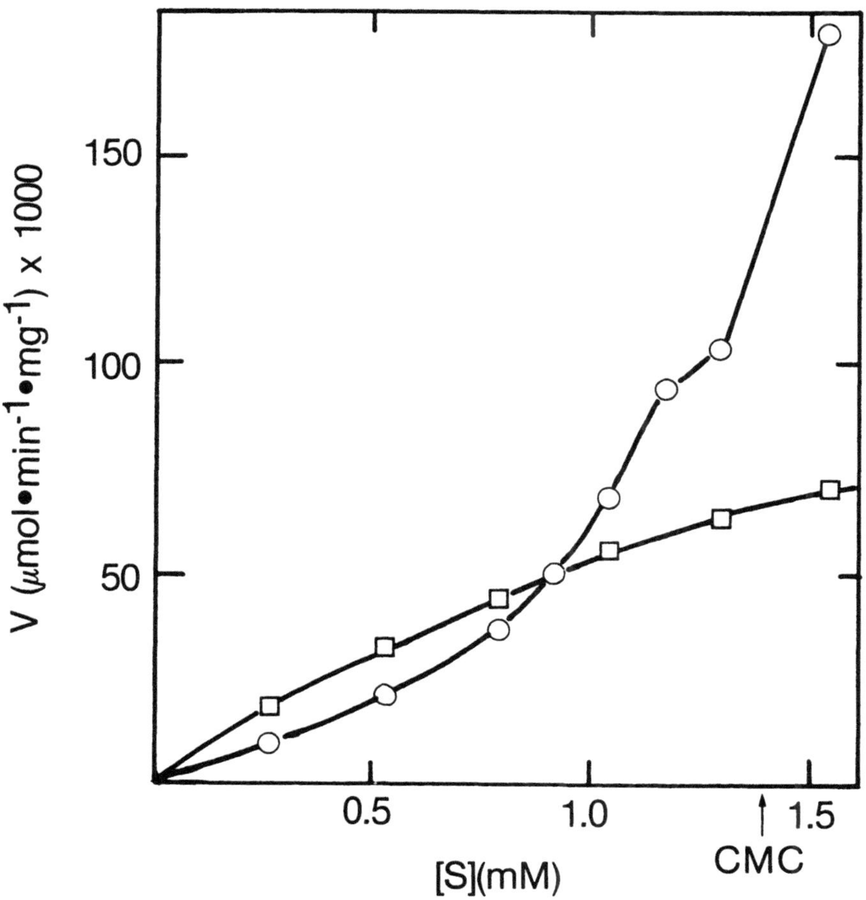

Figure 6. Hydrolysis rates of 2-(decanoylthio)ethyl phosphatidylcholine by phospholipase A_2 (o) and by prophospholipase A_2 (□) at pH 6.0 in the presence of 10 mM CaC12, 100 mM NaCl, and 200 mM sodium acetate at 25°C. (From Volwerk et al., 1979.)

While not often studied, in one series of experiments inverted micelles were used as substrate for PLA_2. These experiments were carried out to more fully understand the basis for the activation of snake venom phospholipases in an ether-aqueous mixture (Wells, 1974b; Poon and Wells, 1974; Misorowski and Wells, 1974). The water content of the inverted phosphatidylcholine micelles can be carefully regulated which allowed Wells and coworkers to investigate the requirements for water structure in hydrolysis. The apparent minimal state of hydration in the inverted micelle that supported hydrolysis was in a structure with a radius of 20Å and approximately 1100 molecules of water (35 or so molecules of phosphatidylcholine). The requirement for Ca^{2+} under these conditions is quite complex in that Ca^{2+} behaved as a competitive inhibitor with water. One criterion for PLA_2 activity is the requirement that the enzyme exist in a soluble form within the micelle. A second and very interesting criterion is that there must be free, unbound water for hydrolysis to occur, that is, the enzyme competes with both Ca^{2+} and substrate for water used in hydrolysis. This series of papers constitute the only study uncovered by this author that describes the nature of water molecules that regulate PLA_2 action.

B. Vesicles

Phospholipid bilayer vesicles are used most often to mimic biological membrane systems. These vesicles are prepared in a wide range of sizes and can be either single- or multishelled structures. The definition of these vesicles and their utility in the study of biological membranes was first described 30 years ago by Alec Bangham and coworkers (Bangham et al, 1965). Accordingly, these vesicles have sometimes been referred to as Bangosomes. Because the commonly used phospholipids for vesicle preparation have very low cmc, consideration of the contribution of monomers or intermediate aggregates usually has been neglected. The vesicles are rather stable and provide an excellent diffusion barrier that prevents entrapped polar solutes from crossing the bilayer. Single shelled vesicles can undergo extensive hydrolysis by PLA_2 without disruption of the vesicle if the products of hydrolysis, lysolipid and free fatty acid, are not removed by proteins such as albumin or fatty acid binding protein, or by metal ions that complex to the freed carboxyl group Jain et al. (1980).

Not all diacyl phospholipids form bilayer structures and consideration of substrate specificity of phospholipases should take that into account. In reality, most have the capacity to form bilayers but the conditions required such as temperature or the pH of the solution are extreme. For example, natural phosphatidylethanolamines form hexagonal II (HII) phases at neutral pH values (de Kruijff, et al., 1985). The tendency of phosphatidylethanolamine to form bilayers versus. HII phases is dependent to a great extent upon its acyl composition. In general, the type of structure formed by phospholipids is related to the relative area of the hydrated polar head group to the area of the cylinder formed by the acyl chains. The closer these are to be equal, the more likely the phospholipid is to forming a bilayer vesicle.

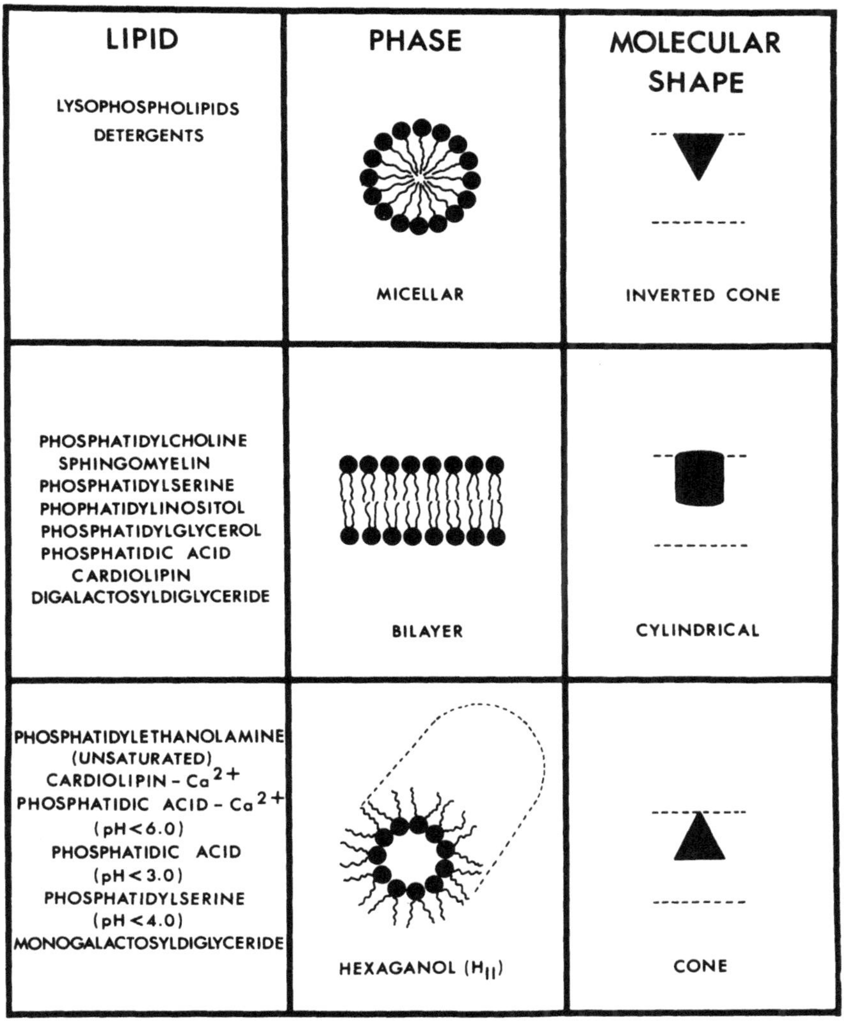

Figure 7. Polymorphic phases and corresponding dynamic molecular shapes of lipids. (From Cullis et al., 1991.)

However, deviation towards a large acyl group cylinder, the more likely the phospholipid is to form a HII phase. Conversely, a large polar head group, relative to the acyl chain cylinder, produces micelles (de Kruijff et al., 1985).

One of the interesting features of the HII phase is the effect of bivalent ions such as Ca^{2+} has upon interaction with anionic lipids. The result of this interaction is to reduce the water of hydration and produce a condensing effect that results in the transition from a bilayer to HII phase. In the case of phosphatidylserine, the interaction with Ca^{2+} produces a related but distinct form termed cochleate (Papahadjapoulos et al., 1975). The action of phospholipases may induce phase changes. For example, hydrolysis of a phospholipid by phospholipase D in the presence of Ca^{2+} can cause disruption of the bilayer vesicle by interaction with the product, phosphatidic acid, with Ca^{2+}.

Temperature has a major effect on the structure formed by phosphatidylethanolamine. Acyl chain motion increases with temperature more than that of other moieties of the molecule which means, in effect, that the area of the acyl chain cylinder increases relative to that of the head group. The consequence of increasing temperature, therefore, is to convert the bilayer to a HII phase via a proposed intermediate inverted micelle or a HI phase. It is possible that a single lipid or appropriate mixtures of lipids can undergo the following with increasing temperature transitions:

bilayer (gel)→bilayer (liquid crystal)→intermediate state (micelle, HexI, cubic)→HexII

Although the relevance of these phase transitions to physiologic systems are still poorly understood, some evidence exists that implicate nonbilayer structures in biological membranes. A prime example of natural lipid that can form nonbilayer structures are those from *A. laidlawii* (Lindbolm and Rilfors, 1989). This organism responds to changes in growth temperature by altering both its acyl composition of the membrane lipid as well as the ratio of the lipid classes. In *A. laidlawii* monoglucosyl-diacylglycerol is the primary lipid to regulate formation of nonbilayer structures. In eukaryotic cells, evidence exists that nonbilayer structures exist as part of the smooth endoplasmic reticulum system. It has been proposed that tubules in the smooth endoplasmic reticulum have a diameter of 30-60nm and have been postulated to be in a "bicontinuous cubic phase" (Lindbolm and Rilfors, 1989).

A major question that is rarely confronted in studies of the regulation of membrane turnover is the extent to which structural transition changes in membranes alter phospholipase action. This type of study is exceedingly complex especially since we are limited in the ability to study transition states of membranes in living biological systems. In static systems of isolated membrane lipid a few examples do show that the transition from one phase to another can drastically alter phospholipase action. It has been well established that phospholipases are more active at the transition of a gel to a liquid crystal phase than on either of the pure phases (op den Kamp, 1974). van den Bosch and coworkers (Lenting et al., 1988) demonstrated that the liver mitochondrial PLA_2 preferentially hydrolyzed HII phase lipid. (See Figure 8.)

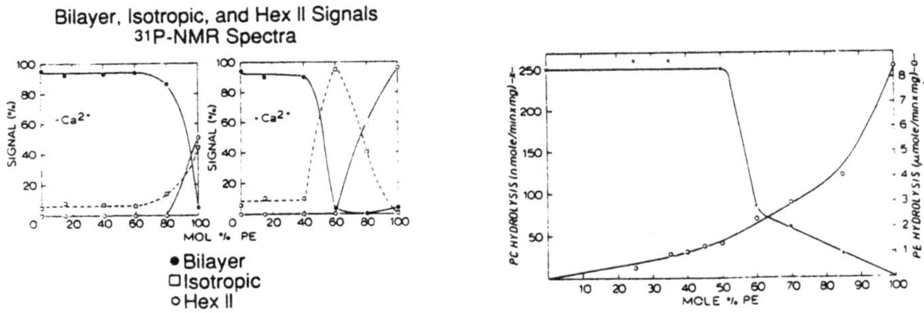

Figure 8. The left panel shows the transition from a bilayer to Hex II phase as the mol percentage of phosphatidylethanolamine is increased as determined by NMR spectroscopy. The right panel shows the activity of the liver PLA_2 as the percentage of phosphatidylethanolamine in phosphatidylcholine is increased. (From Lenting et al., 1988.)

In this study comixtures of phosphatidylcholine and phosphatidylethanolamine were used; these underwent a bilayer to HII phase transition as the percentage of phosphatidylethanolamine was increased. This change was rather abrupt and led to a marked enhancement of phospholipase action. Our studies of the lysosomal phospholipase A_1 from rat liver likewise showed that HII phase lipid was degraded more readily than bilayer vesicles (Robinson and Waite, 1983). The mechanism behind the HII preference exhibited by these two enzymes is not established but could relate to the packing characteristics of lipid in HII versus. bilayer vesicles.

Pure phospholipid vesicles, when used as substrates for phospholipases, have provided a great deal of the knowledge we have about lipolysis. As stated in the beginning of this section, under the appropriate conditions, the entire outer leaflet of the bilayer structure can be hydrolyzed without disruption of the vesicle. Also, transbilayer movement of phospholipid substrate or the products of hydrolysis is negligible unless the vesicle is perturbed. What must be taken into consideration also though, is the number of phospholipid molecules present on the outer leaflet of a single vesicle, therefore, the effective substrate concentration is that of the outer leaflet. Since the transfer rate of natural phospholipid monomers between bilayer vesicles is in the order of 3×10^{-6} per second (Marsh, 1990) phospholipid exchange between vesicles is negligible relative to the catalytic rates of phospholipases (Berg et al., 1991). Under assay conditions with saturating substrate concentrations, the number of vesicles will exceed the number of enzyme molecules several fold. This set of conditions presents an interesting enzymologic challenge, one that has been addressed by many investigators. Multiple steps in the process of substrate binding and catalysis have been considered and a generalized model has appeared. The model described by Jain, Berg and their coworkers focuses on phospholipases A_2 but can be extrapolated to other phospholipases. The contribution of each step in the sequence to the overall hydrolytic rate will vary from one phospholipase to another, however. This is particularly true when an enzyme that has marked interfacial activation (orders of magnitude higher on lipid aggregates then mono-

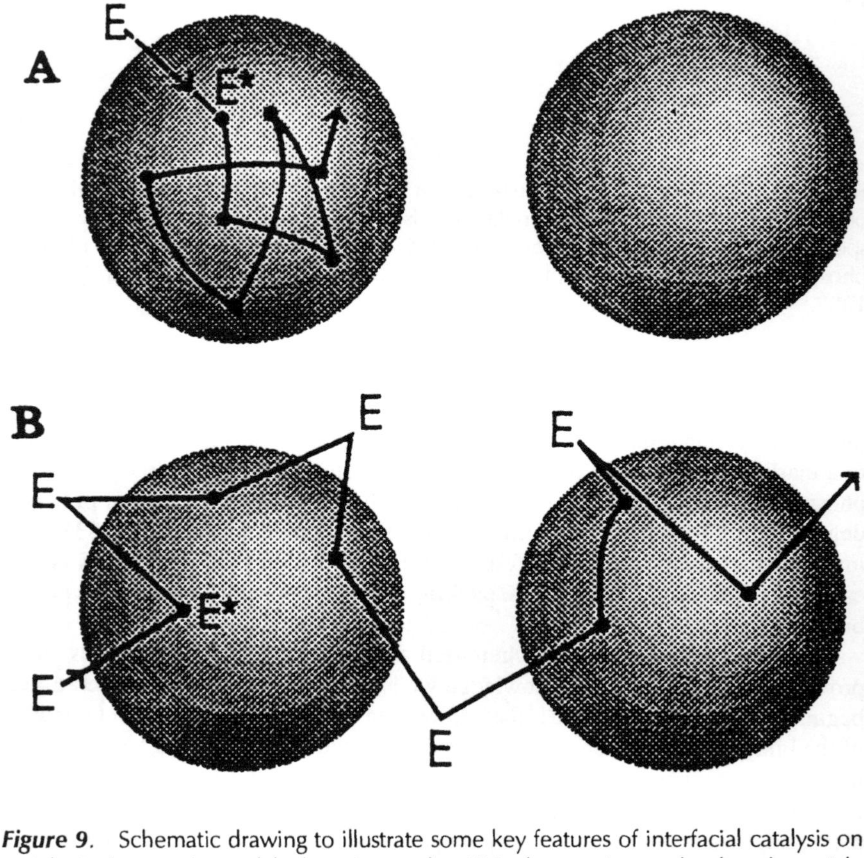

Figure 9. Schematic drawing to illustrate some key features of interfacial catalysis on vesicles in the scooting and the hopping modes. (A) In the scooting mode when the vesicle to enzyme ratio is >5, there is at most one enzyme per vesicle. Due to the high affinity (K_D < 0.1 pM) of PLA_2 for DMPM vesicles, the bound enzyme (E*) does not leave the vesicle even when all of the substrate in the outer monolayer of the target vesicles is hydrolyzed. Therefore, excess vesicles are not hydrolyzed by the enzyme added initially unless the vesicles are allowed to fuse, the bound enzyme is forced to undergo intervesicle exchange, or the excess vesicles are hydrolyzed by adding excess enzyme so that there is a least on enzyme per vesicle. (B) On the other hand, during catalysis in the hopping mode, the enzyme desorbs from the vesicle surface, and thus all vesicles are ultimately hydrolyzed even if the vesicle to enzyme ration is » 1. (From Berg et al., 1991.)

mers) is compared with an enzyme not activated by the interface (activity shows no enhancement of activity when aggregates are present). (See Figure 9.)

The first step in this model is phospholipase binding to the interface. The enzyme is activated (E*) in this binding process by a mechanism(s) that is not yet completely defined. Catalysis then occurs and product is formed. With the intact vesicle system, hydrolysis of the phospholipid on the outer leaflet proceeds until that pool of substrate is exhausted and the reaction ceases. For there to be further hydrolysis, either the enzyme must transfer to a new vesicle or exchange of substrates and products between vesicles must occur. A phospholipase that does not exhibit interfacial activation may in fact be capable of rapid intervesicle exchange. One way this can be studied is to determine if the enzyme hydrolyses the phospholipid present on the outer leaflet of all vesicles or if hydrolysis is limited to the amount of phospholipid present on the number of vesicles equal to the number of enzyme molecules present (number of enzyme molecules times number of phospholipid molecules on the outer leaflet of a vesicle).

Jain, Berg, and their coworkers have studied the kinetics of phospholipases acting on single shelled vesicles in great detail. From these studies they have developed the "scooting" and "hopping" model. The "scooting" mode is represented by part A in Figure 9. Once the enzyme is bound, the activated enzyme "scoots" over the surface of the micelle hydrolyzing substrate en route. A "hopping" event must occur next if the vesicle remains intact and further hydrolysis is to occur. In order to differentiate between these two kinetic mechanisms of phospholipase action, it is necessary to measure the initial burst of activity on the vesicles that contain a molecule of enzyme. These requirements limit the conditions of the assay and as a consequence optimal conditions for the enzyme may not be suitable for the assay. This appears to be the case with venom or pancreatic PLA_2 that require millimolar concentrations of Ca^{2+} for optimal activity, conditions that can lead to vesicle fusion and/or removal of product from vesicles (Yu et al., 1993). The disruption of vesicles is not specific for Ca^{2+} and can occur with other salts. One possible result of vesicle lipid interchange is that following the initial burst of activity (first order kinetics on a single vesicle), vesicle fusion or substrate replenishment becomes rate limiting. This rate of exchange is dependent upon the solution conditions and type of substrate and has little to do with the catalytic rate. Therefore, great caution must be used in studies with bilayer vesicles, as carefully described by Jain, Berg and their colleagues.

Recently there has been great interest in the "cytosolic" (c) PLA_2 that has been purified from a number of mammalian cells (Channon and Leslie, 1990). This enzyme exhibits a marked preference for phospholipids that contain arachidonic acid at position two of the glycerol backbone. Under the usual assay conditions with sonicated phosphatidylcholine vesicles, the enzyme requires micromolar concentrations of Ca^{2+} The Ca^{2+} can be replaced, however, by NaCl and other salts in the millimolar range (Wijkander and Sundler, 1991). It is possible, therefore, that the effect of ions (c) PLA_2 is directed towards the substrate rather than the enzyme.

This possibility is supported by the finding that reduced serum albumin caused a stimulation of activity that was most pronounced after the first burst of hydrolysis (Wijkander and Wykle, unpublished observation).

There are two other interesting considerations when single shell vesicles are used as substrates for phospholipases. First, the size of the vesicle can serve to regulate hydrolysis through the surface packing characteristics. Small vesicles have a higher radius of curvature that corresponds to a loosely packed monomolecular film of the lipid. Large vesicles, on the other hand are more planar and are thought to have closer packing of the polar head group region.

Second, vesicles can be prepared from mixtures of phospholipids that are well suited for substrate specificity studies. This gives a relatively uniform presentation of the substrate and therefore consideration of differences in the physical form of the substrate do not need to be considered. Mixing of lipids in vesicles can change the surface charge and therefore the catalytic properties of the phospholipase under investigation (Robinson and Waite, 1983). This can be seen also by the addition of metal ions to vesicles of lipids. Also, there can be an asymmetric distribution of phospholipids between the inner and outer leaflets of the bilayer (Litman, 1975). This difference can be determined by the acyl composition to some degree but primarily this is most obvious with differences in the hydrated area of the polar head group. This asymmetric distribution was demonstrated with a mixture of phosphatidylcholine and phosphatidylethanolamine in small single-shelled vesicles. As would be expected, phosphatidylethanolamine was found to be preferentially localized in the inner leaflet. It is important therefore to know the relative concentration of the two (or more) lipids on the outer surface when studies are carried out on mixed lipid vesicles.

C. Monomolecular Films

The use of monomolecular (monolayer) films of lipids has been helpful in defining the properties of the lipid-water interface that regulate lipolysis. The utility of monolayer films was recognized some 60 years ago when Hughes (1935) studied the action of snake venom phospholipases on lipid films. Monolayer films can be formed at either an air-water or oil-water interface although for almost all studies of lipolytic enzymes films spread at an air-water interface are used. Both surface pressure and surface potential of monolayers can be readily determined and both have marked influence on enzyme binding to the film (Verger and de Haas, 1976). While the effect of surface charge on lipolysis can be studied with other techniques, effects of surface pressure can only be investigated using monomolecular films. Even though the monolayer system is invaluable in a great number of instances, it is a time consuming approach and has a definite limitation in that the limiting surface area allows only a few percent of the enzyme molecules to bind to the film.

So far as lipid enzymology and the study of phospholipase is concerned, probably the most important single advance made since the original study of

Hughes in 1935 is the introduction of the zero-order kinetic trough (Figure 10-right side) developed by Verger and de Haas in 1973 (Verger and de Haas, 1973).

In this system, the stirred reaction chamber is separated from the lipid reservoir by a teflon partition that confines the enzyme and reaction products. The lipid film, however, is continuous in both chambers (in some configurations, a separate chamber is introduced for the Wilhemy plate attached to the microbalance). As the reaction proceeds, the product desorbs from the film which results in a decrease in the surface pressure that is detected by the surface balance. This change in surface pressure, in turn, is compensated for by the motor that drives the film barrier to maintain constant pressure. The area per molecule, determined by separate force-area studies, is used to quantitate the hydrolytic activity. Computer programs are now available that, when provided with information on the area per molecule at the surface pressure of the experiment, yields real time determination of the hydrolytic rate in molecules hydrolyzed per unit of time. The binding of enzymes to the film can be established using a radiolabeled enzyme and the true specific activity of the bound enzyme can be accurately determined. We've used this approach in our laboratory to establish that the phospholipase action of hepatic lipase is activated by apolipoprotein E (apoE) in a pressure dependent manner and that there is a 1:1

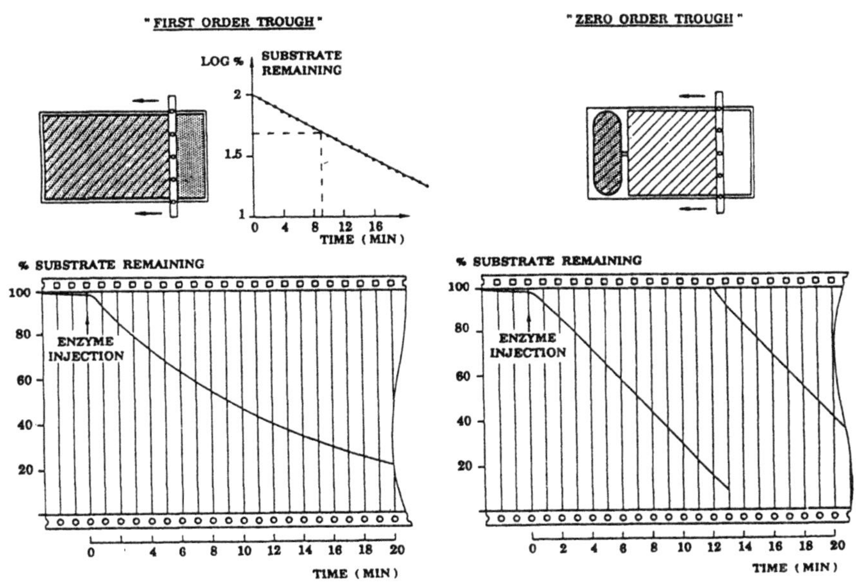

Figure 10. Comparison between the recorded kinetic plots obtained with the first-order trough (left side) and with the zero-order trough (right side). (From Verger and deHaas, 1973.)

stoichiometry between hepatic lipase and apoE under conditions that yield optimal hydrolysis of the film (Thuren et al., 1992).

A great number of lipolytic enzymes have been studied with the monolayer techniques with emphasis on their pressure dependency for film binding and catalysis. The data have been useful in understanding the physiological action of phospholipases, especially with regard to their ability to degrade biological membranes. For example, the porcine pancreatic PLA_2 has a weak capacity to degrade phosphatidylcholine films above 12-13 mN/m and will not degrade erythrocyte membranes (Ransac et al., 1992a). On the other hand, films of phosphatidylglycerol were degraded at pressures as high as 30 mN/m. The PLA_2 purified from rat platelet had roughly comparable activity to that of the pancreatic enzyme on phosphatidylglycerol at 25 mN/m. The platelet enzyme however, failed to hydrolyze phosphatidylglycerol films at 10 or 30 mN/m pressures where the pancreatic enzyme still had nearly full activity. Further, the platelet enzyme failed to degrade monolayers of phosphatidylcholine, at any surface pressure employed. These studies with the monolayer system demonstrate that even though these two enzymes are very closely related structurally, subtle structural variations dictate pronounced differences in their lipid binding capacity and substrate specificity.

The fact is that most of the enzyme in a monolayer system remains in the bulk solution; the groups of de Haas and Verger used this characteristic to advantage in studies of substrate analog inhibitors of phospholipases A_2 (Ransac et al., 1992b). Their approach was to allow [^{3}H]amidated PLA_2 to act on a monolayer film in the presence of an inhibitor, remove the film to determine the amount of enzyme bound, then reinitiate activity by spreading a new lipid film on the remaining subphase that contained the solution phospholipase. They demonstrated that the soluble phospholipase not bound to the film did not interact significantly with monomeric (solution) analog inhibitors. When the PLA_2 bound to the interface and was in the activated state, inhibitors competed with substrate and interacted with the active site. In a formulation analogous to solution kinetic competitive inhibition, the inhibition is expressed relative to the substrate amount in the film as the "inhibitory power," Z. It is interesting to note that Z was pressure dependent, as the surface pressure was increased from 5 to 35 mN/m with phosphatidylglycerol as substrate, the Z value for L-amino-phosphatidylglycerol (nitrogen substitution for oxygen at position-2 of the glycerol) increased nearly four fold. The Z value obtained with a taurocholate micelle system was found to be about four fold higher than that obtained with the same substrates and inhibitors in monolayer films at optimal surface pressures. The reasons for this difference are not yet defined but this finding led to the conclusion that one cannot directly compare the absolute Z values obtained with one interfacial system to another when studying lipophilic analog inhibitors. Physiologically, however, studies on lipophilic inhibitors may be better modeled with the monolayer system since much of the cellular phospholipase does not appear to be membrane bound, similar to the monolayer system.

D. Mixed Micelles

Phospholipid, when mixed with a detergent under the appropriate conditions, will form a population of micelles in which the phospholipid is uniformly dispersed between micelles. This is not to imply, however, that there is a random distribution of phospholipid with a detergent micelle. The micelle structures have characteristics that depend, to a great extent, on the characteristics of the detergent and the mechanism by which it interacts with the phospholipid. The interaction of bile salts with phospholipid can be thought of as a three stage process (Walde, et al., 1987); a process that has pronounced effects on phospholipase action. First, bile salt intercalates into the phospholipid bilayer at concentrations below the bile salt cmc. This perturbs the bilayer and with sufficient bile salt the bilayer becomes leaky. Second, abrupt change occurs at the cmc of the bile salt and mixed micelles are formed. Third, there is a size decrease in the mixed micelle as the bile salt increases. It has been reported that the hydrodynamic radius of bile salt-phosphatidylcholine discoidal micelle increases from 2 up to 30 nMeters as this ratio is decreased and that mixed populations of discs may exist (Mazer et al. 1980). Nonionic detergents such as Triton X-100, however, form spheroid micelles in which the phospholipid is more randomly distributed within the micelle (Robeson and Dennis, 1977). With each type of micelle the actual structure formed is dependent on the conditions of the preparation such as the ratio of detergent to phospholipid, the total concentration of lipid, ionic strength, and temperature.

There is reason to believe that the physical structure and composition of the micellar preparation will influence the interaction of a phospholipase and the substrate within the micelle. For example, the secreted (low molecular weight) PLA_2's were optimally activated by bile salts at concentrations below their cmc and presumably below the conversion of the vesicular to micellar form (Jain et al., 1993). For some phospholipases this activation can be explained as a charge phenomenon since they require a negative surface charge for interfacial interaction. In addition to the cmc, an important factor in determining PLA_2 action on detergent-phospholipid vesicles is the ratio of detergent to phospholipid that defines the intermicelle concentration (imc), the partion coefficient of detergent into the vesicle (Mazer et al., 1980). When the bile salt and phospholipid are between the imc and cmc, the hydrolytic rate decreases dramatically. In this region of phospholipid and bile salt concentrations polydispersed mixed discoidal micelles exist (Mazer et al., 1980). The size and complexity of these discs increase with increasing bile salt to phospholipid ratios that may limit PLA_2 action. The decrease in PLA_2 activity could result from a decrease in phospholipid exchange between micelles. It is thought that the exchange of substrate phospholipid (Nichols, 1988) in these systems is slow relative to the catalytic turnover number of the PLA_2 (Jain et al., 1993). It appears important, therefore, that the exchange of phospholipids substrate and hydrolysis products between the different types of bile salt-phospholipid vesicles and discoidal mixed micelles be established. Nichols and coworkers have addressed the problem of phospholipid exchange between bile salt micelles and reported that both the mechanism and transfer rate is highly dependent on the nature of the micelle (Fullington et al., 1990). More hydrophobic bile salts favor transfer of phospholipid by both collision and diffusion through the aqueous medium mechanisms, primarily through an increase in activation entropy and structured water.

Perhaps the most exhaustive study to define phospholipase kinetics using mixed detergent-phospholipid micelles was carried out by Dennis and coworkers who used Triton X-100-phospholipid mixed micelles as substrate for PLA_2 (Deems et al., 1975). Since Triton X-100 is not a single species of detergent, the pure OPE-9 [p (1,1,3,3 tetramethylbutyl) phenoxypolyoxyethelyene] has been synthesized and studied. They and others examined the physical properties of the mixed micelle and interpreted their data on the basis of the micellar properties. The most straight forward kinetic analysis assumes that there are a defined number of micelle particles with a fixed ratio of phospholipid to Triton X-100 molecules within the micelle. It also is assumed that these two parameters (micelle number and substrate ratio) can be varied independent of each other by first holding the Triton X-100 concentrations constant and varying the phospholipid concentration (variable ratio) and second holding the ratio constant and varying the total concentration of both (variable micelle number). It has been determined by both Dennis' group and ours that the micelle size changes as the ratio of Triton X-100 to phospholipid changes, therefore, with a fixed amount of Triton X-100 the number of particles will probably change. Also, we have found by NMR studies that when the total amount of OPE-9 and phospholipid is increased (in a fixed ratio), the micelle size changes (unpublished data).

Despite these limitations, this model has provided an useful system to study kinetics of phospholipases. Three factors need to be considered when using the Dennis and Dennis model. First, does the enzyme have a sufficiently high turnover number to effectively measure hydrolysis? The cmc of Triton X-100 is about 0.25 mM and a minimal ratio of about 2:1 (Triton X-100 to phospholipid) is required to form mixed micelles. Practically that means that roughly 0.10 mM substrate with 0.45 mM Triton X-100 is the minimal substrate concentration that can be used and have a structure that can be considered defined. From this one would calculate that the concentration of monomeric Triton X-100 should be 0.20 mM and that 0.10 mM Triton would be in a 2:1 mixed micelle with the phospholipid. Given the ambiguities of the premicellar structures mentioned earlier and the accuracy of the determination of the cmc and icm (especially with mixture of OPEs in Triton X-100); even these conditions are poorly defined and a more useful interpretation of data will be obtained at mMolar concentration of substrate and Triton X-100.

The second consideration using mixed micelles is the affinity the phospholipase for the detergent. Often this concern is either ignored or cannot be measured. A binding constant of enzyme for detergent can be measured and incorporated into the kinetic analysis as was done for *B. cereus* phospholipase C (Burns et al., 1982). Since this required an equilibrium binding study, large quantities of enzyme are consumed. We have considered the problem of hepatic lipase binding to Triton X-100 but have not been able to address this problem directly by doing binding studies. However, comparison of maximal velocity and substrate specificity studies using Triton X-100 mixed micelles and monomolecular films in which the number of enzyme molecules bound to the film were determined, we find comparable values (Kucera et al., 1988; Thuren et al., 1992). We conclude, therefore, that hepatic lipase does not bind appreciably to Triton X-100.

The third consideration is the fate of the phospholipase when bound to the micelle. Jain and coworkers pointed out that under conditions that maintain bilayer vesicle stability, a phospholipase bound to the vesicle and functioning in the "scooting" mode is limited to the substrate on the outer leaflet of that vesicle. Most enzymes studied thus far have rather slow disassociation-reassociation rates unless the conditions reaction are altered to force dissociation or vesicle disruption. Is the same true for mixed micelles? The answer appears to be equivocal and dependent upon the detergent. With Triton X-100 (or OPE-9) mixed micelles, we find that a metastable system exists in which there is a very rapid exchange of both OPE-9 and phospholipid between micelles (Kucera et al., 1988; Thomas et al., manuscript in preparation). Using stop-flow florescence measurements and two independent NMR approaches to measure micellar exchange, we estimate the exchange half time of phospholipids between micelles is less then a millisecond. We calculated that the exchange of phospholipid between micelles is at least two orders of magnitude greater than the catalytic rate of hepatic lipase and therefore the extent of hydrolysis was not limited by the surface content of the mixed micelle to which the hepatic lipase was bound. In a typical assay of hepatic lipase activity on phospholipid the number of micelles is in the order of 6×10^{16} whereas the number of enzyme molecules is lower, 1×10^{12}. Assuming that there are on an average 100 phospholipids per micelle and knowing that the turnover number of the enzyme is 90/sec, the hydrolysis would cease after one second and with only 0.01% of the substrate hydrolyzed. Experimentally, we found that hydrolysis was linear up to 35-40% utilization of the substrate. Our measured rapid exchange of Triton X-100 and phospholipid between micelles can therefore account for the observed enzyme kinetics. By comparison, phospholipase A_2's have much higher turnover numbers than hepatic lipase (about 300 molecules/second compared with about 90 for hepatic lipase). Even so, the content of phospholipid in the Triton X-100 micelle should not limit hydrolysis for most known phospholipases.

This conclusion is different from that reached by Jain and coworkers (1991) in their discussion of phospholipase turnover numbers and phospholipid exchange rates. In their study, however, they based their argument on the exchange rate of bile salt micelles which is in the order of 0.2-1000 sec (Fullington et al. 1990). As pointed out by Jain et al., the exchange rate of monomeric phospholipid between vesicles should be in the order of 50 msec if substrate replenishment is not to limit phospholipase A2 action. Our studies show that the exchange rate between OPE-9 mixed micelles is perhaps 3×10^6 times faster than between vesicles (1 h vs. 1 msec).

How can this vast difference be accounted for? One possible, but not likely, possibility is an enhanced collisional effect between micelles, relative to vesicles. It is possible that the tight binding between phospholipids in a vesicle is decreased by the OPE-9 and that a flip from one micelle to another upon collision would be facilitated by OPE-9. The second and more likely explanation is that monomers of phospholipid are transferred through the aqueous medium, not as free molecules but rather with OPE-9 as a "chaperon." This "chaperon exchange" could be short range and appear as pseudo-collisional, that is pseudo-first order. Although this is speculative, two observations support this concept. Our NMR data indicate that OPE-9 exists as multimeres rather than monomer in solution; these multimeres may be as small as dimeres. This is not to imply that monomers do not exist but that under the conditions used for hepatic lipase, multimeres are the predominate nonmicelle form of OPE-9. It is possible then that a phospholipid molecule could associate with this multimere (or related structure) in rather high concentrations that would exchange between micelles with rapid diffusion rates approaching or equal to that of the OPE-9 multimere alone.

There is precedent for postulating that phospholipid exchange is mediated by small numbers of detergent molecules. Shoemaker and Nichols (1992) measured the capacity of bile salts to extract lysophospholipid from bilayer vesicles of diacyl lipid below the cmc of the bile salt. They proposed that this exchange occurred via structures termed subcellular aggregates (SMA) cartooned in Figure 11.

In this case SMA were formed preferentially with monoacyl rather than diacyl phospholipid. Also, the vesicles coexisted with SMA that demonstrates SMA could serve as carriers between vesicles. The enhancement by bile salt of phospholipid transfer through the aqueous medium by hydrophobic bile salts could perhaps be explained by the formation of SMA. The more hydrophobic bile salts may have a greater tendency to form SMA that contain diacyl phospholipid than the more soluble bile salts. Likewise, they would have a lower imc that would favor partitioning into the phospholipid and in both cases increase the order of surrounding water. All these factors should increase phospholipid transfer between micelles. While we cannot predict how the results with lysolipids and bile salts will extrapolate to the Triton X-100 system, the results none the less shed important new insight into the fundamental nature of phospholipid exchange.

Relative to bile salt micelles, it appears that the mixed Triton X-100 micelle system provides a kinetic approach that does not limit the availability of substrate and therefore has certain features of solution kinetics. The Triton X-100 mixed micelle, however, is an entity with defined shape and composition and as such the concepts developed by Dennis and coworkers are valid, given the certain limitations outlined above. However, they are metastatic and undergo rapid exchange of constituents and therefore allow much greater extents by hydrolysis then can be obtained with vesicular substrates. Also, Triton X-100 mixed micelles are not limited in their use to certain conditions of salt concentration, the presence of proteins such as albumin, or multiple lipid substrates.

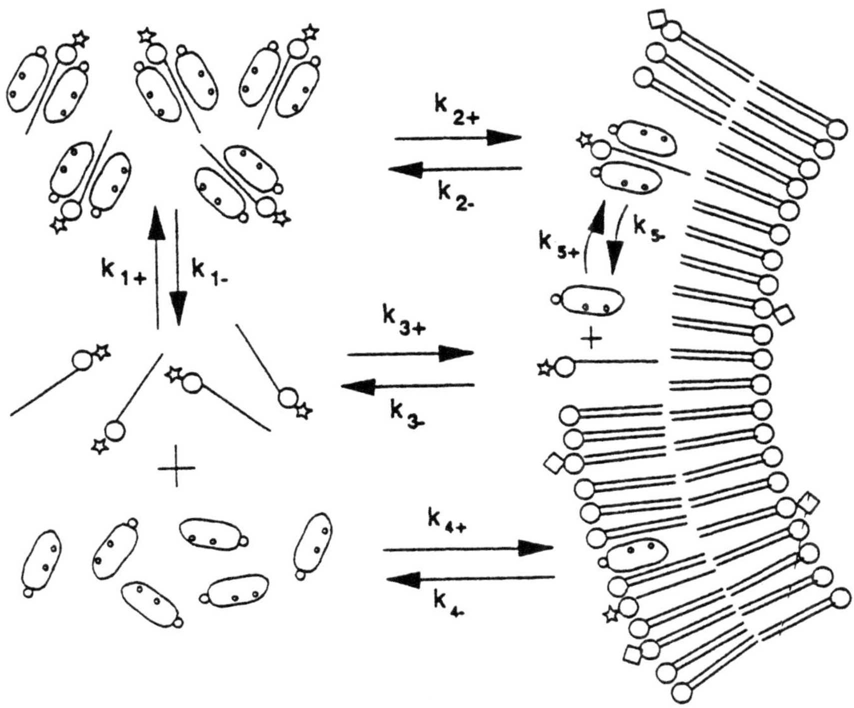

Figure 11. Schematic drawing of the molecular species present under the experimental conditions employed in this study. Phospholipids and lysophospholipids are represented by the traditional ball and stick figures with 19 carbons each on 2 and 1 chain, respectively. The NBD fluorophore attached to the head group of the lysolipids is represented by a star, and the rhodamine attached to the nonextractable diacylated phospholipid is represented by a diamond. The taurodeoxycholate molecule is drawn roughly to scale and is represented by an oval containing the two circular hydroxyl groups with the side chain and taurine represented by a circle on the side of the oval. The SMAs are represented by a single lysolipid coupled with two bile salts for simplicity; however, at the present time, the actual stoichiometry of these submicellar complexes is unknown. (From Shoemaker and Nichols, 1992.)

IV. SUMMARY

The attempt here has been to describe some of the important features of phospholipase model substrates with emphasis on their physical properties. While many features of substrate preparations are not yet defined, much progress has been made in the relevant lipid biophysics. As a consequence, students of phospholipases can better define the mechanism of phospholipase action and therefore, the physiologic role these enzymes play.

REFERENCES

Allgyer, T.T., & Wells, M.A. (1978). Kinetic anomalies associated with phospholipase A_2 hydrolysis of micellar substrates. Adv. Exp. Med. Biol. 101, 153-163.

Arita, H., Hanasaki, K., Nakano, T., Oka, S., Teraoka, H., & Matsumoto, K. (1991). Novel proliferative effect of phospholipase A_2 on Swiss 3T3 cells via specific binding site. J. Biol. Chem. 266, 19139-19141.

Bangham, A.D., Standish, M.M., & Weissmann, G. (1965). Action of steroids and streptolysin S on the permeability of phospholipid structures to cations. J. Mol. Biol. 13, 253-259.

Berg, O.G., Yu, B.Z., Rogers, J., & Jain, M.K. (1991). Interfacial catalysis by phospholipase A_2: Determination of the interfacial kinetic rate constants. Biochemistry 30, 7283-7297.

Birdi, K.S. (1989). Lipid and Biopolymer Monolayers at Liquid Interfaces. Plenum. New York.

Burns, R.A., El-Sayed, M.Y., & Roberts, M.F. (1982). Kinetic model for surface-active enzymes based on the Langmuir adsorption isotherm: Phospholipase C (*Bacillus cereus*) activity toward dimyristoyl phosphatidylcholine/detergent micelles. Proc. Natl. Acad. Sci. USA 79, 4902.

Chachaty, C. (1987). Applications of NMR methods to the physical chemistry of micellar solutions. Prog. NMR Spect. 19, 183-222.

Channon, J.Y., & Leslie, C.C. (1990). A calcium-dependent mechanism for associating a soluble arachidonoyl-hydrolyzing phospholipase A_2 with membrane in the macrophage cell line RAW 264.7. J. Biol. Chem. 265, 5409-5413.

Cullis, P.R., & Hope, M.J. (1991). Physical properties and functional roles of lipids in membranes. In: Biochemistry of Lipids, Lipoproteins and Membranes, (Vance, D.E., & Vance, J., eds.), pp. 1-41. Elsevier, Amsterdam.

Davies, J.T., & Rideal, E.K. (1963). Interfacial Phenomena. Academic Press, NY.

Deems, R.A., Eaton, B.R., & Dennis, E.A. (1975). Kinetic analysis of phospholipase A_2 activity toward mixed micelles and its implications for the study of lipolytic enzymes. J. Biol. Chem. 250, 9013.

van Eijk, J.H., Verheij, H., Dijkman, R., & deHaas, G.H. (1983). Interaction of phospholipase A_2 from *Naja melanolerica* snake venom with monomeric substrate analogs: Activation of the enzyme by protein-protein or lipid-protein interactions. Eur. J. Biochem. 132, 183.

Elsbach, P., Weiss, J., Franson, R.C., Beckerdite-Quagliata, S., Schneider, A., & Harris, L. (1979). The role of intramembrane Ca^{2+} in the hydrolysis of the phospholipids of *Escherichia coli* by Ca^{2+} dependent phospholipases. J. Biol. Chem. 260, 1618.

Fullington, D.A., Shoemaker, D.G., & Nichols, J.W. (1990). Characterization of phospholipid transfer between mixed phospholipid-bile salt micelles. Biochemistry 29, 879-886.

deHaas, G.H., Heemskerk, C.H.T., van Deene, L.L.M., Baker, R.W.R., Galli-Hatchard, J., Magee, W.L., & Thompson, R.H.S. (1963). The mode of action of human pancreatic phospholipase A. In: Biochemical Problems of Lipids (A.C. Frazer, ed.), pp. 244-250. Elsevier, Amsterdam.

Hazlett, T.L., Deems, R.A., & Dennis, E.A. (1990). Activation, aggregation, inhibition and the mechanism of phospholipase A_2. In: Biochemistry, Molecular Biology, and Physiology of Phospholipase A_2 and Its Regulatory Factors. (Mukherjee, A.B., ed.), pp. 49-64. Plenum Press, NY.

Hughes, A. (1935). The action of snake venoms on surface films. Biochem. J. 29, 437.

Jain, M.K., van Echteld, C.J.A., Ramirez, F., deGier, J., deHaas, G.H., & van Deenen, L.L.M. (1980). Association of lysophosphatidylcholine with fatty acids in aqueous phase to form bilayers. Nature 284, 486.

Jain, M.K., Rogers, J., Berg, O.G., & Gelb, J.H. (1991) Interfacial catalysis by phospholipase A_2: Activation by substrate replenishment. Biochemistry, 30, 7340-7348.

Jain, M.K., Rogers, J., Hendrickson, H.S., & Berg, O.G. (1993). The chemical step is not rate-limiting during the hydrolysis by phospholipase A_2 of mixed micelles of phospholipid and detergent. Biochemistry 32, 8360-8367.

op den Kamp, J.A.F., deGier, J., & van Deenen, L.L.M., (1974). Hydrolysis of phosphatidylcholine liposomes by pancreatic phospholipase A_2 at the transition temperature. Biochim. Biophys. Acta 345, 253.

Kinkaid, A., & Wilton, D.C. (1991). Comparison of the catalytic properties of phospholipase A_2 from pancreas and venom using a continuous fluorescence displacement assay. Biochem. J. 278, 843-848.

Krescheck, G.D., Hamori, E., Davenport, G., & Scheraga. H.A. (1966). Determination of the dissociation rate of dodecylpyridinium iodide micelles by a temperature jump technique. J. Am. Chem. Soc. 88, 2, 246-253.

de Kruijff, B., Cullis, P.R., Verkleij, A.J., Hope, M.J., Van Echteld, C.J.A., & Taraschi, T.F. (1985). In: The Enzymes of Biological Membranes (2nd edition), (Martonosi, A.N.,ed.), pp. 131-204. Plenum Press, NY.

Kucera, G.L., Sisson, P.J., Thomas, M.J., & Waite, M. (1988). On the substrate specificity of rat liver phospholipase A,. J. Biol. Chem. 263, 1920-1928.

Lands, W.E.M. (1960). Metabolism of glycerolipids. II. The enzymatic acylation of lysolecithin. J. Biol. Chem, 253, 2233.

Lenting, H.B.M., Nicolay, K., & van den Bosch, H. (1988). Regulatory aspects of mitochondrial phospholipase A_2: correlation of hydrolysis rates with substrate configuration as evidenced by ^{31}P-NMR. Biochim. Biophys. Acta 958, 405.

Lindblom, G., & Rilfors, L. (1989). Cubic phases and isotropic structures formed by membrane lipids—possible biological relevance. Biochim. Biophys. Acta 988, 221-256.

Liscovitch, M. (1992). Crosstalk among multiple signal-activated phospholipases. Trends Biochem. Sci. 17, 393-399.

Litman, B.J. (1975). Surface distribution of the fatty acid side chains of phosphatidylethanolamine in mixed phospholipid vesicles. Biochim. Biophys. Acta 413, 157-162.

Marsh, D. (1990). Handbook of Lipid Bilayers. CRC Press, Boca Raton, FL.

Mazer, N.A., Benedek, G.B., & Carey, M.C. (1980). Quasielastic light-scattering studies of aqueous biliary lipid systems. Mixed micelle formation in bile salt-lecithin solutions. Biochemistry 19, 603-615.

Misiorowski, R. L., & Wells, M.A. (1974). The activity of phospholipase A2 in reversed micelles of phosphatidylcholine in diethyl ether: effect of water and cations. Biochemistry 13, 4921-4927.

Neito, M., Venable, M., & Wykle, R.L. (1990). Role of transacylase activity in the release of arachidonic acid and synthesis of PAF in human leukocytes. Abstr., J. Cell Biol. 111, 257.

Nichols, J.W. (1988). Phospholipid transfer between phosphatidylcholine-taurocholate mixed micelles. Biochemistry 27, 3925-3931.

Papahadjopoulos, D., Vail, W.J., Jacobson, K., & Poste, G. (1975). Cochleate lipid cylinders: Formation by fusion of unilamellar lipid vesicles. Biochim. Biophys. Acta 394, 483-491.

Persson, B., Drakenberg, T., & Lindman, B. (1979). ^{13}C NMR of micellar solutions. Micellar aggregation number from the concentration dependence of the ^{13}C chemical shifts. J. Phys. Chem. 83, 3011-3015.

Poon, P.H., & Wells, M.A. (1974). Physical studies of egg phosphatidylcholine diethyl ether-water solutions. Biochemistry 13, 4928-4936.

Ransac, S., Aarsman, A.J., van den Bosch, H., Gancet, C., de Haas, G.H., & Verger, R. (1992a). Rat platelet phospholipase A_2. Kinetic characterization using the monomolecular film technique. Eur. J. Biochem. 203, 793-797.

Ransac, S., Deveer, A.M.T.J., Riviere, C., Slotboom, A.J., Gancet, C., Verger, R., & de Haas, G. (1992b). Competitive inhibition of lipolytic enzymes. V. A monolayer study using enantiomeric acylamino analogues of phospholipids as potent competitive inhibitors of porcine pancreatic phospholipase A_2. Biochim. Biophys. Acta 1123, 92-100.

Robinson, M., & Waite, M. (1983). Physical-chemical requirements for the catalysis of substrates by lysosomal phospholipase A_1. J. Biol. Chem. 258, 14371-14378.

Robson, R.J., & Dennis E.A. (1977). The size, shape, and hydration of nonionic surfactant micelles. Triton X-100. J. Phys. Chem. 81, 1075-1077.

Schmid, P.C., Johnson, S.B., & Schmid, H.H. (1991). Remodeling of rat hepatocyte phospholipids by selective acyl turnover. J. Biol. Chem. 266, 13690-13697.

Shoemaker, D.G, & Nichols, J.W. (1992). Interaction of lysophospholipid/taurodeoxycholate submicellar aggregates with phospholipid bilayers. Biochemistry 31, 3414-3420.

Small, D.M. (1986). The Physical Chemistry of Lipids. From Alkanes to Phospholipids. In: Handbook of Lipid Research (Hanahan, D.J., ed.). Plenum Publishing Corporation, NY.

Tanford, C. (1980). The Hydrophobic Effect: Formation of Micelles and Biological Membranes. Wiley, New York.

Thuren, T.Y., Weisgraber, K.H., Sisson, P., & Waite, M. (1992). Role of apolipoprotein E in hepatic lipase-catalyzed hydrolysis of phospholipid in high density lipoproteins. Biochemistry 31, 2332-2338.

Verger, R., & deHaas, G.H. (1973). Enzyme reactions in a membrane model. 1. A new technique to study enzyme reactions in monolayers. Chem. Phys. Lipids 10, 127.

Verger, R., & deHaas, G.H. (1976). Interfacial enzyme kinetics of lipolysis. Ann. Rev. Biophys. Bioenerg. 5, 77.

Verheij, H.M., Egmond, M.R., & deHaas, G.H. (1981a). Chemical modification of the a-amino group in snake venom phospholipase A_2: A comparison of the interaction of pancreatic and venom phospholipases with lipid-water interfaces. Biochemistry 20, 94.

Volwerk, J.J., Dedieu, A.G.R., Verheij, H.M., Dijkman, R., & deHaas, G.H. (1979). Hydrolysis of monomeric substrates by porcine pancreatic (pro)phospholipase A_2; the use of a spectrophotometric assay. Recl. Trav. Chim., Pays-Bas 98, 214.

Waite, M. (1987). (The) Phospholipases. In: Handbook of Lipid Research (Hanahan, D.J., ed.) pp. 9-14. Plenum Publishing Corporation, NY.

Walde, P., Sunamoto, J., & O'Connor, C.J. (1987). The mechanism of liposomal damage by taurocholate. Biochim. Biophys. Acta 905, 30-38.

Wells, M.A. (1974a). The mechanism of interfacial activation of phospholipase A_2. Biochemistry 13, 2248-2257.

Wells, M.A. (1974b). The nature of water inside phosphatidylcholine micelles in diethyl ether. Biochemistry 13, 4937-4942.

Wijkander, J., & Sundler, R. (1991). An 100 kDa arachidonate-mobilizing phospholipase A_2 in mouse spleen and the macrophage cell line J774. Purification, substrate interaction and phosphorylation by protein kinase C. Eur. J. Biochem. 202, 873-880.

Yu, B.Z., Berg, O.G., & Jain, M.K. (1993). The divalent cation is obligatory for the binding of ligands to the catalytic site of secreted phospholipase A_2. Biochemistry 32, 6485-6492.

INDEX

www.ingramcontent.com/pod-product-compliance
Ingram Content Group UK Ltd.
Pitfield, Milton Keynes, MK11 3LW, UK
UKHW020234290926
15470UKWH00002B/542

9 780762 302055